Advanced Materials and Techniques for Biosensors and Bioanalytical Applications

Advanced Materials and Techniques for Biosensors and Bioanalytical Applications

Edited by
Pranab Goswami

CRC Press
Taylor & Francis Group
Boca Raton London New York

CRC Press is an imprint of the
Taylor & Francis Group, an **informa** business

Cover image design: Manoharan Sanjay
First edition published 2021

by CRC Press
6000 Broken Sound Parkway NW, Suite 300, Boca Raton, FL 33487-2742
and by CRC Press
2 Park Square, Milton Park, Abingdon, Oxon, OX14 4RN

© 2021 Taylor & Francis Group, LLC
CRC Press is an imprint of Taylor & Francis Group, LLC

Reasonable efforts have been made to publish reliable data and information, but the author and publisher cannot assume responsibility for the validity of all materials or the consequences of their use. The authors and publishers have attempted to trace the copyright holders of all material reproduced in this publication and apologize to copyright holders if permission to publish in this form has not been obtained. If any copyright material has not been acknowledged please write and let us know so we may rectify in any future reprint.

Except as permitted under U.S. Copyright Law, no part of this book may be reprinted, reproduced, transmitted, or utilized in any form by any electronic, mechanical, or other means, now known or hereafter invented, including photocopying, microfilming, and recording, or in any information storage or retrieval system, without written permission from the publishers.

For permission to photocopy or use material electronically from this work, access www.copyright.com or contact the Copyright Clearance Center, Inc. (CCC), 222 Rosewood Drive, Danvers, MA 01923, 978-750-8400. For works that are not available on CCC please contact mpkbookspermissions@tandf.co.uk

Trademark notice: Product or corporate names may be trademarks or registered trademarks, and are used only for identification and explanation without intent to infringe.

Library of Congress Cataloging-in-Publication Data

Names: Goswami, Pranab, editor.
Title: The advanced materials and techniques for biosensors and bioanalytical applications / edited by Pranab Goswami.
Description: First edition. | Boca Raton : CRC Press, 2021. | Includes bibliographical references and index.
Identifiers: LCCN 2020033093 (print) | LCCN 2020033094 (ebook) | ISBN 9780367539658 (hardback) |
   ISBN 9780367539672 (paperback) | ISBN 9781003083856 (ebook)
Subjects: LCSH: Biosensors. | Biosensors--Materials.
Classification: LCC R857.B54 A374 2021 (print) | LCC R857.B54 (ebook) |
   DDC 610.28/4--dc23
LC record available at https://lccn.loc.gov/2020033093
LC ebook record available at https://lccn.loc.gov/2020033094

ISBN: 9780367539658 (hbk)
ISBN: 9781003083856 (ebk)

Typeset in Times LT Std
by KnowledgeWorks Global Ltd.

Contents

Preface vii
About the Author ix
Contributors xi

Chapter 1 Fundamentals of Biosensors..1

    Pooja Rani Kuri, Priyanki Das, and Pranab Goswami

Chapter 2 Emerging Materials and Platforms for Biosensor Applications 29

    Albert Saavedra, Federico Figueredo, María Jesús González-Pabón, and Eduardo Cortón

Chapter 3 Smart Materials for Developing Sensor Platforms 47

    Lightson Ngashangva, Pranab Goswami, and Babina Chakma

Chapter 4 Aptamer: An Emerging Biorecognition System 69

    Ankana Kakoti

Chapter 5 Metal Nanoparticles for Analytical Applications 91

    Priyamvada Jain

Chapter 6 Metal Nanoclusters as Signal Transducing Element 107

    Phurpa Dema Thungon, Pranab Goswami, and Torsha Kundu

Chapter 7 Nanozymes as Potential Catalysts for Sensing and Analytical Applications 143

    Smita Das and Pranab Goswami

Chapter 8 Carbon-Based Nanomaterials for Sensing Applications...163

    Naveen K Singh, Manoharan Sanjay, and Pranab Goswami

Chapter 9 Photoelectrochemical and Photosynthetic Material-Based Biosensors 183

    Mohd Golam Abdul Quadir, Pranab Goswami, and Mrinal Kumar Sarma

Chapter 10 Biofuel Cells as an Emerging Biosensing Device 211

    Sharbani Kaushik, Caraline Ann Jacob, and Pranab Goswami

Chapter 11 Bioelectrochemiluminescence as an Analytical Signal of Extreme Sensitivity 233

    Vinay Bachu and Pranab Goswami

Chapter 12 Paper Electronics and Paper-Based Biosensors 251

    Federico Figueredo, María Jesús González-Pabón, Albert Saavedra, Eduardo Cortón, and Susan R. Mikkelsen

Chapter 13 Strategies to Improve the Performance of Microbial Biosensors: Artificial Intelligence, Genetic Engineering, Nanotechnology, and Synthetic Biology 265

Natalia J. Sacco, Juan Carlos Suárez-Barón, Eduardo Cortón, and Susan R. Mikkelsen

Chapter 14 FET-Based Biosensors (BioFETs): Principle, Methods of Fabrication, Characteristics, and Applications 283

J. C. Dutta

Index 297

Preface

The discipline, analytical science is rapidly growing with the growth of modern civilization. In ancient times, the subject was in a primitive form, mostly limited to a simple measurement and analysis of chemicals and materials used in day-to-day life. The subject has grown in length and breadth over the last century to a robust stream of science, propelled by the continuous demands for an improved living standard with a view for a better future. Clear evidence of this incredible growth is the branching up of this large subject into different independent streams, one being the bioanalytical sciences. Interestingly, the growth of this biology-based analytical science also reached new heights and gave birth to the specialized field, biosensors. Since the invention of the first biosensor in 1962 (amperometric enzyme-electrode for glucose), the field has steadily grown. The growth was highly accelerated later, fueled by the rapid progress in material science, which encompasses a wide range of advanced materials, including ceramics, polymers, semiconductors, magnetic materials, biomaterials, and nanomaterials. The link between biosensors and material science is evident, as described in some chapters of this book. The other driving force that sparks the growth of biosensors is its huge market demand in the field of health care, food and agricultural sectors, environmental monitoring, and the military. Like bioanalytical sciences, the biosensor has also emerged as an academic subject, becoming a part of the course curriculum in undergraduate and postgraduate programs. Of late, this interdisciplinary subject has attracted enormous interest among students from multiple disciplines, including biology, chemistry, electronics, nanotechnology, energy, and environmental studies, as revealed from my own teaching experience at IIT Guwahati.

The current literature on the biosensor and bioanalytical sciences is vast, as evident from the volume of publications in both electronic and printed scientific journals. This developing subject, with ever-increasing new findings, needs compiling into a book form for ready reference for students and researchers. Books on biosensors and bioanalytical sciences are, however, available in the global market, with a couple of recent additions focusing on different aspects of these subjects. While appreciating these books due to their in-depth deliberation emphasizing mostly specific topics suitable for researchers, we realize the need for a more comprehensive book, suitable for both academic and research paradigms. To achieve these objectives, we designed the chapter topics carefully so that the students with modest subject backgrounds could also follow this book, and at the same time, the candidates with a research background find practical information to pursue their research work.

I am thankful to all the authors for their excellent contributions to create this resourceful book.

To start with, we illustrate the first chapter with the fundamentals of biosensors to orientate the readers to the subject. This chapter deals with the definition, classifications, principles, salient features, recent advances, and application potentials of the biosensors. We made an effort to illustrate each class of the biosensors concisely, but at the same time, enriching these classes with underlying principles for a better understanding of the readers. The rest of the chapter deals with the advanced materials and detection techniques used for developing biosensors. The topic "materials" in the context of this subject refers to bioreceptor/recognition elements, smart materials used for developing/amplification of signals, and materials used to construct the sensor platforms. The "techniques" implies here the transduction systems used to generate the bioanalytical signal for the quantitative or qualitative detection of the analyte. Notably, these three components, namely, bioreceptors, signal transduction, and detection platforms, are essential for developing a biosensor. While describing these materials and techniques in length, we consciously restrict our deliberation mostly to the emerging concepts on these topics. For instance, biofuel cells, FET, bioelectrochemiluminescence, and photoelectrochemistry are discussed in chapters as independent topics, whereas the other conventional transduction approaches are discussed along with the other topics. Similarly, we have described the traditional biorecognition elements like enzymes and antibodies alongside the other topics in the chapters.

The development of smart biosensors is incomplete without integrating it into modern computing and communication technology. This book also included a brief description on the topics of artificial intelligence (AI) and machine learning (ML) in a chapter. Additionally, a few contributions included progress on the integration of wireless communication systems to the biosensor devices. Overall, we opine that this book offers a well-balanced theme on the subject of biosensors with an extension to the peripheral aspects of bioanalytical sciences. Nevertheless, a biosensor is a technological endeavor. Prior to this venture, an intensive study in a bioanalytical lab is usually involved in developing a proof-of-concept and its translation to create a simple, portable, and low-cost device for sensitive, selective, and reliable detection of the target analyte of interest.

In this maiden attempt to deliver a good book on the title subject, we may inadvertently commit some mistakes. We therefore request our ardent readers to kindly identify any errors and thereby give us a chance to correct the content in our next edition.

Editor: Prof. Pranab Goswami

About the Author

Prof. Pranab Goswami received an MSc degree in chemistry with a specialization in organic chemistry from Gauhati University, India, and subsequently, an MS degree from BITS Pilani, Rajasthan, India, and finally, a PhD degree from his work at the Regional Research Laboratory (currently, NEIST, CSIR) Jorhat, India, in 1994 in the area of biotechnology. Prof. Goswami was a BOYSCAST fellow of DST, India, at the University of Massachusetts Boston, the USA. He started his professional carrier as a scientist at NEIST, CSIR Jorhat, India, during 1990 and continued there for 12 years. After that, Prof. Goswami moved to the Department of Biotechnology (renamed later as the Department of Biosciences and Bioengineering) at the Indian Institute of Technology (IIT) Guwahati, India, as a faculty member. He attained the level of Professor of Higher Academic Grade in the year 2015. Prof. Goswami also developed administrative experience in the capacity of heads, in the Department of Biosciences and Bioengineering, Centre for Energy, and Centre for Central Instrument Facility at IIT Guwahati.

The primary research area of Prof. Goswami is biocatalysis and biosensors. His research involves the development of nano-biosensors for malaria, myocardial infarction (MI), alcohol, and bilirubin. One of his primary objectives is the development of novel biorecognition systems along with signal transduction platforms for these targets. His group has developed many novel aptamers as recognition molecules for the detection of malaria and MI. Efficient signal transduction through a nanomaterial interface for sensitive and selective detection of the targets is the primary activity in his research lab. In this aspect, his group explores the direct electron transfer from the biorecognition elements (enzyme and bacterial biocatalysts) to the electrode surface through various electrochemical transducing platforms to generate a seamless electrical signal for sensing applications. His contributions to the development of third-generation bioelectrodes following direct electron transfer (DET) principle using FAD-based high-molecular-weight redox enzymes, such as alcohol oxidase, are well recognized. Prof. Goswami has used the concept to develop not only a sensitive amperometric alcohol biosensor but also a prototype on enzymatic-biofuel cell cum biobattery using the enzyme bioelectrode. Among the different electrochemical transducers, Prof. Goswami's lab is known for enzyme- and cyanobacteria-based biofuel cells as signal-transducing platforms for the generation and amplification of signals to detect alcohol and allied applications. A couple of proof of concept his group developed through the intensive basic research has already been translated to portable kits for the detection of malaria, alcohol, methanol, and formaldehyde. Prof. Goswami has published more than 100 peer-reviewed scientific papers and filed many patent applications, mainly in the field of biosensors. He has supervised more than 30 PhD students (20 completed) and received many awards and accolades, including outstanding reviewers' awards from reputed journals and visiting professorships in premiere universities in abroad. He also served as an editorial board member for two international scientific journals.

Contributors

Vinay Bachu
Biosensors and Biofuel Cell Lab (BBL)
Department of Biosciences and Bioengineering
Indian Institute of Technology Guwahati
Guwahati Assam, India

Babina Chakma
Promotional Medical Review Solutions
Indegene Private Limited
Nagawara Bengaluru, India

Eduardo Cortón
Biosensors and Bioanalysis Laboratory (LABB)
Departamento de Química Biológica and
  IQUIBICEN-CONICET
Facultad de Ciencias Exactas y Naturales,
Universidad de Buenos Aires (UBA)
Ciudad Universitaria,
Ciudad Autónoma de Buenos Aires, Argentina

Priyanki Das
Biosensors and Biofuel Cell Lab (BBL)
Centre for Energy, Indian Institute of Technology Guwahati
Guwahati Assam, India

Smita Das
Biosensors and Biofuel Cell Lab (BBL)
Department of Biosciences and Bioengineering
Indian Institute of Technology Guwahati
Guwahati Assam, India

J. C. Dutta
Department of Electronics and Communication Engineering
Tezpur University
Napaam, Tezpur Assam, India

Federico Figueredo
Biosensors and Bioanalysis Laboratory (LABB)
Departamento de Química Biológica and
  IQUIBICEN-CONICET
Facultad de Ciencias Exactas y Naturales
Universidad de Buenos Aires (UBA)
Ciudad Universitaria
Ciudad Autónoma de Buenos Aires, Argentina

María Jesús González-Pabón
Biosensors and Bioanalysis Laboratory (LABB)
Departamento de Química Biológica and
  IQUIBICEN-CONICET
Facultad de Ciencias Exactas y Naturales
Universidad de Buenos Aires (UBA)
Ciudad Universitaria
Ciudad Autónoma de Buenos Aires, Argentina

Pranab Goswami
Biosensors and Biofuel Cell Lab (BBL)
Department of Biosciences and Bioengineering
Indian Institute of Technology Guwahati
Guwahati, Assam, India

Caraline Ann Jacob
Biosensors and Biofuel Cell Laboratory (BBL)
Centre for Energy
Indian Institute of Technology Guwahati
Guwahati, Assam, India

Priyamvada Jain
Department of Applied Mechanics
Indian Institute of Technology Madras
Chennai, Tamil Nadu, India

Ankana Kakoti
Jack H. Skirball Center for Chemical Biology and
  Proteomics
SALK Institute for Biological Studies
La Jolla, California

Sharbani Kaushik
Department of Chemistry and Biochemistry
The Ohio State University
Columbus, Ohio

Torsha Kundu
International Management Institute, Kolkata, India

Pooja Rani Kuri
Biosensors and Biofuel Cell Lab (BBL)
Department of Biosciences and Bioengineering
Indian Institute of Technology Guwahati
Guwahati, Assam, India

Susan R. Mikkelsen
Department of Chemistry
University of Waterloo,
Waterloo, Ontario, Canada

Lightson Ngashangva
Biosensors and Biofuel Cell Lab (BBL)
Department of Biosciences and Bioengineering
Indian Institute of Technology Guwahati
Guwahati, Assam, India

Mohd Golam Abdul Quadir
Biosensors and Biofuel Cell Laboratory (BBL)
Centre for Energy
Indian Institute of Technology Guwahati
Guwahati, Assam, India

Albert Saavedra
Biosensors and Bioanalysis Laboratory (LABB)
Departamento de Química Biológica and
　IQUIBICEN-CONICET
Facultad de Ciencias Exactas y Naturales
Universidad de Buenos Aires (UBA)
Ciudad Universitaria
Ciudad Autónoma de Buenos Aires, Argentina

Natalia J. Sacco
Biosensors and Bioanalysis Laboratory (LABB)
Departamento de Química Biológica and
　IQUIBICEN-CONICET
Facultad de Ciencias Exactas y Naturales
Universidad de Buenos Aires (UBA)
Ciudad Universitaria
Ciudad Autónoma de Buenos Aires, Argentina.

Manoharan Sanjay
Universidad de Buenos Aires
Buenos Aires, Argentina

Mrinal Kumar Sarma
Advanced Biofuel Division,
The Energy and Resources Institute,
New Delhi, India

Naveen K Singh
University of California San Diego
San Diego, California

Juan Carlos Suárez-Barón
Biosensors and Bioanalysis Laboratory (LABB)
Departamento de Química Biológica and
　IQUIBICEN-CONICET
Facultad de Ciencias Exactas y Naturales
Universidad de Buenos Aires (UBA)
Ciudad Universitaria
Ciudad Autónoma de Buenos Aires, Argentina.

Phurpa Dema Thungon
Biosensors and Biofuel Cell Lab (BBL)
Department of Biosciences and Bioengineering
Indian Institute of Technology Guwahati
Guwahati, Assam, India

1 Fundamentals of Biosensors

Pooja Rani Kuri*, Priyanki Das* and Pranab Goswami
Indian Institute of Technology Guwahati, Assam, India (* equal contribution)

CONTENTS

- 1.1 Introduction to Biosensors 2
  - 1.1.1 History of Biosensors 2
- 1.2 General Configuration of a Biosensor 3
- 1.3 Characteristics/Salient Features of a Biosensor 3
  - 1.3.1 Selectivity 3
  - 1.3.2 Sensitivity 3
  - 1.3.3 Reproducibility 4
  - 1.3.4 Stability 4
  - 1.3.5 Response Time 4
  - 1.3.6 Recovery Time 4
  - 1.3.7 Linearity and Dynamic Range 4
  - 1.3.8 Effect of Physical Parameters 4
  - 1.3.9 Additional Factors 4
- 1.4 Classification of Biosensors 5
  - 1.4.1 Classification of Biosensors Based on the Biorecognition Principle 5
  - 1.4.2 Classification of Biosensors Based on the Biorecognition Element 5
    - 1.4.2.1 Enzyme-Based Biosensors 5
    - 1.4.2.2 Immunosensors 6
    - 1.4.2.3 Nucleic Acid/DNA Biosensors 6
    - 1.4.2.4 Cell-Based Biosensors 7
    - 1.4.2.5 Biomimetic-Based Biosensors 7
  - 1.4.3 Classification of Biosensors Based on the Transducer Elements 7
    - 1.4.3.1 Electrochemical Biosensors 7
    - 1.4.3.2 Optical Biosensors 12
    - 1.4.3.3 Calorimetric Biosensors 16
    - 1.4.3.4 Piezoelectric Biosensors 17
- 1.5 Bioreceptor Immobilization Strategies 18
  - 1.5.1 Adsorption 18
  - 1.5.2 Covalent Binding 19
  - 1.5.3 Entrapment 20
  - 1.5.4 Crosslinking 20
  - 1.5.5 Affinity 20
- 1.6 Application of Biosensors 21
  - 1.6.1 Application in the Food Industry 21
  - 1.6.2 Application in the Fermentation Industry 21
  - 1.6.3 Biomedical Applications 21
  - 1.6.4 Environmental Applications 21
  - 1.6.5 Applications in Biodefense 22
- 1.7 Recent Advances in Biosensor Technology 22
  - 1.7.1 Wearable Biosensors 22
  - 1.7.2 Implantable Biosensors 23
  - 1.7.3 Engineered Enzymes for Biosensor Development 24
- 1.8 Market Potential for Biosensors 24
- References 25

1.1 INTRODUCTION TO BIOSENSORS

Today, the importance of monitoring various samples for quality assurance or assessing risk factors or, more seriously, for disease diagnosis cannot be overemphasized. Several analytical techniques, both simple and sophisticated, have made it possible to achieve the same and thereby raise the standards of living of human beings. However, most analytical techniques demand skilled labor and sophisticated instrumentation and are tedious, time-consuming, and expensive. Consequently, there is an aspiration for portable, reliable, fast, and relatively inexpensive detection techniques. With the advent of biosensor technology, such aspirations have begun to come to fruition. A biosensor combines the exquisite selectivity of a biological component with the processing power of a transducer. The biosensor is an interdisciplinary field. It demands an exciting amalgamation of knowledge from several fields viz. the principles of basic sciences (physics, chemistry, biology) with the fundamentals of micro/nanotechnology, electronics, computer sciences, and design. The biosensor as an analytical tool has revolutionized the conventional analyses paradigm and justly finds its applications in the field of clinical diagnostics and biomedical applications, environmental applications, food and processing industries, and much more [1].

As a first step towards understanding a biosensor, it is important to take into account the main difference between a chemical sensor and a biosensor. According to International Union of Pure and Applied Chemistry (IUPAC), a chemical sensor is a device that transforms chemical information, ranging from the concentration of a specific sample component to total composition analysis, into an analytically useful signal. As per IUPAC recommendations from 1999, a biosensor is an independently integrated receptor-transducer device, capable of providing selective, quantitative or semi-quantitative analytical information using a biological recognition element. As a matter of realization, the basic difference between a biosensor and a chemosensor is that a biosensor employs a biological component as a recognition system specific to the analyte of interest and converts the biological signal into a measurable signal. In contrast, a chemosensor converts a chemical or physical property of a specific analyte into a measurable signal. In general, a biosensor is an analytical tool consisting of a biological recognition element, specific to the analyte of interest, in close association with a transduction element, which converts the biochemical signal into a quantitative or semi-quantitative electrical or other suitable signal.

1.1.1 History of Biosensors

The origin of biosensor technology is linked to the development of pH and oxygen electrodes. In this path, M. Cremer demonstrated the proportional relationship between an acid concentration and the rising electric potential of a glass membrane. This led to the concept of pH (hydrogen ion concentration) in 1909 by Søren Peder Lauritz Sørensen. Soon enough, an electrode for pH measurements was realized by W.S. Hughes in 1922. Another significant development was that of Griffin and Nelson, who first demonstrated the immobilization of an enzyme, invertase, on aluminium oxide and charcoal between 1902 and 1922. Thereafter, Leland C. Clark Jr. developed a simple and small amperometric electrode for detection of oxygen and published his definitive paper on the subject in 1956[2]. This followed the development of an amperometric enzyme electrode by Clark and Lyons for the detection of glucose. The concept was illustrated by an experiment in which glucose oxidase was entrapped at a Clark oxygen electrode using a dialysis membrane. The decrease in measured oxygen concentration was proportional to glucose concentration. Clark and Lyons coined the term "enzyme electrode" in their published article [3], and later Updike and Hicks [4] expanded the experimental detail necessary to build functional enzyme electrodes for glucose. This event was followed by the discovery of the first potentiometric biosensor for urea detection in 1969 by Guilbault and Montalvo Jr. Finally, in 1975, with the introduction of a commercial glucose biosensor by Yellow Springs Instruments (YSI), which employed Clark's technology, the biosensor technology marked its first appearance in the market.

Table 1.1 showcases the important milestones in the development of biosensors during the period 1970–1992. Since then, the amalgamation of several fields has led to the emergence of more developed and sophisticated biosensors. The era of the biosensor is an interdisciplinary one. The utility of these sensors is well pronounced in the fields of clinical diagnosis, agriculture, biotechnology, environmental monitoring, military applications, and others.

TABLE 1.1
Important Milestones in the Development of Biosensors During the Period 1970–1992 (Adapted from [5])

1970	Discovery of ion-sensitive field-effect transistor (ISFET) by Bergveld
1975	Fiber-optic biosensor for carbon dioxide and oxygen detection by Lubbers and Opitz
1975	First commercial biosensor for glucose detection by YSI
1975	First microbe-based immunosensor by Suzuki et al.
1982	Fiber-optic biosensor for glucose detection by Schultz
1983	Surface plasmon resonance (SPR) immunosensor by Liedberg et al.
1984	First mediated amperometric biosensor: ferrocene used with glucose oxidase for glucose detection
1990	SPR-based biosensor by Pharmacia Biacore
1992	Handheld blood biosensor by i-STAT

Fundamentals of Biosensors

1.2 GENERAL CONFIGURATION OF A BIOSENSOR

As per the conventional configurations, the biosensor consists of five components, namely, the biocatalysts/biorecognition element, transducer, amplifier, processor, and display (Figure 1.1). The biocatalyst is a biological component, also called a biorecognition element or bioreceptor. This biorecognition element usually interacts specifically with the target analyte of interest present in the samples. Typically, the bioreceptor is incorporated into the system by immobilization. Examples of biorecognition elements include enzyme, antibody, organelle, bacterial cell or other cells, whole slices of mammalian or plant tissues, and nucleic acids. Because the biological component is involved in identifying the target analyte in the sample, it is primarily responsible for conferring the specificity to the developed biosensors.

The next important component is the transducer, which converts the biorecognition signal into a measurable electrical signal. The transducer is generally placed in intimate contact with the recognition layer. It measures the physicochemical changes produced in the biorecognition layer by the presence of the analyte in the sample, generating a signal that is either proportional or inversely proportional to the analyte concentration. The typical transducers used in the system could be electrochemical, piezoelectric, calorimetric, or optical.

The function of the amplifier is to amplify the low electrical signal output from the transducer. Additionally, this unit removes the background noises generated within the electronic components of the transducer by subtracting the "reference" baseline signals.

The main function of a processor is to receive the input signal and provide the appropriate output signal. The signal produced at this level is usually converted to a digital form after being processed through a microprocessor stage, where the data is processed, converted to concentration units, and output to a display device or data store [5].

1.3 CHARACTERISTICS/SALIENT FEATURES OF A BIOSENSOR

1.3.1 Selectivity

Selectivity of a biosensor system refers to the extent to which a bioreceptor can discriminate a particular analyte of interest under given conditions in mixtures – simple or complex – without interference from other components in the mixture. Specificity, on the other hand, refers to the ultimate selectivity, i.e. 100% selectivity and 0% interference. This means that the bioreceptor is specific to a particular analyte of interest, recognizing no other. However, in cases where this ideal performance cannot be realized due to similarity between the analyte of interest and other components in the sample mixture or the lack of tools for specific sensing of the target, a selective sensing approach comes in handy. For example, although sugar-binding proteins – lectins – can be used to detect specifically carbohydrate-containing biomolecules, the individual lectins are not specific to individual glycosylated biomolecules viz. glycoproteins and glycolipids [5].

1.3.2 Sensitivity

Sensitivity refers to the minimum amount of analyte that can be detected by the system. In biosensors, the detection limits are commonly discerned by following two approaches: limit of detection (LOD) and limit of quantification (LOQ). Determination of LOD requires knowledge of limit of blank (LOB). LOB is the highest apparent analyte concentration expected to be found when replicates of a sample containing no analyte (blank) are tested. Although the blank samples are devoid of analyte, it can produce an analytical signal that might otherwise be consistent with a low concentration of analyte. LOB is estimated by measuring replicates of a blank sample and calculating the mean result and the standard deviation (SD).

$$\text{LOB} = \text{mean}_{\text{blank}} + 1.645\,(\text{SD}_{\text{blank}}) \tag{1.1}$$

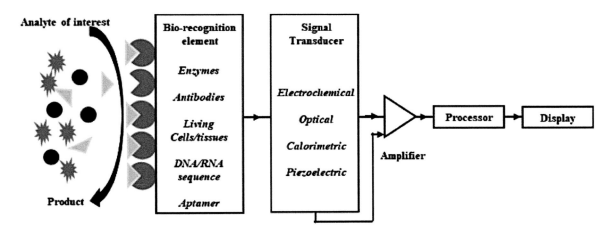

FIGURE 1.1 General configuration of a biosensor.

LOD is the lowest analyte concentration likely to be reliably distinguished from the LOB and at which detection is feasible. It is therefore greater than LOB. LOD is determined by utilizing both the measured LOB and test replicates of a sample known to contain a low concentration of analyte. The mean and SD of the low concentration sample is then calculated and used to calculate LOD.

$$LOD = LOB + 1.645\,(SD_{\text{low concentration sample}}) \quad (1.2)$$

LOQ is the lowest concentration at which the analyte can not only be reliably detected but at which some predefined goals for bias and imprecision are met. The LOQ may be equal to or greater than LOD [6].

Additionally, for a linear calibration curve ($y = a + bx$), it can be assumed that the response (y) is linearly related to the concentration (x) of the analyte in a limited range. This model is used to discern the sensitivity b, LOD, and LOQ from the following relations:

$$LOD = 3 \times \frac{SD_a}{b} \quad (1.3)$$

$$LOQ = 10 \times \frac{SD_a}{b} \quad (1.4)$$

where SD_a is the standard deviation of the response and b is the slope of the calibration curve. The SD_a can be estimated by the standard deviation of y-intercepts, of the regression lines.

1.3.3 Reproducibility

It refers to the property of a system to generate identical responses under identical set-up conditions viz. sample nature, pH, temperature, etc. Reproducibility of a biosensor largely defines the reliability of the biosensor [5].

1.3.4 Stability

Stability refers to the property of a system to be independent of ambient interferences in the form of environmental conditions like humidity, temperature, pH, or other interferences from the sample matrix. These interferences can cause a shift of the output signal to a value that is not accurate or precise. The biorecognition element used in a biosensing system plays a major contributor to stability. For example, the most popularly used biorecognition element – enzymes – are prone to denaturation when subjected to mild changes in ambient conditions. This may demand specific storage conditions for the biosensing system. Stability as a performance factor defines the shelf life of the biosensor system in question [5].

1.3.5 Response Time

This refers to the total time taken by the biosensor to produce the output signal, qualitative or quantitative, that can be recorded and subjected to interpretation. Chemical sensors usually have a very short response time as compared to their counterpart biosensors. However, a few biosensors based on the enzyme catalase for peroxides have exhibited response time of few seconds [5].

1.3.6 Recovery Time

It refers to the interval time required by the biosensor between two subsequent sample analyses. It is desirable to have as low a recovery time as possible. Usually a recovery time of within few minutes is desirable [5].

1.3.7 Linearity and Dynamic Range

Linearity reflects the accuracy of the measured response for a set of measurements with different analyte concentrations to a straight line, mathematically represented as $y = a + bx$, where a is the concentration of the target analyte, y is the output signal, and b is the sensitivity of the biosensor. In this context, the term "linear range" is used to describe the range of analyte concentration for which the biosensor produces a linear response. However, another term, "dynamic range" of the sensor, is used to describe the maximum and minimum values of the analyte concentration that can be measured. Dynamic range can be linear or nonlinear [5].

1.3.8 Effect of Physical Parameters

A biosensor should be independent of physical parameters such as temperature, pH, ionic strength, etc., so as to assure that the performance of the biosensing system is not significantly compromised. This attribute can pave the way for sample analyses with minimal pre-treatment [5].

1.3.9 Additional Factors

A biosensor should possess the following beneficial features for successful commercial applications. (a) If the cofactors or coenzymes involved in the biorecognition reaction are not avidly associated with the biocatalysts, then these should be preferably co-immobilized with the biocatalysts to reduce the operational steps involved in the practical use of the biosensor. (b) For *in vivo* applications, the biosensor probe should be tiny, having no toxic or antigenic effects. (c) The biosensor or its sensing part should be sterilizable without destroying its function for monitoring an analyte in the fermentation setup. (d) The complete biosensor should be inexpensive, small, portable, and capable of being used by semi-skilled operators. (e) Finally, there should be a market for the developed biosensor. It should also fulfil the legislative and other statutory requirements for the country. Further, the biosensor product should meet the overall satisfaction of the customers

Fundamentals of Biosensors

to encourage them to abandon traditional laboratory testing. Moreover, the World Health Organization (WHO) has suggested the ASSURED (Affordable, Sensitive, Specific, User-friendly, Rapid and Robust, Equipment-free, Deliverable to end users) criteria, particularly for the point-of-care (PoC) devices for disease diagnosis in underdeveloped and developing countries. Therefore, any biosensor that complies well with the ASSURED criteria may have better market potential.

1.4 CLASSIFICATION OF BIOSENSORS

Biosensors can be classified based on either of the two important elements i.e. based on the bioreceptor or based on the transducer used in the biosensor. Additionally, biosensors can be classified based on the biological mechanism or biorecognition principle they exploit for sensing – catalytic biosensor or affinity biosensor. However, the classification of biosensor based on the type of transducers is most widely reported in the current literature. Nonetheless, all three modes of classification (Figure 1.2) are briefly described in the following sections.

1.4.1 Classification of Biosensors Based on the Biorecognition Principle

Based on the biorecognition principle, biosensors can be classified into catalytic biosensors and affinity biosensors.

- *Catalytic biosensor*: This refers to the usage of enzymes or may be even living cells as biocatalysts with a transducer to produce a signal proportional to the target analyte concentration. This signal is brought about by the reaction catalyzed by the particular enzyme/cell, which may be heat change (exothermic or endothermic), release/uptake of gases viz. ammonia or oxygen, change in proton concentration, changes in light emission, etc. Enzyme biosensors are widely used biosensors that have been developed due to their substrate specificity and catalytic properties.
- *Affinity biosensor*: This refers to the usage of antibodies or nucleic acids (e.g. aptamers) as bioreceptors that specifically bind to the target analyte. An affinity biosensor works on the principle of receptor-ligand interactions with a high differential selectivity in a nondestructive fashion. Clearly, the structure of the analyte remains chemically unaltered during the detection process using this type of sensor.

1.4.2 Classification of Biosensors Based on the Biorecognition Element

The biorecognition element/bioreceptor/bioelement is largely responsible for conferring selectivity/specificity on the biosensor, and is therefore a distinguishing feature of biosensor technology. Based on the biorecognition element, a biosensor can be classified into different types such as enzyme-based biosensor, immunosensor, nucleic acid biosensor, cell/tissue-based or microbial biosensor, and biomimetic. The enzymes, antibody, nucleic acids, whole cells, biomembrane, and even organelles can be used either in isolation or in combination to act as the bioreceptor that brings about biochemical recognition of the target analyte.

1.4.2.1 Enzyme-Based Biosensors

An enzyme-based biosensor utilizes enzyme(s) as the biorecognition element. An enzyme is a biocatalyst that increases the rate of chemical reaction without itself being changed in the overall process. Virtually, all cellular reactions/processes are mediated by enzymes. Most enzymes are proteins, with

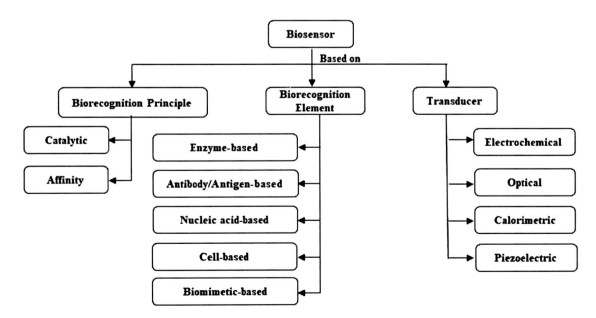

FIGURE 1.2 Classification of biosensors.

the exception of a small group of nucleic acid-based enzymes, such as ribozymes. The catalytic activity of the enzyme depends on the integrity of its native protein conformation. This means that denaturation of the enzyme – dissociation into its constituent subunits or amino acids – renders the enzyme nonfunctional.

Enzymes that are composed entirely of amino acid residues are called simple enzymes, while the enzymes that require a nonproteinaceous component attached to them in order to become functional are called conjugated enzymes, and this nonproteinaceous component is called a cofactor. In the case of conjugated enzymes, the protein component is called the apoenzyme, which when combines with the cofactor to form a fully functional enzyme, which is called the holoenzyme. Cofactors may include metal ions, or complex organic groups, in which case they are called coenzymes. Cofactors that are tightly associated with the apoenzyme are called prosthetic groups.

Enzyme-based biosensors were the earliest developed biosensors, introduced by Clark and Lyons in 1962 for amperometric detection of glucose. Example of the most exploited enzyme bioreceptors include glucose oxidase (GOD) and horseradish peroxidase (HRP). Currently, enzymes are widely used as bioreceptors because of their substrate specificity and usual high turnover rates.

The enzyme is usually immobilized or intimately associated with the transducer surface for signal detection. The catalytic activity of the enzyme accounts for the fast activity and possible signal amplification of the biosensor, thereby contributing sensitivity to the biosensor. This specific catalytic reaction of the enzyme makes enzyme-based biosensors able to detect much lower limits of target analyte than with normal binding techniques.

An enzyme as a bioreceptor can be utilized to detect the analyte of interest mostly by one of the following mechanisms: (a) detection of an analyte that gets converted by the enzyme into a detectable product, (b) detection of an analyte that acts as an enzyme inhibitor or activator, and (c) detection of an analyte that brings about modification of enzyme properties upon interaction.

An enzyme catalyzed reaction, and therefore the performance of the enzyme-based biosensor, is influenced by several factors such as the concentration of the substrate, pH, temperature, and presence of inhibitors or activators [1].

1.4.2.2 Immunosensors

An immunosensor typically utilizes an antibody as a bioreceptor. Antibodies are antigen-binding glycoproteins, synthesized by the body in different amino acid sequences, which confers a binding site for different antigen. They are collectively called immunoglobulins. An antibody is a Y-shaped immunoglobulin, constituting two heavy chains and two light chains. Both the chains have a variable region at their N-terminal end and a constant region at their C-terminal end. The variable region contains the specific amino acid sequence information for binding with the corresponding antigen/target with high affinity. This high affinity of the antibody only for its target antigen confers the property – specificity – which becomes the basis of utilizing antibodies as bioreceptors. By employing antigen-antibody interactions as a biorecognition mechanism, specificity and extremely low detection limits can be attained. Antibodies used in immunosensors can be either monoclonal or polyclonal. Monoclonal antibodies are produced from one type of immune cell and bind to the same epitope of their specific target antigen, making them specific. Polyclonal antibodies can recognize different epitopes on their target antigens, making them highly sensitive but less selective due to possible cross-reactivity. Immunosensors occupy a significant role in clinical diagnostics and health care, where high sensitivity, specificity, and rapid detection are particularly desired. The immunosensors also find applications in food quality and environmental monitoring. Of note, immunosensors developed for bacteria and pathogen detection have captured attention due to their application in POC [1].

1.4.2.3 Nucleic Acid/DNA Biosensors

Nucleic acid sensor utilizes nucleic acids as the bioreceptor, immobilized on the transducer surface. Nucleic acids are biopolymers (oligonucleotides) consisting of nucleotides as the monomeric unit. Each nucleotide comprises three components: a 5-C sugar, a phosphate group, and a nitrogenous base. Ribonucleic acid (RNA) contains ribose sugar, whereas deoxyribonucleic acid (DNA) contains deoxyribose sugar. Nucleic acids typically function as information molecules inside the cell, viz. acting as genetic material and encoding proteins. Single-stranded nucleic acids can form double-stranded structures by forming hydrogen bonds between the appropriate nitrogenous bases of nucleic acid strands. Typically, adenine pairs with thymine (DNA) or uracil (RNA) by forming two hydrogen bonds, and guanine pairs with cytosine by forming three hydrogen bonds. Base paring is sequence specific, employing the complementarity between the two sequences to form the double-stranded structure.

For developing biosensors, nucleic acids are immobilized in the recognition layer using linkers such as thiol or biotin, while the transducer used confers sensitivity. A genosensor consists of an immobilized modified oligonucleotide of known sequence (probe) that can detect the complementary nucleic acid sequence (target) through hybridization. The transducer converts the hybridization signal into a usable detectable signal: optical, electrochemical, or even piezo electrical. DNA can be used as a bioreceptor because base-pairing interactions between complementary sequences are specific and robust, making nucleic acid hybridization the principal basis of nucleic acid biosensors. The important property of DNA as a bioreceptor lies in its ability of thermal denaturation followed by renaturation under suitable conditions, making the sensor easily reusable.

DNA biosensors have been envisioned to play an important role in clinical diagnostics, gene analysis, forensic studies, etc. A few examples include diagnostic tests for mutations, monitoring gene expression, screening for targets associated with a disease, assessment of medical treatment, environmental investigations, and detecting biological warfare agents [1].

1.4.2.4 Cell-Based Biosensors

Cell-based biosensors utilize living cells as the bioelement, immobilized onto the transducer surface. The ability of the living cells to respond to the intracellular and extracellular microenvironment and their corresponding response in terms of certain measurable parameters are exploited in the cell-based biosensors. Microbial cells are particularly explored for use in biosensing for detection of specific molecules in the sample matrix or the overall state of the environment. The enzymes and other proteins present in the microbial cells can produce a response to the analytes selectively. The analyte of interest enters the living cell and undergoes some conversion, into a detectable product, or creates indirect changes in pH, ionic concentration, oxygen level in the biosensing layer, and measurement of other parameters, which could indicate the presence of the analyte in the sample matrix. Since the cell-based biosensors are more tolerant to inhibition by certain solutes and variation of pH and temperature values within a range, they can be more favored as compared to their isolated enzyme counterpart. This means that a longer shelf life can be expected. In addition, proteins and enzymes need not be isolated or purified in the case of cell-based biosensors, thereby reducing the cost of the biosensor.

The environmental conditions in which the microorganisms are immobilized on the transducer surface such that the microbes can stay alive for a longer time determine the limit of detection of such a biosensor. The main challenges faced include maintaining stability of the immobilized cells, which is a function of the narrow range of pH and temperature at which the cells remain tolerant and functional, biocompatibility of the immobilization matrix, and other factors. The cell-based biosensors suffer from poor selectivity as compared to the isolated bioreceptors (enzymes, antibodies, and others) due to the multireceptor behavior of the whole cells. Cell-based biosensors find applications in environmental monitoring, disease diagnosis, drug detection, etc. [1].

1.4.2.5 Biomimetic-Based Biosensors

Natural receptor elements such as enzymes, antibodies, nucleic acids, cells, tissues, etc., have been widely explored, characterized, and optimized to develop biosensors. Although they are robust when judged on some parameters, their scope for biosensing applications is restricted by limited stability under harsh conditions, high production costs, or limited availability. Under such circumstances, exploration of biomimetic receptors that overcome the limitations of natural bioreceptors while retaining their inherent properties of selectivity is of interest. Biomimetic sensors utilize a biomimetic molecule as a biorecognition element. A biomimetic receptor is a synthetic molecule that mimics the natural receptors. Few examples of synthetic receptors include genetically engineered proteins and cells, molecularly imprinted polymers (MIP), synthetic peptides, oligonucleotides viz. locked nucleic acids (LNAs) and peptide nucleic acids (PNAs), and aptamers. Aptamers are of particular interest as they are one of the most widely explored biomimetic receptor so far.

Biosensors that utilize an aptamer as a biorecognition element are referred to as aptasensors. Nucleic acid aptamers are artificial nucleic acid ligands reported for the first time in the early 1990s. These are short, single-stranded oligonucleotides that fold into a well-defined, three-dimensional structure and are capable of binding various molecules (targets) with high affinity and specificity. Aptamers are potential alternatives to antibodies, as they retain the desirable properties of antibodies as a bioreceptor, while overcoming the limitations of thermal stability, have a low cost, and have a wide range of applications [1]. A detailed description of aptamers is discussed in Chapter 4 of this book.

1.4.3 CLASSIFICATION OF BIOSENSORS BASED ON THE TRANSDUCER ELEMENTS

This mode of classifying biosensors is by far the most widely acceptable form. This section discusses various classes of biosensors based on the type of transduction principles being used for their development.

1.4.3.1 Electrochemical Biosensors

According to IUPAC, an electrochemical biosensor is a self-contained integrated device, which is capable of providing specific quantitative or semi-quantitative analytical information using a biological recognition element (biochemical receptor), which is retained in direct spatial contact with an electrochemical transduction element [7]. Electrochemical enzyme biosensors contain a redox enzyme that selectively reacts with the target analyte and produces an electrical signal that is related to the concentration of the analyte being studied. Electrochemical biosensors combine the sensitivity of electrochemical transducers with the specificity of biological recognition processes [8]. The reaction being studied electrochemically typically generates a measurable charge accumulation or potential (potentiometry), alters the conductive properties of the medium between electrodes (conductometry), or produces a measurable current (amperometry). Therefore, electrochemical biosensors are classified as potentiometric, conductometric, and amperometric biosensors.

1.4.3.1.1 Potentiometric Biosensors

Potentiometric measurements entail the determination of the potential difference between either an indicator and a reference electrode or two reference electrodes separated by a permselective membrane, in the absence of significant current between them. The transducer may be an ion-selective electrode (ISE), which is an electrochemical sensor based on thin films or selective membranes as recognition elements [9]. The most common potentiometric devices are pH electrodes; several other ions (Na^+, K^+, Ca^{2+}, NH_4^+, F^-, I^-, CN^-) or gas (CO_2, NH_3) selective electrodes are also available. The potential differences between these indicators and reference electrodes are proportional to the logarithm of the ion activity or gas fugacity (or concentration), as described by Nernst-Donnan. The response of a potentiometric biocatalytic sensor is either

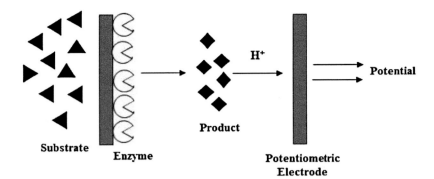

FIGURE 1.3 Schematic of potentiometric biosensor.

steady-state or transient. One significant feature of the ISE-based biosensors, such as pH electrodes, is the large dependence of their response on the buffer capacity of the sample and on their ionic strength [9]. The schematic representation of a potentiometric biosensor is shown in Figure 1.3. In these measurements, concentration and potential are related by the Nernst equation, as given in Equation 1.5.

$$E_{cell} = E^0_{cell} - \frac{RT}{nF} \ln Q \qquad (1.5)$$

where E_{cell} is the observed cell potential at zero current that is called the electromotive force (EMF), E^0_{cell} is a standard potential contribution to the cell, R and T represent universal gas constant and absolute temperature (in degrees Kelvin), respectively, n is the number of charges, F is the Faraday constant, and Q is the ratio of ion concentration at the anode to ion concentration at the cathode [10].

There are many examples of potentiometric biosensors, such as the one developed for the detection of different pesticides [11]. Over the last decade, the ion selective field effect transistor (ISFET)-based potentiometric devices have received intensive interest due to their several advantages over the conventional potentiometric biosensors. A detailed account of the ISFET-based sensors has been illustrated in Chapter 14 of this book.

1.4.3.1.2 Conductometric Biosensors

Conductometry is the measure of the ability of ions in solution to carry current between two inert electrodes under an applied electric field. The ions are formed by the dissociation of the electrolyte. On the application of a potential difference between the two electrodes, a chaotic ion movement occurs, where anions move toward anodes, while cations move toward cathodes in the electrolytic cells (Figure 1.4). The conductivity (S) of the electrolyte solution depends on the ion concentration and mobility. The resistance of electrolyte solution is in direct proportion to the distance L between the immersed electrodes and reciprocal to their area A; therefore, conductivity (S) can be calculated using the following equation:

$$S = \chi \times \left(\frac{A}{L}\right) \qquad (1.6)$$

where χ is specific conductivity. This shows that a conductometric measurement includes determining the conductivity of a solution between two parallel electrodes; it is a sum of all the ions within the solution.

Conductometric biosensors utilize a change in solution conductivity as the transduction mechanism for sensing. These biosensors are based on the fact that almost all biocatalytic reactions involve either consumption or production of charged species and therefore lead to an overall change in the ionic composition of the tested sample. The conductometric transducer is a small two-electrode device fabricated to measure the conductivity of the thin electrolyte layer close to the electrode surface. The interdigitated structure of conductometric electrodes has received increasing interest in sensor and biosensor research. Conductometric biosensors have advantages over other types of transducers – for example, they can be produced through low-cost thin-film standard technology, there is no need for a reference electrode, they

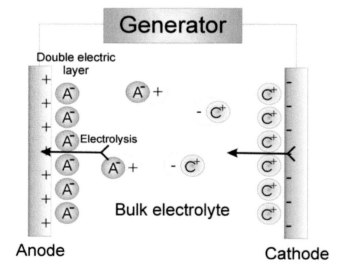

FIGURE 1.4 Ion migration in solution under an applied potential difference between two electrodes [12].

Fundamentals of Biosensors

are insensitive to light, and differential mode measurements permit cancellation of a great number of interferences. The liquids analyzed are generally considered to have significant background conductivity, which is easily modified by different factors; that is why the selectivity of this method is supposed to be low. Interested readers may consult some of the prominent literature to understand the detailed theory and application of this type of electrochemical biosensor [12, 13].

1.4.3.1.3 Amperometric Biosensors

The amperometric transducer-based biosensors work by the production of a current when a potential is applied on the working electrode in an electrochemical setup in response to the analyte of interest [14]. Here we are describing amperometric biosensors citing the widely used biorecognition element enzymes. Normally, the enzymes are immobilized over the electrodes as a selective layer to transduce a biochemical signal into an electrical one under the influence of an appropriate applied potential [15, 16]. If the response arises from the electro-activity, mainly of a co-substrate (oxygen in most of the cases), product, or co-product (H_2O_2 in most of the cases with flavin adenine dinucleotide (FAD)-based redox enzymes) of the enzyme-catalyzed reaction the category of the biosensor is termed first generation. A Clark oxygen electrode-based glucose biosensor is the first example of this type [3]. The requirement of a high operation potential for the measurement of H_2O_2 and limited solubility of oxygen in aqueous samples causing fluctuations in the oxygen tension are some major drawbacks of first-generation biosensors.

Pairing of electrons between the redox active centers of the enzyme and the electrode through some specialized small electroactive molecules to generate the response represents the second-generation biosensors. These specialized molecules are termed "electron transfer mediators" (ETMs), which shuttle electrons between the redox center of the enzyme and the electrode. This approach leads to a considerable reduction of electrochemical interferences due to the involvement of the low redox potential of the enzyme in generating the electrical signal. Many second-generation amperometric biosensors have been developed. One prominent example is the ferrocene-mediated enzyme electrode for the determination of glucose using GOD with FAD in the redox center, which acts as a biorecognition molecule as reviewed in [10]. The reaction involves the following steps, where M/M* are the oxidized and reduced forms of the mediator.

$$Glucose + GOD/FAD \rightarrow Gluconolactone + GOD/FADH_2$$

$$GOD/FADH_2 + 2M \rightarrow GOD/FAD + 2M^* + 2H^+$$

$$2M^* \rightarrow 2M + 2e^- \text{ (transferred to electrode)}$$

The mediator is chosen in such a way that it possess a lower redox potential than the other electrochemically active interferents present in the sample. The redox potential of the mediator should be more positive for oxidative biocatalysts and more negative for reductive biocatalysts than the redox potential of the enzyme-active site. The ETMs should possess some critical characteristics, such as react rapidly with the reduced state of the enzyme, exhibit reversible heterogeneous kinetics, the overpotential for the regeneration of the oxidized mediator should be low and pH independent, stable in oxidized and in reduced forms, and the reduced form should not react with oxygen.

Ferrocene and its few derivatives are widely used as ETMs. Tetrathiafulvalene (TTF), tetracyanoquinodimethane (TCNQ), ferricyanide, N,N,N′,N′ tetramethyl-4-phenylene diamine (TMPD), and benzoquinone are also reported as ETMs in some assays. Organic dyes such as alizarin yellow, azure A and C, methylene blue, methyl violet, phenazines, prussian blue, thionin, and toluidine blue are known for their electron transfer-mediating properties. Many ETMs, however, suffer from a number of problems such as poor stability and pH dependence of their redox potentials (organic dyes). Conversely, many of the inorganic mediators are not easy to tune their redox potentials and solubility by the use of substituents. Mediators are generally added directly to the measuring solution or immobilized on the electrode surface. The first method is easier, though it is not suitable from the technological perspective [10, 17].

The second-generation biosensors, however, suffer from drawbacks, such as poor stability and reproducibility. The diffusion barrier between the enzyme-electrode interface and the leaching susceptibility of the free mediator from the interface to the sample solution are ascribed as the reasons for the poor performance of the second-generation bioelectrode-based biosensors.

In an effort to overcome the drawbacks of first-generation and second-generation biosensing principles, the concept of third-generation biosensors emerged. The principle of third-generation biosensors involves direct electron transfer (DET) between the redox center of the enzyme and the electrode without using any ETMs to generate the response. High selectivity and sensitivity are the main advantages of these biosensors, as they can operate in a potential window closer to the redox potential of the enzyme [18]. There are, however, limited redox proteins with a redox center in the periphery of the protein matrix, such as cytochrome c, a peroxidase that supports the DET principle. Different strategies have been explored to introduce DET-based enzyme electrodes that include nanofabrication of electrodes, immobilization of conductive polymer coupled redox proteins, cofactors tethered by the reconstitution process, etc. [14].

The development of first- to third-generation amperometric biosensors reflects a simplification and improvement of the signal transduction pathway. The electrical signal transduction pathways for the three generations of biosensors are depicted in Figure 1.5, citing the examples with an FAD-based redox enzyme.

The feasibility of electron exchange between the redox proteins and the electrodes can be explained by the electron transfer (ET) theory of Marcus [19]. The ET rate constant (K_{ET}) between a donor and acceptor pair is given by Equation 1.7,

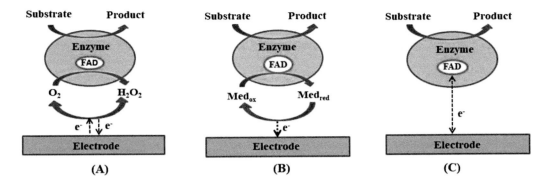

FIGURE 1.5 Three generations of amperometric bioelectrodes: A: first-generation, B: second-generation, and C: third-generation enzyme electrodes.

where d and $d°$ are the distance separating the electron donor and the van der Waals distance, respectively; β is the electron-coupling constant; and $\Delta G°$ and λ are the Gibbs free energy change and the reorganization energy accompanying the ET process, respectively.

$$K_{ET} \propto \exp\left[-\beta(d-d^0)\right].\exp\left[-\frac{(\Delta G^0 + \lambda)^2}{4RT\lambda}\right] \quad (1.7)$$

Thus, the electron exchange between two redox sites depends largely on three factors: the reorganization energy qualitatively representing the conformational rigidity of the redox compound in its oxidized and reduced form, the potential difference between the involved redox entities (since $\Delta G° = \sim -nF\eta$, where, over potential $\eta = E - E°$, the applied and standard potential, respectively), and separation between the redox sites.

The concept of protein film voltammetry (PFV) is beneficial for studying the principle of DET. In PFV a stable mono-/submonolayer film of redox protein on an electrode surface is prepared and studied by a variety of electrochemical techniques.

The thin film protects the chemistry of the active site of the redox proteins and facilitates fast electron transfer due to the proximal distance with the electrode [20, 21]. PFV has several advantages over conventional voltammetry, such as high sensitivity, sharp redox status of the entire sample, requirement of a tiny amount of sample, stoichiometry, and fast reactions.

Some common techniques used for characterizing amperometric biosensors are briefly discussed in the following sections.

1.4.3.1.3.1 Cyclic Voltammetry (CV) CV is a powerful tool for the determination of formal redox potentials, detection of chemical reactions that precede or follow the electrochemical reaction, and evaluation of ET kinetics. It scans a potential window in the forward and reverse directions and measures the resulting current. The rate of change of potential with time is referred to as the scan rate (ϑ). The data from the anodic (I_{pa}) and cathodic (I_{pc}) peak currents versus scan rate plots of the immobilized enzymes is extracted to understand the redox processes on the electrode surface, such as surface-controlled or diffusion-controlled process, electron transfer coefficient (α), reversible or quasi-reversible process, surface coverage area (Γ), ET rate constant (k_s), and number of electrons transferred in the reaction (n). The value of α can be estimated by measuring the anodic (E_{pa}) and cathodic (E_{pc}) peak potentials at various CV scan rates (ϑ) using Equations 1.8 and 1.9 [22]:

$$E_{pa} = E^{0'} + \frac{2.3RT}{(1-\alpha)nF\log\vartheta} \quad (1.8)$$

$$E_{pc} = E^{0'} - \frac{2.3RT}{\alpha nF\log\vartheta} \quad (1.9)$$

where $E^{0'}$ is the formal potential, ϑ is the scan rate, and R is the thermodynamic constant ($R = 8.314$ JK^{-1} mol^{-1}), F is the Faraday constant ($F = 96,500$ C mol^{-1}), T is the temperature in Kelvin, and n and α are the charge transfer number and the charge transfer coefficient, respectively, when $0.5 < \alpha < 1$, in general $n = 1$.

The surface coverage, Γ of the bioelectrode, can be calculated using the following equation:

$$\Gamma = \frac{Q}{n}FA \quad (1.10)$$

where A is the area of the electrode and Q is the charge obtained by integrating the peak current area.

k_s can be estimated using the following equations:

$$\log k_s = \alpha \log(1-\alpha) + (1-\alpha)\log\alpha - \frac{\log RT}{nF\vartheta} - \alpha(1-\alpha)\frac{nF\Delta E_p}{2.3RT}$$

$$(1.11)$$

(When $\Delta E_p > 200$ mV)

$$k_s = \frac{\alpha nF\vartheta}{RT} \quad (1.12)$$

When $\Delta E_p < 200$ mV)

The magnitude of k_s indicates the efficacy of the DET between the immobilized enzymes and electronic unit. The increasing height of the redox peak in CV with increasing concentration of substrate implies the involvement of the DET principle in sensing the substrate of interest by the constructed enzyme electrode [23, 24].

Again, for electrochemically reversible ET processes which involve freely diffusing redox species, the Randles–Sevcik equation (Equation 1.13) explains how the peak current i_p (A) increases linearly with the square root of the scan rate ϑ (V s^{-1}), where n is the number of electrons transferred in the redox process, A (cm^2) is the electrode surface area (generally represented as the geometric surface area), C^0 (mol cm^{-3}) is the bulk concentration of the analyte, and D_0 (cm^2 s^{-1}) is the diffusion coefficient of the oxidized analyte [25].

$$i_p = 0.446 nFAC^0 \left(\frac{nF\vartheta D_0}{RT}\right)^{1/2} \quad (1.13)$$

1.4.3.1.3.2 Chronoamperometry The response curve for biosensors can be determined through a chronoamperometry experiment. Chronoamperometry is a time-dependent technique where a square-wave potential is applied to the working electrode. The current of the electrode, measured as a function of time, fluctuates according to the diffusion of an analyte from the bulk solution toward the sensor surface. Chronoamperometry can therefore be used to measure current-time dependence for the diffusion-controlled process occurring at an electrode. This differs with analyte concentration. The resulting current occurring at the electrode is monitored as a function of time after applying the peak potential. The analysis of chronoamperometry data is based on the Cottrell equation (Equation 1.14), which defines the current-time dependence for linear diffusion control. The Cottrell equation describes the observed current (planar electrode) at any time following a large forward potential step in a reversible redox reaction (or to large overpotential) as a function of $t^{-1/2}$.

$$i = \frac{nFAC_0 \sqrt{D_0}}{\sqrt{(\pi t)}} \quad (1.14)$$

where n = stoichiometric number of electrons involved in the reaction, F = Faraday's constant (96,485 C/equivalent), A = electrode area (cm^2), C_0 = concentration of electroactive species (mol/cm^3), and D_0 = diffusion constant for electroactive species (cm^2/s)

1.4.3.1.3.3 Differential Pulse Voltammetry (DPV) and Square Wave Voltammetry (SWV) DPV and SWV provide a unique alternative to sweeping voltammetry methods such as CV. The main advantage of these pulsed techniques is the higher sensitivity that they offer in terms of both potential and current. This enables the detection of electroactive species at very low concentrations (as low as 10^{-7} M) and facilitates the resolution of overlapping redox features of multiple electroactive species (typically, any redox peaks separated by >50 mV can be resolved). In DPV fixed-magnitude pulses superimposed on a linear potential ramp are applied to the working electrode at a time just before the end of the drop. Again, SWV is a specialized form of DPV, which is a large-amplitude differential technique in which a waveform composed of a symmetrical square wave, superimposed on a base staircase potential, is applied to the working electrode. Readers may consult some works on DPV [26, 27] and SWV [28–30] to understand the application of these techniques for biosensors.

1.4.3.1.3.4 Electrochemical Impedance Spectroscopy (EIS) EIS helps to understand the charge transfer behaviors of the thin-film layers on the electrode surface and to discern the performance of enzyme electrodes. It is commonly employed for analysis of enzymatic electrodes by overlaying a range of alternating current (AC) perturbation signals to an electrode that is under direct current (DC) bias. A Nyquist plot (a plot of the imaginary part of the impedance versus the real part of the impedance for different frequencies) is widely employed, which provides the variations of the frequency response to deduce limiting mechanisms connected with charge transfer [31]. The following example may help to understand the biosensing application of EIS. A DNA aptasensor for sensitive detection of a malaria biomarker, *Plasmodium falciparum* lactate dehydrogenase (PfLDH) using the impedimetric technique was developed. For this a specific aptamer, P38, was immobilized over graphene oxide (GO) on a glassy carbon electrode (GCE). With the help of an external redox probe ([Fe(CN)$_6$]$^{-3}$), the presence of PfLDH in a human serum sample was impedimetrically sensed down to a femtomolar level.

Here, in this paper, EIS experiments were performed in a background of 2.5 mM [Fe(CN)$_6$]$^{-3}$. The obtained spectra were represented in the form of Nyquist plots (Figure 1.6A); the arc radius of the semicircle part represents the magnitude of the charge transfer resistance (R_{ct}). The increase in R_{ct} (more the radius of the semicircle more is the R_{ct} value in the Nyquist plot) with increasing PfLDH concentration infers (Figure 1.6A) that PfLDH was successfully captured on the aptamer electrode surface. This is because the negatively charged DNA phosphate backbone of the aptamers prevented the communication of [Fe(CN)$_6$]$^{-3}$ ions with the electrode, which resulted in a prominent increase in the R_{ct}. The acquired Nyquist plots have been fitted to the Randles–Ershler-type equivalent circuit (Figure 1.6A inset), where R_s is the solution resistance, Z_w is the Warburg impedance, and CPE is the constant phase element. So R_{ct} can be calculated from this Randles–Ershler-equivalent circuit. The response data were generated by plotting the fitting results (Figure 1.6B), i.e. concentration of PfLDH with respect to R_{ct}, from which the concentration of PfLDH was detected up to as low as 0.5 fM [32].

1.4.3.1.4 Biofuel Cells (BFCs)-Based Biosensors BFCs are galvanic cells, can be used to measure potential, current or

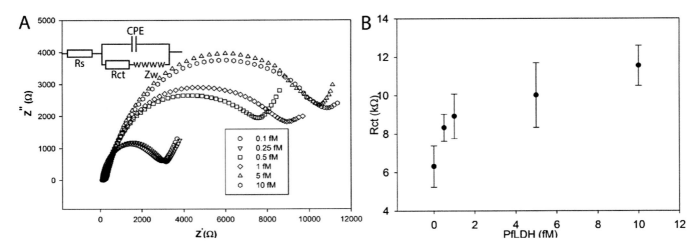

FIGURE 1.6 (A) Nyquist plots on the response of P38-GO-GCE toward increasing PfLDH concentration from 0.1 fM, to 10 fM. (B) Response curve of R_{ct} versus increasing PfLDH concentration [32], Copyright (2016), with permission from Elsevier.

both these signals simultaneously. The bio-catalysts used in the BFC converts chemical energy into these electrical energy forms. The basic differences between these cells and chemical fuel cells are the types of catalysts, fuels, and physical conditions used for their operations. BFCs mostly use enzymes and microorganisms as electrode catalysts. The fuels for BFC are mostly renewable in nature, such as glucose and other carbohydrates, such as alcohols. However, nonrenewable fuels such as hydrocarbons can also be used if the biocatalysts can electrochemically oxidize these compounds or metabolically convert these substances to produce current in the cell. As a whole, BFCs can encompass a wide range of compounds as a fuel substrate. These BFCs function under mild conditions, around physiological pH and room temperature, without using any hazardous or toxic chemicals or materials. Considering these facts, BFCs are acclaimed as green energy technology. The general configuration, principle of operation, and application potential of BFCs are illustrated in Chapters 10 and 13.

The scope of scaling down the size of BFCs to a chip-based platform supported by their self-powered attributes, along with their highly selective responses (potential or current) against a vast array of substrates under physiological operating conditions is the primary stimulus of exploring this energy-transducing technology for biosensing applications. There are two approaches for generating biosensing signal in BFCs: (a) Turn-on sensors. Here the target analyte, which may be a fuel or other substance, activates/initiates the bioelectrocatalytic reactions and generates or increases the electrical output. (b) Turn-off sensors. Here the target analyte, which may be an inhibitor or other substance, de-activates/reduces the bioelectrocatalytic reactions and correspondingly decreases the electrical output signal current [33]. The application of microbial BFCs as a sensor for biochemical oxygen demand (BOD), microorganism load, toxicants, etc., is widely known [34, 35]. Over the last decade, the research on the application of enzymatic BFC-based wearable sensors in the fields of health care and sports have made significant progress.

1.4.3.2 Optical Biosensors

The research and technological developments of optical biosensors have accomplished an exponential growth over the last decade. This is because there is a wide range of optical behaviors, such as fluorescence including various luminescence approaches, ultraviolet (UV)-visible absorption and reflections, internal reflection, and scattering spectroscopy that can be explored to develop the biosensor signals. Moreover, some of the optical behaviors (e.g. fluorescence) offer extremely high sensitivity to the biosensor. Additionally, the color-based detection could be achieved in a simple, low-cost, portable format that greatly boosts their PoC and point of need (PoN) applications relevant to health care, environmental monitoring, and food and agricultural industries, among others [36].

One of the earliest examples of an optical biosensor is a test strip for the detection of glucose in urine [37]. The strip was made of a cellulose pad that consisted of co-immobilized glucose oxidase and peroxidase. The H_2O_2 produced from the enzymatic oxidation of glucose reacts with o-tolidine in the vicinity of peroxidase to form a dye. The color intensity of the dye was proportional to the glucose concentration as envisioned by the naked eye. It represented a semi-quantitative measurement technique for glucose. There has been a parallel growth of technological innovation to process the color intensity of the detection strip to a concentration unit of the target analyte with the help of suitable apps in the interface of modern smart phones. One such example is the quantitative detection of urea in saliva [38]. A further illustration of smart phone-based detection devices has been discussed in a separate chapter of this book. We would like to highlight here working principles of few prominent optical biosensors.

1.4.3.2.1 Absorbance/Reflectance-Based Optical Biosensors

The optical detection principle of absorbance transducers is fundamentally based on the Beer-Lambert law:

Fundamentals of Biosensors

$$\log \frac{I_0}{I} = A = \varepsilon C l \qquad (1.15)$$

where I_o is the intensity of incident light, I is the intensity of transmitted light, A is absorbance, ε is the molar absorbance of the analyte at a specific wavelength, C is the concentration of analyte, and l is the path length of light through solution. The common absorbance transducers utilize a single fiber or fiber bundle that brings light to the analyte-sensitive reagent phase, and the transmitted or reflected light is returned to a measurement instrument or detected through fiber(s), as shown in Figure 1.7A. The optical path length, absorption cross-section of the transducing molecule, and the illumination wavelength determine the extent of the absorption. Changes in the chemical environment can modify the absorption of the biorecognition element, and this modification is monitored as a change in transmitted intensity within the biosensor.

In case of a nontransparent sample, it becomes difficult to measure transmitted light acceptably, and in these cases the intensity of the reflected light may be used as a measure of the color of the recognition element, analyte, or product, as shown in Figure 1.7B [39].

1.4.3.2.2 Surface Plasmon Resonance (SPR) Biosensors

The name itself implies that SPR is a surface phenomenon that occurs on the surface of highly conducting metals, typically gold, which support to generate plasmons. Plasmons are collective charge density oscillations of electron in a metal. Surface plasmons (or surface plasmon polaritons) are surface electromagnetic waves that propagate parallel along a metal-dielectric interface (e.g. metal-air). The excitation of surface plasmons by light is denoted as an SPR for planar surfaces or localized surface plasmon resonance (LSPR) for nanometer-sized metallic structures. The LSPR, on the other hand, is highly sensitive to size, size distribution, and shape of the metal nanostructures, as well as the environment that surrounds them. The LSPR is the fundamental principle behind many color-based biosensor applications and different lab-on-a-chip sensors.

The velocity of the surface plasmons (and hence the light energy for resonance) changes with the change in the refractive index near the metal surface. The refractive index in turn is greatly dependent on the chemical environment of the metal-dielectric medium. In this affinity biosensor, biorecognition elements specific to analyte molecules (in liquid samples) are immobilized on the surface of the metal. The binding of the analyte to the biorecognition layer over the sensor surface gives rise to a refractive index change close to the sensor surface, which can be measured by the optical reader.

The refractive index change (Δn_b) caused by the binding of specific analyte molecules at the sensor surface can be expressed as (often referred to as the de Feijter formula):

$$\Delta n_b = \left(\frac{dn}{dc}\right)_{vol} \Delta c_b = \left(\frac{dn}{dc}\right)_{vol} \frac{\Delta \Gamma}{h} \qquad (1.16)$$

where $(dn/dc)_{vol}$ is the refractive index increment, Δc_b is the wt/vol concentration of bound molecules within the sensitive layer with the thickness h, and $\Delta \Gamma$ is the corresponding surface concentration (mass per surface area). The $(dn/dc)_{vol}$ is a well-characterized property for most of the biochemical species and ranges typically from 0.1 to 0.3 cm^{3}g^{-1}. A change in the effective refractive index of the surface plasmon due to the capture of analyte can be expressed as

$$\Delta n_{ef} = K \Delta \Gamma \qquad (1.17)$$

where K is a constant.

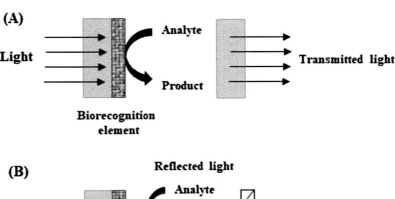

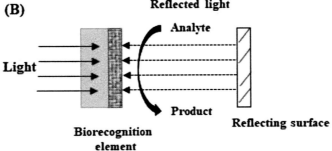

FIGURE 1.7 Absorbance-based optical transducer with **(A)** absorption configuration and **(B)** reflectance configuration [39].

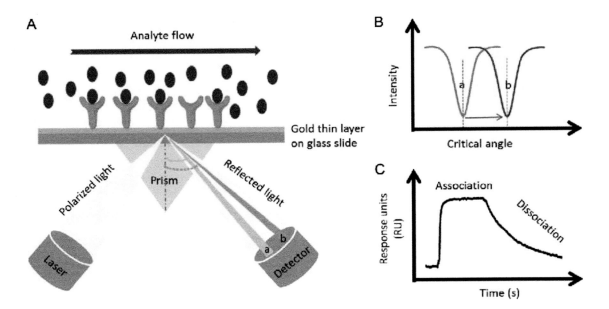

FIGURE 1.8 Schematic representation of SPR phenomenon for the measurement of analyte binding to immobilized ligand. (**A**) The ligand is immobilized on the sensor chip, which is composed of a gold thin layer on a glass slide. (**B**) Reflected light intensity shifts upon a critical angle change from "a" to "b," resulting from a binding interaction event. (**C**) Light intensity shifts are transformed into sensorgrams, a plot of response units (RU) versus time [40]. Copyright (2015), with permission from Elsevier.

The optical reader of the SPR sensor measures changes in a characteristic of a plasmon-coupled light wave resulting from changes in the effective refractive index. Different coupling mechanisms have been developed for coupling a light wave to a surface plasmon, among which the attenuated total reflection in prism couplers and diffraction on diffraction grating couplers are common.

Most of the SPR sensors are designed by using the Kretschmann configuration of the attenuated total internal reflection method. Here, a light wave (usually plane polarized) passed through a high refractive index prism is reflected at the prism base covered with a thin gold film (Figure 1.8). The light evanescently passes through the thin film and excited the plasmon at the outer boundary of the film if the incident light wave and plasmon are closely phase-matched. The phase-matching condition can be written as:

$$n_p \sin\theta = n_{ef} \quad (1.18)$$

where n_p is the refractive index of the coupling prism and θ is the angle of incidence on the metal film (in the prism).

The SPR biosensor's response can be generated by using either (a) wavelength modulation, where the angle of incidence is fixed and the coupling wavelength serves as a sensor output, or (b) angular modulation, where the coupling wavelength is fixed and the coupling angle of incidence serves as a sensor output, or (c) intensity modulation, where both the angle of incidence and the wavelength of incident light are fixed at nearly resonant values and the light intensity serves as a sensor output.

Since the development of the first commercial SPR sensors in 1990, many commercial products are now available in the market. SPR biosensors allow label-free, direct, highly sensitive, and real-time detection of chemical and biological analytes. It is a useful technique for measuring the affinity, stoichiometry, and kinetics of an interaction [41, 42].

1.4.3.2.3 Fluorescent-Based Optical Biosensors

The research on fluorescence-based biosensors and allied analytical techniques is increasing exponentially over the last two decades due to the emergence of various novel and efficient fluorescence probes. The major types of fluorescent probes explored for biosensor applications include organic, nucleic, and cell fusion dyes, fluorescent proteins, and small fluorescent nanoparticles such as quantum dot, metal nanocrystals, and carbon dots. These fluorescence probes absorb (excited with) electromagnetic radiation at their specific wavelengths and they emit fluorescence. The wavelength of the emitted fluorescent light is usually longer than the excitation wavelength. This difference between the absorbance and the emission peak wavelength is termed the Stokes shift after the name of Sir George Stokes [43, 44]. The time interval between absorption and emission in fluorescence is very short, usually on the order of 10^{-9} to 10^{-8} s. The Jablonski diagram, named after the physicist Aleksander Jablonski, commonly depicts the illustration of a single fluorescence event.

Fluorescence biosensor development normally involves coupling the target recognition with a change in fluorescence of the reporter/probe. Based on the nature of the sensing element, different designs of the fluorescent biosensors can be introduced. In single fluorophore-based architecture, the reporter/probe that is sensitive to the microenvironment is attached to the recognition element and then probes the presence of the target analyte in the sample by monitoring the change in fluorescence. As the target molecule binds, the conformation of the recognition element changes, which further changes the microenvironment of the fluorophore. This change in fluorescence intensity may occur due to various mechanisms, such as dynamic quenching caused by the change in accessibility of the solvent, change in polarity caused by the surrounding environment, or a change in fluorophore-protein interactions. However, there can be a change in fluorescence intensity when the fluorophore directly interacts with the bound target molecule and hence can be a basis for designing a biosensor.

A two fluorophore-based design format was also widely used to develop a biosensor following the principle of FRET (Förster resonance energy transfer). FRET takes place by the direct excitation of an acceptor fluorophore by the energy donated by a donor fluorophore, which is excited by electromagnetic radiation in the appropriate wavelength. Transfer of energy takes place when the donor and acceptor are in close proximity (<10 nm distance) and the dipoles of both molecules are oriented appropriately [43, 44]. In these biosensors, fluorophore pairs are normally fused with the recognition element. A change in FRET is observed whenever there is a ligand-induced change in conformation that can alter the distance or relative orientation of the fluorophores of the FRET pair. In many cases, the fluorophore is coupled with a suitable quencher, which dissipates the absorbed energy in the form of heat. The quenchers can absorb energy over a wide range of wavelengths and can also dissipate this absorbed energy in the form of heat, and they remain dark. Due to these properties, quenchers can be very useful molecules for energy acceptors in FRET pairs.

The FRET sensor provides a signal based on the ratio of the acceptor to donor emission that enables the quantitative measurements of the target analyte even in a complex environment. One of the major constraints for developing the FRET-based sensor is to identify a suitable partner for the fluorophores to generate a noise-free effective signal. It may be mentioned that the relationship between the donor acceptor proximity is critical for the FRET phenomenon. FRET-based mechanisms have been employed in the genetically encoded biosensors such as for cyclic adenosine monophosphate (cAMP) and cyclic guanosine monophosphate (cGMP), sugars, phosphate, Ca^{2+}, and adenosine triphosphate (ATP) [44]. The ratio-metric signal-based sensors can also be designed by using an intermolecular charge transfer (ICT) approach. However, the ratio between two relatively broad signal emissions becomes difficult to determine in many ICT-based ion sensors.

Apart from exploiting the normal change in fluorescence intensity and FRET, FCS (fluorescence correlation spectroscopy), and FLIM (fluorescence lifetime imaging) are also explored for generating specific signals for biosensing applications. FCS analyzes small deviations from spontaneous fluorescence intensity of the sample to gain information on the kinetics of thermodynamic processes associated with reversible fluorescence changes, e.g. flow rate, diffusion coefficient, and molecular concentration [45]. FCS is usually useful to deal with low concentrations of molecules.

FLIM is used for imaging biological tissues and reactions taking place in living cells. This method gives information on changes in the local environment of the fluorophore or changes in its energy in response to the interactions with the local environment [45].

Fluorescent nanoparticles are increasingly used to develop various detection methods. As an example, a sol-gel encapsulated CdS quantum dots (QDs)-uricase/horseradish peroxidase (HRP) enzymes hybrid system has been used to detect uric acid. The hybrid system oxidizes uric acid to allantoin, CO_2, and H_2O_2. The produced H_2O_2 has the ability to quench the QDs fluorescence intensity, which is proportional to the uric acid concentration [46].

The information described here is certainly not a comprehensive coverage on the vast subject of fluorescence-based detections, but should provide the reader with at least a general appreciation of the breadth of prominent options available.

1.4.3.2.4 Luminescence Biosensors

Luminescence is defined as the radiation emitted by an atom or a molecule when these species return to the ground state from the exited state. On the basis of the source of excitation, the luminescence phenomenon is mainly classified as photoluminescence (fluorescence and phosphorescence) when the excitation source is energy from absorbed light, chemiluminescence (energy from chemical reactions), and bioluminescence (energy from biologically catalyzed reactions) [47]. When a molecule absorbs a photon in the visible region, exciting one of its electrons to a higher electronic excited state and then radiates a photon as the electron returns to a lower energy state, this process is called photoluminescence. If the molecule experiences internal energy redistribution after the initial photon absorption, the radiated photon may be a longer wavelength than the absorbed photon. Fluorescence and phosphorescence are special forms of photoluminescence. Chemiluminescence occurs on the emission of light by the release of energy from a chemical reaction. Bioluminescence is a type of chemiluminescence occurring in some living forms and involves a protein (enzyme). It is the result of certain oxidation processes (usually enzymatic) in biological systems like fireflies, jellyfish, glow worms, and mostly marine animals. Among different luminescence types, chemiluminescence and electro-chemiluminescence are growingly used in developing biosensors. The principle behind chemiluminescence-based sensors is the combination of light-emitting reactions with sensor capabilities. Optical transducers have been used for the design of these biosensors. They can be used to detect certain biochemical reactions that occur. The immobilized biorecognition element has been marked with a chemiluminescence species, and on reaction with the analyte, it generates light. Generally, photomultiplier tubes (PMTs) are used to detect this light. Chemiluminescence-based sensors have a high sensitivity, a fast dynamic response property, and a wide calibration range. These sensors have been used for immunosensing applications. However, they are expensive and difficult to use for real-time monitoring [47].

A comprehensive discussion on electrochemilunescene biosensors has been included in another chapter of this book.

The assembly of a luminescence-based transducer is similar to that of an absorbance-based transducer. Excitation light is directed to the recognition element, which is exposed to the analyte. The fluorescence is then collected by the detection system. Any change in luminescence intensity, phase, or lifetime can be related to the interaction of recognition element, analyte, and product. The principle of luminescence quenching is normally employed in a luminescence-based transducer. According to Stern and Volmer, the relationship between intensities or lifetimes in the absence and presence of a quencher is given by:

$$\frac{I_0}{I} = 1 + K_{SV}[Q] = \frac{\tau_0}{\tau} \quad (1.19)$$

where I_o and I are the luminescence intensities in the absence and presence of quencher Q, respectively; K_{SV} is the Stern-Volmer constant; $[Q]$ is the quencher concentration; and τ_0 and τ are the luminescence lifetimes in the absence and presence of quencher, respectively [39]. This type of transducer is best exemplified by the oxygen sensors based on luminescence quenching of ruthenium complexes [48, 49].

1.4.3.3 Calorimetric Biosensors

A large number of enzyme-catalyzed reactions are known to be exothermic in nature with a significant evolution of heat (normally $\Delta H \sim -10$ to -200 kJ mol^{-1}). The calorimetric biosensor thus relies on the heat generated in these reactions to produce the response signal for the target analyte of interest. The relations among the heat generated, enthalpy change, and temperature change under adiabatic conditions can be expressed by the following generic equations:

$$Q = -n_p \Delta H \quad (1.20)$$

$$Q = mC_p \Delta T \quad (1.21)$$

$$\Delta T = \frac{-\Delta H n_p}{mC_p} \quad (1.22)$$

where Q is the total heat evolved during a catalytic reaction, ΔH is the molar enthalpy change, n_p is the molar number of the product, ΔT is the temperature change, C_p is the heat capacity, and m is the mass of the system in which the reaction takes place.

The heat generated in the reactions and hence the temperature change, though in principle, could be measured by using the conventional mercury-based thermometer, practically their use is limited due to their low temperature sensitivity. More sensitive different temperature transducers are known and employed in enzyme calorimetric analyzers among which the thermopile (or thermocouple) and thermistor are more popular.

The potential difference (ΔV) of a thermopile transducer depends on the pair number of thermocouples (n), the Seebeck coefficient (ε), and the temperature difference (ΔT):

$$\Delta V = n\varepsilon\Delta T \quad (1.23)$$

Substituting ΔT from Equation 1.22:

$$\Delta V = n\varepsilon \frac{-\Delta H n_p}{mC_p} \quad (1.24)$$

Hence, for the adiabatic environment, the enthalpy change (ΔH) produced by an enzymatic reaction in a biosensor can lead to a direct change in potential (ΔV). Up to 80% of the heat generated in the reaction can be registered as a temperature change in the reaction under such adiabatically controlled conditions. A thermopile transducer has been reported for the detection of many analytes, including organophosphate pesticides such as dichlorvos [50].

The thermistor is a combination of *thermal* and *resistor*, implying that it is a type of resistor whose resistance is dependent on temperature. Hence, thermistors are widely used as temperature sensors, among other applications. Two major types of thermistors are available, depending upon the decrease or increase in resistance with increasing temperature. The thermistor whose resistance decreases as temperature rises (negative temperature coefficient, or NTC, type typically) is commonly used as a temperature sensor. The relationship between resistance and temperature is linear, assuming, as a first-order approximation,

$$\Delta R = k\Delta T \quad (1.25)$$

where ΔR is the change in resistance, ΔT is the change in temperature, and k is the first-order temperature coefficient of resistance. If k is positive, the resistance increases with increasing temperature, and the device is called a positive temperature coefficient (PTC) thermistor. If k is negative, the resistance decreases with increasing temperature, and the device is called an NTC thermistor.

The benefits of using a thermistor are accuracy and stability. Different versions of the thermometric biosensors are known. The conventional ET thermistor system has been progressively modified into mini, micro, and hybrid thermometric devices. A cross-section of a conventional enzyme thermistor instrument with its various components is shown in Figure 1.9.

The instrument contained an immobilized enzyme column for the reaction in a thermostated system, and the temperature at the point of exit was analyzed using a thermistor connected to a Wheatstone bridge. The components were assembled in a compact design for operation of the ET attached to the flow injection analysis. The reference probe contains an immobilized nonenzyme protein. The response signal can be refined by subtracting the signal of the reference column from the signal of the enzyme column [51].

Fundamentals of Biosensors

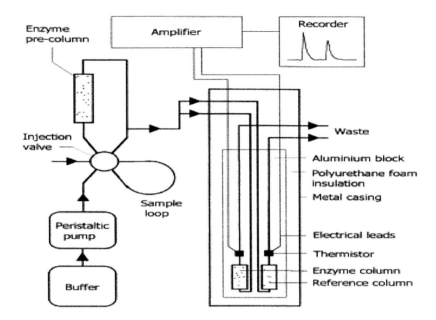

FIGURE 1.9 A cross-section of a conventional enzyme thermistor (ET) instrument showing its various components [51] Copyright (2001), with permission from Elsevier.

The low sensitivity of calorimetric biosensors for the reactions with low enthalpy change could be increased substantially through coupled reactions, which contribute to the heat output of the assay [52]. In the simplest case, this can be achieved by linking together several reactions in a reaction pathway, all of which contribute to the heat output. For instance, the sensitivity of the glucose analysis using glucose oxidase can be substantially increased by the co-immobilization of catalase in the column that exothermically degrades the hydrogen peroxide produced in the oxidase reaction. This highly exothermic reaction doubles the sensitivity of the sensor, reduces the deleterious effects of hydrogen peroxide in the reactor, and enriches the oxygen concentration in the reactor for the oxidase reaction. There have been reports of increasing the enthalpy change by using certain buffers like Tris that can increase the total enthalpy of the proton-producing reactions. Small amounts of organic solvents (around 5%, v/v) present in the aqueous buffer can also be used to increase the registered temperature changes by increasing the total enthalpy change in the reaction, which can be attributed to the lowering of heat capacity of the solvent.

The advantage of the calorimetric biosensors is that their application is not interfered with by the optical properties of the sample, viz. color and turbidity. The extensive application of calorimetry has, however, been restricted by the relatively high cost and complexity of the existing calorimeters. The design fabrication must ensure that the temperature of the sample stream remains constant ($\pm 0.01°C$). Recently numerous inexpensive, less complicated devices for biochemical analysis have been developed merging the universality of calorimeters with the specificity of enzymatic reactions [53].

1.4.3.4 Piezoelectric Biosensors

Pierre Curie discovered the piezoelectric (Greek *piezo* means to squeeze or press) effect in 1880, and later it was used for sensing purposes in the 1950s. The Curie group perceived that anisotropic crystals, i.e. crystals without a center of symmetry, can generate an electric dipole when mechanically squeezed (Figure 1.10). The effect can also work in an opposite way in that an anisotropic crystal becomes deformed when voltage is applied on it [54]. The mechanical deformation is, however, a simple situation, and generally oscillation occurs in the common applications. In the case of oscillation, an alternating voltage is imposed on the crystal and then mechanical oscillation occurs. A piezoelectric sensor is used to measure changes in pressure, acceleration, temperature, strain, or force by converting them to an electrical charge. Many materials, both natural and man-made, exhibit piezoelectricity such as natural crystals: cane sugar, quartz, Rochelle salt, topaz, dry bone, tendon, silk, wood, enamel and man-made crystals: gallium orthophosphate ($GaPO_4$), and langasite ($La_3Ga_5SiO_{14}$).

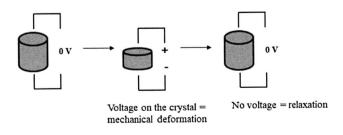

FIGURE 1.10 Piezoelectric effect when mechanical deformation is initiated by an applied voltage [54].

Frequencies of oscillations are determined in standard analytical applications. Analyte can be determined from the electricity produced on the crystal surface through interaction with either crystal alone or electrode. The bound mass on the crystal surface causes slowing of oscillation. For the common quartz crystals, the frequency shift (Δf) is directly proportional to the change of mass (Δm) on the crystal, as described by Sauerbrey [55] (Equation 1.26).

$$\Delta f = \frac{-2f_0^2 \Delta m}{A\sqrt{(\rho_q \mu_q)}} = -2.3 \times 10^6 \, f_0^2 \frac{\Delta m}{A} \quad (1.26)$$

In this equation, f_0 is the fundamental mode of the crystal oscillation (in hertz), A is the piezoelectrically active area (in centimeters), ρ_q means density (2.648 g/cm^3), and μ_q means shear modulus (2.947 × 10^{11} g/cm·s^2) of quartz. The Sauerbrey equation is reliable for calculating frequency shift when the ambient environment is not unaltered. Upon a change in the surrounding environment, since viscosity has an impact on frequency shift, the equation described by Kanazawa and coworkers for quartz crystal should be consulted with [56] (Equation 1.27).

$$\Delta f = f_0^{3/2} \sqrt{\frac{\Delta(\rho_l \eta_l)}{\pi \rho_q \eta_q}} \quad (1.27)$$

The equation states that frequency shift is proportional to an increase of ambient viscosity η. The symbols' meaning is the same as for the Sauerbrey equation – the symbols with index l relate to the ambient liquid and q to quartz crystal.

Piezoelectric biosensors can be of two different types: bulk acoustic wave (BAW) piezoelectric sensors and surface acoustic wave (SAW) piezoelectric sensors. BAW is based on the principle that the wave propagates through the interior of the substrate. Conversely, SAW works on the principle that the wave propagates on the surface of the substrate. As the wave propagates through or on the surface of the substrate, its velocity continually changes. This change can be known by measuring the change in the frequency. This can be related to the physical mass being measured. These sensors apply an electric field that creates mechanical stress (wave). This moves through or on the substrate, and in the last step, is converted back to an electric field before we can measure it.

A quartz crystal microbalance (QCM) is one of the simplest and commonly used BAW devices. SAW sensors are composed of a thick layer of piezoelectric material, like quartz, lithium niobate, or lithium tantalite. Here, Rayleigh waves propagate along the upper surface. The surface transverse wave (STW) sensor is the most commonly used SAW device.

This technique is known for its excellent sensitivity, and hence it has wide applications in the medical, aeronautical, and telecommunications, fields. Quartz is a commonly used piezoelectric material as it is cheap and has the ability to withstand various types of stresses. Other potential materials are lithium niobate and lithium tantalite. These transducers are suitable for label-free and real-time biosensing. They can attained detection limits to the pico level and hence are suitable to measure various gases such as ammonia, hydrogen, methane, and carbon monoxide. Piezoelectric biosensors have been reported to detect various toxins, pathogens in food and water, hepatitis B and C, etc. They have also been used in protein and DNA detection. These sensors have good compatibility with integrated circuits (IC) technology and can be easily manufactured by photolithography, which renders them inexpensive [57].

A QCM-based biosensor was constructed for the determination of organophosphorus and carbamate pesticides in the nM level. The sensor had an immobilized enzyme acetylcholinesterase, which converted 3-indolyl-acetate to insoluble indigo pigment, providing alteration in the oscillations. The biosensor was used for the assay of pesticides, which inhibit the enzyme acetylcholinesterase. When the enzyme became inhibited, the precipitate was not formed [58].

1.5 BIORECEPTOR IMMOBILIZATION STRATEGIES

To efficiently capture the biochemical signal generated as a result of interactions between the bioreceptor and the target analyte, the bioreceptors are usually immobilized on the transducer surface of the biosensor device. This process empowers reusability of the bioreceptor with linked cost reduction of the developed devices/process. There are mainly four different techniques of bioreceptor immobilization, and their choice is based on their intended use. The different immobilization strategies for bioreceptors have been depicted in Figure 1.11 and Table 1.2.

1.5.1 Adsorption

Adsorption refers to the easiest technique of immobilization of the bioreceptor onto a surface by reversible surface interaction between the bioreceptor and the surface. The forces involved in this interaction are weak forces viz. van der Waals forces, ionic bonds, hydrogen bonds, and hydrophobic interactions. In this method the solid support onto which the immobilization is desired is placed in contact with a solution containing the dissolved bioreceptor at optimum pH, ionic concentration, etc., for an appropriate period of time to allow the adsorption to take place. The unbound bioreceptor molecules are removed from the surface by washing with a suitable buffer [59].

Adsorption offers the advantage of retention of bioreceptor activity (native conformation) during immobilization, as this process is not accompanied by chemical changes in the bioreceptor molecule or the surface. It is simple, low cost, and nondestructive to the bioreceptor. However, the disadvantage of this technique is that the immobilized bioreceptor is prone to desorption or leaching under conditions of changed pH, ionic strength, temperature, or polarity of the solvent, as only weak forces are involved to achieve the immobilization. Also, nonspecific interaction of other proteins or molecules

Fundamentals of Biosensors

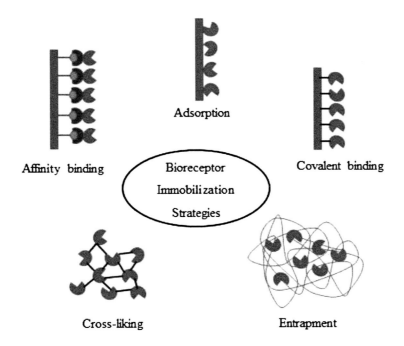

FIGURE 1.11 Bioreceptor immobilization strategies.

on the surface can cause contamination, inhibition, or signal interference. This is why biosensors employing immobilization of the bioreceptor by adsorption usually suffer from poor operational and storage stability. Importantly, the bioreceptor molecules may not exhibit homogeneity in orientation in the immobilized form; hence, there is a possibility of reduction of binding sites/active sites for the analytes of interest. In order to overcome the drawbacks, many strategies are emerged, among which, the immobilization on three-dimensional and/or porous structure of the matrix is widely used. Hydrogel polymers, such as agarose, collagen, gelatin, and polyacrylamide gel, and inorganic materials, such as mesoporous carbon, silica, and zeolite, which have a rigid porous structure for the retention of bioreceptor special enzyme molecules are mostly used as the enzyme immobilizing matrix surface.

1.5.2 Covalent Binding

This involves immobilization of the bioreceptor onto the transducer surface through covalent bonds. The functional groups of bioreceptors, such as amine, hydroxyl, carboxyl, aldehyde, and sulfhydryl, are exploited for the covalent linkages with the solid transducer surfaces. The covalent immobilizations are frequently performed in two steps. First, the surface is activated using linker (also called carrier)

TABLE 1.2
Summary of Bioreceptor Immobilization Strategies

Immobilization Strategy	Characteristic Feature	Advantages	Disadvantages
Adsorption	Weak bonds	• Simple and easy • Limited loss of enzyme activity	• Desorption • Nonspecific adsorption
Covalent binding	Strong chemical bonds	• Stable • No diffusional barrier • Short response time	• High loss of enzyme activity
Entrapment	Confinement of bioreceptors within gel or polymer without chemical reaction	• Enzyme activity not compromised	• Diffusion barrier formed • Enzyme leakage may occurs
Crosslinking	Bond between bioreceptors via crosslinking agents	• Simple	• Likely to suffer from high loss of enzyme activity
Affinity	Affinity bond between affinity tag of bioreceptor and a functional group on the support	• Controlled immobilization	• Requirement of the presence of specific groups on the bioreceptor (e.g. histidine, biotin)

molecules such as glutaraldehyde or carbodiimide that act as the covalent linker between the bioreceptor and the surface. Second, addition of the bioreceptor (viz. enzyme) followed by formation of a covalent bond between the activated substrate and the bioreceptor molecule. The most widely used carrier molecules, particularly for enzyme immobilization, are hydrophilic in nature such as agarose, dextran, cellulose, and starch. These carrier molecules consists of ideal functional groups viz. hydroxyl groups that support the formation of covalent bonds. Also, their ability to form hydrogen bonds with water helps maintain a hydrophilic environment that favors bioreceptor/enzyme stability. The orientation and homogeneity of the linker/carrier molecules can be controlled as they form a self-assemble monolayer (SAM) [59]. This in turn allows for controlled homogeneity of the immobilized bioreceptor on the surface, unlike that of adsorption. Due to the immobilization of bioreceptors through covalent bonds, the leaching problem of the receptor could be overcome. The main disadvantage of this technique lies in the probable denaturation of the bioreceptors accompanied by the loss of their activity; additionally, this method is time consuming and complex as compared to absorption and is poorly reproducible.

1.5.3 Entrapment

Entrapment refers to confinement of the enzyme within a polymeric network without chemically interacting with the support matrix. The pore size of the matrix is controlled to prevent loss of bioreceptor while allowing movement of substrates and products. In this method, a bioreceptor is usually mixed with monomeric units in solution, which is then made to undergo polymerization. The bioreceptor molecules are entrapped within the polymer during the polymerization process. This method offers the advantage of minimizing bioreceptor leakage and denaturation. Additionally it helps to increase stability of the enzyme as the microenvironment of the entrapment matrix can be controlled to have optimal pH, ionic strength, or polarity. However, the disadvantage of this method is that the support matrix polymerization generates mass transfer resistance because of which the binding sites/active sites of the bioreceptor may not be accessible to the substrates efficiently. Special care needs to be taken in controlling the pore size of the matrix to minimize bioreceptor leaching. Also, this method is accompanied with low enzyme loading capacity. All these factors may contribute to serious kinetic implications. A few procedures to achieve entrapment include electrochemical polymerization, photopolymerization, sol-gel process, and microencapsulation.

In electrochemical polymerization, monomer polymerization takes place due to the application of appropriate potential or current. This application of potential or current may generate radical species that facilitate the polymerization process on the electrode surface solution while simultaneously entrapping the bioreceptor. Examples of electropolymerized films used for immobilization include polyaniline (PANI), polypyrrole, thiophenes, etc.

In photopolymerization, monomers are activated to form cross-links or form polymers when exposed to light (photopolymers). In the sol-gel process, a metal alkoxide precursor such as tetramethoxysilane is suspended or dissolved in a suitable acidic solvent for hydrolization to produce silanol groups. This is followed by addition of a base to activate to hydrolized precursor to initiate a condensation reaction to form siloxane polymers. With appropriate time and temperature, gelation of the polymer occurs with bioreceptor entrapment. The sol-gel method offers the advantage of forming a stable nanoporous material with high loading capacity of the enzyme that aids in preserving enzyme activity and increasing biosensor sensitivity. However, this method is comparatively costly and is accompanied with matrix inhomogeneity. Encapsulation refers to entrapment of the bioreceptor inside a closed semi-permeable membrane. The nature of the membrane may be polymeric, lipoidal, or nonionic in nature. The membrane restricts the movement of the enzyme, thereby preventing loss of the enzyme, while allowing the diffusion of small molecules like substrates or products. This method maintains enzyme integrity, as the enzyme remains isolated from the external environment. However, control of membrane pore size is crucial to ensure retention of the enzyme and prevention of enzyme leakage, while allowing the entry and exit of substrate and product molecules [59].

1.5.4 Crosslinking

Immobilization by crosslinking refers to the creation of intermolecular cross-linkages between the bioreceptor molecules by covalent bond formation. This process requires the involvement of a multifunctional reagent like glutaraldehyde or toluene diisocyanate, which links the bioreceptor molecules together. Glutaraldehyde, for example, can crosslink enzyme molecules via the reactions of free amino groups of lysine residues of the individual enzyme molecules. This results in the formation of the three-dimensional complex structure or bioreceptor aggregates. The advantages of immobilization by crosslinking include minimization of bioreceptor leakage due to the involvement of covalent bonds and possible control of the bioreceptor microenvironment to maintain optimum pH, ionic strength, and other factors. The disadvantage of this technique is that the crosslinking agent can bring about certain modifications in the bioreceptor or denaturation that compromises its biological activity to various degrees [60].

1.5.5 Affinity

The strategy of affinity immobilization is employed with the main focus of obtaining a particular orientation of the enzyme or bioreceptor so as to leave its active site/binding site undisturbed and therefore accessible to the substrate/analyte. This can be achieved by creating affinity bonds between an activated support viz. lectin, avidin, metal chelates, and a specific interacting group or sequence of the bioreceptor viz.

carbohydrate residue, biotin, and histidine, respectively. Such groups or affinity tags are sometimes attached to the bioreceptor by genetic engineering methods [61].

1.6 APPLICATION OF BIOSENSORS

1.6.1 Application in the Food Industry

The important concerns relevant to the food industry include quality and safety of the food products, storage and shelf life, and processing. Quality control measures play an important role in approving any food product as suitable for consumption. This involves detection of food spoilage microorganisms or other molecules that may be indicative of degraded food quality and of food-borne pathogens. Because biosensors are specific, sensitive, rapidly responsive, and cost-effective as compared to the traditional methods, these have captured attention for practical application in assessing food quality. The enzyme-based biosensors and immunosensors are widely used for this purpose. Biosensors with different detection principles for detection of foodborne pathogens such as *Staphylococcus aureus* [62], *Salmonella typhimurium* [63], *Salmonella enteritidis* [64], *Escherichia coli* [65], and *Listeria monocytogenes* [66] are some examples.

1.6.2 Application in the Fermentation Industry

Biosensors play a pivotal role in the fermentation industry where continuous monitoring of metabolites, products, enzyme, antibody, biomass, or byproducts is essential. Several commercial biosensors are available to monitor various fermentation processes. A few examples of compounds usually monitored in the fermentation industry are listed in Table 1.3 [67].

1.6.3 Biomedical Applications

The application of biosensors in biomedical field has been rapidly growing since the commercial release of the glucose biosensor. Blood glucose monitoring has been occupying center stage in biosensor research since the first product launched in the market in the year 1975 by Yellow Springs Instruments (YSI). Following this, several new and improved products were introduced in the global market by different companies over the years [68]. Nova's StatStrip has introduced a glucose monitor for use in neonatal testing. StatStrip is the only glucose monitor with 6s analysis time that measures hematocrit on the strip. The company also provides a handheld device for the measurement of blood lactate using a very tiny drop of blood (0.7 µL) with an analysis time of 13s. Nova also commercializes a biosensor that measures creatinine with an analysis time of 30s [68]. Roche Diagnostics markets the Accu-Chek family of products/services for blood glucose monitoring. Its US Patent Number 6,541,216 describes an invention that allows the measurement of ketone levels in blood. In 2001, LifeScan, a part of the Johnson & Johnson companies, launched a glucose measuring device (OneTouch Ultra blood glucose) and the In Duo system, the world's first blood glucose checking and insulin-dosing system. After that, in 2003, LifeScan launched the OneTouch UltraSmart blood glucose monitoring system with a 3,000-record memory for the storage of health, medication, exercise, and meal information [68]. The acceptance and success of biosensors in this field are largely determined by a high level of precision in measuring analytes of clinical importance, capability to sense the analyte in real time, and high sensitivity. A few common applications of biosensors in biomedicine include detection and measurement of disease-specific biomarkers; biomolecules such as glucose, lactate, peroxides, cholesterol, and cytokines; and release of antibodies or other indicator biomolecules in various inflammatory diseases and tumors. For example, biosensors for detection of *Candida* infection [69], circulating tumor cells (CTCs) [70], and antibiotic sensitivity of bacteria [71] are reported.

1.6.4 Environmental Applications

Harmful environmental agents pose a serious threat not only to human health but to the entire ecosystem. The first and foremost step for appropriate treatment of these agents is to detect their levels and their sources. The attainable high sensitivity of the biosensors plays a crucial role in detecting

TABLE 1.3
Analytes of Interest in the Fermentation Industry

Fermentation Process	Compounds to Monitor
Microbial fermentation	Short-chain monocarboxylic and dicarboxylic acids-butyl esters of volatile (C1–C7) and nonvolatile (lactic, succinic, and fumaric) acids
Wine fermentation	Malolactic fermentation compounds
Rice wine	Total sugar content, alcohol, and pH
Wine	L-Lactic acid
Probiotic fermented milk	Oligosaccharides, improved fermentation rates, accelerated lactose hydrolysis
Grapes during yeast fermentation	Volatile flavor chemicals—acetates, ethyl esters, C4–C8 fatty acids
Fermented soybean foods	Proteases and ethanol, ethylene glycol, glucose, isopropanol, and mannitol
Cheese	Tyramine

the otherwise undetectable low level of harmful agents in the environment. Such major pollutants include heavy metals, pesticides, polychlorinated biphenyls, toxic organic wastes, nitrogenous compounds, endotoxin, and several pathogens. Considerable emphasis has been laid on detection of different toxic heavy metals such as Hg, Cd, Ni, Co, Zn, Pb, and Cu, and biosensors with different detection principles have been reported [72–75].

1.6.5 Applications in Biodefense

The world is witnessing increased threats of terrorism, including bioterrorism. Organisms or toxins used for such activities are termed biowarfare agents (BWAs), which include bacteria like *Bacillus anthracis* causing anthrax, toxins such as *Botulinum* neurotoxin and *Cholera* toxin, and viruses. The anthrax bioterrorist attack is predominantly due to the resistant spores, which have been observed to be lethal in 75% of infections. Biosensors hold application in detecting these agents by employing their important attributes of high selectivity. Some of the examples of potential bioterrorism agents against which biosensors have been developed in the literature with their detection limit in parenthesis are detection of *Anthrax* spores (1 μg/mL), *Botulinum* toxin (400 ng/mL), and ricin (400 ng/mL) with an assay time of 15–25 min following flow immunosensor systems, commercialized as a biowarfare agent detection device (BADD); detection of *Anthrax* spores (4,000 spores), *Botulinum* toxin (5 ng), ricin (10 ng), and smallpox (100,000 pfu) with an assay time of 15 min following fluorescent bead immunoassay commercialized by Response Biomedical Corp. [76] detection limits of the noted sensors are shown in the parentheses.

1.7 RECENT ADVANCES IN BIOSENSOR TECHNOLOGY

Recent advancement in biosensor technology comprises miniaturization of the devices, wearable and implantable design using smart materials, and analyzing and transmitting the response signal through modern communication technology, including Internet of Thing (IoT), machine learning, and smart phone technology. In these technology-driven research and developments, application of various nanomaterials for signal transduction and amplifications, and micro-scale domain technology, such as microfluidics and MEMS are increasingly used in biosensing applications to improve the sensitivity, render multiplexing capability in a chip format, and provide cost economy of the devices.

Nanomaterials can be one dimensional viz. surface nanofilms, two dimensional viz. nanofibers, or three dimensional viz. nanoparticles. A common property of the nanomaterials is their tremendously increased surface area as compared to their corresponding bulk materials. The increased surface area supports enhanced immobilization of the bioreceptors. Also, nanoscale quantum effects bring about certain properties viz. unique mechanical, magnetic, electrical, optical, and other properties, which can be used either to amplify or to transduce the biorecognition signal in the biosensor. The most popular nanomaterials used for biosensor applications include gold nanoparticles (AuNPs), carbon nanotubes, graphene, quantum dots, and others, among which, AuNPs are one of the most stable noble metal nanoparticles and most popularly used in biosensors due to their biocompatibility, suitable optical and electrical properties, and relatively simple mode of production and modification [77].

Smart biosensors refer to the coupling of a biosensor with smart technology, especially for application in health care, food safety, and environmental monitoring. An amalgamation of nanotechnology, wireless technology, information technology, machine learning, material sciences, and biomedical sciences has made this concept of smart biosensors possible. One straightforward outcome is the integration of POC biosensors with smart phone technology. This has become possible with the advancements in various technologies viz. deep learning, which is derived from conventional artificial neural networks, wherein a sensory input is received and subjected to an iterative process of training until a desired/accurate output is reported. The applicability of deep learning can be envisioned with the emergence of wearable biosensors, invasive sensors, or embedded sensors in smart devices to collect medical data for disease diagnosis or prognosis. Artificial intelligence is concerned with the creation of intelligent machines that work in analogy with the human brain. It encompasses machine learning, which includes application of pattern-recognition algorithms to improve performance via experience. Pattern-recognition algorithms particularly involve training by using previously acquired data, and its assessment for retrieval of those data with particular characteristics aid in grouping data into classes with similar characteristics and interpretation of the final assignment of the class. In other words, pattern-recognition trains the device to identify the presence or absence of substances in the sample, characteristic composition, analysis, and designation of a reported outcome as readout [78].

1.7.1 Wearable Biosensors

As the name suggests, wearable biosensors are meticulously designed miniaturized biosensors that can be worn on the skin in the form of a temporary skin tattoo or bands, on eyes as contact lenses, or on tooth enamel. Noninvasive monitoring of biofluids (sweat, tears, saliva, and interstitial fluid) is the characteristic feature of such sensors and has great value in the ever-booming health care industry, as they can provide continuous real-time monitoring of targets, management of chronic diseases, and alert the user or medical professionals in case of emergency. In order to achieve this, wearable biosensor platforms must be in direct contact with the noninvasive sample fluid without causing any discomfort to the user. Multidisciplinary research has brought about this possibility by integrating smart materials with the necessary flexibility, stretchability and biocompatibility, and miniaturization technologies which facilitate data processing and transmission for real-time monitoring of the target analyte (biomarkers, metabolites, hormones, etc.). Although several proofs-of-concept

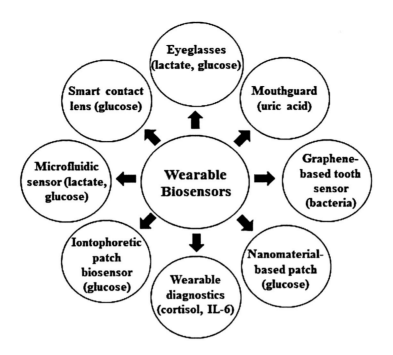

FIGURE 1.12 Representative examples of wearable biosensors [79].

have been forwarded in the literature, a thorough validation for correlation of analyte concentrations in noninvasive body fluids and a gold-standard sample fluid in most cases, i.e., blood, is needed to improve reliability. Additionally, the limited arena of smart materials and miniaturization technologies best suited for the development of wearable biosensors hampers its commercialization so far. Figure 1.12 illustrates a few examples of wearable biosensor, and the target analytes are shown in parentheses [79].

1.7.2 Implantable Biosensors

Most implantable biosensors use the amperometry-based principle, which is envisioned to measure and remotely transmit a record of specific molecular level of a biological analyte within the human body. Implantable biosensors generally use enzymes as the biorecognition molecule to enable the detection of biochemicals of interest within the body under a suitable impressed potential [80]. Trauma management and diagnostics are the foremost aims of research for implantable amperometric biosensor technology [81]. A recently developed dual responsive electrochemical cell-on-a-chip microdisc electrode array (ECC MDEA 5037) transducer was used in a wireless, implantable biosensor system for the continuous measurement of interstitial analytes. Preliminary studies with the MDEA 5037 in a rat hemorrhagic shock model have shown discordance between blood and interstitial lactate levels [82]. Lactate can accumulate more readily in the muscles particularly during periods of compensation and increased peripheral resistance during moderate to severe hemorrhage as blood oxygen delivery will be even further reduced, thus causing a rapid spike in interstitial lactate levels. Now, this lactate will diffuse back into the blood and eventually make its way to the liver. It is hypothesized that under conditions of diminished peripheral perfusion, lactate levels in the tissues will be discordant with systemic lactate levels and that the amount and duration of the tissue lactate levels will be a better indicator of the extent of hemorrhagic shock in the trauma patient. Continued examination of interstitial compartments using biosensors will aid in understanding the temporal relationships among markers of stress in these environments and how they relate to shock-like states. Potentiostat is a vital instrument for an implantable amperometric enzyme biosensor. However, for implantation, the potentiostat is closely connected with two-way telemetry and communications. Three general formats of implantable amperometric biosensors are being followed. The first is an implantable but tethered biotransducer with outwardly located power, electronics and communications components with the external components being mounted outside but on the subject's body [83]. The second is a fully integrated discrete but otherwise fully implanted device [84], and the third is an application-specific integrated circuit (ASIC) where all components are likewise fully implanted [85]. Representative examples of these three formats are shown in Figure 1.13.

Implantable amperometric enzyme biosensors have tremendous technological potential to influence patient management and compliance among diabetics but also to address the management of hemorrhaging victims of trauma. However, they face some major challenges, such as (a) enzyme stability, (b) biomolecular interferences, (c) the performance of molecular mediators used in Generation II biotransducers, and (d) internal calibration. Modern nano-biomedicine approaches such as biomaterials biomimicry, programmed anti-inflammatory drug delivery, and regenerative medicine are examples of tactics being developed to quiet the foreign body response. The

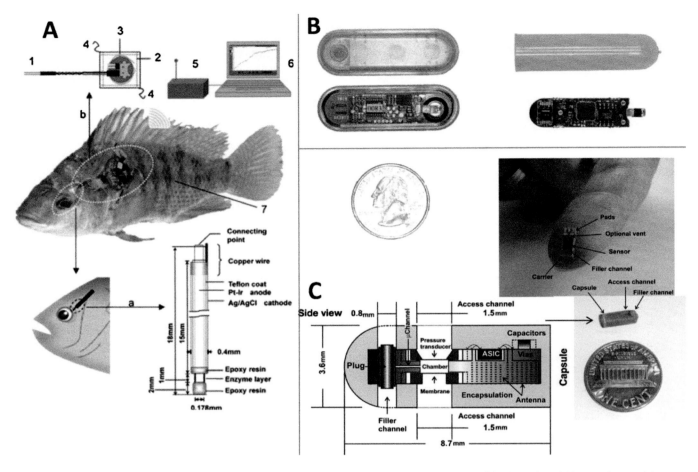

FIGURE 1.13 Examples to illustrate three general formats for implantable biosensor systems. (**A**) The tethered biotransducer with externally located power, electronics, and communications components. (**B**) The fully integrated discrete but otherwise fully implanted biosensor system. (**C**) The application-specific integrated circuit (ASIC) [80] Copyright (2012), with permission from Elsevier.

efforts to address the inherent factors that limit device bioanalytical performance include enzyme engineering to promote stability, hybrid biomaterials to address endogenous interferences, and the emergence of reagentless, third-generation biosensors hold considerable promise. Finally, fully implantable ASIC devices with a small footprint and wireless communication capabilities are being developed [80].

1.7.3 ENGINEERED ENZYMES FOR BIOSENSOR DEVELOPMENT

Genetically engineered acetylcholinesterases (AChEs) have been extensively exploited in enzyme inhibition-based biosensors for the detection of pesticides, like carbamate insecticides and organophosphate [86–88]. Currently great attention is given to protein engineering in order to improve the complete performance of bioelectronics. Specifically, there exist two strategies: (a) improving the biomolecular recognition between enzyme and substrate and (b) increasing the rate of electron transfer between enzyme and electrode. The first strategy can enhance the functional performance of amperometric sensors by increasing their selectivity as well as sensitivity. To address the second strategy, the enzyme is modified, keeping in mind the two key parameters: the turnover number, k_{cat}, and the Michaelis constant, K_M. The first reports on the effect of deglycosylation of redox enzymes were on recombinant horseradish peroxidase (r-HRP) overexpressed in *E. coli* [89]. Bioengineering of GOD extracted from *Penicillium amagasakiense* to make the enzyme less oxygen-dependent demonstrated a promise for making the enzyme more suitable for biosensor applications. Another example of bioengineering of a sugar oxidizing enzyme is the mutation of two ascomycete CDHs to increase the affinity for glucose and at the same time largely decrease the affinity for maltose, which is of vital importance for glucose biosensors. Another recent work in which a FAD-GDH was successfully fused with cyt *c* to mimic CDH and other flavohaemo proteins with direct electron transfer properties to obtain enzymes with modified properties [90]. To make the active site more accessible, eliminating the amino acids that are not vital for the enzyme functionality is another approach that facilitates the electron transfer easy.

1.8 MARKET POTENTIAL FOR BIOSENSORS

The market potential of biosensors is driven by their demand in diverse fields such as medicine, pharmacology, health care, food and agriculture, environmental monitoring, and biodefense. Additionally, with the advantages of biosensors to detect analytes over the established conventional techniques, compounded with possible miniaturization and low production costs, remote monitoring and reduced health care expenses, promising growth in the biosensor market seems visible. According to a recent analysis (Frost & Sullivan), the global biosensor market is expected to grow at a 12% compound annual growth rate (CAGR) during 2018–2023, from revenues of $17.7 billion in 2018, to reach $31.2 billion by the end of 2023. The patent publication in the field is on the rise, with 56% of total patents published by the United States alone from 2016 to 2018. The key areas of innovation include POC diagnostics, wearable biosensors, and noninvasive monitoring. Numerous companies are working worldwide in the area of biosensors for commercialization of their products.

Many factors influence the commercial success of biosensor devices. From the end-user point of view, low cost, simple, and reliable performances of the product are essential factors for penetrating the developed product in the market. It may be mentioned that the WHO prescribed the ASSURED [91] criteria for applications of the diagnostic devices in developing countries. The biosensors should comply with all the performance criteria as discussed elsewhere in this chapter. Addressing these issues may expedite the process of bridging the gap between academia (for proof-of-concept) and industry (to translate the concept to technology) for commercialization of the biosensors. However, few concerns associated with biosensor research need to be identified before developing the product such as the market for biosensors, advantages of biosensors over the existing analytical methods, ease of manufacturing and usage, and last but not the least, hazards and ethics associated with the particular biosensor research in question. Moreover, some other issues, such as government support in terms of ease of doing business and customer perceptions, should encourage for developing and marketing biosensor products for the decentralization of laboratory testing.

REFERENCES

1. V. Perumal, and U. Hashim, "Advances in biosensors: Principle, architecture and applications," *J. Appl. Biomed.*, vol. 12, no. 1, pp. 1–15, 2014, doi: 10.1016/j.jab.2013.02.001.
2. L.C. Clark Jr. "Monitor and control of blood and tissue oxygen tensions," *Trans. Am. Soc. Artif. Intern. Organs.*, vol. 2, pp. 41–48, 1956.
3. L. C. Clark, and C. Lyons, "Electrode systems for continuous monitoring in cardiovascular surgery," *Ann. N. Y. Acad. Sci.*, vol. 102, no. 1, pp. 29–45, 1962, doi: 10.1111/j.1749-6632.1962.tb13623.x.
4. S. J. Updike, and G. P. Hicks, "The enzyme electrode," *Nature*, vol. 214, no. 5092, pp. 986–988, 1967, doi: 10.1038/214986a0.
5. N. Bhalla, P. Jolly, N. Formisano, and P. Estrela, "Introduction to biosensors," *Essays Biochem.*, vol. 60, no. 1, pp. 1–8, 2016, doi: 10.1042/EBC20150001.
6. D. A. Armbruster, and T. Pry, "Limit of blank, limit of detection and limit of quantitation," *Clin. Biochem. Rev.*, vol. 29, Suppl 1, no.August, pp. S49–52, 2008.
7. D. Thevenot, K. Toth, R. A. Durst, and G. S. Wilson, "Technical report: Electrochemical biosensors: recommended definitions and classification," *Biosens. Bioelectron.*, vol. 16, pp. 121–131, 2001.
8. H. Kumar, and Neelam, "Enzyme-based electrochemical biosensors for food safety: A review," *Nanobiosensors Dis. Diagnosis*, p. 29, 2016, doi: 10.2147/ndd.s64847.
9. R. P. Buck, and E. Lindner, "Recommendations for nomenclature of ion-selective electrodes (IUPAC recommendations 1994)," *Pure Appl. Chem.*, vol. 66, no. 12, pp. 2527–2536, 1994, doi: 10.1351/pac199466122527.
10. A. Chaubey, and B. D. Malhotra, "Mediated biosensors," *Biosens. Bioelectron.*, vol. 17, pp. 441–456, 2002.
11. A. N. Ivanov, G. A. Evtugyn, L. V. Lukachova, et al., "New polyaniline-based potentiometric biosensor for pesticides detection," *IEEE Sens. J.*, vol. 3, no. 3, pp. 333–340, 2003, doi: 10.1109/JSEN.2003.814647.
12. N. Jaffrezic-Renault, and S. V. Dzyadevych, "Conductometric microbiosensors for environmental monitoring," *Sensors*, vol. 8, no. 4, pp. 2569–2588, 2008, doi: 10.3390/s8042569.
13. O. O. Soldatkin, V. M. Peshkova, S. V. Dzyadevych, A. P. Soldatkin, N. Jaffrezic-Renault, and A. V. El'skaya, "Novel sucrose three-enzyme conductometric biosensor," *Mater. Sci. Eng. C*, vol. 28, nos. 5–6, pp. 959–964, 2008, doi: 10.1016/j.msec.2007.10.034.
14. P. Das, M. Das, S. R. Chinnadayyala, I. M. Singha, and P. Goswami, "Recent advances on developing 3rd generation enzyme electrode for biosensor applications," *Biosens. Bioelectron.*, vol. 79, pp. 386–397, 2016, doi: 10.1016/j.bios.2015.12.055.
15. A. D. Hirst, and J. F. Stevens, "Electrodes in clinical chemistry," *Ann. Clin. Biochem.*, vol. 22, no. 5, pp. 460–488, 1985, doi: 10.1177/000456328502200503.
16. B. Willner, E. Katz, and I. Willner, "Electrical contacting of redox proteins by nanotechnological means," *Curr. Opin. Biotechnol.*, vol. 17, no. 6, pp. 589–596, 2006, doi: 10.1016/j.copbio.2006.10.008.
17. S. V. Dzyadevych, V. N. Arkhypova, A. P. Soldatkin, A. V. El'skaya, C. Martelet, and N. Jaffrezic-Renault, "Amperometric enzyme biosensors: Past, present and future," *ITBM-RBM*, vol. 29, nos. 2–3, pp. 171–180, 2008, doi: 10.1016/j.rbmret.2007.11.007.
18. L. Gorton, A. Lindgren, T. Larsson, F. D. Munteanu, T. Ruzgas, and I. Gazaryan, "Direct electron transfer between heme-containing enzymes and electrodes as basis for third generation biosensors," *Anal. Chim. Acta.*, vol. 400, nos. 1–3, pp. 91–108, 1999, doi: 10.1016/S0003-2670(99)00610-8.
19. R. a. Marcus, and N. Sutin, "Electron transfers in chemistry and biology," *Biochim. Biophys. Acta – Rev. Bioenerg.*, vol. 811, no. 3, pp. 265–322, 1985, doi: 10.1016/0304-4173(85)90014-X.
20. C. Léger, S. J. Elliott, K. R. Hoke, L. J. C. Jeuken, A. K. Jones, and F. A. Armstrong, "Enzyme electrokinetics: Using protein film voltammetry to investigate redox enzymes and their mechanisms," *Biochemistry*, vol. 42, no. 29, pp. 8653–8662, 2003, doi: 10.1021/bi034789c.
21. J. Hirst, "Elucidating the mechanisms of coupled electron transfer and catalytic reactions by protein film voltammetry," *Biochim. Biophys. Acta – Bioenerg.*, vol. 1757, no. 4, pp. 225–239, 2006, doi: 10.1016/j.bbabio.2006.04.002.
22. E. Laviron, "General expression of the linear potential sweep voltammogram in the case of diffusionless electrochemical systems," *J. Electroanal. Chem. Interfacial Electrochem.*, vol. 101, no. 1, pp. 19–28, 1979, doi: 10.1016/S0022-0728(79)80075-3.

23. M. J. Cooney, V. Svoboda, C. Lau, G. Martin, and S. D. Minteer, "Enzyme catalysed biofuel cells," *Energy Environ. Sci.*, vol. 1, no. 3, p. 320, 2008, doi: 10.1039/b809009b.10.1039/b809009b.
24. P. Vatsyayan, S. Bordoloi, and P. Goswami, "Large catalase based bioelectrode for biosensor application," *Biophys. Chem.*, vol. 153, no. 1, pp. 36–42, 2010, doi: 10.1016/j.bpc.2010.10.002.
25. N. Elgrishi, K. J. Rountree, B. D. McCarthy, E. S. Rountree, T. T. Eisenhart, and J. L. Dempsey, "A practical beginner's guide to cyclic voltammetry," *J. Chem. Educ*, vol. 95, no. 2, pp. 197–206, 2018, doi: 10.1021/acs.jchemed.7b00361.
26. S. R. Chinnadayyala, A. Kakoti, M. Santhosh, and P. Goswami, "A novel amperometric alcohol biosensor developed in a 3rd generation bioelectrode platform using peroxidase coupled ferrocene activated alcohol oxidase as biorecognition system," *Biosens. Bioelectron.*, vol. 55, pp. 120–126, 2014, doi: 10.1016/j.bios.2013.12.005.
27. T. Ghosh, P. Sarkar, and A. P. F. Turner, "A novel third generation uric acid biosensor using uricase electro-activated with ferrocene on a Nafion-coated glassy carbon electrode," *Bioelectrochemistry*, vol. 102, pp. 1–9, 2015, doi: 10.1016/j.bioelechem.2014.11.001.
28. D. Brondani, B. de Souza, B. S. Souza, A. Neves, and I. C. Vieira, "PEI-coated gold nanoparticles decorated with laccase: A new platform for direct electrochemistry of enzymes and biosensing applications," *Biosens. Bioelectron.* vol. 42, no. 1, pp. 242–247, 2013, doi: 10.1016/j.bios.2012.10.087.
29. B. C. Janegitz, R. A. Medeiros, R. C. Rocha-Filho, and O. Fatibello-Filho, "Direct electrochemistry of tyrosinase and biosensing for phenol based on gold nanoparticles electrodeposited on a boron-doped diamond electrode," *Diam. Relat. Mater.*, vol. 25, pp. 128–133, 2012, doi: 10.1016/j.diamond.2012.02.023.
30. S. K. Moccelini, A. C. Franzoi, I. C. Vieira, J. Dupont, and C. W. Scheeren, "A novel support for laccase immobilization: Cellulose acetate modified with ionic liquid and application in biosensor for methyldopa detection," *Biosens. Bioelectron.* vol. 26, no. 8, pp. 3549–3554, 2011, doi: 10.1016/j.bios.2011.01.043.
31. M. J. Moehlenbrock, R. L. Arechederra, K. H. Sjöholm, and S. D. Minteer, "Analytical techniques for characterizing enzymatic biofuel cells," *Anal. Chem*, vol. 81, no. 23, pp. 9538–9545, 2009, doi: 10.1021/ac901243s.
32. P. Jain, S. Das, B. Chakma, and P. Goswami, "Aptamer-graphene oxide for highly sensitive dual electrochemical detection of Plasmodium lactate dehydrogenase," *Anal. Biochem.*, vol. 514, pp. 32–37, 2016, doi: 10.1016/j.ab.2016.09.013.
33. J. Chouler, Á. Cruz-Izquierdo, S. Rengaraj, J. L. Scott, and M. D. Lorenzo, "A screen-printed paper microbial fuel cell biosensor for detection of toxic compounds in water," *Biosens. Bioelectron.*, vol. 102, no. November 2017, pp. 49–56, 2018, doi: 10.1016/j.bios.2017.11.018.
34. Y. Cui, B. Lai, and X. Tang, "Microbial fuel cell-based biosensors," *Biosensors.* vol. 9, pp. 1–18, 2019, doi: 10.3390/bios9030092.
35. J. Z. Sun, G. P. Kingori, R-W. Si, et al., "Microbial fuel cell-based biosensors for environmental monitoring: A review," *Water Sci. Technol.*, vol. 71, no. 6, pp. 801–809, 2015, doi: 10.2166/wst.2015.035.
36. P. Damborský, J. Švitel, and J. Katrlík, "Optical biosensors," *Essays Biochem.* vol. 60, no. 1, pp. 91–100, 2016, doi: 10.1042/EBC20150010.
37. A. H. Free, "Analytical applications of immobilized enzymes," *Ann. Clin. Lab. Sci.*, vol. 7, no. 6, pp. 479–485, 1977.
38. A. Soni, R. K. Surana, and S. K. Jha, "Smartphone based optical biosensor for the detection of urea in saliva," *Sensors Actuators, B Chem.*, vol. 269, pp. 346–353, 2018."
39. M. M. F. Choi, "Progress in enzyme-based biosensors using optical transducers," *Microchim. Acta*, vol. 148, nos. 3–4, pp. 107–132, 2004, doi: 10.1007/s00604-004-0273-8.
40. T. Aristotelous, A. L. Hopkins, and I. Navratilova, *Surface plasmon resonance analysis of seven-transmembrane receptors*, 1st ed., vol. 556. Elsevier Inc., 2015.
41. J. A. D. E. Feijter, J. Benjamins, and F. A. Veer, "Ellipsometry as a tool to study the adsorption behavior of synthetic and biopolymers at the air-water interface," *Biopolymers*, vol. 17, pp. 1759–1772, 1978.
42. M. Piliarik, H. Vaisocherová, and J. Homola, Surface Plasmon Resonance Biosensing in J. M. Walker, *Biosensors and Biodetection Series* Editor. pp. 65–88, 2009, Springer.
43. S. Zadran, S. Standley, K. Wong, E. Otiniano, A. Amighi, and M. Baudry, "Fluorescence resonance energy transfer (FRET)-based biosensors: Visualizing cellular dynamics and bioenergetics," *Appl. Microbiol. Biotechnol.*, vol. 96, no. 4, pp. 895–902, 2012, doi: 10.1007/s00253-012-4449-6.
44. B. T. Bajar, E. S. Wang, S. Zhang, M. Z. Lin, and J. Chu, "A guide to fluorescent protein FRET pairs," *Sensors (Basel).* vol. 16, no. 9, p. 1488, 2016. Published 2016 Sep 14. doi:10.3390/s16091488.
45. R. Cicchi, and F. Saverio Pavone, "Non-linear fluorescence lifetime imaging of biological tissues," *Anal. Bioanal. Chem.*, vol. 400, no. 9, pp. 2687–2697, 2011, doi: 10.1007/s00216-011-4896-4.
46. N. E. Azmi, A. H. A. Rashid, J. Abdullah, N. A. Yusof, and H. Sidek, "Fluorescence biosensor based on encapsulated quantum dots/enzymes/sol-gel for non-invasive detection of uric acid," *J. Lumin.*, vol. 202, no. May, pp. 309–315, 2018, doi: 10.1016/j.jlumin.2018.05.075.
47. T. H. Fereja, A. Hymete, and T. Gunasekaran, "A recent review on chemiluminescence reaction, principle and application on pharmaceutical analysis," *ISRN Spectrosc.*, vol. 2013, pp. 1–12, 2013, doi: 10.1155/2013/230858.
48. J. R. Bacon, and J. N. Demas, "Determination of oxygen concentrations by luminescence quenching of a polymer-immobilized transition-metal complex," *Anal. Chem.*, vol. 59, no. 23, pp. 2780–2785, 1987, doi: 10.1021/ac00150a012.
49. I. Klimant, P. Belser, and O. S. Wolfbeis, "Novel metal-organic ruthenium(II) diimin complexes for use as longwave excitable luminescent oxygen probes," *Talanta*, vol. 41, no. 6, pp. 985–991, 1994, doi: 10.1016/0039-9140(94)E0051-R.
50. Y. H. Zheng, T. C. Hua, D. W. Sun, J. J. Xiao, F. Xu, and F. F. Wang, "Detection of dichlorvos residue by flow injection calorimetric biosensor based on immobilized chicken liver esterase," *J. Food Eng.*, vol. 74, no. 1, pp. 24–29, 2006, doi: 10.1016/j.jfoodeng.2005.02.009.
51. K. Ramanathan, and B. Danielsson, "Principles and applications of thermal biosensors," *Biosens. Bioelectron.*, vol. 16, no. 6, pp. 417–423, 2001, doi: 10.1016/S0956-5663(01)00124-5.
52. S. Salman, S. Soundararajan, G. Safina, I. Satoh, and B. Danielsson, "Hydroxyapatite as a novel reversible in situ adsorption matrix for enzyme thermistor-based FIA," *Talanta*, vol. 77, no. 2, pp. 490–493, 2008, doi: 10.1016/j.talanta.2008.04.003.
53. B. Danielsson, "The enzyme thermistor," *Appl. Biochem. Biotechnol.*, vol. 7, nos. 1–2, pp. 127–134, 1982, doi: 10.1007/BF02798634.
54. M. Pohanka, "The piezoelectric biosensors: Principles and applications, a review," *Int. J. Electrochem. Sci*, vol. 12, no. 1, pp. 496–506, 2017, doi: 10.20964/2017.01.44.

55. G. Sauerbrey, "Verwendung von Schwingquarzen zur Wägung dünner Schichten und zur Mikrowägung," *Zeitschrift für Phys.* vol. 155, no. 2, pp. 206–222, 1959, doi: 10.1007/BF01337937.
56. K. K. Kanazawa, and J. G. Gordon, "Frequency of a quartz microbalance in contact with liquid," *Anal. Chem*, vol. 57, no. 8, pp. 1770–1771, 1985, doi: 10.1021/ac00285a062.
57. B. Drafts, "Acoustic wave technology sensors," *IEEE TRANSACTIONS ON MICROWAVE THEORY AND TECHNIQUES*," vol. 49, pp. 795–802, 2001.
58. J. M. Abad, F. Pariente, L. Hernández, H. D. Abruña, and E. Lorenzo, "Determination of organophosphorus and carbamate pesticides using a piezoelectric biosensor," *Anal. Chem*, vol. 70, no. 14, pp. 2848–2855, 1998, doi: 10.1021/ac971374m.
59. H. H. Nguyen, and M. Kim, "An overview of techniques in enzyme immobilization," *Appl. Sci. Converg. Technol*, vol. 26, no. 6, pp. 157–163, 2017, doi: 10.5757/asct.2017.26.6.157.
60. T. M. Temer, A. M. Omer, and M. Hassan, "Methods of enzyme immobilization," *Int. J. Curr. Pharm. Rev. Res.*, vol. 7, no. 6, pp. 385–392, 2016.
61. H. H. Nguyen, S. H. Lee, U. J. Lee, C. D. Fermin, and M. Kim, "Immobilized enzymes in biosensor applications," *Materials (Basel).* vol. 12, no. 1, pp. 1–34, 2019, doi: 10.3390/ma12010121.
62. H. Yue, Y. Zhou, P. Wang, *et al.* "A facile label-free electrochemiluminescent biosensor for specific detection of *Staphylococcus aureus* utilizing the binding between immunoglobulin G and protein A," *Talanta*, vol. 153, pp. 401–406, 2016, doi: 10.1016/j.talanta.2016.03.043.
63. G. Kim, J. H. Moon, C. Y. Moh, and J. Guk Lim, "A microfluidic nano-biosensor for the detection of pathogenic Salmonella," *Biosens. Bioelectron.*, vol. 67, pp. 243–247, 2015, doi: 10.1016/j.bios.2014.08.023.
64. E. Eser, O. Ö. Ekiz, H. Çelik, S. Sülek, A. Dana, and H. İ. Ekiz, "*Rapid Detection of Foodborne Pathogens by Surface Plasmon Resonance Biosensors*," vol. 5, no. 6, pp. 329–335, 2015, doi: 10.17706/ijbbb.2015.5.6.329-335.
65. N. Hesari, A. Alum, M. Elzein, and M. Abbaszadegan, "A biosensor platform for rapid detection of *E. coli* in drinking water," *Enzyme Microb. Technol.*, vol. 83, pp. 22–28, 2016, doi: 10.1016/j.enzmictec.2015.11.007.
66. H. Sharma, and R. Mutharasan, "Rapid and sensitive immunodetection of Listeria monocytogenes in milk using a novel piezoelectric cantilever sensor," *Biosens. Bioelectron.* vol. 45, no. 1, pp. 158–162, 2013, doi: 10.1016/j.bios.2013.01.068.
67. S. Chandra, J. Chapman, A. Power, J. Roberts, and D. Cozzolino, "The application of state-of-the-art analytic tools (biosensors and spectroscopy) in beverage and food fermentation process monitoring," *Fermentation*, vol. 3, no. 4, 2017, doi: 10.3390/fermentation3040050.
68. J. H. T. Luong, K. B. Male, and J. D. Glennon, "Biosensor technology: Technology push versus market pull," *Biotechnol. Adv.* vol. 26, no. 5, pp. 492–500, 2008, doi: 10.1016/j.biotechadv.2008.05.007.
69. M. A. Pfaller, D. M. Wolk, and T. J. Lowery, "T2MR and T2Candida: Novel technology for the rapid diagnosis of candidemia and invasive candidiasis," *Future Microbiol.* vol. 11, no. 1, pp. 103–117, 2016, doi: 10.2217/fmb.15.111.
70. A. F. Sarioglu, N. Aceto, N. Kojic, *et al.*, "A microfluidic device for label-free, physical capture of circulating tumor cell-clusters," *Nat. Methods.*, vol. 12, no. 7, pp. 685–691, 2015, doi: 10.1038/nmeth.3404.A.
71. J. D. Besant, E. H. Sargent, and S. O. Kelley, "Rapid electrochemical phenotypic profiling of antibiotic-resistant bacteria," *Lab Chip*, vol. 15, no. 13, pp. 2799–2807, 2015, doi: 10.1039/c5lc00375j.
72. S. Magrisso, Y. Erel, and S. Belkin, "Microbial reporters of metal bioavailability," *Microb. Biotechnol.* vol. 1, no. 4, pp. 320–330, 2008, doi: 10.1111/j.1751-7915.2008.00022.x.
73. O. Domínguez-Renedo, M. A. Alonso-Lomillo, L. Ferreira-Gonçalves, and M. J. Arcos-Martínez, "Development of urease based amperometric biosensors for the inhibitive determination of Hg (II)," *Talanta*, vol. 79, no. 5, pp. 1306–1310, 2009, doi: 10.1016/j.talanta.2009.05.043.
74. B. Kuswandi, "Simple optical fibre biosensor based on immobilised enzyme for monitoring of trace heavy metal ions," *Anal. Bioanal. Chem.* vol. 376, no. 7, pp. 1104–1110, 2003, doi: 10.1007/s00216-003-2001-3.
75. A. P. Das, P. S. Kumar, and S. Swain, "Recent advances in biosensor based endotoxin detection," *Biosens. Bioelectron.*, vol. 51, pp. 62–75, 2014, doi: 10.1016/j.bios.2013.07.020.
76. J. J. Gooding, "Biosensor technology for detecting biological warfare agents: Recent progress and future trends," *Anal. Chim. Acta.*, vol. 559, no. 2, pp. 137–151, 2006, doi: 10.1016/j.aca.2005.12.020.
77. X. Zhang, Q. Guo, and D. Cui, "Recent advances in nanotechnology applied to biosensors," *Sensors*, vol. 9, no. 2, pp.1033–1053, 2009, doi: 10.3390/s90201033.
78. S. B. Baker, W. Xiang, and I. Atkinson, "Internet of things for smart healthcare: Technologies, challenges, and opportunities," *IEEE Access*, vol. 5, no. November, pp. 26521–26544, 2017, doi: 10.1109/ACCESS.2017.2775180.
79. J. Kim, A. S. Campbell, B. E. F. de Ávila, and J. Wang, "Wearable biosensors for healthcare monitoring," *Nat. Biotechnol.*, vol. 37, no. 4, pp. 389–406, 2019, doi: 10.1038/s41587-019-0045-y.
80. C. N. Kotanen, F. G. Moussy, S. Carrara, and A. Guiseppi-Elie, "Implantable enzyme amperometric biosensors," *Biosens. Bioelectron.*, vol. 35, no. 1, pp. 14–26, 2012, doi: 10.1016/j.bios.2012.03.016.
81. C. K. Guiseppi-Elie Anthony, "Development of an implantable biosensor system for physiological Status monitoring during Long duration space exploration," *Gravitational Sp. Res.*, vol. 23, pp. 55–64, 2010.
82. A. Guiseppi-Elie, "An implantable biochip to influence patient outcomes following trauma-induced hemorrhage," *Anal. Bioanal. Chem.* vol. 399, no. 1, pp. 403–419, 2011, doi: 10.1007/s00216-010-4271-x.
83. H. Endo, Y. Yonemori, K. Hibi, *et al.*, "Wireless enzyme sensor system for real-time monitoring of blood glucose levels in fish," *Biosens. Bioelectron.*, vol. 24, no. 5, pp. 1417–1423, 2009, doi: 10.1016/j.bios.2008.08.038.
84. C. P. Cheney, B. Srijanto, D. L. Hedden, *et al.*, "In vivo wireless ethanol vapor detection in the Wistar rat," *Sensors Actuators, B Chem.*, vol. 138, no. 1, pp. 264–269, 2009, doi: 10.1016/j.snb.2009.01.052.
85. E. Johannessen, O. Krushinitskaya, A. Sokolov, *et al.*, "Toward an injectable continuous osmotic glucose sensor," *J. Diabetes Sci. Technol.*, vol. 4, no. 4, pp. 882–892, 2010, doi: 10.1177/193229681000400417.
86. B. Bucur, D. Fournier, A. Danet, and J. L. Marty, "Biosensors based on highly sensitive acetylcholinesterases for enhanced carbamate insecticides detection," *Anal. Chim. Acta*, vol. 562, no. 1, p. 115–121, 2006, doi: 10.1016/j.aca.2005.12.060.
87. G. Istamboulie, S. Andreescu, J. L. Marty, and T. Noguer, "Highly sensitive detection of organophosphorus insecticides using magnetic microbeads and genetically engineered acetylcholinesterase," *Biosens. Bioelectron.*, vol. 23, no. 4, pp. 506–512, 2007, doi: 10.1016/j.bios.2007.06.022.
88. S. Sotiropoulou, D. Fournier, and N. A. Chaniotakis, "Genetically engineered acetylcholinesterase-based biosensor

for attomolar detection of dichlorvos," *Biosens. Bioelectron.* vol. 20, no. 11, pp. 2347–2352, 2005, doi: 10.1016/j.bios.2004.08.026.

89. G. Presnova, V. Grigorenko, A. Egorov, *et al.*, "Direct heterogeneous electron transfer of recombinant horseradish peroxidases on gold," *Faraday Discuss.*, vol. 116, no. February, pp. 281–289, 2000, doi: 10.1039/b001645o.

90. P. Bollella, and L. Gorton, "Enzyme based amperometric biosensors," *Curr. Opin. Electrochem.*, vol. 10, pp. 157–173, 2018, doi: 10.1016/j.coelec.2018.06.003.

91. R. W. Peeling, K. K. Holmes, D. Mabey, and A. Ronald, "Rapid tests for sexually transmitted infections (STIs): The way forward," *Sex. Transm. Infect.* vol. 82, Suppl 5, pp. v1–v6, 2006, doi: 10.1136/sti.2006.024265.

2 Emerging Materials and Platforms for Biosensor Applications

Albert Saavedra, Federico Figueredo, María Jesús González-Pabón, and Eduardo Cortón
Universidad de Buenos Aires (UBA), Ciudad Universitaria, Ciudad Autónoma de Buenos Aires, Argentina

CONTENTS

2.1 Hydrogel-Based Biosensors 29
    2.1.1 General Overview: Hydrogels 29
    2.1.2 Hydrogel Materials 30
    2.1.3 Methods to Immobilize the Bioreagent and Increase Biosensor Performance 31
    2.1.4 Bioreagents Used for Biorecognition 31
        2.1.4.1 Natural 31
        2.1.4.2 Artificial (but Bioinspired) 31
    2.1.5 Transducer Used for Hydrogel-Based Biosensors 32
        2.1.5.1 Electrochemical 32
        2.1.5.2 Optical 33
        2.1.5.3 Others 34
    2.1.6 Applications 34
        2.1.6.1 Clinical Chemistry 34
        2.1.6.2 Environmental Analysis 34
2.2 Nanoporous Anodic Alumina-Based Biosensor 35
    2.2.1 General Overview: Porous Anodic Alumina 35
    2.2.2 Fabrication and Structure of PAA 35
    2.2.3 Properties for Biosensing Applications 36
    2.2.4 Methods to Immobilize the Bioreagent and Increase Biosensor Performance 36
    2.2.5 Bioreagents Used for Biorecognition 37
        2.2.5.1 Natural Biorecognition Molecules 37
        2.2.5.2 Artificial (but Bioinspired) Biorecognition Molecules 38
    2.2.6 Transducers Used for Anodic Alumina-Based Biosensors 38
        2.2.6.1 Electrochemical Transducers 38
        2.2.6.2 Optical Transducers 40
    2.2.7 Applications 43
2.3 Perspectives 43
References 43

2.1 HYDROGEL-BASED BIOSENSORS

2.1.1 General Overview: Hydrogels

Hydrogels are three-dimensional (3D) networks of cross-linked polymers. They are interesting materials capable of absorbing large quantities of water without dissolving. In fact, hydrogels are similar to a container of a solvent made of a three-dimensional mesh. In a dried state, it acts as a solid material; however, when hydrated it swells without dissolution (in the short term) up to 99% (w/w) of the dry weight. The ability to absorb water is related to the hydrophilic functional groups attached to the polymer backbone; their resistance to dissolution is related to the cross-links between network chains. Hydrogels present a variety of unique properties in addition to swelling behavior, such as presenting a solvated surface and super-permeation, which are not present in either pure liquids or solids [1, 2].

Since the early discovery in 1968 [3, 4], hydrogel materials have been increasingly studied for applications in diagnostics [5, 6], drug delivery [7], and tissue regeneration [8], among others, for a number of reasons: They provide a semi-wet, 3D environment for molecular-level biological and chemical interactions; some of them can provide inert surfaces to prevent nonspecific adsorption of proteins or other compounds;

they can control the diffusion behavior of molecules through the polymer networks; biological molecules can be incorporated into their structures; their mechanical properties are tunable; and they can be designed to change properties in response to external factors.

Hydrogels can undergo a volume-phase transition when they are exposed to an external stimulus resulting in changes in the network such as swelling, collapse, or solution-to-gel transitions. In general, the so-called stimulus-sensitive hydrogels [2, 9, and 10] are those that are designed to respond to small changes in the environment by altering their swelling properties.

Hydrogel biosensors are those that are designed to detect a biochemical or biological interaction by means of an immobilized bioreceptor. The interaction between the bioreceptor and the analyte produces a swelling alteration or/and another measurable signal that may be translated by optical, conductometric, amperometric, or mechanical readouts for biosensing [6]. Hydrogel biosensors are highly tunable systems (Figure 2.1), since a wide variety of bioreceptors from different origins (natural or artificial) can be immobilized, the porous structure can be controlled in order to limit the diffusion of large molecules, and depending on the network nature, they can bring a protected environment for the bioreceptors, thereby preserving their activity and structure [11].

2.1.2 Hydrogel Materials

Depending on the network properties and their origin, hydrogels can be classified as natural, synthetic, and conducting. Natural hydrogels are those that are constructed with materials from a natural origin, such as natural gums, proteins, peptides, cellulosic materials [11], collagen [12], hyaluronic acid [13], and polysaccharides [14] (chitosan, chitin, agarose, etc.), among others. Natural hydrogels offer some advantages for biosensor design and construction, such as high biocompatibility, heavy metal ion chelation, high protein affinity, low cost, and the presence of a diversity of functional groups that allow further surface chemical modification [15–17].

Synthetic hydrogels are those that are fabricated with artificial materials, such as polyvinyl alcohol (PVA), polyacrylate-based compounds, and polyethylene glycol (PEG), among others. In particular, PVA hydrogels showed high biocompatibility and hydrophilic properties. Moreover, one of the main advantages is that they can be used for implantable sensors and biosensors due to their flexibility and stability in several environmental conditions. On the other hand, PEG is a hydrophilic material that displays antifouling properties due to the low interfacial energy preventing protein and cell adhesion [18–20]. PEG can repel large molecules such as high-molecular-weight proteins. Polyacrylate-based compounds are composed mainly of polyacrylic acid, polyhydroxyethyl methacrylate, polyacrylamide, and poly (n-isopropylacrylamide). Most of the monomers used for hydrogel polymerization can be modified with methacrylic or cinnamyl side groups to become photo-crosslinkable, making them compatible with microfabrication techniques such as photolithography. Other techniques for hydrogel fabrication using photo-crosslinking methodologies were recently reviewed [21]. With these approaches, interesting studies were focused on label-free detection of deoxyribonucleic acid (DNA) using polyacrylate-based compounds to photopolymerize hydrogel plugs in microfluidic channels [22, 23]. Hydrogel materials have attracted the attention of scientists, since polymerization can be finely controlled, so very small 2D and 3D structures can be obtained. These structures can be incorporated as an interface between the solution to be analyzed and the biosensor transducer, for example, over microelectrodes or inside microfluidic devices as well. Hydrogels can be displayed in at least four geometries, such as a 3D bulk hydrogel material, as small particles, as brush layers attached to solid supports, and as thin hydrogel network films.

Conductive hydrogels, a combination of gel structures with electronic functionality, have attracted attention for electrochemical biosensor development. They are created by adding a conducting polymer to the hydrogel to enhance the current conductivity from the reaction site (e.g., redox enzyme) to the transducer (e.g., electrode). The conducting polymers most used are polyenes and polyaromatic compounds, such as polypyrrole, polyaniline, and polythiophene derivatives. The conductive behavior of the conducting polymer arises from the conjugated backbones (simple and double bonds). The polymers are synthesized in a preferable oxidized form in the presence of a "dopant" (a negative charge/anion in most cases); therefore, the backbones are stabilized and the charge neutralized. Conductive hydrogels have been used at the surface of the electrodes as an interface. Their doping/dedoping mechanism produces a change in surface resistance, current, or electrical potential, which appears as a response

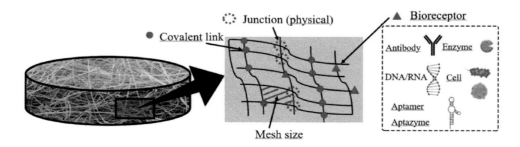

FIGURE 2.1 Network properties of hydrogels, including the types of bioreceptors used to construct hydrogel biosensors.

related to the analyte concentration. Polypyrrole has shown good *in vitro* and *in vivo* biocompatibility, chemical stability, and high conductivity. Polyaniline (PANI) is also one of the most used conducting polymers after polypyrrole. PANI has been investigated for its application in biosensing devices but not for implantable devices, since it showed lack of flexibility, nonbiodegradability, and chronic inflammation issues once implanted [24]. Finally, polythiophene derivatives such as poly (3, 4-ethylenedioxythiophene) (PEDOT) are a conducting polymer commonly used in biosensing and bioengineering applications.

2.1.3 Methods to Immobilize the Bioreagent and Increase Biosensor Performance

To obtain a hydrogel-based biosensor, the bioreceptor materials (enzymes, antibodies, nucleic acids, etc.) are subjected to an immobilization process. There are several strategies to do that – the choice should take into account the bioreceptor viability in time, the availability of reactive groups, the binding type (covalent or noncovalent), and the risk of the bioreceptor degradation (ultraviolet [UV] exposure, free radicals, temperature, and organic solvents) [25]. The hydrogel porosity, rigidity, and swelling behavior are also important parameters to take into account for the hydrogel biosensor design. The target analyte (and cofactors, if any) should have access through the porous structure, and by-products may have to diffuse out, both processes maintaining the structural integrity of the hydrogel.

Bioreceptors can be immobilized by means of physical absorption and entrapment, covalent binding, crosslinking methods, or a combination of some of these techniques. Similarly to other materials, covalent binding and crosslinking methods are preferred due to the stability achieved and bioreceptor leakage to the bulk solution (bleeding) is minimized. However, if covalent bonds interact with the active sites, the bioreceptors can lose functionality. On the other hand, the bioreceptors can be entrapped when the hydrogel is polymerized if the pore size (intermolecular distance) is less than the size of the bioreceptors. The main disadvantage of the entrapment method is the bioreceptor leakage over time. Other immobilization methods such as physical absorption or affinity ligand binding are not efficient enough, since they are sensitive to environmental conditions, leading to weak reproducibility, random orientation of bioreceptors, and weak attachment. In all situations, the hydrogel material, hydrophobicity, water solubility, and porous structure need to be controlled depending on the analyte to be determined.

In living biosensors, cells or tissue are in close contact with the hydrogel, since physiological parameters can be monitored, making it possible to detect a broad range of known and unknown chemical and physical agents. In particular, hydrogels used for living biosensors need to accomplish high porosity of the matrix as a necessary feature to allow the rapid diffusion and transport of nutrients, cell wastes, and analyte molecules.

2.1.4 Bioreagents Used for Biorecognition

2.1.4.1 Natural

2.1.4.1.1 Antibodies, Enzymes, and Nucleic Acids

PVA hydrogels have been used for the fabrication of several biosensors based on antibodies [26] and enzymes [27, 28]; more rarely, peptides have been used [29]. PEG-based materials have been used for the fabrication of enzymatic biosensors as well [30–34]. Polyacrylate hydrogels have been mostly used for DNA biosensors because of their high compatibility with DNA molecules. In addition, most of the polyacrylic acid can be incorporated into microfluidic devices when photopolymerization is done.

DNA hybridization-based biosensors are commonly constructed by immobilizing small DNA fragments over 2D rigid surfaces (electrodes for electrochemical biosensors or solid substrates for surface plasmon resonance and light scattering biosensors), but this is a suboptimal solution given that molecular recognition is affected and leads to less sensitivity. Meanwhile, when immobilization of DNA fragments is done over 3D hydrogel structures, it results in enhanced signal intensity. This is related to a higher immobilization capacity and an undisturbed molecular structure, given the hydrophilic nature and flexibility of the hydrogel matrix. Similar effects are found when other sensing moieties such as enzymes and antibodies are used.

2.1.4.1.2 Living Cells

One of the most important features related to living biosensors is that they have to maintain life for the intended biosensor working time (hours to weeks) [35]. Hydrogels can achieve this goal if nutrients and wastes can move through the network at a rate compatible with cellular needs. Cells can be encapsulated within the hydrogel, for example, in alginate crosslinked with divalent cations forming a highly porous structure. When microorganisms such as *Escherichia coli* were encapsulated in alginate hydrogels, the cell activity was maintained for several weeks [17]. Similar results were obtained with macroporous PVA hydrogel produced by cryogelation (at −20°C), which was used for living biosensor construction [36]. Some conducting hydrogels such as polypyrrole can be synthesized with a controlled porous structure, making them a good candidate for living biosensors. Cell adhesion and growth were achieved with polypyrrole films showing potential application for tissue engineering and implantable biosensors as well [37].

2.1.4.2 Artificial (but Bioinspired)

Molecular imprinted hydrogels have great potential for biosensor applications, since they provide a robust means for target analyte recognition, being stable over time. The molecular imprinting method can be performed by polymerization of the monomer(s) in the presence of the target analyte and the crosslinking agent. When the polymerization process is concluded, the target analyte can be washed away, and then molecular imprinted recognition sites can be employed for the detection of the target analyte present in the sample to be measured (Figure 2.2).

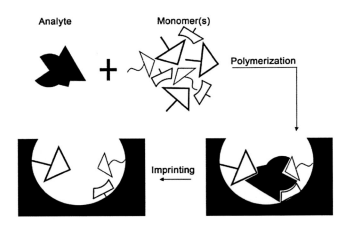

FIGURE 2.2 Steps needed to produce a molecularly imprinted polymer with sensing capabilities. Adapted, under the Creative Commons Attribution License which permits unrestricted use, distribution, and reproduction, from Tavakoli et al. [6].

2.1.5 Transducer Used for Hydrogel-Based Biosensors

2.1.5.1 Electrochemical

Electrochemical transducers have been widely employed for the signal readout of hydrogel biosensors, since they can be miniaturized and manufactured with low-cost materials, as well being an affordable technology for point-of-need or point-of-care applications. A conductive hydrogel matrix is an ideal environment for bioreceptors, maintaining their biological activity and conducting the electrical signal to the transducer (electrode). The 3D structure provides better results to maintain the bioreceptors and analytes in their innate structure in comparison to 2D films, as was previously shown by other authors [38]. In this sense, a whole variety of electrochemical biosensors containing a conductive polymer hydrogel matrix over the electrode have been studied in recent years. In contrast, there are some interesting approaches regarding to conducting polymer hydrogels coupled to various conductive materials, such as metal nanoparticles, carbon nanotubes, graphene oxide, and conducting polymers that are still being studied for practical applications. One of the most promising applications of conductive polymer hydrogels is the development of multianalyte electrochemical biosensors, such as those presented by the Revzin group [39]. They photopolymerized PEG diacrylate prepolymer containing glucose and lactate oxidase enzymes (GOx and LOx, respectively) to obtain the hydrogel over an array of gold electrodes (Figure 2.3A). In a similar way a recent study shows the possibility to use photolithography techniques to pattern the conductive polymer PEDOT:PSS on thin solid [40] or flexible [41] substrates, as

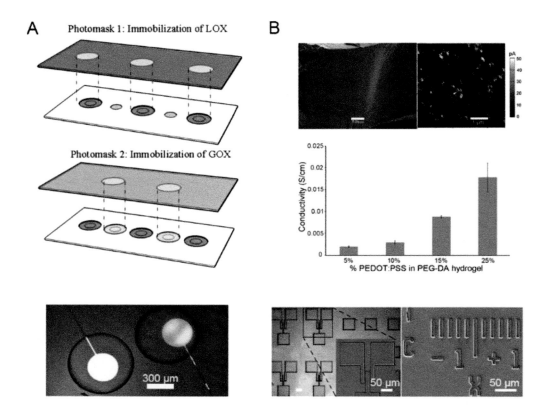

FIGURE 2.3 Electrochemical hydrogel biosensors. (A) Enzymatic biosensors constructed by photopolymerization over metal electrodes. (B) PEG:DA hydrogel containing PEDOT:PSS, SEM pictures and electrochemical microscopy characterization. Effect of the PEDOT:PSS amount incorporated to PEG-DA hydrogel on the conductivity and finally SEM pictures showing the photopolymerized structures. (LOX = LOx; GOX = GOx). Reprinted with permission from Yan, et al. [39]; copyright (2017) American Chemical Society (panel A). Reprinted with permission from Pal et al. (panel B) [40].

Emerging Materials for Biosensor Applications

can be seen in Figure 2.3B. These fabrication techniques open the possibilities to produce low-cost electrodes with different geometries and biosensing functionalities.

2.1.5.2 Optical

The high transparency and optical quality of hydrogels make them an ideal material for the design of optical biosensors based on absorbance, fluorescence, surface plasmon resonance (SPR), and light scattering (LS) techniques. An interesting approach was recently done by employing a multiplexed enzyme-based assay using shape-coded PEG hydrogel particles inside a Y-shaped microfluidic device [42]. Hydrogel particles containing GOx and alcohol oxidase (AOx) were collected in a chamber to detect fluorescence changes in the presence of glucose and ethanol. In this sense, simultaneous detection of glucose and ethanol was possible (1–10 mM) in the same detection chamber, as can be seen in Figure 2.4A. Later, the same research group developed micropatterned fibrous membranes consisting of polystyrene (PS)/poly (styrene-alt-maleic anhydride) (PSMA) nanofibers and PEG hydrogels microstructures fabricated by combination of electrospinning and photolithography [43]. These microstructures were synthetized first with PEG hydrogels polymerized in the presence of enzymes, and finally they were modified with antibodies that were selectively immobilized over PS/PSMA nanofibers, since the PEG hydrogel has a protein-repellent nature; in this way, simultaneous detection of anti-IgG and glucose was performed. Another strategy based on a multianalyte biosensor was achieved fabricating a PEG acrylate hydrogel array by using soft lithography [44]. Oligonucleotides containing a methacrylamide functional group and fluorescent marker were copolymerized within the hydrogel matrix. The biosensor was developed in order to detect three different target DNA sequences producing three different colors. Another interesting study showed the possibility of using hydrogel materials coupled with microfluidic paper-based analytical devices to mediate fluidic flow and signal readout [45]. In brief, an aptamer-crosslinked hydrogel was used as an analyte-responsive valve. In the absence of the analyte, the hydrogel is formed and stops the flow; meanwhile, when the analyte is present, the solution flows through the device, allowing the color indicator to appear at the observation spot for visible detection (Figure 2.4B).

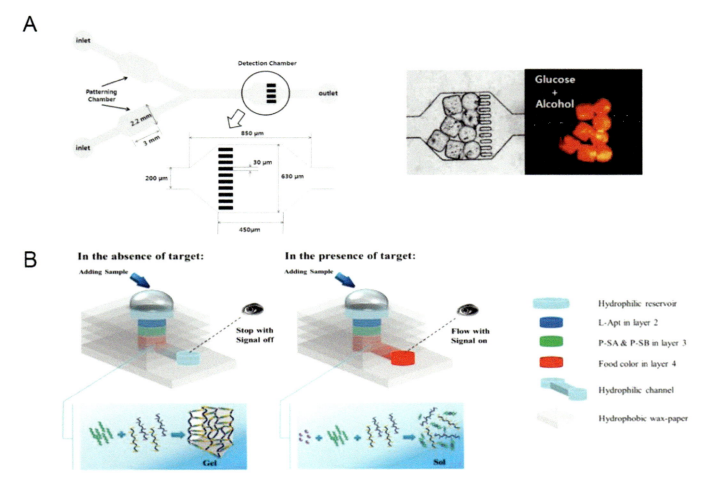

FIGURE 2.4 Biosensing strategies by using hydrogel biosensors. (A) Y-shaped microfluidic device used for the injection and detection of glucose and alcohol. (B) Microfluidic paper-based analytical device constructed with hydrogel. Reprinted with permission from Jang et al. (panel A) [42]. Reprinted with permission from Wei et al. (panel B); copyright (2015) American Chemistry Society [45].

2.1.5.3 Others

A wide range of transduction methods, including pressure, capacitive, cantilever based, bending plates, and microgravimetric transducers, can be used for the design and operation of hydrogel biosensors. Capacitive and pressure biosensors involve several fabrication steps and can suffer from insensitivity [6]; however, changes in mass, temperature, and stress produce swelling variations that can be determined – for example, with a quartz crystal microbalance for the detection of DNA, glucose, protease, and virus [38].

2.1.6 Applications

Most of the hydrogel biosensor development and optimization studies have been focused on small metabolite detection, such as glucose, lactate, uric acid, and cholesterol. On the other hand, there is a lot of research for nucleic acid biosensors and antibody biosensors for pathogen detection. Recent developments involving living biosensors used for clinical chemistry and environmental applications will be discussed in the following section.

2.1.6.1 Clinical Chemistry

Biosensors that are composed of cells or tissue as the bioreceptor are usually called living biosensors. They are developed in order to acquire measurable signals related principally to cell metabolism or metabolic pathways. Other names or definitions can be found in bibliography, such as whole-cell biosensors or microbial biosensors (which include only cells from a microbial origin), among others. Living biosensors can be constructed with cells from any origin, but given that microbial cells are easy to culture and maintain, most of the published work replies on them. In this section we will discuss living biosensors that use animal cells and tissue, since they are intensely studied for their application in clinical chemistry such as drug screening, toxicology assays, cell-biomaterial interaction screening, and the identification of unknown pathogens and toxins [46].

In the first studies, biosensors were constructed by direct attachment of animal cells or tissue onto a biotic or abiotic substrate surface forming a 2D cell monolayer. Despite the advances achieved with 2D cell culture, they do not reproduce accurately the characteristics of cells' behavior in an *in vivo* environment such as the human body. In the past few decades, tremendous progress has been made in the development of 3D cell culture systems. In a 3D growing structure, cells promote many biologically relevant functions not observed in a 2D cell culture. Significant differences in cell morphology, protein expression, differentiation, migration, functionality, and viability have been reported for 3D cultures in comparison to 2D cultures. Many studies have focused on the transition of 2D to 3D cell cultures using hydrogel materials as scaffolds for cell proliferation in 3D architectures, similarly to the native extracellular matrix that mimics the structural, biochemical, and mechanical properties of an *in vivo* environment [47]. 3D culture models involve spheroids and cells embedded in 3D matrixes, such as multilayered tissue models or scaffolds (or supported matrix) models. The hydrogel materials commonly used are collagen, collagen-chitosan, Matrigel, fibrinogen, alginate, Puramatrix scaffolds, and PEG [46]. Live-cell imaging, viability, metabolism, impedance change, motility, and adhesion, among others, can be detected from living biosensors using electrical and optical transducers, depending on the nature of the biological reactions. The potential toxicity of a number of drugs was tested with a 3D cellular microarray containing human cells entrapped in collagen-alginate by means of the optical detection of cytochrome P450-produced metabolites (Figure 2.5A). In comparison to multiwell assays, the detection time was shortened from 24 to 7 h [48]. On the other hand, an interesting approach was made by Inal and colleagues [47], since they report a monitoring device constructed with a macroporous conducting scaffold made of poly(3,4-ethylene dioxythiophene):poly(styrenesulfonate) (PEDOT:PSS). These structures supported 3D cell cultures and co-cultures too. They integrated a perfusion tube inside the scaffold for cell spreading and inclusion inside the hydrogel (Figure 2.5B). The conducting polymer hydrogel allows the impedimetric detection of cell growth, but with the incorporation of dodecylbenzenesulfonic acid (DBSA) and collagen, both electrical and mechanical properties were enhanced. Future work can focus on the optimization of impedance parameters and hydrogel materials containing conducting polymers, nanoparticles, nanotubes, microparticles, and others.

2.1.6.2 Environmental Analysis

The detection of chemical compounds, toxins, and pathogens in the environment (soil, water, air) can be determined selectively by means of traditional analytical techniques such as liquid gas chromatography combined with mass spectrophotometry. However, these tools are not suitable for continuous monitoring systems that need, for example, to be incorporated in several locations along a river in order to detect a specific toxic compound (as needed with complete early-warning systems). Biosensors have several characteristics that make them an excellent tool for environmental monitoring. Enzymes and antibodies bioreceptors have already been used for a variety of environmental analysis. However, these biomolecules are sensitive to most environmental conditions, since the optimal working conditions are similar to biological fluids. In particular, living biosensors can be constructed with bioreceptors such as environmental bacteria or fungi cells that avoid most of the problems related with the sample matrix, as microorganisms are adapted to survive to a broad range of environmental conditions (when compared to enzymes). Living biosensors detect responses of cells after exposure to a sample, and this response can be correlated with the sample toxicity or other sample characteristics. A hydrogel can be incorporated as the biosensor component that provides a suitable environment for the cells used as bioreceptors, maintaining their integrity and viability. Many hydrogel materials have been used for cell entrapment such as gellan/xanthan, PVA/alginate, alginate, and polypyrrole, among others. In recent

Emerging Materials for Biosensor Applications

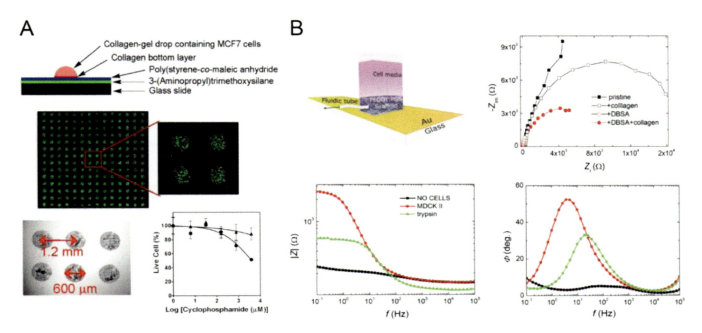

FIGURE 2.5 Living biosensors contructed with hydrogel materials. (A) Collagen hydrogel containing MCF7 cells for the construction of a toxicity biosensor array. Materials and construction details (up) and a fluorescence microscopy picture showing the array (middle). Optical picture showing drop size and distance between them and a typical dose response curve obtained for cyclophosphamide (down). (B) Conducting scaffold constructed over a gold electrode. Biosensor scheme and details (upper-left corner). Nyquist plot for the pristine scaffold and those modified with collagen and DBSA (upper-right corner). Electrochemical impedance spectrosopy results (down) of the biosensor after (black) and before cell incubation (red); trypsin treatment is also show (green). Freely available online through the PNAS open access option, from Lee et al. (panel A) [48]. Reprinted with permission from Inal et al. (panel B) [47].

years, alginate electropolymerization techniques were used to produce tunable hydrogels. Recently, Vigués et al. [49] developed an electrochemical living biosensor fabricated with an electropolymerized conductive alginate hydrogel (alginate + graphite) containing bacteria cells (bacterial electrotrapping). The authors showed that the procedure was reproducible (CV < 0.5%); meanwhile, cell integrity and activity are maintained at an acceptable level (cell viability = 56%) [49].

2.2 NANOPOROUS ANODIC ALUMINA-BASED BIOSENSOR

2.2.1 General Overview: Porous Anodic Alumina

Aluminum is an interesting metal because of its lightness, abundance, and anticorrosive properties, given its ability to form passive compounds, such as inert oxides that remain over the surface as a protective layer [50]. These features have propelled this metal to be used in the industry as an anticorrosive layer. Although it was not until 1990s that researchers discovered that highly ordered nanoporous structures can be achieved by properly controlling anodization conditions, such as electrolyte composition and concentration, temperature, and anodization voltage. The research and development that focused on the formation of protective oxide layers resulted in techniques such as anodization, an electrochemical method where a controlled potential (or current) is applied [51]. The resulting materials are known today as porous anodic alumina (PAA), anodic aluminum oxide (AAO), nanoporous alumina membrane (NAM), nanoporous anodic alumina (NAA), and porous aluminum oxide (PAO). These nanostructured materials have attracted great interest in nanotechnology due to their physical and chemical properties, low manufacturing cost, and potential applications [52, 53]. At present, they are interesting platforms for nanofilters and biosensors development, among other applications [54]; key developments in this area are discussed in Figure 2.6.

2.2.2 Fabrication and Structure of PAA

The manufacture of PAA is carried out on smooth surfaces of aluminum, which are initially corroded electrochemically in a controlled manner with the passage of current [54]. Corrosion generates ordered and vertically aligned nanochannels [55]. The manufacture is usually carried out in several stages. First, the aluminum is electropolished in a mixture of perchloric acid and ethanol (1:4) by galvanostatic (small surfaces) and potentiostatic (relatively large surfaces) anodizing processes, obtaining an aluminum oxide layer. Second, the aluminum oxide barrier cracks. Third, nucleation of the pores occurs, and finally steady-state growth [56, 57]. The thickness ratio of inner to outer layers can be controlled by the electrolyte type used, in this order: chromic acid > phosphoric acid > oxalic acid > sulfuric acid [56, 57]. After the anodization process, the PAA material will ideally present a continuous and regular pore distribution [58], as shown in Figure 2.7.

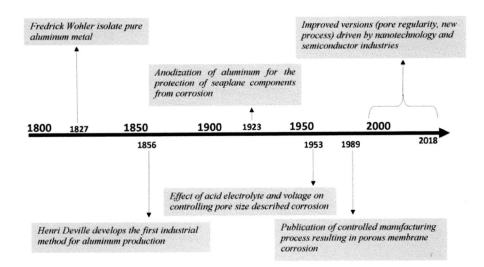

FIGURE 2.6 Key developments in the PAA manufacturing process.

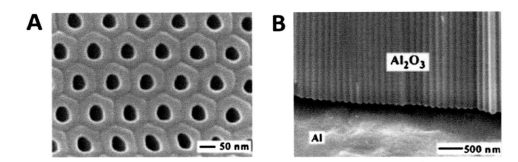

FIGURE 2.7 FESEM pictures of nanoporous anodic alumina films formed in 10% sulfuric acid at 70 V and 0°C for 50 min. (A) transverse fracture sections, (B) vertical fracture section. Reproduced, with permission, from Chu et al. [59]; copyright 2006, The Electrochemical Society.

2.2.3 Properties for Biosensing Applications

The physical and chemical properties of PAA's porous structure make it an excellent platform for the development of biosensors; most of the transduction systems assayed have been optical (absorbance, photoluminescence, reflectivity, transmittance) or electrochemical (conductance, resistance, impedance). As a bonus, miniaturization and integration are easy, and the nanopores can be used as containers for bioreagents [60].

It is interesting to note that not only the geometry of the pores and the surface chemistry of PAAs can be customized but also multifunctional properties can be introduced (e.g., optical and electrochemical activities), providing flexibility to develop advanced sensors for the detection of multiple analytes [61]. Another notable feature of PAA is its ability to control functionalization within the pores with lipids, antibodies, DNA, proteins, and enzymes, among others. Thus, the large specific surface of the PAA can be activated to interact or capture a wide range of molecules, which can later be analyzed by optical or electrochemical techniques. These modifications can also improve the properties of PAA (e.g., reflectivity, hydrophobicity or hydrophilicity, housing and maintenance of recognition biomolecules) for specific and high-throughput analysis [62].

PAAs could become a complementary alternative to porous silicon systems, providing some advantages in terms of chemical and mechanical stability, controllable pore geometry, scalable, and manufacturing process with competitive costs.

2.2.4 Methods to Immobilize the Bioreagent and Increase Biosensor Performance

In the last decade, many strategies have been developed to immobilize recognition biomolecules in PAA pores, such as carbohydrates, proteins (antibodies, enzymes), glycoproteins (lectins), and DNA. In general, these biomolecules can bind to the pores by covalent chemical bonds or noncovalent immobilization (physical-chemical adsorption) [63, 64]. As the ordered PAA surface suffers from chemical instability, the functionalization is advantageous, avoiding any passivation or fouling effects and making it more stable in the long term. Several PAA surface functionalization strategies have been described, including self-assembly and layer deposition. In general, chemical compounds are known to form monolayers

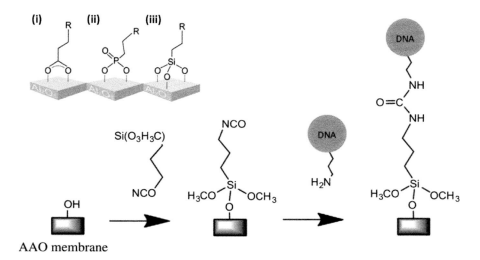

FIGURE 2.8 Monolayer formation over PAA. Silanization of the hydroxylated PAA surface with triethoxysilane and subsequent immobilization of amine-terminated DNA. In the insert: Chemical modification by using (i) carboxylic acid, (ii) phosphonic acids, and (iii) organosilanes (AAO = PAA). Modified, with permission, from Ingham et al. [54] (inset) and Jani et al. [60] (main figure).

on aluminum oxide, such as carboxylic acids, organosilanes, and phosphonic acids; as they are very thin, usually the structural properties of the PAA remains unchanged (Figure 2.8).

Most studies used organosilanes to functionalize the PAA with organic molecules. This is due to its packing density and stability over aluminum oxide; it has been applied to change the surface load, to create superhydrophobic materials, and to reduce the fouling effect [65]. Silanization is widely used for biofunctioning, being the main problem for the entire functionalization of the pore (bottom and top); in contrast, strategies based on the anodization were developed to obtain selective functionalization of the pore. To do this, after the first short anodization, the porous aluminum oxide generated is eliminated, and the remaining layer (70–90 nm) is treated with (3-aminopropil) triethoxysilane (APTES). Then, a subsequent higher anodization starts at the bottom of the first porous layer, resulting in an increase in the internal volume of the pores. The newly generated pores are functionalized with hydrophobic pentafluorophenyldimethylchlorosilane (PFPTES) with different properties than the APTES present in the pore openings. In this way, two specific areas of functionalization can be achieved [66].

Protein immobilization can be done through a nonspecific (random) or site-specific union. The most common method for covalent protein immobilization on surfaces employs links with amine groups of lysine's lateral reactive chain, such as N-hydroxysuccinimide, epoxy, aldehyde, and others. These are simple and well-established methodologies, but they often lead to random orientations in protein immobilization due to the abundance of amine groups in the protein structure, which can impair protein function or stability. Amine groups have also been used for the immobilization of DNA [61]. Another approach to nonspecific protein and DNA immobilization employs the strong and stable interaction between biotin and avidin/streptavidin. To use this method, the surface is initially modified with either of the ligands, and the molecule to be joined is previously modified to present the complementary molecule. Although the biotin-streptavidin system shows a very strong noncovalent interaction, the main disadvantage is that it requires the modification of biotin or streptavidin to label biomolecules (e.g., biotin oligo modification) in most cases [61].

Another strategy is the adsorption of polymers, as the formation of poly-L-lysine monolayers in PAA results in a positively charged surface, capable of joining DNA fragments through electrostatic interactions. This strategy has been used for the manufacture of FRET-based sensors with single-pore resolution [67].

2.2.5 Bioreagents Used for Biorecognition

The PAA provides a suitable environment for different biorecognition molecules, whether natural or artificial (imprinted polymer, aptamer, and others). It should be noted that nanoporous membranes not only act as platforms for the detection of biomolecular targets but also serve as filters to minimize the interfering effect of substances present in complex biological or environmental samples [68].

2.2.5.1 Natural Biorecognition Molecules

2.2.5.1.1 Enzymes

A very interesting advantage of using enzymes is that when they are inside the pores of the PAA, the stability (and catalytic activity) seems to be better than in the soluble form, as noted by several authors [69]. Li et al. [70] reported an interesting device based on a nanomachined flow channel biosensor for real-time electrochemical detection of glucose. Pt was sputtered on one side of the membrane and then used as a working electrode to detect the formation H_2O_2 that was produced by GOx covalently attached to the internal walls of the PAA nanochannels. Once glucose solutions flow through the nanochannels, the electrochemical signal is monitored using the Pt-modified electrode. The biosensor (flow-through operated) showed high reproducibility (±5.8%, n = 8) and sensitivity (86.62 mA mM^{-1} cm^{-2}). The electrochemical response demonstrated the stability of the immobilized enzymes, when it was measured with different rates of substrate flow, presenting good linearity [70].

2.2.5.1.2 Antibody

Antibodies are specific natural molecules, which are used in multiple types of diagnostic tests or immunoassays, such as the enzyme-linked immunosorbent assay (ELISA). The direct detection of the antibody-antigen reaction is usually made by optical detection (ELISA plate readers), but simple, miniaturizable, and disposable devices are highly needed; some work using PAA membranes has been presented [71]. Rajeev et al. reported an interesting strategy, taking advantage of the optical characteristics presented by the PAA membranes. They used interferometric reflectance spectroscopy (IRS) for the detection of tumor necrosis factor alpha (TNF-α), a proinflammatory cytokine important for wound healing (Figure 2.9). TNF-α determination has a predictive value about the healing process. To reach this goal, PAA nanopore walls were functionalized with anti-TNF antibodies using the silanization chemistry; after exposure to the analyte, the ligand reaction was followed by measuring the effective optical thickness (EOT), which is an IRS parameter. The study presented a detection limit of 0.13 μg mL^{-1} [72].

The detection of bacteria (total number and identification) is of special interest, since numerous epidemic outbreaks throughout the world are caused by bacterial pathogens, which have become a growing threat to public health. Current methods for the detection of food-borne bacteria include colony plate count and other microbiological methods, such as metabolic characterization, ELISA, and nucleic acid amplification. Most of them are labor intensive, time consuming, and require multiple tests to detect multiple types of pathogens in the samples. Currently, there is a need to develop new methods; some of them use PAA-based biosensors, which show great sensitivity and simplicity in terms of fabrication. Usually, nanopore-based methods are used to detect a single type of bacteria; however, the objective today is the simultaneous identification of various types of bacteria. Therefore, the integration between the microfluidics and the adequate functionalization of the PAA pores is necessary to provide a multifunctional platform for the simultaneous detection of multiple bacteria in a sample.

Tian et al. [73] used functionalized PAA membranes integrated with PEG designed to detect quickly, sensitively, and simultaneously two pathogens transmitted by food: *E. coli* O157:H7 and *Staphylococcus aureus*. For this purpose, bacteria are transported by a microfluidic system and captured using antibodies immobilized in PAA membranes. The results with samples of pure bacteria cultures showed specific responses and only a low degree of cross-response. The measurements were made using electrochemical impedance spectroscopy. The detection range was 10^2–10^5 CFU mL^{-1} and the limit of detection (LOD) was 10^2 CFU mL^{-1} [73]. The principles of detection and analytical characteristics of the method are presented in Figure 2.10.

2.2.5.1.3 Nucleic Acids

DNA and ribonucleic acid (RNA) are not only hereditary materials that codify genetic information; they are also ideal candidates for the recognition of complementary sequences given their high specificity. Moreover, other similar but bioinspired molecules (such as aptamers) have further increased the interest in nucleic acid biosensing applications [71]. A new strategy for the detection of unlabeled oligonucleotides has been proposed, where PAA channels are chemically modified with a morpholino (a DNA analogue) that can hybridize with the target DNA [74]. When hybridization occurs, the electrostatic field produced by the complex act as a barrier decreasing the diffusion of Fe(CN)$_6^{3-}$ through the channel. The electrochemical detection of ferricyanide was done with a Au electrode sputtered at the end of the nanochannel. The flux of the probe is strongly dependent on the ionic force, the channels diameter and the quantity of complex formed indicating a synergistic effect of both steric and electrostatic phenomena (Figure 2.11).

2.2.5.2 Artificial (but Bioinspired) Biorecognition Molecules

2.2.5.2.1 Aptamers

In the last decade, the use of aptamers on PAA platforms has gained importance in the development of biosensors, given its relatively easy design (SELEX or others), reproducible fabrication (given that they are just a short single-chain DNA or a RNA sequence), and specificity in the recognition of molecules such as DNA and proteins. Cancer diagnosis is one of the important fields where new and fast technologies are needed. An interesting strategy was reported recently by [75] (Figure 2.12), where the capture of circulating tumor cells (CTC) was performed by using a matrix of a hybrid nanochannel and ionchannel, which connected two chambers. Aptamer probes were immobilized on the surface of the ionic channel to react with the CTCs' membrane proteins. After that, mass transfer properties were modified, and then the CTCs were detected using a linear sweep voltammetry technique. The results showed that one type of acute leukemia denominated CCRF-CEM can be detected at a concentration as low as 100 cells mL^{-1} [75].

2.2.6 Transducers Used for Anodic Alumina-Based Biosensors

2.2.6.1 Electrochemical Transducers

The PAA-based electrochemical biosensors are designed considering the electrochemical parameter to modify and the signal to be determined; some examples include voltammetry, amperometry, and impedance spectrometry methods. In particular, PAA displays insulating characteristics, and several methods have been used to partially or completely coat the PAA with conductive surfaces, including the deposition of

Emerging Materials for Biosensor Applications

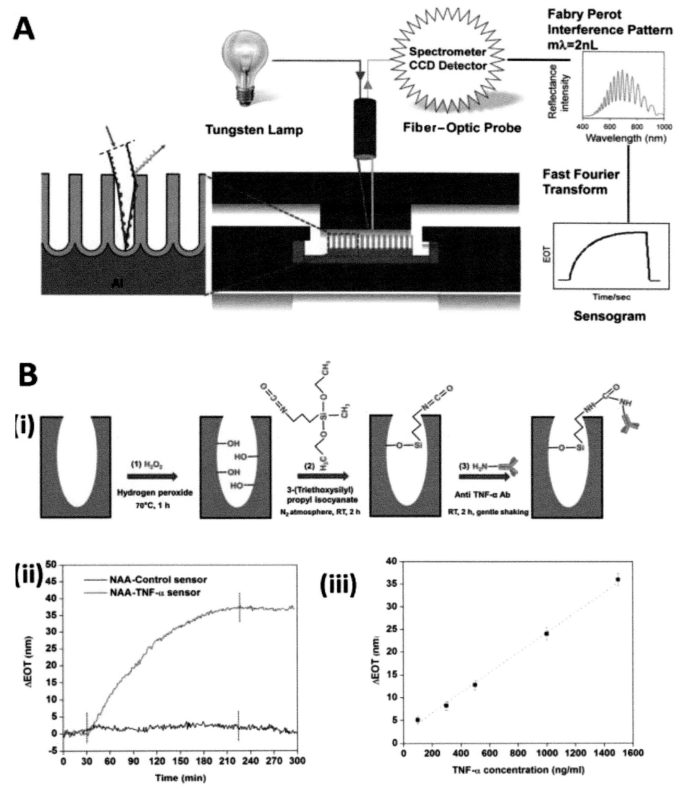

FIGURE 2.9 PAA-immunobiosensor, where the TNF-α is the analyte. (A) Schematic representation of IRS optical sensing setup and working principle. (B) Immobilization procedure and analytical response: (i) surface modification of PAA, (ii) sensing response, (iii) calibration curve. Reproduced, with permission from Rajeev et al. [72].

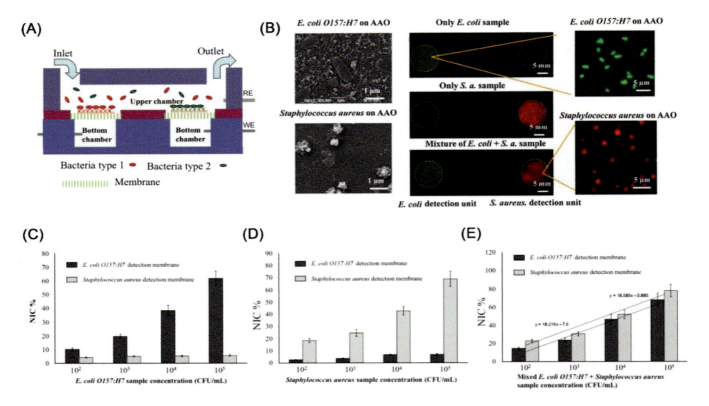

FIGURE 2.10 Detection of bacteria using PAA membranes. (A) Principle of simultaneous detection for two types of bacteria using the microfluidic device integrated with nanoporous membranes. (B) FESEM pictures of *E. coli* O157:H7 and *S. aureus* captured on specific antibody immobilized nanoporous alumina membranes; fluorescent images of nanoporous membrane detection units for samples of only *E. coli* O157:H7 sample, only *S. aureus*, and a mixture of the two bacteria samples; fluorescence images for rod-shaped *E. coli* O157:H7 and round-shaped *S. aureus* captured on nanoporous alumina membrane. Impedance sensing for simultaneous detection of *E. coli* O157:H7 and *S. aureus* for (C) only *E. coli* O157:H7 sample, (D) only *S. aureus* sample, (E) mixed bacteria sample with ratio 1:1. AAO = PAA. Reproduced, with permission, from Tian et al. [73].

Au, Pt, SnO$_2$, growth of carbon nanotubes (CNT), conductive polymers (polypyrrole), or by using other electrochemically active molecules, such as Prussian blue [76].

2.2.6.1.1 Voltammetry and Amperometry

The PAAs are excellent platforms for electrochemical biosensors development, considering their large surface area for immobilization within the pores and taking advantage of their filtering capacity to reject possible interfering substances (by electrostatic repulsion or size). Signals are usually obtained from a redox mediator reacting over the PAA-modified surface by means of differential pulse voltammetry (DPV), cyclic voltammetry (CV), or other related methods [61]. Some authors demonstrated this detection principle when biosensors for detection of dengue virus type 2 [77], DNA [78], and glucose [79]; LODs were of 1 pfU mL^{-1} (plaque-forming units), 9.55 pM, and 1 µM, respectively.

2.2.6.1.2 Impedance Spectroscopy

This technique measures the dielectric properties of a material as a result of the interaction between the applied electrical field (frequency) and the electrical dipole moment of the material. The electrical behavior of a PAA can be expressed by an equivalent circuit that will depend on the model used. Measurements are based on the electrical resistance of PAA pores, which can vary depending on the components reacting within the pore; in this way, it is possible to characterize and quantify biological events [54]. This principle has been used for the detection of *E. coli* [80] and dengue virus type 2 [81], with LODs of 10^2 CFU mL^{-1} and 1 pfU mL^{-1}, respectively.

2.2.6.2 Optical Transducers

Because of their dimensions, geometry, and chemical composition, the structure of PAAs show specific responses when interacting with light. This makes them an attractive material for developing optically active devices. Many studies have demonstrated the applications of PAA for optical filters, waveguides, reflective surfaces, resonators, or microcavities [82]. The PAA has proven to be a particularly outstanding platform for the development of detection devices with a unique set of optical properties, including reflectance, transmittance, absorbance, photoluminescent, chemiluminescent, and waveguide. These features were the main key for the development of optical sensors using techniques discussed next.

2.2.6.2.1 Surface Plasmon Resonance (SPR)

SPR methods are based on the excitation of surface plasmons generated by an evanescent electromagnetic wave

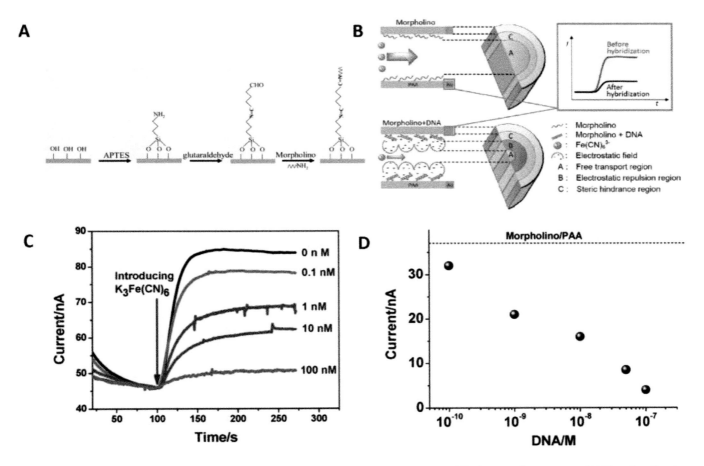

FIGURE 2.11 Development of an unspecific DNA biosensor by measuring ferricyanide diffusion trough nanopores. (A) Functionalization of the inner walls of PAA nanochannels with aminated morpholino. (B) Schematic representation of the working principle for the label-free DNA electrochemical detection. (C) The steady-state current response of the morpholino-functionalized nanochannel array after hybridizing in different concentrations of cDNA. (D) Calibration curve form data presented in C. Reprinted with permission from Li et al.; copyright (2010) American Chemical Society [74].

and have been introduced for characterization of thin films [83]. SPR-PAA biosensors are integrated into a prism-based configuration of a thin PAA film. The plasmon properties of this surface allow interrogating the changes in the refractive index within distances of about 200 nm from the metal film, allowing the detection of biological ligand-ligand events [83]. Different studies using this technique show the possibility to detect invertase [84], BSA [85], and avidin [86], with LODs of 10 nM, 60 nM, and 10 μg mL^{-1}, respectively.

2.2.6.2.2 Surface-Enhanced Raman Spectroscopy (SERS)

Many studies define SERS as a powerful analytical technique capable of detecting a wide range of analytes with implications in many fields of research. The exact mechanism of SERS is still not completely known, but it has been associated with the enhancement of the electromagnetic field surrounding small objects near intense and localized SPR phenomena [87]. SERS scattering consist of using the large local field enhancements that can exist at metallic surfaces to boost Raman scattering signals of adsorbed biomolecules. SERS is not only a quantitative technique but also qualitative, since biomolecules present characteristic signals of SERS, which can be used as fingerprints for identification. Interesting results using this kind of transducer include the detection of 3-mercaptobenzoic [88] and benzenethiol [89], with LODs of 3 mM and 500 ppb, respectively. Moreover, the system has been applied to follow the DNA amplification process [62].

2.2.6.2.3 Reflectometric Interference Spectroscopy (RIfS)

Different studies have shown promising potential for this interrogation technique in PAA platforms. The RIfS spectrum of PAA structures presents an interference pattern produced by the Fabry-Pérot effect, defined by $2n_{eff}L_p = m\lambda$, where n_{eff} is the effective refractive index of PAA, L_p is the physical thickness of the PAA film, and m is the RIfS strip order, of which the maximum is presented in the wavelength λ [90]. RIfS is a label-free technique, allowing the ability to follow biological events in real time and *in situ* without the use of fluorescent or radioactive labels. RIfS systems based on a PAA can perform an ultra-sensitive detection of analytes, such as organic gases and biomolecules. The principle was used to demonstrate the detection of DNA [91], CTCs [90],

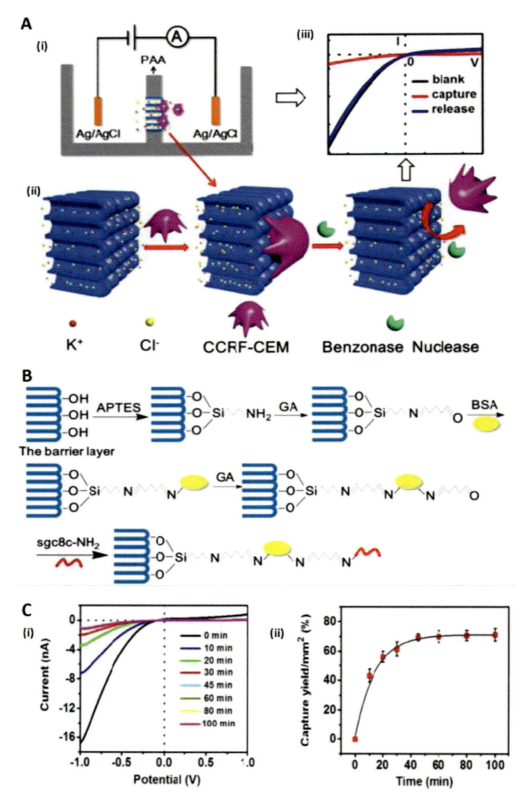

FIGURE 2.12 Method based on immobilized aptamers to detect cancer cells (CTCs). (A)(i), Schematic illustration of the setup for CTC capture, detection, and release, (A)(ii) illustration of CTC capture and release process on the nanochannel-ionchannel hybrid, (A)(iii) schematic representation of the varied electrochemical responses toward CTC capture and release. (B) Chemistry involved in the surface process of linking the aptamer probe on the nanochannel-ionchannel hybrid. (C)(i) Effect of the CCRF-CEM cells (a T lymphoblastic line) incubation time on I-V properties of PAA. (C)(ii) Capture yield of CCRF-CEM cell at different incubation times. Cell concentration was 2×10^5 cell mL^{-1}. Reprinted with permission from Cao et al.; copyright 2017 American Chemical Society [75].

and human IgG [92], with LODs of 2 nmol cm^{-2}, 10^3 cells mL^{-1}, and 600 nM, respectively.

2.2.6.2.4 Photoluminescence (PL)

A very interesting characteristic of PAAs studied during the last decade and very useful to develop new biosensors is their photoluminescence properties, the origin of which is still a question of debate [93] that we will not discuss here, but can be reviewed elsewhere [94]. This technique, unlike RIfS, amplifies the effect of Fabry-Pérot in wavelengths corresponding to the optical modes of the cavity formed by the air-PAA-Al system. Thus, the PAA PL spectrum shows an abundance of oscillations, and the number, intensity, and position of these oscillations can be tuned by modifying the length and diameters of the pores. The detection method enables the generation of a wide range of PL barcodes (like the barcode used in consumer products), which are suitable for intelligent development of optical biosensors for a wide range of analytes. Different studies showed the possibilities of these techniques to fabricate biosensors to detect trypsin [95], DNA [96], and glucose [97], with LODs of 40 μg mL^{-1}, 100, and 10 mM, respectively.

2.2.7 Applications

NAA-based biosensors can be useful in several areas. There are many reports showing applications in clinical chemistry (many of which were cited earlier), environmental analysis, food industry such as the identification of *Listeria monocytogenes* [98], and drug delivery systems [99], among others. Most of these uses have been reviewed and presented in tabular form by Santos et al. [69]. Usually, PAA-based chemical sensors and biosensors have been used in a wide range of applications for the detection of gases, organic molecules, biomolecules (DNA, proteins, antibodies), viruses, and cells (bacteria, cancer cells) in air, water, and biological samples. The availability of small and economic optical or electrochemical transducers may allow the development of biosensors able to work in point-of-care and point-of-need conditions.

2.3 PERSPECTIVES

Recent progress and findings in the application of hydrogel-based and PAA-based biosensors have been summarized. PAA and hydrogel manufacturing processes are based on well-established and scalable technologies that use low-cost fabrication methods. Both sensing platforms prove to possess a unique set of optical and electrochemical properties, which make them suitable materials to be coupled to many transducers. By matching them with different biochemical receptors, thousands of different biosensors can be developed. Some of the relevant properties of hydrogels are related to their biocompatibility and capability to allow rapid diffusion and transport of ions and molecules. The outstanding properties of PAAs are related to the large surface area, the easy control of nanopore size, adjustable geometries, and chemical resistance, as well as mechanical and thermal stability. Therefore, there are still excellent opportunities for future developments of optical and electrochemical biosensors based on PAA and hydrogel systems. In particular, interest lies in the implementation of lab-on-a-chip systems and the multiplexed capabilities for multiple analyte detection. In addition, the development of implantable biosensors for *in vivo* applications and the development of medical and environmental diagnostic systems is a goal.

REFERENCES

1. Annabi N, Tamayol A, Uquillas JA. Rational design and applications of hydrogels in regenerative medicine. Adv Mater 2014; 26: 85–124.
2. Miyata T, Uragami T, Nakamae K. Biomolecule-sensitive hydrogels. Adv Drug Deliv Rev 2002; 54: 79–98.
3. Dušek K, Patterson D. Transition in swollen polymer networks induced by intramolecular condensation. J Polym Sci Part A-2 Polym Phys 1968; 6: 1209–1216.
4. Shibayama M, Tanaka T. Volume phase transition and related phenomena of polymer gels. In Responsive gels: Volume transitions I. Springer, Berlin, Heidelberg. 1993; pp. 1–62.
5. Langer R, Tirrell DA. Designing materials for biology and medicine. Nature 2004; 428: 487–492.
6. Tavakoli J, Tang Y. Hydrogel based sensors for biomedical applications: An updated review. Polymers 2017; 9: 364.
7. Culver HR, Clegg JR, Peppas NA. Analyte-responsive hydrogels: Intelligent materials for biosensing and drug delivery. Acc Chem Res 2017; 50: 170–178.
8. Pina S, Oliveira JM, Reis RL. Natural-based nanocomposites for bone tissue engineering and regenerative medicine: A review. Adv Mater 2015; 27: 1143–1169.
9. Stuart MAC, Huck WTS, Genzer J. Emerging applications of stimuli-responsive polymer materials. Nat Mater 2010; 9: 101–113.
10. Hoffman AS. Stimuli-responsive polymers: Biomedical applications and challenges for clinical translation. Adv Drug Deliv Rev 2013; 65: 10–16.
11. Feng L, Wang L, Hu Z, Tian Y, Xian Y, Jin L. Encapsulation of horseradish peroxidase into hydrogel, and its bioelectrochemistry. Microchim Acta 2009; 164: 49–54.
12. Chattopadhyay S, Raines RT. Review collagen-based biomaterials for wound healing. Biopolymers 2014; 101: 821–833.
13. Highley CB, Prestwich GD, Burdick JA. Recent advances in hyaluronic acid hydrogels for biomedical applications. Curr Opin Biotechnol 2016; 40: 35–40.
14. Krajewska B. Application of chitin- and chitosan-based materials for enzyme immobilizations: A review. Enzyme Microb Technol 2004; 35: 126–139.
15. Kotanen CN, Moussy FG, Carrara S, Guiseppi-Elie A. Implantable enzyme amperometric biosensors. Biosens Bioelectron 2012; 35: 14–26.
16. Wang X, Han M, Bao J, Tu W, Dai Z. A superoxide anion biosensor based on direct electron transfer of superoxide dismutase on sodium alginate sol–gel film and its application to monitoring of living cells. Anal Chim Acta 2012; 717: 61–66.
17. Polyak B, Geresh S, Marks RS. Synthesis and characterization of a biotin-alginate conjugate and its application in a biosensor construction. Biomacromolecules 2004; 5: 389–396.
18. Sun C, Miao J, Yan J. Applications of antibiofouling PEG-coating in electrochemical biosensors for determination of glucose in whole blood. Electrochim Acta 2013; 89: 549–554.

19. Lowe S, O'Brien-Simpson NM, Connal LA. Antibiofouling polymer interfaces: Poly(ethylene glycol) and other promising candidates. Polym Chem. 2015; 6(2): 198–212.
20. Muñoz EM, Yu H, Hallock J, Edens RE, Linhardt RJ. Poly(ethylene glycol)-based biosensor chip to study heparin–protein interactions. Anal Biochem 2005; 343: 176–178.
21. Yao H, Wang J, Mi S. Photo processing for biomedical hydrogels design and functionality: A review. Polymers 2017; 10: 11.
22. Olsen KG, Ross DJ, Tarlov MJ. Immobilization of DNA hydrogel plugs in microfluidic channels. Anal Chem 2002; 74: 1436–1441.
23. Zangmeister RA, Tarlov MJ. DNA Displacement assay integrated into microfluidic channels. Anal Chem 2004; 76: 3655–3659.
24. Huang L, Zhuang X, Hu J. Synthesis of biodegradable and electroactive multiblock polylactide and aniline pentamer copolymer for tissue engineering applications. Biomacromolecules 2008; 9: 850–858.
25. Le Goff GC, Srinivas RL, Hill WA, Doyle PS. Hydrogel microparticles for biosensing. Eur Polym J 2015; 72: 386–412.
26. Omidfar K, Dehdast A, Zarei H, Sourkohi BK, Larijani B. Development of urinary albumin immunosensor based on colloidal AuNP and PVA. Biosens Bioelectron 2011; 26: 4177–4183.
27. Guascito MR, Chirizzi D, Malitesta C, Mazzotta E. Mediator-free amperometric glucose biosensor based on glucose oxidase entrapped in poly(vinyl alcohol) matrix. Analyst 2011; 136: 164–173.
28. Pundir CS, Sandeep Singh B, Narang J. Construction of an amperometric triglyceride biosensor using PVA membrane bound enzymes. Clin Biochem 2010; 43: 467–472.
29. Bertok T, Gemeiner P, Mikula M, Gemeiner P, Tkac J. Ultrasensitive impedimetric lectin based biosensor for glycoproteins containing sialic acid. Mikrochim Acta 2013; 180: 151–159.
30. Dey P, Adamovski M, Friebe S, et al. Dendritic polyglycerol–poly(ethylene glycol)-based polymer networks for biosensing application. ACS Appl Mater Interfaces 2014; 6: 8937–8941.
31. Ahmad N, Abdullah J, Yusof N, Ab Rashid A, Abd Rahman S, Hasan M. Amperometric biosensor based on zirconium oxide/polyethylene glycol/tyrosinase composite film for the detection of phenolic compounds. Biosensors 2016; 6: 31.
32. Yan J, Pedrosa VA, Enomoto J, Simonian AL, Revzin A. Electrochemical biosensors for on-chip detection of oxidative stress from immune cells. Biomicrofluidics 2011; 5: 107–121.
33. Lafleur JP, Jönsson A, Senkbeil S, Kutter JP. Recent advances in lab-on-a-chip for biosensing applications. Biosens Bioelectron 2016; 76: 213–233.
34. Yadavalli VK, Koh W-G, Lazur GJ, Pishko M V. Microfabricated protein-containing poly(ethylene glycol) hydrogel arrays for biosensing. Sensors Actuators B Chem 2004; 97: 290–297.
35. Bjerketorp J, Håkansson S, Belkin S, Jansson JK. Advances in preservation methods: Keeping biosensor microorganisms alive and active. Curr Opin Biotechnol 2006; 17: 43–49.
36. Philp JC, Balmand S, Hajto E. Whole cell immobilised biosensors for toxicity assessment of a wastewater treatment plant treating phenolics-containing waste. Anal Chim Acta 2003; 487: 61–74.
37. Ghasemi-Mobarakeh L, Prabhakaran MP, Morshed M. Application of conductive polymers, scaffolds and electrical stimulation for nerve tissue engineering. J Tissue Eng Regen Med 2011; 5: 17–35.
38. Song HS, Kwon OS, Kim J-H, Conde J, Artzi N. 3D hydrogel scaffold doped with 2D graphene materials for biosensors and bioelectronics. Biosens Bioelectron 2017; 89: 187–200.
39. Yan J, Pedrosa VA, Simonian AL, Revzin A. Immobilizing enzymes onto electrode arrays by hydrogel photolithography to fabricate multi-analyte electrochemical biosensors. ACS Appl Mater Interfaces 2010; 2: 748–755.
40. Pal RK, Turner EE, Chalfant BH, Yadavalli VK. Mechanically robust, photopatternable conductive hydrogel composites. React Funct Polym 2017; 120: 66–73.
41. Pal RK, Pradhan S, Narayanan L, Yadavalli VK. Micro-patterned conductive polymer biosensors on flexible PDMS films. Sensors Actuators B Chem 2018; 259: 498–504.
42. Jang E, Koh W-G. Multiplexed enzyme-based bioassay within microfluidic devices using shape-coded hydrogel microparticles. Sensors Actuators B Chem 2010; 143: 681–688.
43. Yeol L, Hyun JL, Kyung JS, Won-Gun K. Fabrication of hydrogel-micropatterned nanofibers for highly sensitive microarray-based immunosensors having additional enzyme-based sensing capability. J Mater Chem 2011; 21: 4476–4483.
44. Meiring JE, Schmid MJ, Grayson SM. Hydrogel biosensor array platform indexed by shape. Chem Mater 2004; 16: 5574–5580.
45. Wei X, Tian T, Jia S. Target-responsive DNA hydrogel mediated "stop-flow" microfluidic paper-based analytic device for rapid, portable and visual detection of multiple targets. Anal Chem 2015; 87: 4275–4282.
46. Edmondson R, Broglie JJ, Adcock AF, Yang L. Three-dimensional cell culture systems and their applications in drug discovery and cell-based biosensors. Assay Drug Dev Technol 2014; 12: 207–218.
47. Inal S, Hama A, Ferro M. Conducting polymer scaffolds for hosting and monitoring 3D cell culture. Adv Biosyst 2017; 1: 1700052.
48. Lee M-Y, Kumar RA, Sukumaran SM, Hogg MG, Clark DS, Dordick JS. Three-dimensional cellular microarray for high-throughput toxicology assays. Proc Natl Acad Sci 2008; 105: 59–63.
49. Vigués N, Pujol-Vila F, Marquez-Maqueda A, Muñoz-Berbel X, Mas J. Electro-addressable conductive alginate hydrogel for bacterial trapping and general toxicity determination. Anal Chim Acta. 2018; 1036:115–120.
50. McCullough JC. The Aluminum Industry. J Chem Educ 1930; 7:1977.
51. Keller F, Hunter MS, Robinson DL. Structural features of oxide coatings on aluminum. J Electrochem Soc 1953; 100: 411.
52. Ferré-Borrull J, Pallarès J, Macías G, Marsal L. Nanostructural engineering of nanoporous anodic alumina for biosensing applications. Materials 2014; 7: 5225–5253.
53. Warkiani ME, Bhagat AAS, Khoo BL. Isoporous micro/nano-engineered membranes. ACS Nano 2013; 7: 1882–1904.
54. Ingham CJ, ter Maat J, de Vos WM. Where bio meets nano: The many uses for nanoporous aluminum oxide in biotechnology. Biotechnol Adv 2012; 30: 1089–1099.
55. Santos A. Nanoporous anodic alumina photonic crystals: Fundamentals, developments and perspectives. J Mater Chem C 2017; 5: 5581–5599.
56. Sulka GD, Stępniowski WJ. Structural features of self-organized nanopore arrays formed by anodization of aluminum in oxalic acid at relatively high temperatures. Electrochim Acta 2009; 54: 3683–3691.
57. Zaraska L, Kurowska E, Sulka GD, Senyk I, Jaskula M. The effect of anode surface area on nanoporous oxide formation

58. Zaraska L, Wierzbicka E, Kurowska-Tabor E, Sulka GD. Synthesis of nanoporous anodic alumina by anodic oxidation of low purity aluminum substrates. Nanoporous Alumina 2015:61–106. during anodizing of low purity aluminum (AA1050 alloy). J Solid State Electrochem 2014; 18: 361–368.
59. Chu SZ, Wada K, Inoue S, Isogai M, Katsuta Y, Yasumori A. Large-scale fabrication of ordered nanoporous alumina films with arbitrary pore intervals by critical-potential anodization. J Electrochem Soc 2006; 153: B384.
60. Md Jani AM, Losic D, Voelcker NH. Nanoporous anodic aluminium oxide: Advances in surface engineering and emerging applications. Prog Mater Sci 2013; 58: 636–704.
61. Sriram G, Patil P, Bhat MP. Current trends in nanoporous anodized alumina platforms for biosensing applications. J Nanomater 2016; 2016: 1–24.
62. Tran BM, Nam NN, Son SJ, Lee NY. Nanoporous anodic aluminum oxide internalized with gold nanoparticles for on-chip PCR and direct detection by surface-enhanced Raman scattering. Analyst 2018; 143: 808–812.
63. Donczo B, Kerekgyarto J, Szurmai Z, Guttman A. Glycan microarrays: New angles and new strategies. Analyst 2014; 139: 2650.
64. Trilling AK, Beekwilder J, Zuilhof H. Antibody orientation on biosensor surfaces: A minireview. Analyst 2013; 138: 1619.
65. Liakos IL, Newman RC, McAlpine E, Alexander MR. Comparative study of self-assembly of a range of monofunctional aliphatic molecules on magnetron-sputtered aluminium. Surf Interface Anal 2004; 36: 347–354.
66. Mutalib Md Jani A, Anglin EJ, McInnes SJP, Losic D, Shapter JG, Voelcker NH. Nanoporous anodic aluminium oxide membranes with layered surface chemistry. Chem Commun. 2009; 21: 3062.
67. Matsumoto F, Nishio K, Masuda H. Flow-through-type DNA array based on ideally ordered anodic porous alumina substrate. Adv Mater 2004; 16: 2105–2108.
68. De la Escosura-Muñiz A, Chunglok W, Surareungchai W, Merkoçi A. Nanochannels for diagnostic of thrombin-related diseases in human blood. Biosens Bioelectron 2013; 40: 24–31.
69. Santos A, Kumeria T, Losic D. Nanoporous anodic alumina: A versatile platform for optical biosensors. Materials. 2014; 7: 4297–4320.
70. Li S-J, Xing Y, Tang M-Y, Wang L-H, Liu L. A novel nanomachined flow channel glucose sensor based on an alumina membrane. Anal Methods 2013; 5: 7022.
71. Dusan L, Abel S. Nanoporous alumina: Fabrication, structure, properties and applications. Springer International Publishing Switzerland, Cham, Switzerland, 2015, 219.
72. Rajeev G, Xifre-Perez E, Prieto Simon B, Cowin AJ, Marsal LF, Voelcker NH. A label-free optical biosensor based on nanoporous anodic alumina for tumour necrosis factor-alpha detection in chronic wounds. Sensors Actuators B Chem 2018; 257: 116–123.
73. Tian F, Lyu J, Shi J, Tan F, Yang M. A polymeric microfluidic device integrated with nanoporous alumina membranes for simultaneous detection of multiple foodborne pathogens. Sensors Actuators B Chem 2016; 225: 312–318.
74. Li S-J, Li J, Wang K. a nanochannel array-based electrochemical device for quantitative label-free DNA Analysis. ACS Nano 2010; 4: 6417–6424.
75. Cao J, Zhao X-P, Younis MR, Li Z-Q, Xia X-H, Wang C. Ultrasensitive capture, detection, and release of circulating tumor cells using a nanochannel–ion channel hybrid coupled with electrochemical detection technique. Anal Chem 2017; 89: 10957–10964.
76. Chik H, Xu JM. Nanometric superlattices: Non-lithographic fabrication, materials, and prospects. Mater Sci Eng R Reports 2004; 43: 103–138.
77. Parkash O, Shueb R. Diagnosis of dengue infection using conventional and biosensor based techniques. Viruses 2015; 7: 5410–5427.
78. De la Escosura-Muñiz A, Mekoçi A. Nanoparticle based enhancement of electrochemical DNA hybridization signal using nanoporous electrodes. Chem Commun 2010; 46: 9007.
79. Xian Y, Hu Y, Liu F, Xian Y, Feng L, Jin L. Template synthesis of highly ordered Prussian blue array and its application to the glucose biosensing. Biosens Bioelectron 2007; 22: 2827–2833.
80. Tan F, Leung PHM, Liu Z. A PDMS microfluidic impedance immunosensor for *E. coli* O157:H7 and *Staphylococcus aureus* detection via antibody-immobilized nanoporous membrane. Sensors Actuators B Chem 2011; 159: 328–335.
81. Peh AEK, Li SFY. Dengue virus detection using impedance measured across nanoporous alumina membrane. Biosens Bioelectron 2013; 42: 391–396.
82. Santos A, Balderrama VS, Alba M. Nanoporous anodic alumina barcodes: Toward smart optical biosensors. Adv Mater 2012; 24: 1050–1054.
83. Green RJ, Frazier RA, Shakesheff KM, Davies MC, Roberts CJ, Tendler SJ. Surface plasmon resonance analysis of dynamic biological interactions with biomaterials. Biomaterials 2000; 21: 1823–1835.
84. Nagaura T, Takeuchi F, Yamauchi Y, Wada K, Inoue S. Fabrication of ordered Ni nanocones using a porous anodic alumina template. Electrochem Commun 2008; 10:681–685.
85. Lau KHA, Tan L-S, Tamada K, Sander MS, Knoll W. Highly sensitive detection of processes occurring inside nanoporous anodic alumina templates: A waveguide optical study. J Phys Chem B 2004; 108: 10812–10818.
86. Hiep HM, Yoshikawa H, Tamiya E. Interference localized surface plasmon resonance nanosensor tailored for the detection of specific biomolecular interactions. Anal Chem 2010; 82: 1221–1227.
87. Han XX, Huang GG, Zhao B, Ozaki Y. Label-free highly sensitive detection of proteins in aqueous solutions using surface-enhanced Raman scattering. Anal Chem 2009; 81: 3329–3333.
88. Velleman L, Bruneel J-L, Guillaume F, Losic D, Shapter JG. Raman spectroscopy probing of self-assembled monolayers inside the pores of gold nanotube membranes. Phys Chem Chem Phys 2011; 13: 19587.
89. Kodiyath R, Malak ST, Combs ZA. Assemblies of silver nanocubes for highly sensitive SERS chemical vapor detection. J Mater Chem A 2013; 1: 2777.
90. Kumeria T, Losic D. Controlling interferometric properties of nanoporous anodic aluminium oxide. Nanoscale Res Lett 2012; 7: 88.
91. Pan S, Rothberg LJ. Interferometric sensing of biomolecular binding using nanoporous aluminum oxide templates. Nano Lett 2003; 3: 811–814.
92. Dronov R, Jane A, Shapter JG, Hodges A, Voelcker NH. Nanoporous alumina-based interferometric transducers ennobled. Nanoscale 2011; 3: 3109.
93. Santos A, Alba M, Rahman MM. Structural tuning of photoluminescence in nanoporous anodic alumina by hard anodization in oxalic and malonic acids. Nanoscale Res Lett 2012; 7: 228.

94. Du Y, Cai WL, Mo CM, Chen J, Zhang LD, Zhu XG. Preparation and photoluminescence of alumina membranes with ordered pore arrays. Appl Phys Lett 1999; 740: 2951–2953.
95. Jia RP, Shen Y, Luo HQ, Chen XG, Hu ZD, Xue DS. Enhanced photoluminescence properties of morin and trypsin absorbed on porous alumina films with ordered pores array. Solid State Commun 2004; 130: 367–372.
96. Feng C-L, Zhong XH, Steinhart M, Caminade A-M, Majoral J-P, Knoll W. Graded-bandgap quantum- dot-modified nanotubes: A sensitive biosensor for enhanced detection of DNA hybridization. Adv Mater 2007; 19: 1933–1936.
97. Santos A, Kumeria T, Losic D. Optically optimized photoluminescent and interferometric biosensors based on nanoporous anodic alumina: A comparison. Anal Chem 2013; 85: 7904–7911.
98. Zhou C-X, Mo R-J, Chen Z-M. Quantitative label-free *Listeria* analysis based on aptamer modified nanoporous sensor. ACS Sensors 2016; 1: 965–969.
99. Kang H-J, Kim DJ, Park S-J, Yoo J-B, Ryu YS. Controlled drug release using nanoporous anodic aluminum oxide on stent. Thin Solid Films 2007; 515: 5184–5187.

3 Smart Materials for Developing Sensor Platforms

Lightson Ngashangva and Pranab Goswami
Indian Institute of Technology Guwahati, Assam, India

Babina Chakma
Indegen Private Limited, Nagawara, Bengaluru, India

CONTENTS

3.1 Introduction 48
3.2 Sensor Platforms and Smart Materials 48
    3.2.1 Sensor Platforms 48
    3.2.2 Smart Materials 49
        3.2.2.1 Chromogenic Materials 49
        3.2.2.2 Piezoelectric Materials 49
        3.2.2.3 Magnetic Materials 50
        3.2.2.4 Shape Memory Materials 50
        3.2.2.5 Luminescent Materials 51
        3.2.2.6 Plasmonic Materials 51
        3.2.2.7 Self-Healing Materials 51
3.3 Sensor Platforms Based on Smart Materials 51
    3.3.1 Organic-Based Smart Materials 51
        3.3.1.1 Carbon Nanomaterials 52
        3.3.1.2 Organic Semiconductors 52
        3.3.1.3 Organic Thin Film Transistors (OTFTs) 52
        3.3.1.4 Polymer Materials 53
        3.3.1.5 Hydrogels 53
    3.3.2 Inorganic-Based Smart Materials 54
        3.3.2.1 Metal Oxides 54
        3.3.2.2 Noble Metal Nanomaterials 54
        3.3.2.3 Transition Metal Dichalcogenides 55
        3.3.2.4 Liquid Metals 55
        3.3.2.5 Ionic Liquids (ILs) 55
    3.3.3 Biomolecule-Based Smart Materials 55
        3.3.3.1 Oligonucleotides 56
        3.3.3.2 Peptides 56
        3.3.3.3 Silk Fibroin (SF) 56
        3.3.3.4 Cellulose Materials 56
    3.3.4 Hybrid Smart Materials 56
        3.3.4.1 Sol-Gel Materials 57
        3.3.4.2 Bionanomaterials 57
3.4 Techniques for Developing Sensor Platforms 57
    3.4.1 Synthesis of Smart Materials 57
        3.4.1.1 Synthesis of Polymers 57
        3.4.1.2 Synthesis of Nanomaterials 58
    3.4.2 Deposition Techniques 58
    3.4.3 Modification Techniques 58
        3.4.3.1 Molecular Self-Assembly 60
    3.4.4 Fabrication Techniques 60
        3.4.4.1 Microfabrication Techniques 60

	3.4.4.2 Nanofabrication Techniques	62
3.5	Advanced Application of Smart Materials	64
	3.5.1 Wireless Sensors	64
	3.5.2 Wearable Sensing Technologies	64
	3.5.2.1 Composition of Wearable Sensors	64
3.6	Flexible Conducting Electrodes	65
3.7	Smartphone-Based Sensing Device	66
	3.7.1 Detector	66
	3.7.2 Interface	66
3.8	Conclusion	66
References		66

3.1 INTRODUCTION

The progress made over the last two decades in the field of material sciences, electronics and communication technology, and biotechnology has revolutionized the technology domain [1–4]. One of the significant advantages that could be accrued through this advancement is the scope of scaling down the size of the product. This miniaturized product development has been achieved in many areas like medical and health care, environmental science, and consumer electronics, among others. One of the significant contributions of scientific innovation to scaling down the sizes of the products is the development of *smart materials*. These advanced materials may be used to design the platform for developing next-generation sensors and biosensor. Smart materials are also called "responsive materials," as they can sense any external stimuli without human intervention. The external stimuli may be of different types, such as temperature, pressure, pH, and light. Smart materials are designed to have one or more properties that can be responsive to external changes. These materials have been incorporated in various devices ranging from simple to sophisticated technology-based products.

Since the inception of developing smart, intelligent, and adaptive materials in the mid-1980s, there has been an enormous advancement of technological application of these materials into our daily lives. The interdisciplinary approach of research and development has played an important role in achieving this progress. The emerging research area of integrating electroactive functional materials can be cited as an example to demonstrate the development of smart material-based technology. There are many bioprocesses, which inspire the design of smart and high-performance biointerface materials. Weak interactions like hydrogen bonding (H-bonding) in biological systems remains one of the most important interaction forces for numerous bioprocesses. One of the highly specific H-bonding interactions can be seen in our genetic system – the interaction of the complementary base pair of nucleotides A, T, C, G, and U. The secondary and tertiary structure formation of biomolecules is also very much driven by the multiple H-bondings, resulting in various activities and functionalities. There are many such examples where biological systems and nature can inspire to develop next-generation high-performance materials. There may be some difficulties in harnessing the biological properties, but the physical properties and some of the chemical properties could be obtained by introducing an intelligent design and process.

Many smart materials have been employed for designing sensing platforms. Metallic and nonmetallic nanomaterials, bimetallic, polymer, and chemical compounds as coating agent on the materials and biomolecules like deoxyribonucleic acid (DNA), ribonucleic acid (RNA), aptamers, oligonucleotides, peptides, and proteins are some of the generally used smart materials. One material may not be sufficient to exhibit the desired sensing outcome, so hybrid materials are also explored to enhance the effects from the two or more materials combined together. Several emerging techniques that aid in achieving the miniature technologies through smart materials are nanoelectrical mechanical systems (NEMS), microelectrical mechanical systems (MEMS), and bio-NEMS/MEMS [5]. This chapter describes some commonly available smart materials along with the various techniques for designing smart platforms mainly for biosensing applications.

3.2 SENSOR PLATFORMS AND SMART MATERIALS

3.2.1 SENSOR PLATFORMS

The sensor platform is one of the main components of the sensor device, as discussed elsewhere. It is also called a "sensor element" or "receptor" in some of the literature. The platform may be defined as the specific area or location where various reactions are performed between analytes and the sensing platform/element. There are various platforms known to develop sensing devices as per the requirement and demands. Some of the material parameters considered for sensor platforms are as follows: (a) *Inertness or stable:* The materials should not be influenced by any minor physical change, including temperature or chemical environment. (b) *Reactive sites or binding sites:* The materials should possess reactive sites or binding sites for immobilization of compounds. (c) *Reactivity:* The materials should be reactive, and any chemical change or reaction on the platform should be easily detected even for a small change (physically or chemically). (d) *Robustness:* The materials should be strong and sturdy, so even in harsh environmental conditions, the working of the sensor should not be affected. (e) *Electrochemically*

Smart Materials for Developing Sensor Platforms

active: Materials with electrochemical activity can promote fast electron transfer. (f) *Biodegradable:* Biodegradability of the platform is desired to address the global concerns with environmental pollution.

The platform materials are chosen based on the desired sensing applications. The materials which possess physicochemical properties, such as catalytic ability, electron/ion transport, thermal and optical properties, plasticity, and quantum mechanical properties, may have additional advantages for multiplex sensing applications. The natural or other materials that do have sensing ability can be harnessed for the application in sensing technology. In fact, the next-generation technology has been focused on the natural smart materials. However, some of the naturally occurring smart materials could not fulfil the parameters that are mentioned earlier. Nevertheless, some materials that do not fall within the category of smart materials could be modified with some techniques, such as functionalization, to introduce the smart material property. Various physical, chemical, and biological attributes of the target analytes are probed by the sensors and biosensors. The specificity of the biosensors and sensors depend on the specific interaction of the attributes with the sensing platforms and elements. Some of the commonly targeted physical, chemical, and biological attributes for biosensors and sensors are as follows: (a) *Physical:* temperature, heat, pressure, acoustics, touch, light, colors, etc. (b) *Chemical:* volatile organic compounds, organic compounds, inorganic compounds, heavy metals (Hg, Cu, Pb, Ag, etc.), toxic gases, etc. (c) *Biological:* antibody, nucleic acids, proteins, pathogens, biomolecules, amino acids, etc.

3.2.2 Smart Materials

A material may be termed a smart material simply if one or more of its properties (mechanical, physicochemical, biological) are controllably altered in response to the predetermined stimulus. Hence, the material may also be called "stimulus responsive material." The smart materials are increasingly used in a wide range of products such as children's toys that change color with change in temperature, high-throughput industrial applications using shape memory alloys, and in various sensing applications. Smart materials can be divided depending on how the material is responding to the stimulus. Light, temperature, pH, redox reaction (in chemical systems or biological systems), etc., are some of the known stimuli which are harnessed for the high-end product technology applications. Some of the smart materials based on the stimulus are explained in the following sections.

3.2.2.1 Chromogenic Materials

These materials can undergo a change in optical properties by applying an external stimulus like heat/temperature, light, and voltage. Some of the commonly known chromogenic properties are thermochromic, photochromic, electrochromic, and gasochromic materials, which reversibly change the optical property, like color or transparency, in response to heat, light, applied voltage, and redox gases. For instance, the characteristics of electrochromic materials arise from the reversible redox reactions of the transition metal ions under applied voltage. The performance of the materials will depend on the amount of reduced/oxidized metal ions and the switching kinetics. The phenomenon of electrochromism in polymeric electrochromic material is the change of π-electrons of the polymer backbone, which is followed by a reversible change in redox potential. Hence, chromogenic materials have been used for various applications by incorporating the advanced materials like nanoparticles, polymers, and other chemical compounds. Hybrid electrochemical materials like a polymer-metal nanomaterial can be fabricated to produce plasmonic resonance, as shown in Figure 3.1. The rapid switching property of the chromogenic materials along with other characteristics features make the materials a potential candidate for sensing applications as well [6].

3.2.2.2 Piezoelectric Materials

These materials are usually embedded in electronic devices and used in many technological applications, including sensors and biosensors. The primary requirement of the piezoelectric effect is that the materials should be anisotropic crystals. The materials generate piezoelectricity under an oscillation frequency, which is dependent on the analyte interaction on the crystal surface. The mass bound on the crystal surface is related to the oscillation frequencies. The change in

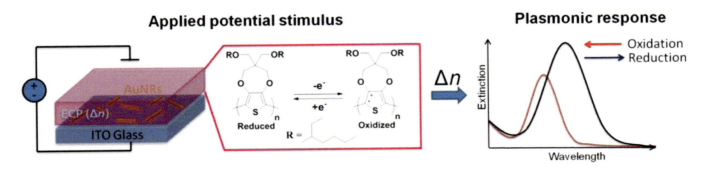

FIGURE 3.1 The plasmonic response can be observed by monitoring the absorption and scattering of light resulting from the hybrid electrochromic polymer with metal nanomaterial. (Reused with permission from the publisher [12].)

frequency (Δf) is directly proportional to the change in mass (Δm) bound on the crystal, as given by Sauerbrey:

$$\Delta f = -\frac{2 f_0^2 \Delta m}{A\sqrt{\rho_q \mu_q}} = -2.3 \times 10^6 f_0^2 \frac{\Delta m}{A}$$

where

f_0: The fundamental mode of the crystal in hertz
A: Piezoelectrically active area in centimeters
ρ_q: Density (2.648 g/cm^3 for quartz)
μ_q: Shear modulus (2.947 × 10^{11} g/cm·s^2 of quartz)

If there is an interaction other than physical or chemical, such as ambient viscosity, which can also drive the frequency shift, the equation described by Kanazawa and co-workers for a crystal should consider:

$$\Delta f = f_0^{3/2} \sqrt{\frac{\Delta(\eta_l \rho_l)}{\pi \eta_q \rho_q}}$$

Where symbols with index l relates to the ambient liquid and q to quartz crystal. There are many anisotropic materials that produce piezoelectric effect that are inorganic, organic, and even biomolecules like nucleic acid. Some of the examples of piezoelectric materials are as follows: *Inorganic anisotropic materials*: aluminum phosphate (also known as berlinite), aluminum nitride, zinc oxide, crystallized topaz, crystallized tourmaline, barium titanate, gallium orthophosphate, lead titanate, quartz SiO$_2$, etc. *Organic materials*: sodium potassium tartrate tetrahydrate (or Rochelle salt), organic polymers like polyvinylidene fluoride, Piezo1, collagen, cellulose, etc. [7].

3.2.2.3 Magnetic Materials

The shape of the materials can be changed if an external magnetic field is applied. The phenomenon involved in these material properties, like magnetostriction and magnetorheology, may be explained as follows:

Magnetostrictive Materials: A small change in the magnetic domain results in the change in length of the materials, which may be due to the internal strain. The strain thus leads to stretching of the materials in the direction of the magnetic field. This phenomenon is called *magnetostriction*. In general, magnetostriction is a reversible energy exchange between magnetic and mechanical forms which allows the material to be used in actuators and sensors. The "Joule effect" is one of the most understood effects of magnetostriction. In this effect, the expansion or contraction of a ferromagnetic rod is related to the longitudinal magnetic effect. The applied magnetic field is roughly proportional to the increase in length, which is also called longitudinal strain or the contraction of diameter, which is also called lateral strain. Another effect is the "Villari effect" related to the mechanical stress of the sample. This effect takes place when there is a change in the magnetic flux density. Here, the level of applied stress is proportional to the change in flux density. The effect is reversible in nature and is used for various sensor applications.

Magnetorheological Materials: These smart materials are controlled by an external magnetic field to alter mechanical properties of the materials. In the presence of a magnetic field, the reversible transition of the liquid state to a nearly solid state occurs, which is called *magnetorheological effect*. The materials could exhibit changes in viscosity of several orders of magnitude, which make them very good candidates for application in mechanical systems. The use of magnetic fields to produce a magnetorheological effect on the materials can be used for various other applications such as isothermal magnetic advection, chemical sensors, and biomedical applications, among others. Traditionally, the magnetorheological (MR) fluids are of two-phase fluids, which are prepared in nonmagnetizable liquid with highly magnetizable micronsized particles. Iron particles are commonly used because of the large saturation magnetization and minerals, silicone oils, polyesters, polyethers, etc., are typical carrier liquids. Since, the MR effect is attributed to the magnetic field induced on the suspended particles, the suspensions have low viscosity if there is no magnetic field. However, when a magnetic field is applied, the particles attract each other because of the magnetization along the field lines and form anisometric aggregates, which results in large yield stress and thus enhances the viscoelasticity. Another MR fluid is the inverse ferrofluids (or magnetic holes), which are formed by micron-sized, nonmagnetizable particles in the fluid. One advantage is that the mechanical properties of the ferrofluid can be controlled by varying the magnetic field strength. Since, many types of nonmagnetizable particles are available, particle size, shape, and functionality can be easily tunable. Magnetic gels, elastomers, magnetic elastomers, or field-responsive composite materials such as superparamagnetic polymer-based particles are made by dispersing the magnetic particles in a polymer solution before crosslinking in a polymerization process. These have been used in colloidal aggregation, optical trapping studies, and magneto-electrorheological applications. Another example of application is the electroconductive composite by introducing graphite microparticles. Fabrication of various sensors and transducers, magnetoresistors, etc., can be exploited with this property [8].

3.2.2.4 Shape Memory Materials

When a particular stimulus is applied, shape memory materials can recover their original shape, and this is called the shape memory effect. Many materials can undergo the shape memory effect under certain conditions such as superelasticity in the case of alloys and viscoelasticity in the case of polymers [9].

Shape Memory Alloys: The material transforms back to its original state by heating above the critical temperature, which follows the reversible austenite-martensite phase transformation. By rearrangement processes during the high temperature phase (which is austenite) and the low temperature phase (which is martensite), there is a change in the structure of the crystal lattice resulting in two distinct stress-strain curves.

TABLE 3.1
The Characteristic Features of Self-Healing Polymers

Extrinsic Self-Healing Polymer	Intrinsic Self-Healing Polymer
• Self-healing property triggered by external agents which are encapsulated in capsules/networks.	• No external triggered mechanism is required for self-healing.
• Prevents the damage propagation and healing the materials by releasing the healing agents from the capsules or networks once the polymer composite is damaged.	• Self-healing is achieved without external healing agents through reversibility of molecular interaction of the matrix polymer.

Shape Memory Polymer (SMP): Compared to metals/alloys, the designing and tailoring of the material properties of polymers is much easier and cost-effective. Additionally, the SMPs are lighter, have a higher recovery strain, and shape recovery can be triggered by many stimuli, which allows the polymers to be used in various application, including sensing. The underlying mechanism of the SMPs is that they have a dual segment/domain system as a hard/elastic segment and the other as the transition segment. With the advancement of technology and material design, shape memory composite and hybrid materials have been developed for smart sensing applications, mini-actuators, MEMS, etc.

3.2.2.5 Luminescent Materials

The luminescent materials have been widely studied for developing smart materials for sensing applications. In response to changes in various environmental conditions, some of the physicochemical properties on the materials surface are dynamically altered. The materials can also be controlled by a trigger mechanism at the interface of the solid/liquid phase. One such prominent example is Lanthanide ions, which have been extensively studied due to their high color purity. The intraconfigurational *f–f* transitions and abundant ladder-like energy levels the ions possess long lifetimes of the excited state. Additionally, the optical property of lanthanides is governed by the *4f–5d* transitions of divalent and trivalent lanthanides, which enables the lanthanide compounds to have a tunable emission and luminescent lifetime. Hence, the optical properties of lanthanide ions could be harnessed to develop advanced sensing devices with respect to various external stimuli like light/photons, pH, thermal, magnetic, electric field, or multiple stimuli [10].

3.2.2.6 Plasmonic Materials

Plasmonic resonances have been exploited and used for various sensing applications because the sensors can provide label-free, signal-enhanced, and real-time monitoring and sensing at the molecular level [11]. In most of the cases, plasmonic materials are metallic in nature, and in general gold (Au) and silver (Ag) are widely used substrates for the applications. Classically, plasmonic materials support the electromagnetic oscillations (known as surface plasmon polaritons) at the metal-dielectric interface. The excitation of surface plasmons by light is known as surface plasmon resonance (SPR) for planar surfaces or localized surface plasmon resonance (LSPR) for metallic nanoparticle surface. The LSPR is greatly sensitive to size, size distribution, shape, and the surrounding environment of the nanoparticles. Many standard tools including biosensors for measuring the adsorption of material onto planar metal surfaces or onto the metal nanoparticles surface have been developed following this optical phenomenon.

3.2.2.7 Self-Healing Materials

Self-healing is defined as the ability to repair and heal damages automatically without any external influences. The materials should have the property of self-healing, self-repairing, or self-recovery [13]. However, there are certain materials where an external triggering mechanism is required to undergo self-healing. There are two mechanisms of self-healing: (a) *autonomic* (without any external influence) and (b) *nonautonomic* (needs external triggering). There are materials which have their own self-healing mechanisms like plastics/polymers, paints/coatings, metals/alloys, etc. Over the last decade, many self-healing materials have been introduced successfully in organic electronics such as wearable and flexible devices. For self-healing behavior, polymeric materials remain one of the widely studied materials. In a polymer system, self-healing is achieved by functionalization and polymerization even at low temperature. Self-healing can be quantified in terms of efficiency, which measures the restoration or recovery of the lost property/performance metric. Generally, self-healing polymers are categorized into two types that are based on the integration of the healing functionalities into the bulk materials: extrinsic and intrinsic self-healing polymers, as shown in Table 3.1.

3.3 SENSOR PLATFORMS BASED ON SMART MATERIALS

In this section, we will extend our discussion to the use of smart materials for developing sensor platforms. As a whole, smartness can be introduced in the devices by using smart materials or stimulus-responsive materials, as they response to one or more external stimuli.

3.3.1 Organic-Based Smart Materials

In most of the cases, the skeleton structure of the organic matter is made up of carbon, which is found in different forms of allotropes. In its fundamental state, the electronic configuration of carbon is $1s^2\,2s^2\,2p^2$. Carbon can form different types of bonds through rearrangement of the electronic configuration. In these rearrangements, three different hybridizations are possible, such as *sp*, *sp²*, and *sp³* hybridization. These hybridizations define the chemical bonds such as simple bond (σ bond) in *sp³* hybridization, double bond (one σ bond and one π bond) in *sp²* hybridization, and triple bond (one σ bond and two π bonds) in *sp* hybridization. Due to these inherent properties, carbon as a material can be modified and engineered for various applications.

3.3.1.1 Carbon Nanomaterials

Nanomaterials from carbon have been extensively used for developing various advanced technology/devices. Easy synthesis protocols, less expensive, and diverse optical and electronic properties have been cited as reasons to use carbon nanomaterial for developing various products and processes. The properties of carbon-based materials can be transformed by simple modification and functionalization. For example, by changing the hybridization from *sp²* to *sp³* carbon, the electrical and thermal conductivities of the carbon nanotubes (CNTs) could be changed. The technique has been widely used to design materials for various sensing applications [14], particularly with the emerging nanoscale carbon materials such as CNTs, graphene, and nano/mesoporous carbon [15]. Some of the characteristic features of the carbon and its types are shown in Table 3.2.

3.3.1.2 Organic Semiconductors

The biosensing applications of organic semiconductors are a comparatively recent development. Organic semiconductors are nonmetallic. This section focuses on organic semiconductor single crystals.

Organic semiconductor single crystals are small-molecule and polymer organic semiconductors that form the basis of the emerging field of organic electronics and photonics. These are promising materials used for the development of new electronic devices such as organic solar cells, organic light emitting displays, flexible inexpensive circuits, and molecular sensors. Even with the commendable progress of organic electronics in the applied research, our understanding of the fundamental properties of this important class of materials remains limited. This impediment is mainly rooted in considerable disorder in polycrystalline and amorphous organic thin films that governs their electronic properties. However, in recent past, electronic devices have been developed that are based on single crystals of organic semiconductors with unprecedented structural order and chemical purity. This achievement helped to advance experimental access to the intrinsic fundamental electronic properties of organic semiconductor materials and devices. Following this progress, the organic single crystals have received wide recognition as an important tool for various fundamental investigations such as excitonic energy transfer, polaronic charge transport, and interfacial phenomena. Eventually, the optical properties of organic semiconductors lead to a better understanding of organic electronic materials and devices.

3.3.1.3 Organic Thin Film Transistors (OTFTs)

These are three-terminal electronics based on metal electrodes with a thin organic semiconductor layer with an insulator. The active organic materials in OTFTs have been fabricated with a plastic substrate producing low-cost, lightweight, and flexible devices. In general, OTFT consists of the gate electrode, which is separated by the dielectric (insulator

TABLE 3.2
Carbon Nano/Materials and Their Typical Characteristics

Material Type	Typical Characteristics
Single wall carbon nanotubes (SWNT)	a. Excellent mechanical and electrochemical properties. b. A wall of one atom thick and diameters typically 1.4 nm (range 0.3–2 nm). c. Grown by chemical vapor deposition (CVD).
Multiwall carbon nanotubes (MWNTs)	a. Multiple walls with diameters of 10 nm and larger. b. Good electrochemical properties, moderate cost, and can grow arrays of nanotubes. c. Easier to process due to large diameter of 10–50 nm and 1–50 mm length, grown by CVD; for example, bamboo MWNT 20–40 nm diameter, 1–20 mm length, internal closeouts.
Carbon nanofibers (CNFs)	a. Moderate electrochemical properties and ease of incorporation to polymers due to the fibrous properties. b. Multiple concentric nested tubes with walls angled 20° to the longitudinal axis. The surface shows steps at the termination of each tube wall. c. The nanofibers include PR-24 (~65 nm diameter) and the PR-19 (~130 nm diameter).
Graphene	a. sp^2 hybridized C, where the in-plane σ_{c-c} bond is one of the strongest bonds in materials and the out-of-plane π bond is responsible for the electron conduction. b. Inherent properties include highest specific surface area; excellent mechanical, electrical, optical, thermal, chemical, and electrochemical inertness; easy and simple surface modification is possible; etc. c. Very sensitive to a range of stimuli such as gas molecules, pH value, mechanical strain, electrical field, and thermal or optical excitation.
Porous carbon	a. Three classifications based on pore size: microporous <2 nm, mesoporous 2–50 nm, and macroporous >50 nm. b. High surface area, surface chemistry functionality, and shorter pathway.

layer) from the organic semiconductor. The source-drain electrode is fabricated on the top of the semiconductor organic materials. The semiconductor may be of small molecules, polymers, carbon nanotubes, and others as well. A number of contact layers for the OTFTs are used besides the typical gold layers such as conductive polymers, metallic nanoparticles, etc. The dielectric layer is usually composed of organic or inorganic materials like polymers, composites, and inorganic oxide. This dielectric layer is crucial in terms of delivering fast, low-power-enhancing stability.

3.3.1.4 Polymer Materials

The response of many polymer membranes can be controlled by an external stimulus as the membrane demonstrates the gating function. Whenever they are exposed to certain changes in environmental conditions such as temperature, pH, ionic strength, light, and concentration in solution, the pore size of the membrane also changes. Sometimes, a change in pore size of the membrane can be followed by a change in the surface charges. Stimuli-responsive polymers are used to produce polymer membranes such as light-/thermos-responsive polyelectrolyte (PE) hydrogels, thermos-responsive brushes, PE films, PE brushes, etc. In all the cases, the response may be due to the conformational changes within the polymer chain itself and in the polymer solution with phase segregation. For instance, in the case of thermos-responsive polymer hydrogels, as the name suggests, they are prepared from the polymers which are sensitive to the change in temperature. So, during the specific temperature, the transition state of interaction from the polymer-water to the polymer-polymer state occurs. This transition is followed by the conformational change in polymer chain as well like from a swollen to a shrunken coil. This particular transition results in the phase separation, one being polymer enriched and the other being water enriched. Thus, the stimuli-responsive polymer as smart membrane depends on the chemical structure of the polymer. Some of the polymers which are triggered by external stimuli are given in Table 3.3.

Poly(N-isopropylacrylamide) (PNIPAM) is one of the most attracted stimuli responsive polymers due to the fact that it has a lower critical solution temperature (LCST) at 32°C, which is close to the human body temperature. Moreover, there is always a scope to modify and functionalize (at the time of polymerization) PNIPAM to incorporate more desired property. The polymer is often used to make thermosensors [16].

3.3.1.5 Hydrogels

A hydrogel is a water-swollen and crosslinked polymeric network produced by the simple reaction of one or more monomers [17]. The material has the ability to swell and retain a large amount of water within its structure. The ability to absorb water may be due to the hydrophilic functional groups present on the polymeric backbone, whereas the crosslinks between the network chains of hydrogels make them nondissolvable in water.

TABLE 3.3
A Comparative Account of Some Commonly Used Polymers for Sensing Applications

Types	Commonly Used Polymer	Mechanism
pH sensitive	a. Weak polyacids: Poly (acrylic acid); poly (methacrylic acid); poly (aspartic acid); alginic acid; poly (itaconic acid); poly (4-vinylbenzoic acid); poly (L-glutamic acid) b. Weak polyalkaline: Poly (allylamine); polyacrylamide; poly (4-vinylpyridine); poly (ethylene imine); poly (N-vinylimidazole); poly (N-isopropylacrylamide)	a. Protonation/deprotonation of polymer, inducing property change. b. Polymers are usually weak acids or weak bases containing carboxylic acids and amino functional groups. c. Changes in pH induce change in net charge of the polymer.
Ion sensitive	Calix[n]arene; tridodecyl-methyl-ammonium chloride; 8-hydroxyquinoline; crown ethers; sodium tetraphenyl borate; tris(2-ethylhexyl) phosphate; ethyl-2-benzoyl-2-phenyl-carbamoylacetate; bis [4-(1,1,3,3-tetramethylvutyl) phenyl] phosphoric acid.	a. Ionic strength of the solution affects the charge density of the polyelectrolytes chains. b. Ionic concentration changes lead to change of the hydrophobic property of polyelectrolytes.
Humidity sensitive	Nafion; poly(sodium-p-styrene sulfonate); poly(dially-dimethyl ammonium chloride); poly(2-hydroxy-3-methacryloxypropyl-trimethylammonium chloride); poly(ethylene oxide); polyimide; poly(z-hydroxylethyl methacrylate); poly(vinyl alcohol); poly(N,N-dimethyl-3,5-dimethylene piperidinium chloride)	a. Hydrophilic property of polymers is sensitive to humidity. b. Co-polymers which are insoluble in water can be incorporated or grafted with hydrophilic segment. c. Hydrophobic polymers can be modified with hydrophilic polymers.
Gas sensitive	Conducting polymers: Poly(aniline); poly(pyrrole); poly(thiophene); polydimethyl-siloxane; poly(vinyl chloride); polyethylene; poly(3,4-ethylenedioxythiophene); poly (methylmethacrylate); poly(4-vinylphenol); poly(N-vinylpyrrolidone); polysulfone.	a. Doping and de-doping process by redox reaction between conducting polymers and the gases. b. Besides, conducting polymers, hydrogels are also employed for the gas sensing.
Bio-sensitive	Proteins; nucleic acids such as DNA, RNA, and peptide nucleic acid; enzymes; polysaccharides,	Detection mechanism depends on the specific individual applications.

a. *Thermochromic Hydrogels:* The hydrogels undergo a hydrophilic to hydrophobic transition when the temperature is below the LCST. At LCST, the intermolecular hydrogen bonds are prevalent between the polymer chains and surrounding water molecules. When the temperature is increased above the LCST, the intermolecular hydrogen bonds are broken, which collapses the polymer chains resulting in phase separation and polymer aggregation. These aggregated polymers thus reduce the transparency of hydrogels by scattering the incident light. The extensively studied thermochromic hydrogels include polyampholyte, poly(N-isopropylacrylamide), and hydroxypropyl cellulose.

b. *Conductive Hydrogels*: Conductive hydrogels have unique properties, including good electronic properties, tunable mechanical flexibility, and simple to process. The hydrogel materials also have good biological characteristics such as self-healing, self-adhesive, antimicrobial activity, and biocompatibility. These cumulative properties of the hydrogels and conductive hydrogels have incited tremendous interest to exploit them for developing flexible electronics like flexible electrodes, flexible mechanical sensors, and flexible displays. Further, the conductive hydrogel materials promote high carrier mobility, mechanical and environmental stability, and overall electrical performance. Conductive polymers, metal nanoparticles/nanowires, carbon-based materials, etc., have been explored for constructing conductive hydrogels.

3.3.2 Inorganic-Based Smart Materials

3.3.2.1 Metal Oxides

Inorganic oxides are a good choice for many industrial applications because of the advantages including good chemical and mechanical stability, low cost, and photoelectric properties. Some of the inorganic oxides have the capability of switching the surface from one stable state to another functional group (i.e., oxygen vacancies and hydroxyl groups). Some semiconductor oxides like WO_3, TiO_2, ZnO, SnO_2, V_2O_5, and Ga_2O_3 have been studied for responsive surfaces.

The electronic properties of the semiconductors and their comparison with the conducting and nonconducting materials could be best described by using the diagram based on the Band theory, as shown in Figure 3.2. When a molecule interacts with the surface of metal oxide film, the concentration of charge carrier of the material can alter. This change in the charge carrier concentration can alter the conductivity or resistivity of the materials. The semiconducting materials have few free electrons in the crystal lattice (i.e., valence band [VB]) that can flow under certain conditions. By controlling the amount of impurities (the process is called doping), the conductivity of the intrinsic semiconductor materials could be controlled. Semiconductors are of two types: n-type and p-type (n and p refer to negative and positive, respectively). The resistivity of the n-type could be increased or decreased by doping it with an oxidizing or reducing agent, respectively. The opposite is the case for a p-type semiconductor.

While designing sensor platforms, metal oxide nanomaterials with a large surface area, electrochemical activity, high absorptive capacity, and stability are extremely helpful. By tailoring the properties of metal oxide nanomaterials, the performance of the sensors can be well tuned. To synthesize metal oxide nanomaterials, chemical methods involving a reduction of metal ions with controlled separation of the formed nanoparticles from the bulk solution can be followed. Nanomaterials of various types such as nanorods, nanofibers, nanobelts, nanocombs, and nanotubes are obtained. SnO_2, In_2O_3, ZnO, TiO_2, WO_3, PdO, CuO, NiO, MgO, and V_2O_5 have been commonly used for sensing application as metal oxide nanomaterials [18].

3.3.2.2 Noble Metal Nanomaterials

Noble metals such as gold (Au), silver (Ag), platinum (Pt), and palladium (Pd) have been extensively used to produce nanomaterials due to their unique size and shape-dependent physical, chemical, and electrochemical properties. Their metallic alloys and core shell have also been studied extensively for various sensing application. The metallic nanomaterials offer the following advantages while designing sensors and biosensors: (i) ease of surface functionalization, (ii) stress-free sensor fabrication approach, (iii) catalysis of electrochemical reactions, and (iv) the enhancement of an electron transfer process [19].

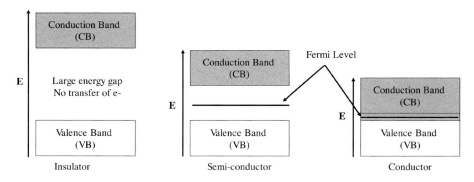

FIGURE 3.2 Schematic band diagram of insulator, semiconductor, and conductor. The larger the gap between CB and VB, the more difficult it is to transfer electrons.

3.3.2.3 Transition Metal Dichalcogenides

Transition metal dichalcogenides (TMDCs) are 2D nanomaterials consisting of a monolayer of transition metal atoms sandwiched between two layers of chalcogen atoms in a hexagonal lattice. The chalcogen atoms are usually selenium, sulfur, or telluride. TMDCs are considered one of the 2D graphene analogues. These materials offer a variety of promising technologies in the field of optoelectronics, nanoelectronics, energy storage, photonics, sensing, etc. The widely studied TMDCs include MoS_2, $MoSe_2$, hexagonal boron nitride (h-BN), borophene (2D boron), silicone (2D silicon), germanene (2D germanium), WS_2, WSe_2, etc. Depending on their chemical composition and structural configurations, TMDs can be categorized into metallic, semimetallic, semiconducting, insulating, or superconducting. They are as thin, transparent, and flexible as a graphene material. However, unlike graphene, many TMDs are semiconductors, which have the potential to develop ultra-small and low-power transistors for miniaturized technology. Additionally, TMDs can be deposited on the flexible substrate without compromising the chemical composition and structural conformation. Since the materials are scaled down to monolayers, these 2D materials exhibit unique electrical and optical properties. This property of 2D TMDs may be from the quantum confinement and also surface effects arising due to the transition from indirect to direct bandgap. Thus, the tunability in TMDs would make them a strong photoluminescent material. The typical structure of the TMDs is shown in Figure 3.3 [20].

3.3.2.4 Liquid Metals

It is a liquid state of a metal at room temperature, usually consisting of a eutectic alloy with a low melting point. Gallium-based alloys include galinstan (an alloy of gallium, indium, tin), which is one of the widely used metals of this kind in various applications due to their high conductivity and nontoxic properties. Galinstan exists in a liquid phase at room temperature and has a melting point of −19°C. Due to these properties, gallium-based alloys have been introduced to stretchable and wearable electronics. Another advantage of using the liquid metals is that under mechanical deformation or cracking, they can reshape or regrow in an ambient atmosphere which is sometimes known as native oxide skin. Thus, the features enable the fabrication methods including microcontact printing, stencil printing, 3D printing, direct writing, and spray deposition. Liquid metals are used for physical on-skin sensors because of their stretchability and mechanical stability [21].

3.3.2.5 Ionic Liquids (ILs)

It is defined as salt in a liquid state without any decomposition or vaporization. It consists of ions and short-lived ion pairs instead of neutrally charged molecules. An ionic liquid-based complex film is a mixture of transition metal compounds and ionic liquids [21]. Since IL is entirely ionic in composition, the intrinsic conductivity and negligible vapor pressure make them unique materials to develop stable electrochemical gas sensors for oxygen, carbon dioxide, sulfates, halides, etc. When IL is incorporated in electrodes, some common attributes such as conductivity, catalytic ability, stability, sensitivity, and selectivity are improved. An additional feature of IL is the compatibility with biomolecules, and whole cells while developing biosensors. For instance, the property of thermochromic ionic liquids is based on an octahedral-tetrahedral configuration of metal complexes with solvent molecules. Octahedral complexes are formed at low temperature, whereas at high temperatures, tetrahedral complexes are formed leading to the light transmittance variation. So ionic-based complex materials show high sensitivity with temperature change. The organic approach hybridizes an IL-based complex system with thermochromic hydrogels known as ionogels or ion gels [22].

3.3.3 Biomolecule-Based Smart Materials

The use of biomolecules for designing sensor platforms is a well-developed concept due to various suitable properties associated with different biomolecules. One of the most attractive biomolecules is DNA, which has surpassed any

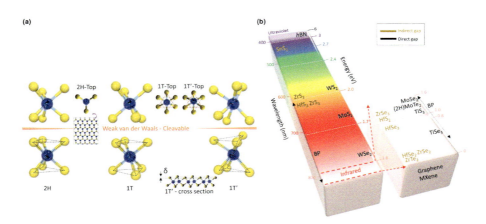

FIGURE 3.3 (a) Typical structure of layered TMDCs. Cleavable 2H, 1T, and 1T′ structures in layered TMD are shown. (b) Different bandgap of 2D layered materials like graphene (white color) to bandgap of hBN [20]

synthetic analog with remarkable molecular recognition properties. DNA oligonucleotides can be synthesized with an unlimited range of sequences and functionalized to design various shapes, sizes, and functions to develop programmable and other biosensors. The unique properties of DNA thus aid in developing new sensor technology with unmatched functionalities. With the advent of nanotechnology, designing of nanoparticle-oligonucleotide conjugates to develop a framework of simple dimers to complex 3D heterostructures has been possible. The smart material properties of DNA have also been realized from DNAzymes and synthetic oligonucleotides analogues, such as DNA aptamers for application in sensor technology. Apart from the nucleic acids, other biomolecules such as protein, polysaccharides, peptides, and composite biomaterials are also found to exhibit interesting material properties useful for sensing applications [23].

3.3.3.1 Oligonucleotides

Oligonucleotides are short, single-stranded polymers of nucleic acids used for various biosensing applications, [24] including diagnostics, environmental hazard detection, detection of small molecules, biomolecules, peptides, etc. The broad application of the oligonucleotides may be because of their intelligent material properties, mostly the hybridization-based recognition of the target oligonucleotides. Oligonucleotides readily bind to the respective complementary oligonucleotides, DNA, or RNA, or hybrids in a sequence-specific manner. On the laboratory scale, specific oligomers can be synthesized following the standard chemical protocol like solid phase peptide synthesis.

Nucleic acid aptamers are oligonucleotides with single-stranded DNA or RNA. The attractions on aptamers are due to the fact that they can bind with high affinity to a broad range of targets, including proteins, peptides, amino acids, drugs, metal ions, whole cells, etc. The specificity of the aptamers is conferred mainly by their tertiary structure. Aptamers are synthesized through an *in vitro* chemical process known as systematic evolution of ligands by exponential enrichment (SELEX). A detailed discussion on aptamers has been included in another chapter of this book.

3.3.3.2 Peptides

Protein-based biomaterials are interesting because of their various attributes such as mechanical, electrical, electromagnetic, and optical [25]. With the advent of protein engineering technology, different proteins are studied such as reflectins, amyloids, and plant proteins. One of the major drawbacks of protein-based biorecognition systems and biosensing platforms is their instability under harsh environmental conditions. However, there are reports that the functionality and the specificity of the protein can be retained by using artificial (or synthetic) peptides. Some of the problems associated with the biomolecules can be minimized by employing short peptides. The peptides can be synthesized following solid-phase peptide synthesis. The sensitivity of the biosensor can also be achieved using a short specific peptide sequence, which can be further modified structurally and functionally to improve some functionality. The chemical versatility of the peptides thus can be exploited for various multiplexing biosensor technologies [26].

3.3.3.3 Silk Fibroin (SF)

Silk is a natural biopolymer, which is generally obtained from silkworms, *Bombyx mori*. Silk is composed of hydrophilic (fibroin) and hydrophobic (sericin) proteins. SF has attracted a great deal of interest as sensing materials due to its various intrinsic properties such as biocompatibility, biodegradability, thermal and electrical properties like piezoelectric, excellent mechanical strength, and nontoxicity [27]. It also offers a variety of morphological structures for a wide range of applications. These biomaterials can be transformed to different forms such as fibers, films, and powder for immobilization and stabilization of enzymes. The fibrous nature of the SF protein with a dominant hydrophobic property may also serve as an excellent support for microorganisms for biosensing applications. The SF can be electrospun to nanofibers, microfibers, self-assembled to hydrogels, cast into films or emulsified into capsules, etc. With these features, silk-based electronics have great potential for advanced technological applications like e-skins, e-bandages, biosensors, wearable displays, implantable devices, artificial muscles, and many others. The innovative design of sensing materials can go hand-in-hand with the SF materials, which fulfills the concept of "materials by design." This can be done because SF can be deposited, prepared, and processed using diverse techniques. Silk-fibroin can be co-polymerized with other biological materials such as elastin, collagen, and DNA, which can be further processed with numerous techniques and methods of polymerization [28].

3.3.3.4 Cellulose Materials

Cellulose-based materials such as paper have some advantages over glass, silica, etc., as a biosensing platform, as the fluid moves in these biomaterials through capillary action and hence, no external pump or pressure is required to transport aqueous analyte solutions from the sample zone to detection zone. Additionally, the hydroxyl functional groups of cellulose can be modified to change the surface properties according to the need of the particular detection device. Further, paper is cheap and easily available, disposable, and portable. Hence, paper as a biosensing platform is preferred for PoC diagnostic sensing devices. Different papers like Whatman filter paper, glossy paper, and nitrocellulose (NC) membrane have been exploited for sensor platforms [29].

3.3.4 Hybrid Smart Materials

Sometimes a mixture of materials is used to design a smarter sensing platform. In this way, the limitation of one material can be overcome by complementing the property of another material. For example, a simple PDMS/paper hybrid system for various multiplexing biosensing can be developed [30], where the porous paper substrate offers a simple 3D storehouse for reagents and reactions.

3.3.4.1 Sol-Gel Materials

Sol-gel materials are the intermediate materials between glasses and polymers that open up new areas for bioimmobilization. These materials with the interface of nanoscience, novel sol-gel nanocomposites have been prepared and used to design different sensors and biosensors. The method has the ability to form solid or semisolid metal oxide through aqueous processing of hydrolytically labile precursors. So, using this method, various antibodies or antigens, RNA or DNA, etc., can attach to various supports such as silica, metal oxide, organosiloxane or sol-gel hybrid polymers. Briefly, the method involves two steps: (a) sol soluble hydroxylated oligomers are generated by the hydrolysis of the precursor in aqueous medium, (b) followed by formation of hydrogel of hydrate oxide by polymerization and phase extraction. Some of the precursors are (a) metallic salts of Al(III), Ti(IV), V(V), Sn(IV), (b) esters of silicic (VI) or polysilicic (IV). In majority of the sensor materials, sol-gel materials derived from glasses are mostly used because of chemical inertness, thermal stability, photochemical, and structural stability with excellent optical transparency. Incorporation of metal salts, organic dyes, and complexes before the hydrogel formation is one of the crucial features of sol-gel polymerization. Another advantage of the sol-gel process is that it can be performed at room temperature.

3.3.4.2 Bionanomaterials

Biomolecules can be integrated with nanoscale objects forming new platforms with combined characteristics of nanomaterials and catalytic function of biomolecules. The bionanomaterials inspire new strategy for sensor platform design and applications. Bionanomaterials involving enzymes, antibodies (antigens), or DNA or naturally occurring polymer (biopolymer) can be coupled with an inorganic moiety on nanometer scale. Due to the properties like transport, electronic, and catalytic function, bionanomaterials-based sensors are highly selective and sensitive. Nanobiocomposite like L-phenylalanine ammonia-lyase enzyme is used for biosensing capsaicin, an extract from chilies. To increase the absorption capacity, stability, and retaining activity, the self-assembled monolayers (SAMs) technique has been employed widely to functionalize the platform.

3.4 TECHNIQUES FOR DEVELOPING SENSOR PLATFORMS

There are numerous approaches one can follow to design smart materials. As discussed in the earlier section, smart materials are those that respond immediately to an external stimuli/influence. Upon exposure to external stimuli, the target molecules should be responsive and may have the capability of changing their conformational structure and properties. The physical or chemical properties of the smart materials or molecules change if polarity, conformation, or functional group of the molecule changes accordingly. So, the building block of the smart materials should be responsive molecules. Single-responsive and even dual/multiresponsive materials could be developed by a combination of the building blocks and different functional groups [31].

The chemical and physical effect that produces useful transduction is crucial in selecting and designing sensors. Numerous methods and techniques are used to design smart materials, and hence methods cannot be generalized for the production of various smart materials. For instance, to produce smart nanomaterials, four main production methods are employed, including (i) physical strategy (such as physical vapor deposition and laser ablation); (ii) chemical strategy (such as chemical vapor deposition, sol-gel processes, thermal decomposition, and hydrothermal method); (iii) electrochemical (such as anodic oxidation and electrodeposition); and (iv) photochemical strategy (such as photodeposition). In sensor applications the, absorption and interaction of nanomaterials with chemicals or biomolecules play a very significant role. Nanomaterials have increased the sensitivity of sensors, as they are used as carriers or mediators and they can also increase or accelerate electron transfer when used as functional materials on electrode surfaces in case of electrochemical-based sensors.

3.4.1 SYNTHESIS OF SMART MATERIALS

Many methods are available to synthesize materials, but not every method is used to produce smart materials. Depending on the types of platforms, smart materials can be synthesized and further surface modification could be done. Due to the demand of highly sensitive and selective sensors, original materials which have been used for the fabrication of a platform couldn't meet certain demands. To meet the criteria, the surface of the materials can be modified and enhance its functionalities. The synthesis procedure and methods may be different for inorganic- and organic-based smart materials. Some of the commonly used methods of synthesis are discussed in detail next.

3.4.1.1 Synthesis of Polymers

Usually, polymers are produced from their constituent monomers through a process called polymerization. There are different types of polymerizations such as addition and condensation polymerization depending on the types of linking of monomers through physical or chemical processes. The molecular weight of the polymers can be gradually built by step reactions or by chain reaction, depending on the chemical structure of the monomer. The polymerization generally proceeds by following three steps: initiation, propagation, and termination. So, most of the chain polymerization methods depend on the type of initiation such as free radical chain polymerization, ionic chain polymerization, and coordinating chain polymerization. With these classical methods, various polymers have been designed and synthesized. The methods and techniques of polymerization have improved and modified to synthesize smart polymers for sensing application. One such example would be the newly emerging type of polymer, porous ionic polymers (PiPs), where ionic moieties are incorporated into a polymer backbone or are covalently attached to the framework. Various physicochemical properties, functional groups, and active sites of PiPs can be tuned in for various range of applications.

3.4.1.2 Synthesis of Nanomaterials

Nanomaterials with better properties, more functionality, and lower cost than the existing ones are the major motivations in technology-driven research. Several techniques have been developed to improve the properties of nanomaterials by controlling the size, dimension, and distribution [14]. Generally, there are three methods and approaches to nanomaterials synthesis: (i) *top-down:* size reduction from bulk materials, (ii) *bottom-up*: materials synthesis from the atomic level, and (iii) *hybrid*: combining top-down and bottom-up to obtain nanomaterials. In the top-down approach, bulky material is converted to a smaller one by using physical processes like crushing, milling, or grinding. This approach is not suitable for preparing uniformly shaped materials, and it is also difficult to realize small particles even with high energy consumption. The bottom-up approach, as the name suggests, refers to the building up of a material from the bottom, which is from the atomic level to molecule level or cluster, and then to the nanoscale dimension. This method has been used to prepare a uniform size, shape, and distribution. This could effectively control the chemical reaction and particle growth. The hybrid approach is applied when the top-down and bottom-up approaches could not be used due to technical problems related to both the approaches. A thin film sensor like magnetic sensor is usually developed using the hybrid approach. A thin film is grown via the bottom-up approach, whereas etching to sensing circuit is done through a top-down approach [32]. Some of the precursors, reducing agents, and polymeric stabilizers which are used in metallic nanoparticle preparation are given in Table 3.4.

The size of the nanomaterials can be controlled and engineered by using different types of reducing agents. If smaller nanomaterials are required, a strong reducing agent could be used, as it promotes a fast reduction reaction leading to a high super saturation of the grown species. Thus, a large number of nuclei are formed, which results in a small size of nanomaterials. On the other hand, a weak reducing agent causes a slow reaction leading to the formation of large particles. The quantum dots (QDs) are another class of nanomaterials, which can be further designed and modified chemically by using various ligands, as shown in Figure 3.4. A detailed discussion on QDs has been included in another chapter of the book.

3.4.2 Deposition Techniques

The method is used generally in the semiconductor industry to develop smart material based on silicon. There are many deposition techniques employed in microelectronic fabrication. Four methods are commonly used as deposition techniques: (a) physical vapor deposition (PVD), (b) chemical vapor deposition (CVD), (c) electrochemical deposition (ECD), and (d) spin-on coating. The physical and chemical deposition processes are summarized in Table 3.5.

3.4.3 Modification Techniques

Many materials can be transformed into smart materials through surface modification for various applications. The modification of the surface will enhance the adaptability of the materials as well. The performance of biosensors and sensors greatly depend on the chemistry of the materials platform surface including the chemistry used to immobilize the components such as antibodies and enzymes on the surface. Various surfaces are used for the immobilization of biomolecules such as silicon, glass (silicon dioxide), nitrocellulose, gold, silver, polystyrene, and graphene. The end terminal functional groups are involved in such modification methods. Some of the commonly used modification methods are the (1) *physisorption-based modification,* (2) *chemical or covalent modification,* (3) *electrochemical modification,* and (4) *electrostatic modification,* as shown in Scheme 3.1. The chemical or covalent modification is widely used. Generally, biomolecules like DNA, proteins, and carbohydrates are immobilized on the platform surface. The common platforms used for sensor devices are silicon, glass slides, glass membranes, carbon, nitrocellulose, polystyrene, silver, gold, and others. The surface modification focuses on three main criteria: (i) the target of interest must bind with high affinity and selectivity to the sensitive area; (ii) the target molecules must be efficiently transported from the bulk solution to the sensor; and (iii) the transducer should be sufficiently sensitive to detect low coverage of captured molecules within reasonable time scales. The physisorption and chemisorption are summarized in Table 3.6.

TABLE 3.4
Precursors, Reducing Agents, and Polymeric Stabilizers Used in the Preparation of Metallic Nanoparticles

Category	Name
Precursor	Palladium chloride, potassium, silver nitrate, chloroauric acid, rhodium chloride, Metal anode, tetrachloroplatinate II
Reducing agent	Sodium citrate, carbon monoxide, methanol, formaldehyde, hydrogen peroxide, sodium tetrahydraborate, hydrogen, citric acid
Polymeric stabilizer	Poly(vinylpyrrolidone), sodium polyphosphate, sodium polyacrylate, polyvinyl alcohol

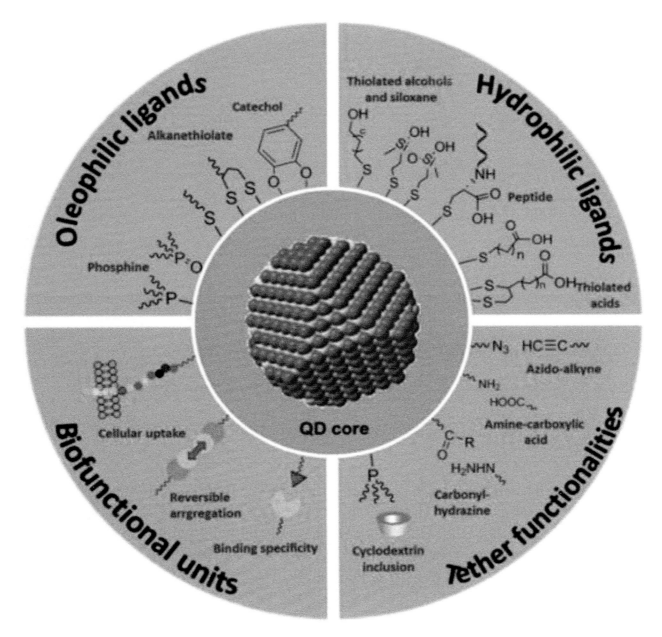

FIGURE 3.4 Surface engineering strategies of quantum dots for various applications using various ligands like (a) oleophilic ligands, (b) hydrophilic ligands, (c) tether functionalities, and (d) biofunctionalities. (adapted and reused with permission from the publisher [33].)

TABLE 3.5
Physical and Chemical Deposition Processes

Physical Deposition		Chemical Deposition	
Evaporation techniques	Vacuum thermal evaporation	Sol-gel techniques	
	Electron beam evaporation	Chemical bath deposition	
	Laser beam evaporation	Spray pyrolysis technique	
	Arc evaporation	Plating	Electroplating technique
	Molecular beam epitaxy		Electroless deposition
	Ion plating evaporation	Chemical vapor deposition (CVD)	Low pressure (LPCVD)
Sputtering techniques	Direct current sputtering		Plasma enhanced (PECVD)
	Radio frequency sputtering		Atomic layer deposition

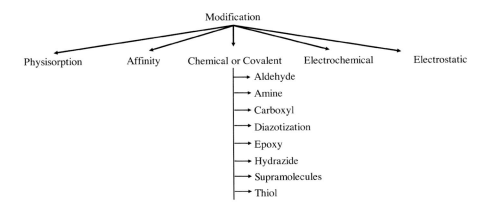

SCHEME 3.1 Some commonly used modification techniques/methods for smart sensing materials.

TABLE 3.6
Surface Modification via Physisorption and Chemisorption Methods

Type of Bonding	Bonding Molecule/Moiety	Surface Materials
Weak bonding (physisorption)	Amino acids; proteins; polyelectrolytes; hydrophobic chains; hydrophilic lipid head group	Flexible but nonspecific metal oxides; charged surfaces; hydrophobic surfaces; hydrophilic surfaces
Strong bonding (chemisorption)	Silanes	SiO_2 (metal and nonmetal oxides/hydroxides)
	Thiols	Au, Ag, Cu, Pt, ITO
	Phosph(on)ates	TiO_2, Al_2O_3, TiO_2, and many other transition metal oxides
	Catechols	TiO_2, Fe_2O_3 (metal oxides)

3.4.3.1 Molecular Self-Assembly

In this technique, molecules such as peptides, proteins, and DNA and organic or inorganic molecules are spontaneously brought together by noncovalent forces like hydrophobic, van der Waals, and electrostatic interactions and hydrogen bonding for thermodynamically favored conformations and interactions. Using this technique, one can arrange the molecules in a nanometer scale and can be an alternative to lithographic techniques for fabrication. In this technique, three-dimensional structures are fabricated, and they have the capability to control the pattern formation and assembly to the subnanometer scale. One of the widely accepted methods is layer-by-layer assembly, which utilizes the electrostatic attraction between the molecules. Multiple thin poly-ion films are deposited consecutively and stabilized by the electrostatic attractive force generated by the oppositely charge molecules from the solution, which allows the use of materials like biomolecules such as proteins, DNA, etc. The film formed by this technique provides great control over structure, thickness, and function.

3.4.4 Fabrication Techniques

The principle of nanofabrication and microfabrication are the same with a difference of dimensional aspects between these two techniques. The general principle of these techniques includes lithography (photo- or soft-), deposition process (film), etching, and bonding. The processes like molecular self-assembly, nanopatterning, rapid prototyping, x-ray, and focused ion beam lithography are also involved according to the requirements. Due to the advancement in micro- and nano-fabrication, miniaturization of bulky technologies is feasible. The miniaturization of the device offers many advantages such as portability, handheld, implantable, or even injectable device.

3.4.4.1 Microfabrication Techniques

Bulk micromachining is the process when the structures are built within the bulk of substrate materials, and if the structure is built on the surface of the substrate, it is called surface micromachining. However, both surface and bulk micromachining may be utilized to obtain the desired structure in the fabrication process. Material selection is one of the crucial steps in the microfabrication process for successful application of advanced and low-cost microfluidic systems. Silicon, metal, silica, polymeric, or paper are some of the commonly used materials for a microfluidic system. The general methods for rigid structures and for polymer-based microfluidic structures are given in Tables 3.7 and 3.8, respectively.

3.4.4.1.1 Photolithography

Photolithography is one of the most widely employed techniques to create a pattern on a material. The desired pattern is generated on the surface of a substrate by exposing the regions of a light-sensitive material to ultraviolet (UV) light.

TABLE 3.7
Fabrication Methods for Rigid Structures

Material	Fabrication Techniques	Fabrication Methods	Interface
Silicon	Bulk micromachining	Wet etching	Fluid inlet/outlet interface
Glass		Surface micromachining	Electrical interface
		Wafer bonding	Analytical interface
	High-aspect ratio MEMS	Reactive ion etching	
		Deep reactive ion etching	
	Other micromachining technologies	XeF_2 dry phase etching	
		Electro-discharge	
		Micromachining	
		Laser micromachining	
		Focused ion beam	

TABLE 3.8
Microfabrication Methods for Polymer-Based Microfluidic Devices

Polymer Material	Fabrication Methods	Bonding Methods	Interface
Elastomer	Soft lithography	Oxygen plasma	Fluid inlet/outlet interface
	Replica micromolding	Thermal compression	Electrical interface
	Microcontact printing	Adhesive	Analytical interface
	Microtransfer molding	Localized welding	
	Microcapillary molding		
Thermoplastics	Hot microembossing	Direct bonding	
	Injection molding	Thermal compression	
	3D printing	Adhesive	
	CNC micromilling		

The general steps of photolithography techniques may be explained as follows: (a) A substrate material such as silicone or glass is spin-coated with a photoresist, or light-sensitive polymer. (b) A mask (or photomask), made by designing the desired shape or pattern beforehand with an opaque material on glass dish or transparent material, is placed on top of the substrate and photoresist. (c) The assembly is then irradiated with UV light, which exposes the section of photoresist of the mask. The photoresist can be either positive or negative, depending on which the transformation happens upon exposure to light. (d). After exposure, the substrate is treated with developing solution. If irradiated on positive photoresist, the exposed regions would break down, as it is soluble in the developing solution. A negative photoresist on the other hand, becomes crosslinked upon exposure to light, which makes it insoluble in the developing solution. Hence, the parts which are not exposed to light will be removed consequently. The photolithography is depicted schematically in Scheme 3.2 [5].

3.4.4.1.2 Soft Lithography
It is a method to transfer a designed pattern onto a surface. In many cases, this technique utilizes a replica of microstructure produced by molding a polymer such as PDMS to a master, which is manufactured by photolithography or other microfabrication techniques. The mold obtained from soft lithography is reusable and does not require clean room manipulation. The steps involved in soft lithography include microstamping, stencil patterning, and microfluidic patterning.

3.4.4.1.3 Film Deposition
In this step, layers of materials, or films, are applied or grown on the microstructure during microfabrication. The films play a crucial role in designing the microchip due to their structural and functional capabilities. The most commonly used films for microfabrication are plastics, silicon-containing compounds, metals, and biomolecules.

3.4.4.1.4 Etching
After deposition of the film, the surface could be modified by creating various topographical features using selective removal either by a physical or chemical approach. There are different types of etching: isotropic etching, anisotropic etching, wet etching, and dry etching which is explained in Table 3.9. The characteristic slanted profile for wet anisotropic etching may be because of the interaction of the reagents with the crystalline structure of the materials. So, the rate of etching occurring at the crystal plane can be determined with the crystal structures.

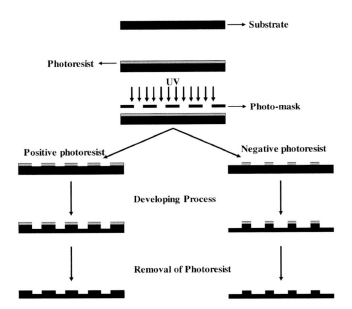

SCHEME 3.2 Schematic diagram of the general photolithography technique. The procedure proceeds from raw substrate to desired imprinted substrate following the standard protocol in a clean room facility.

TABLE 3.9
Isotropic and Anisotropic Etching Processes

Etching	Features	Profile
Isotropic	Etching equally in all directions. It doesn't occur only in the direction of depth, but also laterally as well.	Curved profile.
Anisotropic	Etching in one specific direction. It occurs in only one direction and thus increases the depth of the cavity.	Flat profile is obtained if dry anisotropic etching is used, whereas inclined sidewalls are obtained if wet anisotropic etching is used.

3.4.4.1.5 Bonding

In order to form tight seals and to obtain the desired structures, a reversible and irreversible bonding step is usually performed between the microstructures. Since numerous bonding methods are available, one has to choose the methods specific for the material of interest. For instance, irreversible anodic bonding between silicone substrate and non-pure glass film.

3.4.4.1.6 Microfabrication of Paper

Cellulose in general and paper in particular is very attractive for sensors for the following reasons: (a) biodegradable, abundantly available, flexible and lightweight, economical, and easy to process; (b) surface can be functionalized to manipulate properties like hydrophilicity, permeability, or reactivity, etc., of cellulose for specific purposes; (c) cellulose has piezoelectric properties; and (d) cellulose has the natural capillary action. With these inherent properties of paper, a wide range of paper-based sensors and devices have been developed with low-cost and simple fabrication methods. Sensors like chemical, gas, and other electronic devices or microfluidic-based biosensor devices have been recently designed using paper as the main substrate or component. In this chapter, paper-based devices for biosensing applications are introduced to provide some knowledge, but more details are provided in another chapter. One of the greatest assets of using paper as a sensing platform is the inherent property like capillary action where no external power source is required for liquid transport. To capitalize the property, paper/cellulose has been focused to integrate in advance technology. So this makes the fabrication and device cost effective and small which is suitable to develop PoC devices [34]. Hydrophilic and hydrophobic channels or barriers are created on the paper using simple patterning methods, some of the fabrication methods are (a) ink-jet printing; (b) photolithography; (c) plasma treatment; (d) wax printing; (e) flexography printing; (f) screen printing; (g) laser treatment; and (h) one-step plotting. A hydrophobic agent is used to create a hydrophobic and hydrophilic microchannel on the paper platform, as explained in Table 3.10.

3.4.4.2 Nanofabrication Techniques

There are some difficulties associated with photolithography: (a) nanostructures <100 nm are difficult to produce

TABLE 3.10
Various Fabrication Processes of the Paper-Based Analytical System

Fabrication Method	Microchannel Fabrication	Patterning Agent
Ink-jet printing	Incorporate digital ink-jet printing technique for fabrication. Hydrophobic polymer as ink is printed on the platform with precision followed by drying.	Hydrophobic polymer like polystyrene, alkyl ketene dimer (AKD), etc.
Photolithography	Hydrophobic patterning agent is activated and allowed to impregnate within the paper substrate with the help of light energy. Mostly UV lights are used but normal sunlight can also be harvested for the purpose.	Photoresist SU-8; hydroxyl propyl cellulose
Plasma treatment	The paper is hydrophobized wholly using hydrophobic polymer. Hydrophilic microchannels are created by plasma treatment on a mask with channel network.	AKD, octadecyltrichlorosilane
Wax printing	Printing solid wax pattern on the paper substrate which is followed by heating.	Solid wax
Flexography printing	The ink is transferred onto anilox roll, then doctor blade removes the excess ink. The ink is transferred to paper substrate which affixed to impression roll.	Polystyrene in toluene
Screen printing	Solid wax is printed on the paper substrate using a screen.	Solid wax
Laser treatment	Hydrophobic paper substrate is created using photopolymer. The hydrophilic microchannel is created using laser treatment.	Hydrophobic paper like parchment paper, wax paper or palette paper
One-step plotting	Formed permanent ink markers on the paper substrate	Commercial permanent ink; Hydrophobic resin
3D device fabrication	(i) Placing patterned paper and double-sided adhesive tape with through holes in ordered manner; (ii) using origami technique.	PDMS

with photolithography because of the diffraction effects, (b) masks and the pattern on the wafer substrate should be perfectly aligned, (c) the density of defects needs to be carefully controlled, and (d) photolithographic tools are expensive. Therefore, alternative methods have been developed to overcome such limitations. Electron-beam lithography and x-ray lithography techniques have been developed to overcome the limitations of photolithography techniques [35]. The technique utilizes similar principle as the microfabrication method to generate patterns or devices at the nanoscale level (1–100 nm). Some of the lithographic techniques used to accomplish miniaturization are briefly discussed next.

3.4.4.2.1 Electron Beam (E-beam) Lithography
The pattern in e-beam lithography is written in a polymer film with a beam of electrons. This technique mainly utilizes the electron beam to scan the material and form the desired pattern. In this technique, the magnetic lenses are positioned to focus the beam. Another characteristic feature of the technique is to use thermionic emitters, electron sources, which usually gives an output in the range of 1–200 keV. To achieve the desired nanoscale resolution, characteristics of electron beam and the position of the specimen are controlled electronically, and as such less than 10 nm resolution has been demonstrated. Few issues encountered are the cost associated with the instrument and the high maintenance of the system.

3.4.4.2.2 Focused Ion Beam Lithography
In this technique, instead of using electrons, ions are used to pattern the resist substrate. In general, ions are generated from a liquid metallic tip, which contain elements like gallium. The operational energy levels are usually in the range of 10–200 keV. The strength of the technique is that it can pattern the features directly to the substrate by selective material removal or deposition while eliminating the need of the photomask. The resolutions obtained from the technique are in the range of 5–20 nm with 5 nm lateral feature size.

3.4.4.2.3 Colloid Monolayer Lithography
In this technique, self-organization of one- or two-dimensional colloidal systems is used as layers of nanofabrication. The electron or ionic lithographic techniques are expensive and sophisticated therefore, this technique can be used as an economical alternative to produce nanoscale patterns. The colloidal particles are deposited on the surface of a substrate before solvent evaporation, using spin coating, or through electrophoresis techniques. Theoretically, if the size and the geometry of the colloidal particles are controlled, the spatial distribution of the colloids can be also controlled. The resulting array is influenced by a number of parameters including colloidal concentration, rate of solvent evaporation, wetting characteristics of the substrate, and competition with multiple layer formation. So far, it is reported that as low as 5-nm resolution in all three dimensions can be achieved with this technique.

3.4.4.2.4 X-Ray Lithography
In this technique, electromagnetic radiation is employed with the wavelength range of 0.5–4 nm. This is commonly known as "soft x-rays," and is used for transferring a pattern from a mask to a substrate material. Laser-induced plasma generators or synchrotrons are the common sources of soft x-rays. The x-rays masks are composed of silicon carbide, which are few microns in thickness. To perform lithography, the mask

is placed at a distance of several microns to the wafer for good resolution. The larger gap of distances between mask and wafer proportionally reduces the resolution. The other two parameters are Fresnel diffraction and photoelectron diffusion or photoelectron blur. Photoelectron blur is indirectly proportional to the wavelength of energy being used, whereas the Fresnel diffraction is directly proportional. Thus, these two variables have to be balanced to achieve the pattern with small resolutions.

3.5 ADVANCED APPLICATION OF SMART MATERIALS

Over the decades, the use of Internet of Things (IoT) is exponentially growing in the field of defense, health care, environmental uses, etc., due to its potential in real-time data acquisition and analysis. Continuous feedback and information allows us taking precautions to prevent health abnormalities and prevent any severe medical situations. The demand of instant information can be materialized in IoT by the advancement of mobile communication technologies, which are integrated with wireless chemical sensors and biosensors. For health care, wearable sensors can play a major role as they can provide vital health-related information [36, 37]. The wearable sensors are not only limited to on-body applications; this technology can be integrated with other surfaces like vehicles or buildings as well. But the obstacle of the application is that the sensors progressively lose function with time. Another issue for developing the reliable implantable sensors is the biocompatibility. In this section, advanced application of smart materials, wireless sensors, wearable sensors, and smartphone sensors are covered even though there are resemblance, as well as overlapping technology, that can be seen.

3.5.1 WIRELESS SENSORS

Wireless technology has a significant advantage to transfer observed data from one system to another system over a wired system, particularly in monitoring chemical analytes that are not easily accessible and hazardous in nature [36]. As such, wireless chemical sensors are used to monitor environmentally hazardous gases or chemicals, and in health care, implantable or swallowable sensors are employed. These are hybrid devices which involve in the collection of biochemical data from the local environment and then process and transmit the analytical information to remote devices by wireless technology like radio communications. Both wired and wireless sensors are used in PoC and PoN in the field of agriculture and food production industries, chemical process monitoring, environment monitoring, defense and homeland security, etc. [38, 39]. The advantage offered specifically by wireless technology is that the (bio)chemical analytes can be monitored over long distances for long-term *in situ* such as hazardous areas or sites which are not easily accessible. Additionally, the sensing can be achieved and monitored over large areas.

In most of the wireless sensors, the devices contain radios capable of receiving, generating, and transmitting radio-frequency waves. The sensors usually have an energy source for transmitting and receiving information over the distance. Thus, active wireless sensors usually have at least a power supply or a battery. For instance, wireless sensors which contain active transceiver (transmitter-receiver) such as Bluetooth or ZigBee technology can communicate with a master controller/reader device over quite some distance. On the other hand, passive transponders do not require a battery, which may be because of the fact that they can communicate with the reader without radio. Instead the readers generate the radio-frequency electromagnetic field which is reflected or modulated with transponders to communicate. Thus, the sensors based on passive operation such as radio-frequency identification (RFID) and near-field communication (NFC) are attractive in terms of compact size and zero energy consumption. But the passive transponders are operated in a very short range, like a few centimeters for NFC and several meters for RFID.

3.5.2 WEARABLE SENSING TECHNOLOGIES

There are many limitations associated with the traditional centralized health care systems, such as the patients have to travel to clinical centers or hospitals for any treatment and observation. However, there are many medical cases which cannot be addressed, especially in emergency cases with these centralized medical facilities. Many sensor/biosensor devices for diagnosis and real-time monitoring have been developed. Some of the examples can be illustrated as follows: to monitor intracranial pressure (ICP), and implanted catheter with external transducer is used, and for diabetes patients, pressure sensors are integrated at the soles of the feet. But these cause huge concern because of the poor biocompatibility and discomfort. Now, a concept to improve comfortability and real-time health monitoring includes a new wearable and attachable health monitoring system based on flexible electronics, as shown in Scheme 3.3. Many flexible sensors have been developed for monitoring the physiological signals such as piezo-resistive/electrical sensors, capacitive, and field effect transistor-based sensor devices. Common existing wearable sensors usually track the user's physical activities and various physiological signals like heart rate to develop sensors that can monitor electrocardiogram (ECG), body motion, or body temperature, etc. And in addition to these, there are many crucial chemical parameters which are released as biofluid at the molecular level such as sweat, saliva, and tears. Currently, wearable chemical sensing is one of the key application areas for wireless chemical sensors, which may be either an electrochemical or optical transducer-based system. For electrochemical wearable sensors, wireless communication may be essential, but the output signals for optical wearable sensors can be read or seen by the naked eye [40].

3.5.2.1 Composition of Wearable Sensors

The word "wearable" should be associated with "flexibility" and "stretchability." The wearable sensor is an electronic

Smart Materials for Developing Sensor Platforms

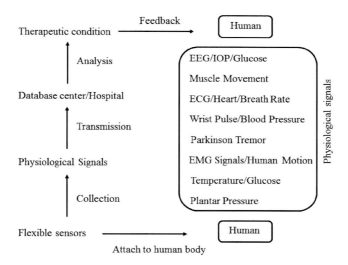

SCHEME 3.3 Wearable health device for humans where physiological signals are collected and transmitted to the database center or hospital for analysis. Then, feedback related to therapeutic conditions or body conditions are provided for further assessment and care.

TABLE 3.11
The Physicochemical Properties of Flexible Substrates

Materials	Stretchable or Bendable	Thermal Stability	Chemical Resistance
Polyethylene Terephthalate (PET)	Bendable	Resist temperature (<100°C)	Dissolvable in acetone
Silicone (PDMS, Ecoflex)	Stretchable	Resist temperature (<100°C)	Ethanol and acetone (in short time)
Polyimide (PI)	Bendable	Resist temperature (<450°C)	Weak acids and alkali
			Ethanol and acetone
Polyethylene naphthalate (PEN)	Bendable	Resist temperature (<180°C)	Easily permeated by oxygen and water
Self-healing	Stretchable	Thermal healing	Moisture oxygen
		Thermal expansion	Ethanol and acetone (in short time)
Paper	Bendable	Resist temperature (<100°C)	No
Metal foil	Bendable	Resist temperature (≈250°C)	Ethanol and acetone

device which can be attached to or worn on the target body/surface. This sensor comprises a substrate which should be flexible/stretchable, a conducting electrode, a sensing material, and an encapsulation material occasionally. Normally two approaches have been employed to obtain stretchability/flexibility: the material should have elasticity property and the material should have a flexible/stretchable structure. In addition to these approaches, incorporation of special materials such as micro/nanostructures or a serpentine/helix structure for better functionalities and applications have been introduced.

One of the crucial parts in the wearable device is the substrate material. These materials should possess physical and chemical properties like thermal stability, chemical resistance, and mechanical flexibility. Therefore, the materials such as polymers, rubbers, or metal foils are frequently employed for such cases that exhibit flexibility, high tensile strength, and temperature stability. These materials should also be endowed with resistivity to weak acids and alkalis and stable in common solvents. Some of the flexible substrates and their physicochemical properties are summarized briefly in Table 3.11. In a recent development, fibers and textiles have been integrated with multifunctional flexible electronics to realize both the nature and properties together. Sometimes the textile is integrated with the sensor, or the textile is composed of a conducting property with a sensing fiber for sensor configuration bringing about smart, multifunctional sensing platforms [41].

3.6 FLEXIBLE CONDUCTING ELECTRODES

Conducting electrodes are usually embedded in or on top of the flexible substrate to design sensing devices. Traditionally, materials like Au, Pt, and indium tin oxide (ITO) films demonstrate high sensitivity. However, because of the rigidity and brittleness, the electrode materials prepared from these metals show cracks during deformation. Nanowire and nanotube networks can overcome this limitation, as they are transparent, stretchable, electrically stable, and durable [42].

Semiconducting polymers and nanowires are used in flexible electrodes as sensing materials. Some of the polymers employed in constructing wearable electrodes for field effect transistors (FETs) include poly(3,4-ethylenedioxythiophene)-poly(styrenesulfonate) (PEDOT: PSS), polyaniline (PANI), polypyrrole (Ppy), and poly(3-hexylthiophene) (P3HT). Even though they have drawbacks like low carrier mobility and instability compared to inorganic semiconductors such as nanowires of SnO_2, In_2O_3, and ZnO, these polymers are preferred because of their tunability in terms of their physical and chemical properties. The mechanical property of semiconducting polymers can be compared with an insulated polymer substrate, and their Young's modulus is also comparable to human skin (0.34 MPa), which is not possible with inorganic nanowires. Carbon-based materials like carbon nanotubes and graphene are incorporated in wearable electrodes because of their amazing physical and chemical properties as discussed elsewhere.

3.7 SMARTPHONE-BASED SENSING DEVICE

An area called "diagnostic and communication technology" (DTC) is emerging because of the mobile phone-based biosensing application [43]. As such, mobile phone contains many attractive components and features suitable for biosensing applications such as an analytical reader: the screen to display and control, high-quality camera to capture input signals, memory to store data, and others like physical sensors with wired or wireless connectivity modes. Moreover, smartphones are comparatively cheaper, easily available in the public domain, and easily portable [44]. There are two types of smartphone-based devices used for sensing technology: (i) smartphone as a detector and (ii) smartphone as an interface. Briefly, these two categories will be discussed for a better understanding.

3.7.1 Detector

In this category, the phone camera should be of high quality to identify signal outputs. A commonly used method by smartphone-based sensing devices is the colorimetric method. In this, the pixel intensity captured by the smartphone is determined by an image processing algorithm.

3.7.2 Interface

Smartphones have many features built in, which can be exploited as an interface device through Wi-Fi, Bluetooth, and microUSB with analytical instruments. They can be integrated with sensors like test strips, handheld detectors, sensor chips, etc. Numerous smartphone apps are available, and so more physical quantities can be measured. The data can be easily shared with several wired as well as wireless methods after the measurement.

One of the advantages of smartphone interfaced based sensors is that the data can be obtained instantaneously and saved on a secure server automatically. It can perform rapid colorimetric analysis both quantitatively and qualitatively. Additionally, no trained personnel are required to use a smartphone-based sensing device. So, this could be one of the powerful PoC and PoN devices for the next generation of sensor technology. Besides health care, there are many other areas such as environmental monitoring, food processing and distribution chain, and bioterrorism where smartphone-based sensing devices can be employed.

3.8 CONCLUSION

In this chapter, we discussed smart materials and their applications for designing various sensing platforms. Thus, various methods and techniques employed to design and develop smart materials have been elaborated. Among the techniques, nanotechnology and microtechnology are emerging as the most efficient techniques. As the size of the material decreases, various significant properties of the materials suitable for biosensing applications are exhibited. Designing smart materials for sensing applications can be achieved due to the ever-growing interdisciplinary approach in science and technology and its allied fields. The basic understanding of the materials properties could be helpful in designing the smart materials. This will not only enhance the technology and development but also further exploring science as well. Understanding and modulating the properties of multilayered sensing platform may be challenging, but the approach of using multilayered platform for sensors and biosensors has been greatly explored. The hybrid approach of exploiting the various properties of the sensing materials needs further exploration to develop next generation of biosensors and sensors. There is also a growth in the development of sensor platforms for wearable sensors using flexible biocompatible materials including textiles and fibers. There are many potential areas where sensing technologies could be tremendously helpful. Since the technology is developing and advancing day by day, the scope of improvement and exploring the unexplored dimension are encouraged. Thus in this chapter, we have modestly explored the smart materials for sensor applications.

REFERENCES

1. P. D. Thungon, A. Kakoti, L. Ngashangva, and P. Goswami, "Advances in developing rapid, reliable and portable detection systems for alcohol," *Biosens. Bioelectron.*, vol. 97, no. March, pp. 83–99, 2017, doi: 10.1016/j.bios.2017.05.041.
2. L. Ngashangva, V. Bachu, and P. Goswami, "Development of new methods for determination of bilirubin," *J. Pharm. Biomed. Anal.*, vol. 162, pp. 272–285, 2019, doi: 10.1016/j.jpba.2018.09.034.
3. T. Vo-Dinh and B. Cullum, "Biosensors and biochips: Advances in biological and medical diagnostics," *Fresenius J. Anal. Chem.*, vol. 366, no. 6–7, pp. 540–51, 2008.
4. C. R. Ispas, G. Crivat, and S. Andreescu, "Review: Recent developments in enzyme-based biosensors for biomedical analysis," *Anal. Lett.*, vol. 45, nos. 2–3, pp. 168–186, 2012, doi: 10.1080/00032719.2011.633188.
5. M. L. Kovarik, D. M. Ornoff, A. T. Melvin, *et al.*, "Micro total analysis systems: Fundamental advances and applications in the laboratory, clinic, and field," *Anal. Chem.*, vol. 85, no. 2, pp. 451–472, 2013, doi: 10.1021/ac3031543.

6. C. G. Granqvist, "Recent progress in thermochromics and electrochromics : A brief survey," *Thin Solid Films*, vol. 614, pp. 90–96, 2016, doi: 10.1016/j.tsf.2016.02.029.
7. M. Pohanka, "The piezoelectric biosensors: Principles and applications, a review," *Int. J. Electrochem. Sci.*, vol. 12, pp. 496–506, 2017, doi: 10.20964/2017.01.44.
8. F. T. Calkins, A. B. Flatau, and M. J. Dapino, "Overview of magnetostrictive sensor technology," *J. Intell. Mater. Syst. Struct.*, vol. 18, no. 10, pp. 1057–1066, Oct. 2007, doi: 10.1177/1045389X06072358.
9. W. M. Huang, Z. Ding, C. C. Wang, J. Wei, Y. Zhao, and H. Purnawali, "Shape memory materials," *Mater. Today*, vol. 13, nos. 7–8, pp. 54–61, 2010, doi: 10.1016/S1369-7021(10)70128-0.
10. Y. Yang, P. Su, and Y. Tang, "Stimuli-responsive lanthanide-based smart luminescent materials for optical encoding and bio-applications," *ChemNanoMat.*, pp. 1–25, 2018, doi: 10.1002/cnma.201800212.
11. S. Barizuddin, S. Bok, and S. Gangopadhyay, "Plasmonic sensors for disease detection – A review," *J. Nanomedicine Nanotechnol.*, vol. 7, no. 3, pp. 1–10, 2016, doi: 10.4172/2157-7439.1000373.
12. P. A. Ledin et al., "Design of hybrid electrochromic materials with large electrical modulation of plasmonic resonances," *ACS Appl. Mater. Interfaces*, vol. 8, pp. 13064–13075, 2016, doi: 10.1021/acsami.6b02953.
13. Y. J. Tan, J. Wu, H. Li, and B. C. K. Tee, "Self-healing electronic materials for a smart and sustainable future," *ACS Appl. Mater. Interfaces*, vol. 10, pp. 15331–15345, 2018, doi: 10.1021/acsami.7b19511.
14. J. Wang, "Carbon-nanotube based electrochemical biosensors: A review," *Electroanalysis*, vol. 17, no. 1, pp. 7–14, Jan. 2005, doi: 10.1002/elan.200403113.
15. J. P. Bahamonde, H. N. Nguyen, S. K. Fanourakis, and D. F. Rodrigues, "Recent advances in graphene-based biosensor technology with applications in life sciences," *J. Nanobiotechnology*, pp. 1–17, 2018, doi: 10.1186/s12951-018-0400-z.
16. K. C. Persaud, "Polymers for chemical sensing," *Mater. Today*, vol. 8, no. 4, pp. 38–44, Apr. 2005, doi: 10.1016/S1369-7021(05)00793-5.
17. S. J. Buwalda, K. W. M. Boere, P. J. Dijkstra, J. Feijen, T. Vermonden, and W. E. Hennink, "Hydrogels in a historical perspective : From simple networks to smart materials," *J. Controlled Release*, vol. 190, pp. 254–273, 2014, doi: 10.1016/j.jconrel.2014.03.052.
18. A. Dey, "Semiconductor metal oxide gas sensors: A review," *Mater. Sci. Eng. B*, vol. 229, pp. 206–217, Mar. 2018, doi: 10.1016/j.mseb.2017.12.036.
19. U. Saxena and P. Goswami, "Electrical and optical properties of gold nanoparticles: Applications in gold nanoparticles-cholesterol oxidase integrated systems for cholesterol sensing," *J. Nanoparticle Res.*, vol. 14, no. 4, p. 813, Mar. 2012, doi: 10.1007/s11051-012-0813-9.
20. W. Choi, N. Choudhary, G. H. Han, J. Park, D. Akinwande, and Y. H. Lee, "Recent development of two-dimensional transition metal dichalcogenides and their applications," *Mater. Today*, vol. 20, no. 3, pp. 116–130, 2017, doi: 10.1016/j.mattod.2016.10.002.
21. M. Varga, C. Ladd, S. Ma, J. Holbery, and G. Tröster, "On-skin liquid metal inertial sensor," *Lab. Chip*, vol. 17, no. 19, pp. 3272–3278, 2017, doi: 10.1039/C7LC00735C.
22. V. V. Singh, A. K. Nigam, A. Batra, M. Boopathi, B. Singh, and R. Vijayaraghavan, "Applications of ionic liquids in electrochemical sensors and biosensors," *Int. J. Electrochem.*, vol. 2012, pp. 1–19, 2012, doi: 10.1155/2012/165683.
23. S. Hollingshead, C.-Y. Lin, and J. C. Liu, "Designing smart materials with recombinant proteins," *Macromol. Biosci.*, vol. 17, no. 7, p. 1600554, Jul. 2017, doi: 10.1002/mabi.201600554.
24. R. Bhardwaj, N. Lightson, Y. Ukita, and Y. Takamura, "Development of oligopeptide-based novel biosensor by solid-phase peptide synthesis on microchip," *Sens. Actuators B Chem.*, 2014, doi: 10.1016/j.snb.2013.10.086.
25. Q. Liu, J. Wang, and B. J. Boyd, "Peptide-based biosensors," *Talanta*, vol. 136, pp. 114–127, 2015, doi: 10.1016/j.talanta.2014.12.020.
26. L. Ngashangva, Y. Ukita, and Y. Takamura, "Development of programmable biosensor using solid phase peptide synthesis on microchip," *Jpn. J. Apllied Phys.*, vol. 53, pp. 05FA09-1-05FA09-6, 2014.
27. D. N. Rockwood, R. C. Preda, T. Yücel, X. Wang, M. L. Lovett, and D. L. Kaplan, "Materials fabrication from Bombyx mori silk fibroin," *Nat. Protoc.*, vol. 6, no. September, pp. 1612–1631, 2011, doi: 10.1038/nprot.2011.379.
28. L.-D. Koh, J. Yeo, Y. Y. Lee, Q. Ong, M. Han, and B. C.-K. Tee, "Advancing the frontiers of silk fibroin protein-based materials for futuristic electronics and clinical wound-healing (Invited review)," *Mater. Sci. Eng. C*, vol. 86, pp. 151–172, May 2018, doi: 10.1016/j.msec.2018.01.007.
29. K. Mahato, A. Srivastava, and P. Chandra, "Paper based diagnostics for personalized health care : Emerging technologies and commercial aspects," *Biosens. Bioelectron.*, vol. 96, no. May, pp. 246–259, 2017, doi: 10.1016/j.bios.2017.05.001.
30. A. Fahmi, T. Pietsch, C. Mendoza, and N. Cheval, "Functional hybrid materials," *Mater. Today*, vol. 12, no. 5, pp. 44–50, 2009, doi: 10.1016/S1369-7021(09)70159-2.
31. M. D. Sonawane and S. B. Nimse, "Surface modification chemistries of materials used in diagnostic platforms with biomolecules," *J. Chem.*, vol. 2016, pp. 1–19, 2016, doi: 10.1155/2016/9241378.
32. W. Wen, Y. Song, X. Yan, et al., "Recent advances in emerging 2D nanomaterials for biosensing and bioimaging applications," *Mater. Today*, vol. 21, no. 2, pp. 164–177, Mar. 2018, doi: 10.1016/j.mattod.2017.09.001.
33. J. Zhou, Y. Liu, J. Tang, and W. Tang, "Surface ligands engineering of semiconductor quantum dots for chemosensory and biological applications," *Biochem. Pharmacol.*, vol. 20, no. 7, pp. 360–376, 2017, doi: 10.1016/j.mattod.2017.02.006.
34. D. D. Liana, B. Raguse, J. Justin Gooding, and E. Chow, "Recent advances in paper-based sensors," *Sens. Switz.*, vol. 12, no. 9, pp. 11505–11526, 2012, doi: 10.3390/s120911505.
35. P. Das, M. Das, S. R. Chinnadayyala, I. M. Singha, and P. Goswami, "Recent advances on developing 3rd generation enzyme electrode for biosensor applications," *Biosens. Bioelectron.*, vol. 79, pp. 386–397, 2016, doi: 10.1016/j.bios.2015.12.055.
36. E. Ghafar-zadeh, "Wireless integrated biosensors for point-of-care diagnostic applications," *Sensors*, vol. 15, pp. 3236–3261, 2015, doi: 10.3390/s150203236.
37. Kenry, J. C. Yeo, and C. T. Lim, "Emerging flexible and wearable physical sensing platforms for healthcare and biomedical applications," *Microsyst. Nanoeng.*, vol. 2, no. October 2015, p. 16043, 2016, doi: 10.1038/micronano.2016.43.
38. P. Kassal, M. D. Steinberg, and I. Murkovi, "Wireless chemical sensors and biosensors : A review," *Sens. Actuators B: Chem.*, vol. 266, pp. 228–245, 2018, doi: 10.1016/j.snb.2018.03.074.
39. M. S. Arefin, J. Redoute, and M. R. Yuce, Wireless biosensors for POC medical applications, In Narayan, Roger J. (ed), Medical biosensors for point of care (POC) applications, Woodhead Publishing Limited, Duxford, pp. 151–180, 2017.

40. A. J. Bandodkar and J. Wang, "Non-invasive wearable electrochemical sensors: A review," *Trends Biotechnol.*, vol. 32, no. 7, pp. 363–371, 2014, doi: 10.1016/j.tibtech.2014.04.005.
41. A. J. Bandodkar, W. Jia, C. Yard, X. Wang, J. Ramirez, and J. Wang, "Tattoo-based noninvasive glucose monitoring: A proof-of-concept study," *Anal. Chem.*, vol. 87, pp. 394–398, 2015, doi: 10.1021/ac504300n.
42. L. Donaldson, "Conducting nanowires: Electronic materials," *Mater. Today*, vol. 14, no. 10, p. 459, Oct. 2011, doi: 10.1016/S1369-7021(11)70204-8.
43. S. Kanchi, M. I. Sabela, P. S. Mdluli, and K. Bisetty, "Smartphone based bioanalytical and diagnosis applications : A review," *Biosens. Bioelectron.*, vol. 102, pp. 136–149, 2018, doi: 10.1016/j.bios.2017.11.021.
44. Y. Jung, J. Kim, O. Awofeso, H. Kim, F. Regnier, and E. Bae, "Smartphone-based colorimetric analysis for detection of saliva alcohol concentration," *Appl. Opt.*, vol. 54, no. 31, p. 9183, 2015, doi: 10.1364/AO.54.009183.

4 Aptamer
An Emerging Biorecognition System

Ankana Kakoti
SALK Institute for Biological Studies, La Jolla, California, USA

CONTENTS

4.1 Introduction 69
4.2 Aptamer Selection 70
    4.2.1 Nitrocellulose Membrane Filtration-Based SELEX 71
    4.2.2 Capillary Electrophoresis SELEX 71
    4.2.3 Affinity and Magnetic Bead-Based SELEX 72
    4.2.4 Cell SELEX 73
    4.2.5 *In vivo* SELEX 73
    4.2.6 SELEX on a Chip 74
4.3 Sequence Analysis 74
4.4 Post-Selex Modifications 75
    4.4.1 Truncation 75
    4.4.2 Chemical Modifications 75
4.5 Affinity Measurements 76
    4.5.1 Dialysis and Ultrafiltration 76
    4.5.2 Gel Electrophoresis 77
    4.5.3 Capillary Electrophoresis 78
    4.5.4 High Performance Liquid Chromatography (HPLC) 78
    4.5.5 Fluorescence Intensity and Fluorescence Anisotropy/Polarization 78
    4.5.6 UV-Vis Absorption 78
    4.5.7 Circular Dichroism (CD) 78
    4.5.8 Flow Cytometry 78
    4.5.9 Surface Plasmon Resonance (SPR) 78
    4.5.10 Isothermal Titration Calorimetry (ITC) 79
    4.5.11 Microscale Thermophoresis (MST) 79
    4.5.12 Biolayer Interferometry (BLI) 80
4.6 Biosensing Platforms 80
    4.6.1 Optical Detection 80
        4.6.1.1 Fluorescence Detection 80
        4.6.1.2 Colorimetric Detection 81
        4.6.1.3 Chemiluminescence Detection 82
    4.6.2 Electrochemical Detection 82
    4.6.3 Mass Sensitive Detection 84
4.7 Aptamers Commercialized or En Route To Commercialization 84
4.8 Concluding Remarks 84
References 85

4.1 INTRODUCTION

Biological recognition elements constitute one of the most important parts of a biosensor. As being a central part of the system, biological recognition elements are responsible for specific interactions with the target of interest. The choice of biorecognition element mainly depends upon the analyte, its affinity towards the analyte, and its overall stability in the sample solution. Since the development of the first enzyme-based biosensor, various different types of biorecognition elements have been utilized. Most of these systems take inspiration from biomolecular pairs found in biological systems, like antibodies and antigens, enzymes and substrates, hormones and hormone receptors, etc. These systems are highly selective and sensitive but are not always suited for use under the confines of a biosensor due to their labile structures and sensitivity to changes in their environment. This underlines the importance of exploring new synthetic ligands that can function in different environmental (biological/chemical) conditions and with comparable performance.

Aptamers (Latin *aptus* means "to fit") are one such synthetic ligand (oligonucleotide or peptide) that are emerging as an alternative biorecognition element. It was in 1990 that three independent labs reported the *in vitro* evolution of single-stranded nucleic acid molecules that can specifically interact with targets with high affinity [1–3]. The *in vitro* evolution process widely used to generate nucleic acid aptamers is known as SELEX (systematic evolution of ligands by exponential enrichment). The aptamers are also sometimes referred to as "chemical antibodies" due to their high selectivity and affinity for their target, which if not superior, is comparable to that of antibodies. Apart from this, aptamers have several other inherent characteristics that make them promising candidates for use in biosensors. First, aptamers can be chemically synthesized *in vitro* as opposed to antibodies that require the use of animals. They can also be selected targeting nonimmunogenic and toxic molecules, as the process does not rely on inducing the animal's immune system, as in the case of raising antibodies. Second, they are thermally stable and can be easily modified for incorporating reporter molecules and functional groups. Third, SELEX can be carried out under both physiological and nonphysiological conditions. Fourth, aptamer-target complex formation can be easily reversed using a chaotropic agent, rendering the aptasensor reusable. These positive traits have incited a wide interest in using aptamers as a biorecognition element in biosensors. This chapter will detail the different types of SELEX procedures used, post-SELEX characterization of aptamer sequencing, their binding affinity measurement techniques, and finally a brief introduction to the biosensing platforms that have been utilized for developing aptasensors.

4.2 APTAMER SELECTION

Aptamers are isolated through the process of SELEX (Figure 4.1). The first step of the selection process involves generation of a combinatorial oligonucleotide library with a typical sequence diversity of 10^{12}–10^{15} single-stranded DNA or RNA sequences. Generally, each sequence in the library has a central randomized region (20–90 nucleotides) flanked by fixed primer binding sites for polymerase chain reaction (PCR) amplification. The basic steps involved in SELEX include binding, partition, elution, and amplification. Initially, the target is incubated with the randomized oligonucleotide library, during which the target molecules interact with the aptamer library either in a free form or in a form that is immobilized on a solid support substrate surface (e.g., on magnetic beads, microtiter plates, membranes, columns, etc.). Following aptamer-target complex formation, the next critical step is the removal of unbound or weakly bound oligonucleotide candidates from the bound forms. This can be achieved through various methods like filtration, affinity elution, magnetic isolation, washing of microtiter plates, capillary electrophoresis, etc. The target-bound oligonucleotides are then eluted (using heat or a chaotropic agent) followed by PCR or reverse transcription PCR (RT-PCR) amplification. The enriched selected oligonucleotide pool so generated is then used for the next SELEX selection cycle. Typically 6–20 selection cycles are carried out, including counter-SELEX cycles to select specific and high-affinity aptamers. The enriched aptamer pool obtained at the last step of SELEX is cloned and sequenced for further analysis. In the case of RNA SELEX, the additional steps of *in vitro* transcription and reverse transcription are introduced. Briefly, RNA sequences bound to the target are eluted, reverse-transcribed into DNA, and amplified with PCR. The amplified DNA is

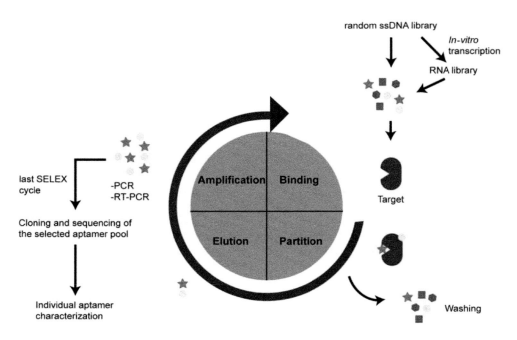

FIGURE 4.1 General process of SELEX.

then transcribed *in vitro* to yield the RNA sequences, which are then used for the next selection cycle.

The conventional form of SELEX, although well-established, is labor intensive and time consuming, which has necessitated the development of alternative SELEX strategies, as discussed next.

4.2.1 Nitrocellulose Membrane Filtration-Based SELEX

Nitrocellulose membranes, due to their high protein binding affinity, are routinely used in Western blots for protein immobilization. This property is exploited in the SELEX method, where proteins are immobilized on the membrane and then incubated with the initial library. Following this, the membrane is washed to remove the unbound species, and finally the bound sequences are eluted and amplified for use in the next SELEX cycle. This method was initially developed by Pristoupil and Kramlova [4] to separate protein from RNA, which was subsequently used by Tuerk and Gold [3] for selection of aptamers against the T4 DNA polymerase. This strategy is of convenience for protein targets but suffers from drawbacks, for instance, the incapability of nitrocellulose membranes to bind small molecules, peptides, etc.

4.2.2 Capillary Electrophoresis SELEX

One of the most successful variations of the SELEX process is capillary electrophoresis (CE) (Figure 4.2). It enables separation of aptamer-target complexes from unbound oligonucleotides based on their electrophoretic mobility, which in turn is influenced by the charge, hydrodynamic radius, and frictional forces [5]. In principle, oligonucleotides bound to the target have lower mobility as compared to free oligonucleotides and thus can be collected after the free nucleotides have passed out of the capillary. It has the advantage of improved resolution of aptamer-target complexes from free nucleic acids, thus requiring fewer cycles of selection for isolation of high affinity aptamers. Additionally, CE-based selection circumvents the problem of nonspecific binding of the library with stationary supports, as it is performed in free solution. Several modifications of CE-SELEX were introduced to increase the partitioning efficiency in comparison to the traditional CE-SELEX methods.

One such format is non equilibrium capillary electrophoresis of equilibrium mixtures (NECEEM) [6]. In this approach, the target is mixed with the oligonucleotide library to obtain the equilibrium mixture, which is then injected into the capillary and separated by applying an electric field. During separation, free DNA and the target migrate as a single electrophoretic zone, while the target-DNA complex dissociates slowly during migration creating a nonequilibrium production of DNA and protein. Based on the fractions collected during this time frame, aptamers with high affinity can be isolated as well as determine their dissociation constant (K_d) and the off rate-constant (K_{off}).

Unlike NECEEM, ECEEM allows selection of smart aptamers with predefined K_d values [7]. It involves incubation of the target with the oligonucleotide library, equilibration, and then application to a capillary prefilled with a running buffer with the target at the same concentration as present in the equilibrium mixture. Finally, the different components in the

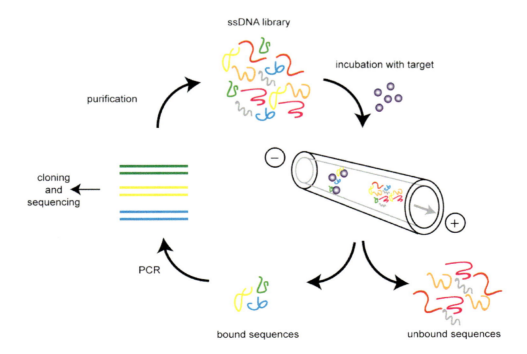

FIGURE 4.2 Schematic representation of CE-SELEX. An oligonucleotide pool is first incubated with the target and then the mixture is separated on the basis of mobility shift. The bound sequences are then collected, PCR amplified, cloned, sequenced, and characterized. (Adapted from Yang and Bowser [159].)

equilibrium mixture are separated by CE, while maintaining equilibrium between the target and aptamers. In ECEEM, aptamers with different K_d values migrate with different mobility. Thus, collecting fractions with different mobility results in smart aptamers with different and predefined K_d values.

Another variation of kinetic CE is Sweep CE (sweeping CE) which not only provides the K_d and K_{off} values but also the K_{on} values, thus extending the available information of aptamer-target interactions [8]. Briefly, a capillary filled with a solution of DNA is used for electrophoresis of the protein. Due to its greater electrophoretic mobility, the protein travels toward DNA forming the aptamer-protein complex, which in turn migrates with a velocity higher than that of unbound DNA.

Though highly advantageous, CE-SELEX is not without its limitations. For example, only very low reaction volumes can be applied to the capillary, which limits the overall number of sequences that can be analyzed at a given time. As a means of evading this, microfluidic free flow electrophoresis (µFFE) was developed [9]. It allows the continuous application, separation, and collection of an oligonucleotide library, thus enabling screening and collection of a higher number of aptamer sequences. µFFE, however, requires the fabrication of a special µFFE device, which might limit its application in a regular laboratory.

4.2.3 Affinity and Magnetic Bead-Based SELEX

Affinity bead-based SELEX capitalizes on the principle of affinity chromatography, wherein biological interactions between the target and ligand are utilized for purification of the target from a biochemical mixture. Target molecules are immobilized on the stationary phase beads of the column, thus ensuring selection of oligonucleotide sequences that specifically binds to the target. With regard to proteins, various purification tags such as glutathione S-transferase (GST) and His tags are used, while for small molecules covalent immobilization on beads is the preferred method [10–12]. A similar approach is used for magnetic bead-based SELEX, where the target is immobilized on the magnetic beads via a chemical or biological interaction between the affinity tag and the substrate on the beads (Figure 4.3). Use of magnetic beads renders the SELEX process more easy and rapid, as separation of the target immobilized magnetic beads from the unbound oligonucleotide library can be achieved by applying a magnet [13]. A modified version of this method is FluMag-SELEX, which incorporates fluorescence labeling of the bound ssDNA after the first cycle of SELEX [14]. This allows quantification of bound and unbound fractions of the library, as well as calculation of the dissociation constants of the enriched aptamer sequences. Usually bead-based SELEX predominantly uses immobilization of target on the beads followed by incubation with the oligonucleotide library. Recently, a new method called capture SELEX was generated, which involves immobilization of the DNA library on the magnetic beads [15]. A docking sequence is incorporated into the random region of the library, which in turn can hybridize to the complementary strand fixed on the magnetic beads. The aptamer sequences that bind specifically to the target are released from the beads

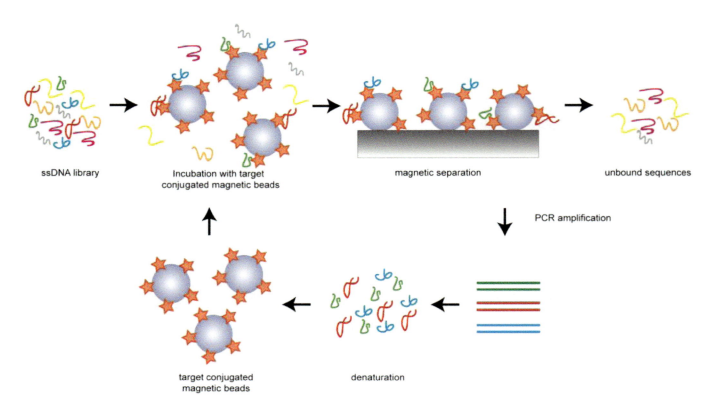

FIGURE 4.3 Basic steps of magnetic bead-based SELEX.

and are then amplified, purified, and immobilized to the magnetic beads for the next round of SELEX.

4.2.4 Cell SELEX

In cell SELEX, instead of using highly purified proteins or small molecules, whole cells, either prokaryotic or eukaryotic, are used as the target [16, 17]. As in conventional SELEX, the cells are incubated with a starting library and the unbound sequences are separated by washing (Figure 4.4). The bound sequences are then collected and amplified by PCR for the next round. Negative rounds of SELEX using control cells are an important step in the cell SELEX procedure to avoid sequences that bind to normal cells from being enriched during the process. This is because many cancer cells express surface proteins that are also expressed by normal healthy cells. Based on this method, a number of specific aptamers have been generated against different cancer cell lines like lymphocytic leukemia [18], myeloid leukemia [19], liver cancer [20], small-cell lung cancer [21], lung adenocarcinoma [22], prostate cancer [23], and colorectal cancer [24]. Apart from cancer cell lines, aptamers have also been generated against somatic cells and stem cells [25, 26]. A substantial amount of aptamers have also been reported for pathogenic organisms like *Salmonella typhimurium* [27], *Escherichia coli* [28], *Trypanosoma brucei* [17], *Trypanosoma cruzi* [29], etc.

Cell-SELEX has also been applied for the generation of aptamers specifically binding to virus-infected cells. As a result of viral infection, the host cell displays various viral proteins on its surface, which acts as specific aptatopes for recognition by the selected aptamers. Cell-SELEX against *Vaccinia virus*-infected adenocarcinomic epithelial cells (A549) resulted in the selection of DNA aptamers that could bind to different types of cell lines infected by the virus, thus indicating that the aptamers were able to recognize specific viral proteins expressed on the host cell membrane [30].

4.2.5 In vivo SELEX

The recombinant proteins or cells that are routinely used for conventional SELEX procedures may not perfectly mimic their natural antigenic state or folding *in vivo*. This could be due to influence of the microenvironment in which the cells are found, varying antigenic display, or other interactions. As a result, aptamers selected for these cells or proteins *in vitro* may fail to perform in the same way *in vivo*. To overcome this difficulty, *in vivo* SELEX was introduced that allows for selection of specific aptamers in living organisms. In

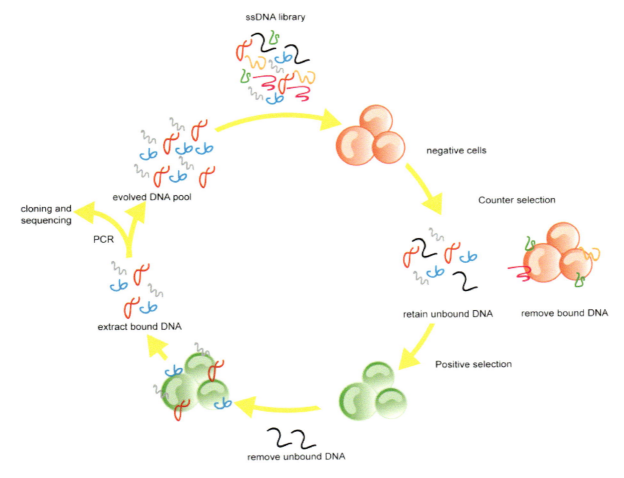

FIGURE 4.4 Schematic illustration of DNA aptamer selection using cell SELEX.

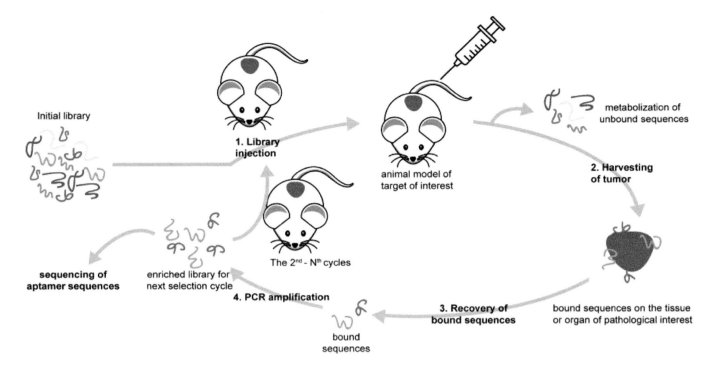

FIGURE 4.5 *In vivo* selection of aptamers: tumor-bearing mice are intravenously injected with a random nucleic acid library, followed by harvesting of the tumor, extraction, and amplification of the bound nucleic acids. The enriched sequences are then used for the next round of *in vivo* SELEX.

2010, Mi et al. selected aptamers against intrahepatic tumors inside live mice [31]. For this they injected 2′-fluoropyrimidine-modified RNA aptamers in to the tail vein of intrahepatic tumor-bearing mice (Figure 4.5). Following this, RNA aptamers were extracted from the liver tumors, amplified, and reinjected into mice bearing the same kind of tumor. In this way, high affinity aptamers were selected against p68 and RNA helicase with K_d values in the low nanomolar range. In the light of this procedure, a similar method was applied to generate aptamers that could penetrate the blood-brain barrier in mice [32]. The selection scheme was the same as noted earlier, except for an additional negative selection cycle wherein aptamers obtained from the previous rounds were incubated with mouse serum. The aptamers thus selected could bind to brain capillary endothelia and penetrate parenchyma.

4.2.6 SELEX ON A CHIP

SELEX, when performed manually, usually takes from a few weeks to months for completion. Therefore, to facilitate high throughput selection of aptamers, it was consequently automated using a robotic platform in 96 well formats. The first automated system integrated a PCR thermocycler, a magnetic bead separator, reagent trays, and a pipette tip station. Automated selection has been successfully performed for generating aptamers against various targets like hen egg white lysozyme [33, 34], U1A [35], CYT-18 [36], MEK-1 [36], and Rho protein [36].

Besides ease of automation, the use of magnetic bead-based separation has enabled miniaturization of the SELEX process, thus decreasing the consumption of reagents as well as selection time [34]. Microfluidics-based SELEX thus reduces average selection times by applying highly stringent conditions, usually through the use of a very small amount of targets. They utilize target immobilized magnetic beads inside microchannels to capture specific aptamers and then ensure separation from unbound sequences by applying a magnetic field within the microchannel. These chip-based *in vitro* selection platforms have been applied for the successful selection of aptamers against CRP [37], AFP [38], A/H1N1 virus [39], bovine serum albumin [40], BoNT/A-rLC protein [41], etc.

4.3 SEQUENCE ANALYSIS

The enriched library at the end of SELEX is subjected to cloning, colony picking, and then sequencing of a small number of colonies. Usually Sanger sequencing is used to obtain the sequences, which are then aligned and grouped on the basis of their similarities. This process is slowly being replaced by high-throughput next-generation sequencing (NGS) as it renders several advantages over conventional Sanger sequencing. It allows the analyses of millions of sequences from each round of SELEX, thus enabling continuous monitoring of enriched sequences by tracking their frequency distribution over the entire SELEX procedure. As a result, enriched sequences can be observed with fewer selection cycles, thereby reducing the time required for an aptamer selection process. Finally, the sequences obtained are clustered into similar families based on their primary sequence motifs or secondary structure motifs. Some of the programs used for NGS data analysis for SELEX experiments are included in Table 4.1.

TABLE 4.1
Comparison of Several Programs Dedicated to NGS Analysis for SELEX Experiments (Ducongé et al. [160].)

Name	System	Rounds Analyzed	Clustering Based on Primary Sequence	Clustering Based on Secondary Predicted Structure	References
IniMotif	Not determined	Several	Clustering of the most enriched subsequences using Hamming distance.	-	[42]
TFAST	Mac/Linux/PC	Several	Alignment on a genome	-	[43]
FASTAptamer	Mac/Linux; Galaxy web platform	2	Levenshtein distance on sequence "seeds"; possibility to look for a known motif	-	[44]
AptaCluster	Linux	Several	LSH method followed by k-mer distance on sequence "seeds"	-	[45]
PATTERNITY-seq	Available through services	Several	Levenshtein distance on sequence "seeds"	Look for predicted structure motifs shared by several primary clusters	[46]
COMPAS	Available through services	Several	k-mer distance or Shannon's information entropy	Can detect stem loops shared by several primary clusters	[47]
AptaTrace	Mac/Linux/PC	Several	Look for enrichment of k-mer with a predicted structure; then, primary alignment of k-mers with the same predicted structure is realized to form clusters		[48]
APTANI	Linux	Several	Look for four kinds of substructures in each sequence; primary alignment of the substructures to form clusters		[49]
MPBind	Mac/Linux	2	Rank sequences based on k-mer enrichment	-	[50]
SEWAL	Mac	Several	Hamming distance on two sequences (seed and control) to obtain (x,y) coordinates of all sequences to build the empirical "landscape"	-	[51]
AptaMut	Linux	2	Rank variants of a primary sequence family based on their enrichment	-	[52]

4.4 POST-SELEX MODIFICATIONS

In contrast to all the advantages that aptamers present over antibodies, only a few of them have entered clinical trials, and only one (pegaptanib) has been approved so far by the U.S. Food and Drug Administration (FDA) for clinical use. The practical applications of aptamers are restricted primarily due to inadequate stability to nuclease digestion, reduced thermal stability, fast renal clearance, and sometimes reduced affinity, among others. Therefore, modifications of aptamers are required to enhance their performance in real-life applications. Some commonly used strategies are discussed in the following sections.

4.4.1 Truncation

Aptamers generated through SELEX usually consist of a 30- to 50-nucleotide random region and fixed primer-binding sites on both ends. However, studies have indicated that constant regions generally do not contribute to aptamer binding and are minimally involved in maintaining the overall aptamer structure [53]. Therefore, minimizing the length of aptamers could reduce overall synthesis costs. To deduce the consensus high affinity-binding motif, multiple sequence alignments of the enriched library can be performed using software like CLUSTALW/Omega [54, 55] and MEME Suite [56], or secondary structure prediction tools like Mfold [57] and RNA structure [58] could be used to predict the secondary structures of aptamers. Based on the consensus motif and the secondary structure, aptamers can be truncated to their minimal binding motifs without the loss of binding affinity.

4.4.2 Chemical Modifications

Chemical modifications of aptamers can be either incorporated in the sugar ring, bases, or internucleotide linkage. RNA aptamers are usually modified at the 2′ position of nucleosides, while DNA aptamers are modified in their backbone. Modifications of the 2′ position include 2′-F, 2′-O-CH$_3$, and 2′-NH$_2$ nucleotides as substitutes for RNA aptamers, which increases their resistance to nuclease

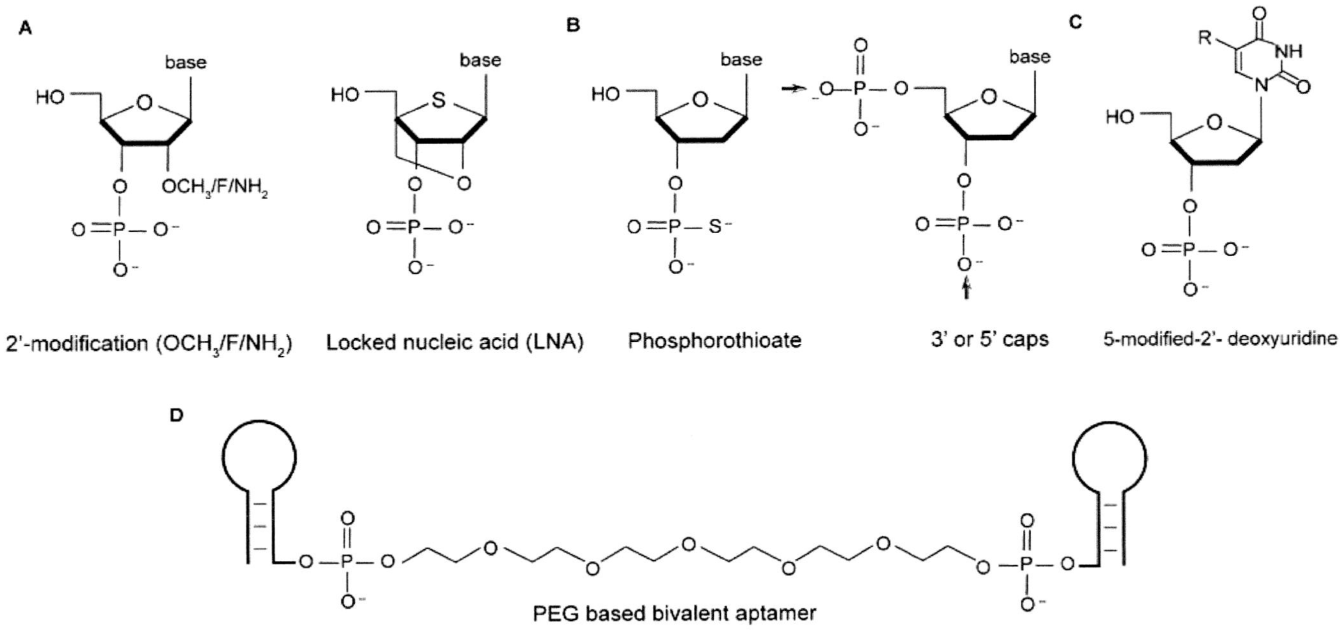

FIGURE 4.6 Chemical modifications to improve aptamer properties. (A) Modification of the sugar ring (2′ replacement, LNA). (B) Modification of linkage (phosphorothioate replacement, 3′ or 5′ capping). (C) Modification of bases (5′ position of pyrimidine). (D) Bivalent or multivalent modifications (PEG modification).

digestion. Locked nucleic acids are also used due to their high nuclease stability, thermal stability, and low cytotoxicity. Bases could be modified by incorporating a wide variety of hydrophobic, hydrophilic, and charged groups to increase the interactions between the aptamer and its target. The most common modification of the backbone that is used is substitution with thiophosphate. However, using a thiophosphate backbone can lead to decreased thermal stability as well as nonspecific interactions with nontarget proteins. Moreover, to enhance the pharmacokinetic properties of aptamers, 3′ or 5′ capping and PEG conjugation are also frequently included (Figure 4.6).

4.5 AFFINITY MEASUREMENTS

Understanding the interactions necessary for aptamer-target binding is important for their application in the real world. This calls for detailed studies on binding kinetics, stoichiometry, and measurement of equilibrium constants. In its simplest binding equilibrium with 1:1 stoichiometry, aptamer-target interactions can be expressed as:

$$A + T \rightleftharpoons C$$

where A is the aptamer, T is the target, and C is the aptamer-target complex. The equilibrium can be represented in the form of the dissociation constant (K_d) or association constant (K_a), as shown in the following equations:

$$K_d = \frac{[A][T]}{[C]}$$

$$K_a = \frac{1}{K_d} = \frac{[C]}{[A][T]}$$

The equilibrium constants are determined by titrating an increasing concentration of target against a fixed concentration of the aptamer. Therefore, the fraction of bound aptamer is calculated from the following equation:

$$f_a = \frac{[T]}{K_d + [T]}$$

There are several platforms for detection and characterization of aptamer-target biomolecular interactions. The major techniques are briefly described in the next sections.

4.5.1 DIALYSIS AND ULTRAFILTRATION

Both these methods separate free aptamer from aptamer-protein complex mixtures based on size. During dialysis different concentrations of target are titrated against fixed concentrations of aptamer in one compartment and buffer in the other compartment, separated by a membrane. Only free aptamers, due to their size, can pass through the membrane to the other compartment. After equilibrium is reached, the amount of free aptamer in the second compartment is determined. In ultrafiltration, the aptamer-target mixture is passed through the membrane using pressure or a vacuum, allowing only the free aptamer to pass through the membrane, while the aptamer-target complex is trapped on the membrane [59, 60]. The aptamer-target complex on the membrane can then be detected if radioactively labeled aptamers were used (Figure 4.7A).

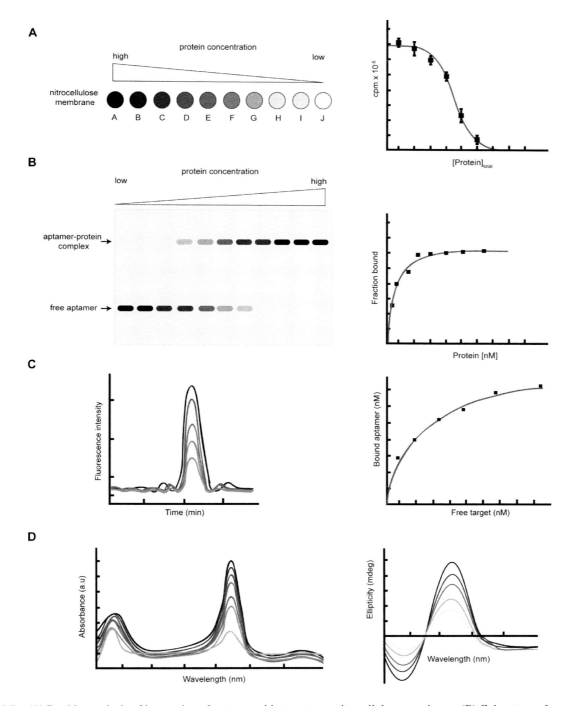

FIGURE 4.7 (A) Dot blot analysis of interaction of aptamer with target on a nitrocellulose membrane. (B) Gel pattern of a mobility shift assay with aptamer and target. (C) Fluorescence intensities of free aptamer at different concentrations of target in CE electropherograms. (D) UV-Vis absorption and CD spectra at a fixed aptamer concentration and increasing concentrations of the target.

4.5.2 Gel Electrophoresis

Nondenaturing PAGE or agarose gels are routinely used for studying nucleic acid-protein binding kinetics. It is used to separate nucleic acid-protein complexes based on their charge and mobility. Nucleic acids with strong negative charge travel faster, while proteins possess lower charge and therefore lower mobility. The protein-aptamer complex possesses an intermediate electrophoretic mobility between free aptamer and target protein, making it easier to separate protein-aptamer complexes from free aptamer. Different concentrations of the target are incubated with the aptamer and then separated on a gel and then quantified using ultraviolet (UV) absorbance, radioactive labeling, fluorescent staining, or blotting [61–67]. Finally, the binding curve is constructed and the equilibrium constants determined (Figure 4.7B).

4.5.3 Capillary Electrophoresis

As already mentioned, capillary electrophoresis can be used to determine dissociation constants by calculating the bound aptamer fraction from the decrease in intensity of the free ligand peak with the addition of increasing amounts of protein. This is because in CE two peaks corresponding to the complex and the unbound aptamer are observed in the electropherogram [68–70]. Therefore any decrease in the peak intensity of unbound aptamer with increasing amount of target indicates the amount that is bound (Figure 4.7C).

4.5.4 High Performance Liquid Chromatography (HPLC)

HPLC can be used to separate free aptamers from protein-aptamer complexes on the basis of their size, and these zone separations can be used to estimate the K_d [71–79].

4.5.5 Fluorescence Intensity and Fluorescence Anisotropy/Polarization

Change in fluorescence intensity of a labeled aptamer on binding to its target can be used to estimate binding affinity. Quenching of fluorescence, as well as shift in the fluorescent emission profile, can also be used based on individual systems under study. Specifically fluorescence signals of the aptamer-target complex can be subtracted from that of free ligand and background signal to obtain intensity changes and thereby calculate the binding fractions. Fluorescence polarization can also be used for determining binding kinetics if binding of the labeled aptamer to its target causes an increase in polarization of the fluorescent dye attached [80–83].

4.5.6 UV-Vis Absorption

Due to its simplicity, UV-Vis absorption is a widely used method for studying aptamer-target interactions. Both DNA and proteins are known to show absorption maxima in the UV range; therefore, any change in the intensities of their absorption maxima can be used for estimation of their binding constants [84–86] (Figure 4.7D).

4.5.7 Circular Dichroism (CD)

Proteins and nucleic acids absorb left and right circularly polarized light differently due to the presence of asymmetric carbons in amino acid residues and sugars. This difference is expressed in the form of absorption (ΔA), differential molar extinction coefficient ($\Delta \varepsilon$), or degree of ellipticity (θ). Since DNA and protein CD spectra are observed at different absorption ranges (for protein under 250 nm and above 250 nm for DNA), changes in their intensities on titrating with the target protein can be used to calculate the binding fraction [87–92] (Figure 4.7D).

4.5.8 Flow Cytometry

Flow cytometry is widely used to detect the binding of aptamers to whole cells. It allows monitoring of aptamers to their targets in their native conformational states on the cell surface. In this method, the aptamer is labeled with a fluorescent dye, like FITC, and incubated with the target cell. Binding of the aptamer to the cells is determined from the fluorescence intensity of the labeled cells. The higher the intensity, the higher the binding capacity of the aptamer toward the cell [93, 94]. Apart from cells, targets that can be immobilized on magnetic/affinity beads can also be analyzed for their binding to their respective aptamers [95].

4.5.9 Surface Plasmon Resonance (SPR)

SPR measures the effect of changes in electromagnetic surface waves on reflected light. These waves are sensitive to the dielectric properties of the metal surface, as well as the sample solution in contact with the sensor tip. Usually, aptamer is immobilized on the sensor tip and allowed to interact with the target in solution. When no target is bound by the aptamer, there is total internal reflection of the incident polarized light. However, upon binding, the electromagnetic surface waves are altered, which leads to a shift in the angle of the reflected light. This change in resonance angle can directly be related to the amount of target bound on the sensor tip. This method allows for label-free detection of both thermodynamic and kinetic parameters in addition to K_d [96–100] (Figure 4.8).

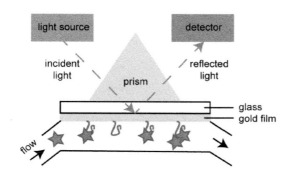
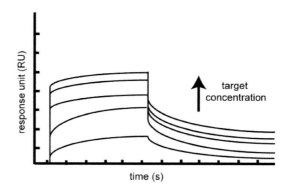

FIGURE 4.8 Schematic of SPR technique for studying aptamer-protein binding affinity. A typical sensorgram of aptamer-target interaction at different target concentrations is shown in the right.

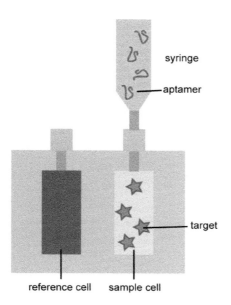
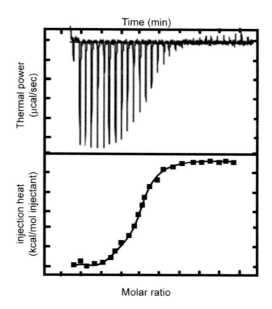

FIGURE 4.9 Schematic of ITC cells and injection syringe. On the right are the representative raw ITC data from titration experiment and the binding isotherm, respectively.

4.5.10 Isothermal Titration Calorimetry (ITC)

ITC is a label-free, immobilization-free method that allows for the simultaneous determination of K_d, stoichiometry, and thermodynamic parameters by measuring the absorption or dissipation of heat when aptamer-protein complexes are formed. Ligand in small amounts is added to the target in a calorimetric cell, and the amount of electric power required to maintain the temperature in the sample cell same as the control cell is measured. The amount of power required is directly proportional to the amount of target bound [101–104] (Figure 4.9).

4.5.11 Microscale Thermophoresis (MST)

MST measures the movement of molecules through a temperature gradient. For studying aptamer target binding, multiple capillaries containing a fixed concentration of ligand and different concentrations of targets are analyzed. For this, intrinsically fluorescent or labeled ligands are used, and their movement is measured in the presence or absence of an infrared laser. Any change in size, charge, or hydration caused by binding affects the movement of molecules, which is recorded for determination of affinity constants [105, 106] (Figure 4.10).

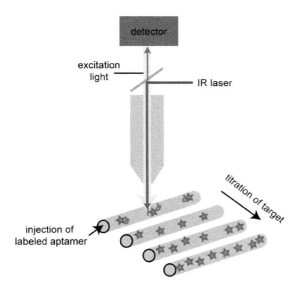
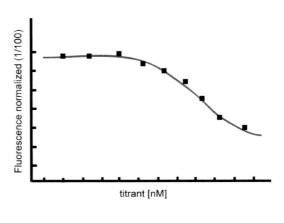

FIGURE 4.10 Schematic showing target titration in capillaries and injection of labeled aptamers. Plot of normalized fluorescence vs titrant concentration is shown in the right.

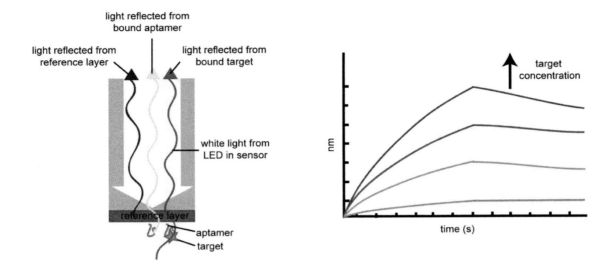

FIGURE 4.11 Schematic showing changes in reflected light with binding of aptamer with target. The typical association and dissociation graph for a target binding to an aptamer is depicted in the right.

4.5.12 Biolayer Interferometry (BLI)

BLI measures changes in reflected light when the target binds to ligand on the sensor tip. White light emitted by the sensor is reflected back from the biosensor tip and a reference layer, and based on the changed interference pattern, the amount of target bound is determined. This change in interference is reported as a change in wavelength over time [107, 108] (Figure 4.11).

4.6 BIOSENSING PLATFORMS

A typical biosensor consists of a biological recognition element (enzyme, antibody, receptors, whole cells, etc.), transducer (electrochemical, thermal, optical, etc.), signal amplifier, and display. When the biorecognition element used is an aptamer, then it is called an aptasensor. The application of aptamers as the biorecognition element renders advantages beyond the obvious benefits of aptamers over antibodies. Aptamers have high affinities for their targets, are easily tunable, and can be immobilized at high density on sensor surfaces. Additionally, they render the important property of reusability to sensors. Based on the signal-transduction method used, aptasensors can be classified into the types discussed in the following sections.

4.6.1 Optical Detection

The most popular optical detection techniques using aptamers as the biorecognition element are fluorescence, colorimetry, and chemiluminescence.

4.6.1.1 Fluorescence Detection

Aptamers are known to undergo significant changes in their conformation on binding to their targets [109, 110]. This property can be used to incorporate fluorophores in regions of the aptamer that are prone to conformational disruption, so that upon ligand binding there is in an associated change in the fluorescence property (intensity/anisotropy) of the fluorophore, induced by changes in its chemical environment. Fluorescent labels can be introduced either at the sugar moiety, the backbone, or the bases, depending on the nature of the label and the site of introduction, the mechanism of signal transduction is affected. In addition to labeling with fluorescent dyes, replacement of nucleotides with their fluorescent counterparts is also used for fluorescent sensors. The fluorescence of these analogs when remaining in a base-stacked structure is quenched, but on any disruption to their base stacking, the fluorescence intensity increases. For example, modification to the aptamers for human alpha-thrombin, immunoglobin E, and platelet-derived growth factor B with fluorescent analogs in positions that undergo conformational changes on ligand binding resulted in an increase in fluorescence signal of up to 30-fold [111]. However, to apply this strategy for signal transduction, information about the tertiary structure of the aptamer is desired to enable incorporation of fluorescent labels/nucleotides at positions that can induce the maximum change in fluorescent intensity on target binding.

Apart from single reporter systems, two reporter systems utilizing a quencher and either one or two fluorophores are also commonly used. The simplest format is aptamer-based molecular beacons, wherein the aptamer is placed in a molecular beacon hairpin structure with one end labeled with a fluorophore and the other end with a quencher. Binding of the target can either bring the two closer, resulting in quenching, or can distance them, resulting in increased fluorescence (light-up) [112–114]. Another frequently used strategy is to use a complementary antisense DNA strand labeled with a quencher to disassemble the aptamer (labeled with fluorophore) in such a way that on addition of the target, the complementary strand can no longer bind to the aptamer, thus

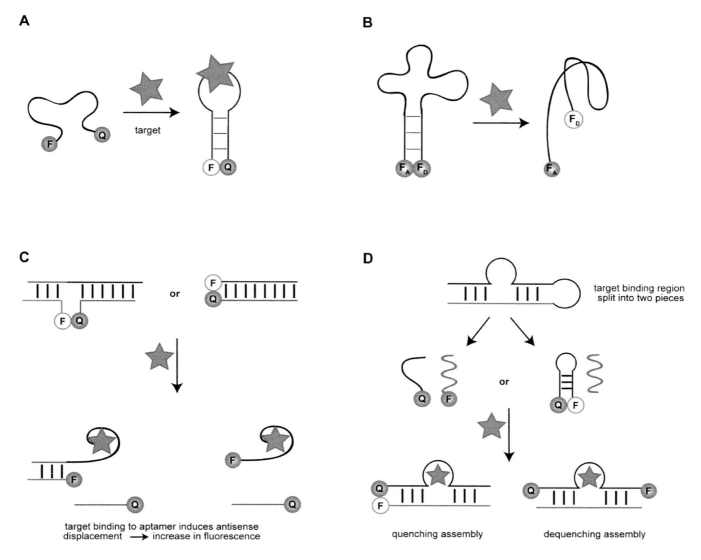

FIGURE 4.12 Two reporter optical aptasensors. (A) Quenching aptamer. (B) Fluorescence resonance energy transfer (FRET) aptamer beacon. (C) Disassembly aptamer beacon. (D) Assembly aptamer beacon. F, fluorophore; Q, quencher; F_A, acceptor fluorophore; F_D, donor fluorophore. (Adapted from Cho et al. [161]).

resulting in an increase in fluorescence [114, 115]. Aptamers can also be split rationally into two parts, such that on target binding, the two fragments can self-assemble, resulting in a change in fluorescence signal [116–119]. This has been successfully used for detection of HIV-1 Tat protein, cocaine, and adenosine (Figure 4.12).

However, quenching-based strategies are susceptible to interference by certain solvents and ligands, resulting in false-positive signals. As a means to avoid this hindrance, beacons can be constructed based on FRET, which relies on energy transfer between two fluorescent molecules – the donor and acceptor [120, 121]. Apart from organic fluorescent dyes, quantum dots have been employed offering increased photostability, longer fluorescent lifetimes, sharper emission bands, and the possibility to control the wavelength of emitted light by changing the size and composition of the materials [122–124].

4.6.1.2 Colorimetric Detection

Colorimetric detection, though not as sensitive as fluorescence reading, is a convenient technique for target estimation without the use of complex analytical instruments. Gold nanoparticles (AuNPs) are the most commonly used reagents wherein changes in their absorption properties, depending upon their aggregation state, are exploited. Dispersed AuNPs smaller than 100 nm in solution appear red, but on aggregation, change to blue due to the shift of surface plasmon resonance to a higher wavelength. This property of AuNPs has been used to develop simple colorimetric formats using aptamers that have distinctive adsorptive properties toward functionalized AuNPs. Several approaches have been used successfully for the detection of analytes using AuNPs and aptamers. DNA aptamers in solution can tightly bind to unmodified AuNPs and stabilize them against salt-induced aggregation. However, upon binding to its target, the aptamers

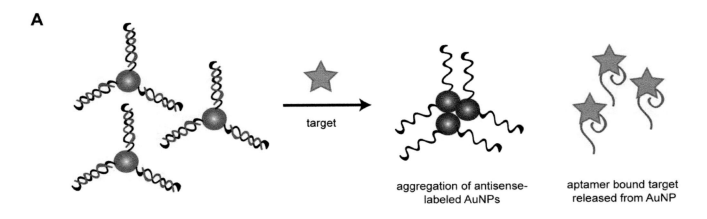

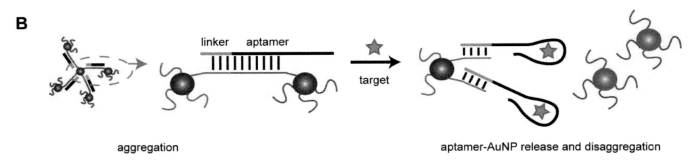

FIGURE 4.13 Gold nanoparticle-based optical sensors. (A) Aptamer release and AuNP aggregation. (B) Aptamer release and disaggregation. (Adapted from Cho et al. [161]).

detach from the nanoparticle surface, thus exposing them to aggregation. This leads to a change in color from red to blue, which can be easily recorded with the naked eye [125, 126]. In another approach, aptamers have been used as linkers to assemble AuNPs into purple aggregates, which in the presence of the target are separated into red-colored individual NPs [127, 128] (Figure 4.13). Additionally, the nanoparticle interface can be enlarged in a growth solution containing $HAuCl_4$ and reducing agents, thus further enhancing the sensitivity of these assays [129, 130].

4.6.1.3 Chemiluminescence Detection

Chemiluminescence detection has several advantages over other optical methods, as it does not need an external light source like fluorescence to produce a light signal and is highly sensitive compared to colorimetric assays. For this, the catalytic activity of AuNPs has been frequently used. The aggregation of AuNPs upon binding of the target with its aptamer can enhance the luminal-H_2O_2 chemiluminescence reaction. This assay format has been used for sensitive detection of thrombin, α-fetoprotein, Pb^{2+}, adenosine, cocaine, and severe acute respiratory syndrome (SARS) coronavirus nucleocapsid protein [131–136].

4.6.2 Electrochemical Detection

Electrochemical transduction in aptasensors is finding greater preference due to its fast response, high sensitivity, low cost, ease of miniaturization, and compatibility with turbid samples. So far, a number of electrochemical detection methods have been utilized to create aptasensors, including amperometry, cyclic voltammetry, electrochemical impedance spectroscopy, potentiometry, electrogenerated chemiluminescence, and field effect transistors (Figure 4.14). The most common strategy utilizes the structural changes of the aptamer on binding to its target. Upon binding, the aptamer is capable of folding into well-defined three-dimensional rigid structures from their single-stranded flexible forms. Therefore, when an aptamer is bound to a conducting surface and the other end is tethered to a redox-active moiety, formation of the aptamer-target complex can be monitored based on the electron transfer property of the tethered redox moiety. Any change in the distance between the electroactive group and the conductive surface leads to a change in the signal recorded. Different reporters have been used for this purpose such as methylene blue, ferrocene, ferrocene-bearing polymers, ruthenium complexes, etc. [137–140]. Redox-active reporting molecules can also be employed in electrochemical formats without covalent

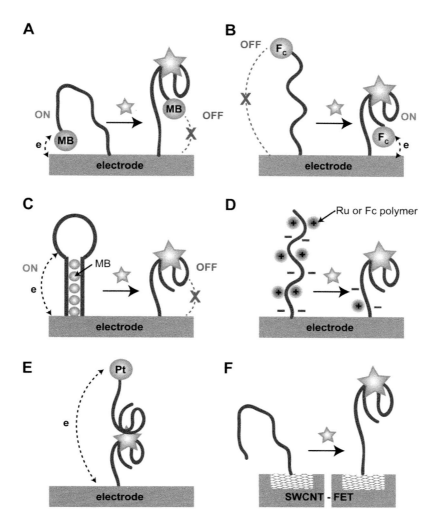

FIGURE 4.14 Schemes for electrochemical sensors. (A) After binding to target, the aptamer self-assembles into a compact structure shielding methylene blue (MB) from transferring the electron to the electrode, leading to a negative signal. (B) Aptamer-target complex formation leads to a rigid conformation of the aptamer, bringing the ferrocene moiety close to the electrode, leading to easy electron transfer and producing a positive signal. (C) Binding of target to the target disrupts the hairpin structure of the aptamer, thus releasing the intercalated MB and decrease in electrochemical signaling. (D) Binding of the aptamer to the target, thereby blocking the binding of cationic reporters to the aptamer, thereby resulting in depleted signal. (E) Aptamer functionalized PtNPs were utilized to catalyze the electrochemical reduction of H_2O_2 for the amplified detection of targets in a sandwich format. (F) Binding of the target by the aptamer causes in a change in conductance through the device, thus enabling the detection of targets. (Adapted from Song et al. [162].)

attachment to the aptamer. For example, they can be used to intercalate the double-stranded aptamer like methylene blue [141], or could be bound to the aptamer phosphate backbone via electrostatic interaction [142, 143]. On binding to the target, the aptamer in turn loses interaction or releases the redox reporters, leading to a change in the electrochemical signal.

To increase the specificity as well as sensitivity of electrochemical bioassays, sandwich configurations are often employed utilizing the catalytic properties of redox enzymes. These formats use one or two aptamers or in combination with antibodies. The enzyme conjugated to the aptamer, in the presence of the target and redox mediator, catalyzes reduction of the substrate leading to amperometric detection of the target. For example, glucose dehydrogenase was employed for the amplified amperometric detection of thrombin at levels low as 1 μM [144]. Apart from enzymes, nanoparticles capable of catalyzing the electrochemical reduction of H_2O_2 have also been used for the same purpose [145, 146]. For the detection of target-aptamer interactions, impedance spectroscopy measuring the change in resistance in the presence of the redox mediator has also been used. In the absence of the target, the negatively charged aptamer repulses the redox mediator away from the sensor surface, thus resulting in increased resistance. In this way, detection of thrombin and potassium ions has been reported [138, 139, 147].

The essential features that make electrochemical sensors desirable are their size and fast response, both of which can be found in field effect transistors (FETs). Aptamers are smaller in size (1–2 nm) and therefore when interactions with the target occurs, they take place within the confines of Debye length (3 nm for 10-mM ionic concentrations), thus enabling perturbations of the gate potential. This type of sensor has been used for the detection of thrombin and IgE with high sensitivity [148, 149].

4.6.3 Mass Sensitive Detection

Mass sensitive biosensors measure properties that are directly proportional to the differential change in mass on the sensor surface. Aptamer-based mass sensitive biosensors are mainly of the following types: SPR, surface acoustic wave (SAW), quartz crystal microbalance (QCM), and microchannel cantilever sensors (Figure 4.15). These label-free methods are, however, most suitable for large analytes either of the same or a larger size than the aptamer itself.

In SPR, the sensor chip is immobilized with aptamers and the analyte is injected at a constant flow rate, during which the instrument measures changes in the resonance angle that occur at the surface. This change in the resonance angle is proportional to the amount of bound analyte. For example, an SPR-based sensor for detection of thrombin was constructed by immobilizing an aptamer on the gold plasmon resonance surface. This instrument was capable of detecting thrombin in the range of 0.1–150 mM in human plasma [150–152].

SAW devices detect analyte on the basis of changes in the propagation velocity of acoustic waves. Any change in mass on the sensor surface leads to a reduction in resonance frequency or an alteration of the phase shift between the input and output signals. Schlensog et al. in 2004 designed a love-wave biosensor for label free real-time measurements of thrombin and HIV-1 Rev peptide by coupling aptamers to the surface of the biosensor [153]. This sensor was capable of detecting thrombin and HIV-1 Rev peptide at a detection limit of 72 pg/cm^2 and 77 pg/cm^2, respectively [154].

QCM uses piezoelectric quartz crystals, where the frequency of the quartz crystal is controlled by changes in the mass associated with the crystal. Therefore, any interaction of the target on the aptamer-immobilized crystals increases the mass of the transducer, resulting in a decrease in the resonance frequency of the crystal. QCM aptasensors have been used for the detection of thrombin and HIV-1 Tat protein by measuring the change in gravimetric resonance frequency [151, 155].

In microcantilever-based aptasensors, the aptamers are bound to the top surface of the microcantilever. Binding of aptamers to their targets leads to an increase in mass that forces the cantilever to bend. The difference in bending between a reference (random DNA) and sensor (aptamer) cantilever is used to measure binding events. Using these sensors Taq DNA polymerase and hepatitis C virus have been detected using DNA and RNA aptamers [156, 157].

4.7 APTAMERS COMMERCIALIZED OR EN ROUTE TO COMMERCIALIZATION

A vast number of aptamers have been reported to be successfully used for various diagnostic applications in literature. Some of them have now been translated for use in commercial diagnostic products/reagents. A few of them are listed in Table 4.2 [158].

4.8 CONCLUDING REMARKS

Aptamers as an emerging class of biorecognition element in biosensing have found a broad range of application in analytical as well as diagnostic platforms. This chapter is aimed at presenting a comprehensive general idea about the methodologies used for aptamer generation and development finally culminating in their use for biosensing purposes. Use of aptamers in biosensing has evolved from simple colorimetric assays to more specific and sensitive sensors. However, challenges still remain. First, the number of aptamers for practical use is still limited to only a handful. Increased automation of aptamer selection needs to be carefully planned with a focus on the final use of the aptamer. There is no single universal protocol for aptamer generation; therefore, careful consideration should be made of the conditions used for aptamer selection, such as buffer condition, temperature, pH, salt

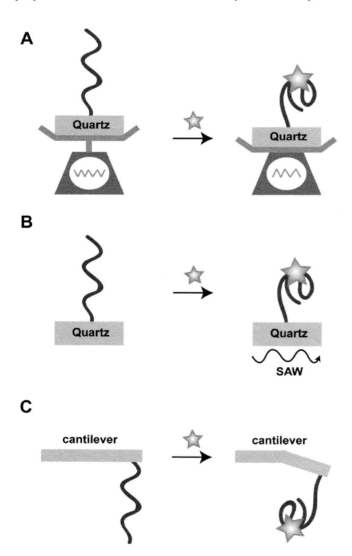

FIGURE 4.15 (A) QCM-based aptasensor. (B) SAW-based aptasensor. (C) Micromechanical cantilever-based aptasensor. (Adapted from Song et al. [162].)

TABLE 4.2
Aptamers in Some Selected Diagnostic Applications

Product	Company	Mode of Detection	Application	Limitation
OTA-Sense and AflaSense	NeoVentures Biotechnology, Inc.	Fluorescence-based assay	Detection of Mycotoxins (produced by *Aspergillus* and *Penicillium* species) in food	Require extraction of toxin from the sample
AptoCyto	AptSci, Inc.	Aptamer-based flow cytometry	Aptamer and magnetic bead-based pulldown of biomarker-positive cells and protein isolation	Dependence on expensive instrument (flow cytometer) to visualize the outcome
AptoPrep	AptSci, Inc.	Fluorescence-based assay and polyacrylamide gel electrophoresis (PAGE)	A highly efficient platform with multiplex capability for biomarkers discovery and diagnostics	4°C storage is recommended for long-term storage
SOMAscan	Somalogic	SOMAmer-based detection and quantification of biomarkers	Detection of foodborne pathogens	Multistep process
CibuxDx	CibusDx	Electrochemical sensing	Detection of active thrombin	Instrument-based technology
OLIGOBIND	Sekisui Diagnostics	Fluorogenic activity assay	Hot start PCR	Platelet contamination in plasma sample may interfere with assay
Hot Start *Taq* DNA polymerase	NEB	Aptamer-based reversible inhibition of *Taq* DNA polymerase		In comparison to normal polymerase this is relatively expensive

concentrations, competitors, and other SELEX parameters, to generate an aptamer of desired functionality. Aptasensors have been mainly developed for clinical applications; however, while exploring into a different arena like food or environmental contamination, it will take conscious effort to use the same assay conditions as would be found in real samples as opposed to spiked samples. Second, only aptamers that have been reported should be used, and special care should be taken to check and validate the sequences (e.g., if the aptamer is an RNA or DNA, importance of the primer binding sites in target binding, etc.). Fourth, aptamer-based assays are still at their infancy as compared to immunoassays due to a limited availability of aptamers as well as knowledge of their surface immobilization techniques. A rapid automation of the aptamer selection process, along with development of improved surface functionalization and transduction strategies, will be crucial for enhanced performance of aptasensors. Overall, aptamers, with their unique properties, will continue to be used as a biosensing element; however, it will be newer and innovative strategies with respect to aptamer selection and transducing mechanisms that will be crucial for the real-world applications of aptasensors.

REFERENCES

1. Robertson DL, Joyce GF. Selection in vitro of an RNA enzyme that specifically cleaves single-stranded DNA. Nature 1990; 344(6265):467–8.
2. Ellington AD, Szostak JW. In vitro selection of RNA molecules that bind specific ligands. Nature 1990; 346(6287):818–22.
3. Tuerk C, Gold L. Systematic evolution of ligands by exponential enrichment: RNA ligands to bacteriophage T4 DNA polymerase. Science 1990; 249(4968):505–10.
4. Pristoupil TI, Kramlova M. Microchromatographic separation of ribonucleic acids from proteins on nitrocellulose membranes. J Chromatogr 1968; 32:769–70.
5. Mendosa SD, Bowser MT. *In vitro* evolution of functional DNA using capillary electrophoresis. J Am Chem Soc 2004; 126:20–1.
6. Berezovski M, Krylov SN. Nonequilibrium capillary electrophoresis of equilibrium mixtures-a single experiment reveals equilibrium and kinetic parameters of protein-DNA interactions. J Am Chem Soc 2002; 124:13674–5.
7. Drabovich A, Berezovski M, Krylov SN. Selection of smart aptamers by equilibrium capillary electrophoresis of equilibrium mixtures (ECEEM). J Am Chem Soc 2005; 127:11224–5.
8. Okhonin V, Berezovski M, Krylov SN. Sweeping capillary electrophoresis: a non-stopped-flow method for measuring bimolecular rate constant of complex formation between protein and DNA. J Am Chem Soc 2004; 126:7166–7.
9. Jing M, Bowser MT. Isolation of DNA aptamers using micro free flow electrophoresis. Lab Chip 2011; 11:3703–9.
10. Song KM, Cho M, Jo H, Min K, Jeon SH, Kim T, Han MS, Ku JK, Ban C. Gold nanoparticle-based colorimetric detection of kanamycin using a DNA aptamer. Anal Biochem 2011; 415:175–81.
11. Vianini E, Palumbo M, Gatto B. *In vitro* selection of DNA aptamers that bind L-tyrosinamide. Bioorg Med Chem 2001; 9:2543–8.

12. Lévesque D, Beaudoin JD, Roy S, Perreault JP. In vitro selection and characterization of RNA aptamers binding thyroxine hormone. Biochem J 2007; 403:129–38.
13. Bruno JG. In vitro selection of DNA to chloroaromatics using magnetic microbead-based affinity separation and fluorescence detection. Biochem Biophys Res Commun 1997; 234:117–20.
14. Stoltenburg R, reinemann C, Strehlitz B. FluMag-SELEX as an advantageous method for DNA aptamer selection. Anal Bioanal Chem 2005; 383:83–91.
15. Stoltenburg R, Nikolaus N, Strehlitz B. Capture-SELEX: selection of DNA aptamers for aminoglycoside antibiotics. J Anal Methods Chem 2012; 2012:415697.
16. Morris KN, Jensen KB, Julin CM, Weil M, Gold L. High affinity ligands from in vitro selection: complex targets. Proc Natl Acad Sci USA 1998; 95(6):2902–7.
17. Homann M. Goringer HU. Combinatorial selection of high affinity RNA ligands to live African trypanosomes. Nucleic Acids Res 1999; 27:2006–14.
18. Tang Z, Shangguan D, Wang K, et al. Selection of aptamers for molecular recognition and characterization of cancer cells. Anal Chem 2007; 79:4900–7.
19. Sefah K, Tang Z, Shangguan D, et al. Molecular recognition of acute myeloid leukemia using aptamers. Leukemia 2009; 23:235–44.
20. Ninomiya K, Kaneda K, Kawashima S, Miyachi Y, Ogino C, Shimizu N. Cell-SELEX based selection and characterization of DNA aptamer recognizing human hepatocarcinoma. Bioorg Med Chem Lett 2013; 23:1797–802.
21. Kunii T, Ogura S, Mie M, Kobatake E. Selection of DNA aptamers recognizing small cell lung cancer using living cell-SELEX. Analyst 2011; 136:1310–2.
22. Jiménez E, Sefah K, López-Colón D, Van Simaeys D, Chen HW, Tockman MS, et al. Generation of lung adenocarcinoma DNA aptamers for cancer studies. PLoS One 2012; 7:e46222.
23. Wang Y, Luo Y, Bing T, Chen Z, Lu M, Zhang N, et al. DNA aptamer evolved by cell-SELEX for recognition of prostate cancer. PLoS One 2014; 9(6):e100243.
24. Li WM, Bing T, Wei JY, Chen ZZ, Shangguan DH, Fang J. Cell-SELEX-based selection of aptamers that recognize distinct targets on metastatic colorectal cancer cells. Biomaterials 2014; 35(25):6998–7007.
25. Guo KT, Schäfer R, Paul A, et al. Aptamer-based strategies for stem cell research. Mini Rev Med Chem. 2007; 7:701–5.
26. Iwagawa T, Ohuchi SP, Watanabe S, et al. Selection of RNA aptamers against mouse embryonic stem cells. Biochimie 2012; 94:250–7.
27. Dwivedi HP, Smiley RD, Jaykus LA. Selection of DNA aptamers for capture and detection of Salmonella typhimurium using a whole-cell SELEX approach in conjunction with cell sorting. Appl Microbiol Biotechnol 2013; 97:3677–86.
28. Kim JW, Kim EY, Kim SY, Byun SK, Lee D, Oh KJ. Isolation and characterization of DNA aptamers against Escherichia coli using a bacterial cell-systematic evolution of ligands by exponential enrichment approach Anal Biochem 2013; 436(1):22–8.
29. Nagarkatti R, Bist V, Sun S, Fortes de Araujo F, Nakhasi HL, Debrabant A. Development of an aptamer-based concentration method for the detection of Trypanosoma cruzi in blood. PLoS One 2012; 7:e43533.
30. Tang Z, Parekh P, Turner P, et al. Generating aptamers for recognition of virus-infected cells. Clin Chem. 2009; 55:813–22.
31. Mi J, Liu Y, Rabbani ZN, Yang Z, Urban JH, Sullenger BA, et al. In vivo selection of tumor targeting RNA motifs. Nat Chem Biol 2010; 6:22–4.
32. Cheng C, Chen YH, Lennox KA, Behlke MA, Davidson BL. In vivo SELEX for identification of brain-penetrating aptamers. Mol Ther Nucleic Acids 2013; 2:e67.
33. Cox JC, Ellington AD. Automated selection of anti-protein aptamers. Bioorg Med Chem 2001; 9(10):2525–31.
34. Hybarger G, Bynum J, Williams RF, Valdes JJ, Chambers J.P. A microfluidic SELEX prototype. Anal Bioanal Chem 2006; 384: 191–8.
35. Cox JC, Rajendran M, Riedel T, Davidson EA, Sooter LJ, Bayer TS, Schmitz-Brown M, Ellington AD. Automated acquisition of aptamer sequences. Comb Chem High Throughput Screen 2002a; 5: 289–99.
36. Cox JC, Hayhurst A, Hesselberth J, Bayer TS, Georgiou G, Ellington AD. Automated selection of aptamers against protein targets translated in vitro from gene to aptamer. Nucleic Acids Res 2002b; 30(20):e108.
37. Huang CJ, Lin HI, Shiesh SC, Lee, GB. Integrated microfluidic system for rapid screening of CRP aptamers utilizing systematic evolution of ligands by exponential enrichment (SELEX). Biosens Bioelectron 2010; 25:1761–6.
38. Huang CJ, Lin HI, Shiesh, SC, Lee, GB. An integrated microfluidic system for rapid screening of alpha-fetoprotein-specific aptamers. Biosens Bioelectron 2012; 35:50–5.
39. Lai HC, Wang CH, Liou TM, Lee GB. Influenza A virus-specific aptamers screened by using an integrated microfluidic system. Lab Chip 2014; 14(12):2002–13.
40. Oh SS, Qian J, Lou X, Zhang Y, Xiao Y, Soh HT. Generation of highly specific aptamers via micromagnetic selection. Anal Chem 2009; 81:5490–5.
41. Lou X, Qian J, Xiao Y, Viel L, Gerdon AE, Lagally ET, et al. Micromagnetic selection of aptamers in microfluidic channels. Proc Natl Acad Sci USA 2009; 106:2989–94.
42. Jolma A., Kivioja T, Toivonen J, Cheng L., Wei G, Enge M, Taipale M, Vaquerizas JM, Yan J, Sillanpaa MJ, et al. Multiplexed massively parallel SELEX for characterization of human transcription factor binding specificities. Genome Res 2010; 20:861–73.
43. Reiss DJ, Howard FM, Mobley HL. A novel approach for transcription factor analysis using SELEX with high-throughput sequencing (TFAST). PLoS One 2012; 7:e42761.
44. Alam KK, Chang JL, Burke DH. FASTAptamer: A bioinformatic toolkit for high-throughput sequence analysis of combinatorial selections. Mol Ther Nucleic Acids 2015; 4:e230.
45. Hoinka J, Berezhnoy A, Sauna ZE, Gilboa E, Przytycka TM. AptaCluster—a method to cluster HT-SELEX aptamer pools and lessons from its application. Res Comput Mol Biol 2014; 8394:115–28.
46. Nguyen QN, Bouvier C, Lelandais B, Ducongé F. PATTERNITY-seq v.1.0.: High-throughput analysis of sequence patterns and paternity relationship between them during molecular evolution processes. In Presented at Aptamers in Bordeaux, Bordeaux, France, 24–25 June 2016.
47. Blank, M. Next-generation analysis of deep sequencing data: bringing light into the black box of SELEX experiments. Methods Mol Biol 2016; 1380:85–95.
48. Dao P, Hoinka J, Takahashi M, Zhou J, Ho M, Wang Y, Costa F, Rossi JJ, Backofen R, Burnett J, et al. AptaTRACE elucidates RNA sequence-structure motifs from selection trends in HT-SELEX Experiments. Cell Syst 2016; 3:62–70.
49. Caroli J, Taccioli C, De La Fuente A, Serafini P, Bicciato S. APTANI: a computational tool to select aptamers through sequence-structure motif analysis of HT-SELEX data. Bioinformatics 2016; 32:161–4.
50. Jiang P, Meyer S, Hou Z, Propson NE, Soh HT, Thomson JA, Stewart R. MPBind: a meta-motif-based statistical framework

50. and pipeline to predict binding potential of SELEX-derived aptamers. Bioinformatics 2014; 30, 2665–67.
51. Pitt JN, Rajapakse I, Ferre-D'Amare AR. SEWAL: an open-source platform for next-generation sequence analysis and visualization. Nucleic Acids Res 2010; 38:7908–15.
52. Hoinka J, Berezhnoy A, Dao P, Sauna ZE, Gilboa E, Przytycka TM. Large scale analysis of the mutational landscape in HT-SELEX improves aptamer discovery. Nucleic Acids Res 2015; 43:5699–707.
53. Cowperthwaite MC, Ellington AD. Bioinformatic analysis of the contribution of primer sequences to aptamer structures. J Mol Evol 2008; 67(1):95–102.
54. Larkin MA, Blackshields G, Brown NP, Chenna R, McGettigan PA, McWilliam H, Valentin F, Wallace IM, Wilm A, Lopez R, Thompson JD, Gibson TJ, Higgins DG. Clustal W and Clustal X version 2.0. Bioinformatics 2007; 23:2947–48.
55. Sievers F, Wilm A, Dineen DG, Gibson TJ, Karplus K, Li W, Lopez R, McWilliam H, Remmert M, Söding J, Thompson JD, Higgins DG. Fast, scalable generation of high-quality protein multiple sequence alignments using Clustal Omega. Mol Syst Biol 2011; 7:539. doi: 10.1038/msb.2011.75
56. Bailey TL, Bodén M, Buske FA, Frith M, Grant CE, Clementi L, Ren J, Li WW, Noble WS. MEME SUITE: tools for motif discovery and searching. Nucleic Acids Res 2009; 37:W202–W208.
57. Zuker M. Mfold web server for nucleic acid folding and hybridization prediction. Nucleic Acids Res 2003; 31:3406–15.
58. Reuter JS and Mathews DH. RNA structure: software for RNA secondary structure prediction and analysis. BMC Bioinformatics 2010; 11:129.
59. Carey J, Cameron V, Dehaseth PL, Uhlenbeck OC. Sequence-specific interaction of R17 coat protein with its ribonucleic acid binding site. Biochemistry 1983; 22(11):2601–10.
60. Wong I, Lohman TM. A double-filter method for nitrocellulose-filter binding: application to protein-nucleic acid interactions. PNAS 1993; 90 (12):5428–32.
61. Fried M, Crothers DM. Equilibria and kinetics of lac repressor-operator interactions by polyacrylamide gel electrophoresis. Nucleic Acids Res 1981; 9:6505.
62. Garner MM, Revzin A. A gel electrophoresis method for quantifying the binding of proteins to specific DNA regions: application to components of the Escherichia coli lactose operon regulatory system. Nucleic Acids Res 1981; 9:3047.
63. Jing D, Agnew J, Patton WF, Hendrickson J, Beechem JM. A sensitive two-color electrophoretic mobility shift assay for detecting both nucleic acids and protein in gels. Proteomics 2003; 3:1172.
64. Jing D, Beechem JM, Patton WF. The utility of a two-color fluorescence electrophoretic mobility shift assay procedure for the analysis of DNA replication complexes. Electrophoresis 2004; 25:2439.
65. Shcherbakov D, Piendl W. A novel view of gel-shifts: analysis of RNA-protein complexes using a two-color fluorescence dye procedure. Electrophoresis 2007; 28:749.
66. Westermeier, R. Protein Purification. New York: John Wiley & sons. 1998.
67. Chen G, Kelly C, Chen H, Leahy A, Bouchier-Hayes D. Thermotolerance protects against endotoxin mediated microvascular injury. J Surg Res 2001; 95:79–84.
68. Mendonsa SD and Bowser MT. In vitro evolution of functional DNA using capillary electrophoresis. JACS 2004a; 126:20.
69. Mendonsa SD, Bowser MT. In vitro selection of high-affinity DNA ligands for human IgE using capillary electrophoresis. Anal Chem 2004b; 76:5387.
70. Mendonsa SD, Bowser MT. In vitro selection of aptamers with affinity for neuropeptide Y using capillary electrophoresis. JACS 2005; 127:9382.
71. Deng Q, German I, Buchanan D, Kennedy RT. Retention and separation of adenosine and analogues by affinity chromatography with an aptamer stationary phase. Anal Chem 2001; 73:5415.
72. Deng Q, Watson CJ, Kennedy RT. Aptamer affinity chromatography for rapid assay of adenosine in microdialysis samples collected in vivo. J Chromatogr A 2003; 1005:123.
73. Michaud M, Jourdan E, Villet A, Ravel A, Grosset C, Peyrin E. A DNA aptamer as a new target-specific chiral selector for HPLC. JACS 2003; 125:8672.
74. Kotia RB, Li LJ, Mcgown LB. Separation of nontarget compounds by DNA aptamers. Anal Chem 2000; 72:827.
75. Connor AC, McGown LB. Aptamer stationary phase for protein capture in affinity capillary chromatography. J Chromatogr A 2006; 1111:115.
76. Clark SL, Remcho VT. Electrochromatographic retention studies on a flavin-binding RNA aptamer sorbent. Anal Chem 2003; 75:5692.
77. Zhao Q, Li XF, Le XC. Aptamer-modified monolithic capillary chromatography for protein separation and detection. Anal Chem 2008a; 80:3915.
78. Zhao Q, Li XF, Shao YH, Le XC. Aptamer-based affinity chromatographic assays for thrombin. Anal Chem 2008b; 80:7586.
79. Rupcich N, Nutiu R, Li Y, Brennan JD. Solid-phase enzyme activity assay utilizing an entrapped fluorescence-signaling DNA aptamer. Angew Chem Int Ed 2006; 45:3295.
80. Gokulrangan G, Unruh JR, Holub DF, Ingram B, Johnson CK, Wilson GS. DNA aptamer-based bioanalysis of IgE by fluorescence anisotropy. Anal Chem 2005; 77:1963.
81. Nutiu R, Li Y. Aptamers with fluorescence-signaling properties. Methods 2005; 37(1):16–25.
82. Klymchenko AS and Mely Y. Fluorescent environment-sensitive dyes as reporters of biomolecular interactions. ProgMolBiolTranslSci 2013; 113:35–58.
83. Cho EJ, Rajendran M, Ellington AD. Aptamers as Emerging Probes for Macromolecular Sensing. In: Geddes CD, Lakowicz JR (Eds.), Advanced Concepts in Fluorescence Sensing. Topics in Fluorescence Spectroscopy, vol 10. Springer, Boston, MA, 2005.
84. Basu S. Ultraviolet absorption studies on DNA. Biopolymers 1967; 5:876.
85. Donovan JW. Physical principles and techniques of protein chemistry. Phys Principles Tech Protein Chem 1969; 101.
86. del Toro M, Gargallo R, Eritja R, Jaumot J. Study of the interaction between the G-quadruplex-forming thrombin-binding aptamer and the porphyrin 5,10,15,20-tetrakis-(N-methyl-4-pyridyl)-21,23H-porphyrin tetratosylate. Anal Biochem 2008; 379:8.
87. Moss T, Leblanc B. (Eds.). DNA-Protein Interactions: Principles and Protocols, third edition, vol. 543. Methods in Molecular Biology, Totowa, NJ, 2009.
88. Johnson WC Jr. Circular dichroism and its empirical application to biopolymers. Methods Biochem Anal 1985; 31:61.
89. Johnson WC Jr. Secondary structure of proteins through circular dichroism spectroscopy. Ann Rev Biophy Biophy Chem 1988; 17:145.
90. Johnson BB, Dahl KS, Tinoco I Jr, Ivanov VI, Zhurkin VB. Correlations between deoxyribonucleic acid structural parameters and calculated circular dichroism spectra. Biochemistry 1981; 20:73.

91. Gray DM, Hung S-H, Johnson KH. Absorption and circular dichroism spectroscopy of nucleic acid duplexes and triplexes. Methods Enzymol 1995; 246:19.
92. Kakoti A, Goswami P. Multifaceted analyses of the interactions between human heart type fatty acid binding protein and its specific aptamers. BBA General Subject 2017; 1861:3289–99.
93. Quang NN, Miodek A, Cibiel A, Duconge F. Selection of aptamers against whole living cells: from cell-selex to identification of biomarkers. Methods Mol Biol 2017; 1575: 253–72.
94. Shangguan D, Li Y, Tang Z, Cao ZC, Chen HW, Mallikaratchy P, Sefah K, Yang CJ, Tan W. Aptamers evolved from live cells as effective molecular probes for cancer study. Proc Natl Acad Sci USA 2006; 103:11838–43.
95. Tolle F, Brändle GM, Matzner D, Mayer G. A versatile approach towards nucleobase-modified aptamers. Angew Chem Int Ed 2015; 54:10971–74.
96. Fagerstam LG, Frostell-Karlsson A, Karlsson R, Persson B, Ronnberg I. Biospecific interaction analysis using surface plasmon resonance detection applied to kinetic, binding site and concentration analysis. J Chromatogr 1992; 597:397.
97. Jonsson U, Fagerstam L, Ivarsson B, Johnsson B, Karlsson R, Lundh K, Lofas S, Persson B, Roos H, Ronnberg I, et al. Real-time biospecific interaction analysis using surface plasmon resonance and a sensor chip technology. Biotechniques 1991; 11:620.
98. Di Primo C, Lebars I. Determination of refractive index increment ratios for protein-nucleic acid complexes by surface plasmon resonance. Anal Biochem 2007; 368:148.
99. Balamurugan S, Obubuafo A, Soper SA, McCarley RL, Spivak DA. Effect of linker structure on surface density of aptamer monolayers and their corresponding protein binding efficiency. Langmuir 2006; 22:6446.
100. Tang Q, Su X, Loh KP. Surface plasmon resonance spectroscopy study of interfacial binding of thrombin to antithrombin DNA aptamers. J Colloid Interface Sci 2007; 315:99.
101. Freire E, Mayorga OL, Straume M. Isothermal titration calorimetry. Anal Chem 1990; 62:950A.
102. Doyle ML. Characterization of binding interactions by isothermal titration calorimetry. Curr Opin Biotechnol 1997; 8:31.
103. Holdgate GA. Making cool drugs hot: isothermal titration calorimetry as a tool to study binding energetics. Biotechniques 2001; 31:164.
104. Mueller M, Weigand JE, Weichenrieder O, Suess B. Thermodynamic characterization of an engineered tetracycline-binding riboswitch. Nucleic Acids Res 2006; 34:2607.
105. Jerabek-Willemsen M, Wienken CJ, Braun D, Baaske P, Duhr S. Molecular interaction studies using microscale thermophoresis. Assay Drug Dev Tech 2011; 9(4):342–53.
106. Jerabek-Willemsen M, Andre T, Wanner R, MarieRoth H, Duhr S, Baaske P. Microscale thermophoresis: interaction analysis and beyond. J Mol Struct 2014; 1077:101–13.
107. Chang AL, McKeague M, Liang JC, Smolke CD. Kinetic and equilibrium binding characterization of aptamers to small molecules using a label-free, sensitive, and scalable platform. Anal Chem 2014; 86:3273–78.
108. Gao S, Zheng X, Hu B, Sun M, Wu J, Jiao B, Wang L. Enzyme-linked, aptamer-based competitive biolayer interferometry biosensor for polytoxin. Biosens Bioelectron 2016; 89(2):952–8.
109. Hermann T, Patel DJ. Adaptive recognition by nucleic acid aptamers. Science 2000; 287:820–25.
110. Patel DJ, Suri AK. Structure, recognition and discrimination in RNA aptamer complexes with cofactors, amino acids, drugs and aminoglycoside antibiotics. Rev Mol Biotechnol 2000; 74:39–60.
111. Katilius E, KatilieneZ, Woodbury NW. Signaling aptamers created using fluorescent nucleotide analogs. Anal Chem 2006; 78:6484–89.
112. Stojanovic MN, de Prada P, Landry DW. Aptamer-based folding fluorescent sensor for cocaine. J Am Chem Soc 2001; 123:4928–31.
113. Fang X, Sen A, Vicens M, Tan W. Synthetic DNA aptamers to detect protein molecular variants in a high-throughput fluorescence quenching assay. Chem Bio Chem 2003; 4:829–34.
114. Yamamoto R, Baba T, Kumar PK. Molecular beacon aptamer fluoresces in the presence of tat protein of HIV-1. Genes Cells 2000; 5(5):389–96.
115. Stojanovic MN, de Prada P, Landry DW. Fluorescent sensors based on aptamer self-assembly. J Am Chem Soc 2000; 122:11547–48.
116. Nutiu R, Li Y. Structure-switching signaling aptamers. J Am Chem Soc 2003; 125:4771–78.
117. Nutiu R, Li Y. Structure-switching signaling aptamers: transducing molecular recognition into fluorescence signaling. Chemistry 2004; 10(8):1868–76.
118. Elowe NH, Nutiu R, Allali-Hassani A, Cechetto JD, Hughes DW, et al. Small-molecule screening made simple for a difficult target with a signaling nucleic acid aptamer that reports on deaminase activity. Angew Chem Int Ed 2006; 45:5648–52.
119. Li N, Ho CM. Aptamer-based optical probes with separated molecular recognition and signal transduction modules. J Am Chem Soc 2008; 130:2380–81.
120. Li W, Yang X, Wang K, Tan W, Li H, Ma C. FRET-based aptamer probe for rapid angiogenin detection. Talanta 2008; 75:770–74.
121. Yang CJ, Jockusch S, Vicens M, Turro NJ, Tan W. Light-switching excimer probes for rapid protein monitoring in complex biological fluids. Proc Natl Acad Sci USA 2005; 102(48):17278–83.
122. Chu TC, Shieh F, Lavery LA, Levy M, Richards-Kortum R, et al. Labeling tumor cells with fluorescent nanocrystal-aptamer bioconjugates. Biosens Bioelectron 2006; 21:1859–66.
123. Ikanovic M, Rudzinski WE, Bruno JG, Allman A, Carrillo MP, et al. Fluorescence assay based on aptamer–quantum dot binding to Bacillus thuringiensis spores. J Fluoresc 2007; 17:193–99.
124. Choi JH, Chen KH, Strano MS. Aptamer-capped nanocrystal quantum dots: a new method for label-free protein detection. J Am Chem Soc 2006; 128:15584–85.
125. Wang L, Liu X, Hu X, Song S, Fan C. Unmodified gold nanoparticles as a colorimetric probe for potassium DNA aptamers. Chem Commun (Cambridge) 2006; 3780–82.
126. Wei H, Li B, Li J, Wang E, Dong S. Simple and sensitive aptamer-based colorimetric sensing of protein using unmodified gold nanoparticle probes. Chem Commun (Cambridge) 2007; 3735–37.
127. Liu J, Lu Y. Fast colorimetric sensing of adenosine and cocaine based on a general sensor design involving aptamers and nanoparticles. Angew Chem Int Ed Engl 2005; 45(1):90–94.
128. Liu J, Lu Y. Preparation of aptamer-linked gold nanoparticle purple aggregates for colorimetric sensing of analytes. Nat Protoc 2006; 1(1):246–52.
129. Pavlov V, Xiao Y, Shlyahovsky B, Willner I. Aptamer-functionalized Au nanoparticles for the amplified optical detection of thrombin. J Am Chem Soc 2004; 126:11768–69.

130. Li YY, Zhang C, Li BS, Zhao LF, Li XB, Yang WJ, Xu SQ. Ultrasensitive densitometry detection of cytokines with nanoparticle-modified aptamers. Clin Chem 2007; 53(6):1061–66.
131. Qi YY, Li BX. A sensitive, label-free, aptamer-based biosensor using a gold nanoparticle-initiated chemiluminescence system. Chem–Eur J 2010; 17:1642–8.
132. Li T, Wang E, Dong S. Lead(II)-induced allosteric G-quadruplex DNAzyme as a colorimetric and chemiluminescence sensor for highly sensitive and selective Pb2+ detection. Anal Chem 2010; 82:1515–20.
133. Yan XL, Cao ZJ, Kai M, Lu JZ. Label-free aptamer-based chemiluminescence detection of adenosine. Talanta 2009; 79:383–7.
134. Yan XL, Cao ZJ, Lau CW, Lu JZ. DNA aptamer folding on magnetic beads for sequential detection of adenosine and cocaine by substrate-resolved chemiluminescence technology. Analyst 2010; 135:2400–7.
135. Ahn DG, Jeon IJ, Kim JD, Song MS, Han SR, Lee SW, Jung H, Oh JW. RNA aptamer-based sensitive detection of SARS coronavirus nucleocapsid protein. Analyst 2009; 134:1896–1901.
136. Zhang SS, Yan YM, Bi S. Design of molecular beacons as signaling probes for adenosine triphosphate detection in cancer cells based on chemiluminescence resonance energy transfer. Anal Chem 2009; 81:8695–701.
137. Xiao Y, Lubin AA, Heeger AJ, Plaxco KW. Label-free electronic detection of thrombin in blood serum by using an aptamer-based sensor. Angew ChemInt Ed Engl 2005; 44(34):5456–9.
138. Radi AE, Acero Sanchez JL, Baldrich E, OSullivan CK. Reagentless, reusable, ultrasensitive electrochemical molecular beacon aptasensor. J Am Chem Soc 2006; 128(1):117–24.
139. Radi AE, OSullivan CK. Aptamer conformational switch as sensitive electrochemical biosensor for potassium ion recognition. Chem Commun (Cambridge) 2006; (32):3432–4.
140. Mir M, Katakis I. Aptamers as elements of bioelectronic devices. Mol Biosyst 2007; 3(9):620–2.
141. Bang GS, Cho S, Kim BG. A novel electrochemical detection method for aptamer biosensors. Biosens Bioelectron 2005; 21(6):863–70.
142. Le Floch F, Ho HA, Leclerc M. Label-free electrochemical detection of protein based on a ferrocene-bearing cationic polythiophene and aptamer. Anal Chem 2006; 78(13):4727–31.
143. Cheng AK, Ge B, Yu HZ. Aptamer-based biosensors for label-free voltammetric detection of lysozyme. Anal Chem 2007; 79(14):5158–64.
144. Ikebukuro K, Kiyohara C, Sode K. Novel electrochemical sensor system for protein using the aptamers in sandwich manner. Biosens Bioelectron 2005; 20(10):2168–72.
145. Polsky R, Gill R, Kaganovsky L, Willner I. Nucleic acid-functionalized Pt nanoparticles: catalytic labels for the amplified electrochemical detection of biomolecules. Anal Chem 2006; 78(7):2268–71.
146. He P, Shen L, Cao Y, Li D. Ultrasensitive electrochemical detection of proteins by amplification of aptamer-nanoparticle bio bar codes. Anal Chem 2007; 79(21):8024–9.
147. Rodriguez MC, Kawde AN, Wang J. Aptamer biosensor for label-free impedance spectroscopy detection of proteins based on recognition-induced switching of the surface charge. ChemCommun 2005; 4267–9.
148. So HM, Won K, Kim YH, Kim BK, Ryu BH, Na PS, Kim H, Lee JO. Single-walled carbon nanotube biosensors using aptamers as molecular recognition elements. J Am Chem Soc 2005; 127(34):11906–7.
149. Maehashi K, Katsura T, Kerman K, Takamura Y, Matsumoto K, Tamiya E. Label-free protein biosensor based on aptamer-modified carbon nanotube field-effect transistors. Anal Chem 2007; 79(2):782–7.
150. Luzi E, Minunni M, Tombelli S, Mascini M. New trends in affinity sensing: aptamers for ligand binding. Trends Anal Chem 2003; 22(11):810–18.
151. Tombelli S, Minunni M, Luzi E, Mascini M. Aptamer-based biosensors for the detection of HIV-1 Tat protein. Bioelectrochemistry 2005; 67(2):135–41.
152. Li Y, Lee HJ, Corn RM. Detection of protein biomarkers using RNA aptamer microarrays and enzymatically amplified surface plasmon resonance imaging. Anal Chem 2007; 79(3):1082–8.
153. Schlensog MD, Gronewold TMA, Tewes M, Famulok M, Quandt E. A love-wave biosensor using nucleic acids as ligands. Sens Actuators B 2004; 101(3):308–15.
154. Gronewold TM, Glass S, Quandt E, Famulok M. Monitoring complex formation in the blood-coagulation cascade using aptamer-coated SAW sensors. Biosens Bioelectron 2005; 20(10):2044–52.
155. Hianik T, Ostatna V, Zajacova Z, Stoikova E, Evtugyn G. Detection of aptamer-protein interactions using QCM and electrochemical indicator methods. Bioorg Med Chem Lett 2005; 15(2):291–5.
156. Savran CA, Knudsen SM, Ellington AD, Manalis SR. Micromechanical detection of proteins using aptamer-based receptor molecules. Anal Chem 2004; 76(11):3194–8.
157. Hwang KS, Lee SM, Eom K, Lee JH, Lee YS, Park JH, Yoon DS, Kim TS. Nanomechanical microcantilevers operated in vibration modes with use of RNA aptamer as receptor molecules for label-free detection of HCV helicase. Biosens Bioelectron 2007; 23(4):459–65.
158. Kaur H, Bruno JG, Kumar A, Sharma TK. Aptamers in the therapeutics and diagnostics pipelines. Theranostics 2018;8(15):4016–32.
159. Yang J, Bowser MT. Capillary electrophoresis–SELEX selection of catalytic DNA aptamers for a small-molecule porphyrin target. Anal Chem 2013; 85(3):1525–30.
160. Quang NN, Perret G, Ducongé F. Applications of high-throughput sequencing for in vitro selection and characterization of aptamers. Pharmaceuticals 2016; 9:76. doi: 10.3390/ph9040076.
161. Cho EJ, Lee JW, Ellington AD. Applications of aptamers as sensors. Ann Rev Anal Chem 2009; 2:241–64.
162. Song S, Wang L, Li J, Zhao J, Fan C. Aptamer-based biosensors. Trends Anal Chem 2008; 27(2):108–17.

5 Metal Nanoparticles for Analytical Applications

Priyamvada Jain
Indian Institute of Technology Madras, Chennai, India

CONTENTS

5.1 Introduction 91
5.2 Optical Properties of Metallic Nanoparticles 91
    5.2.1 Studying Light-Matter Interactions 91
    5.2.2 Nanoparticle Geometry and LSPR 93
5.3 Metallic Nanoparticles 95
    5.3.1 Gold Nanoparticles 95
    5.3.2 Silver Nanoparticles 95
        5.3.2.1 Physical Methods 97
        5.3.2.2 Chemical Methods 98
        5.3.2.3 Biological Methods 98
5.4 LSPR-Based Sensors 98
    5.4.1 Nucleic Acid Detection Assays 98
    5.4.2 Immunoassays 100
    5.4.3 Cellular Detection Assays 101
    5.4.4 Other Biomolecular Interactions 101
5.5 Conclusion 102
Acknowledgment 103
References 103

5.1 INTRODUCTION

Metal nanoparticles, especially gold and silver, have been vigorously studied for their characteristic optoelectronic and thermal properties. The unique properties of these nanoparticles are attributed to localized surface plasmon resonance (LSPR), a nanoscale phenomenon responsible for sharp spectral absorption and scattering peaks, as well as strong electromagnetic near-field enhancements [1]. Studies on LSPR revealed a strong correlation between this phenomenon and the size, shape, environment, and surface chemistry of the nanoparticles. This knowledge led to significant improvements in the fabrication of nanomaterials, their bioconjugation, and application over the past decade. In this chapter, we discuss the fundamentals and physical basis of LSPR and its dependence on nanoparticle geometry. We then introduce the two most applicable nanoparticles in biomedicine (i.e., gold and silver nanoparticles), their properties, and synthesis approaches. Finally, we highlight the recent elegant examples in signal amplification strategies using biofunctional nanomaterials, constructed by conjugation of gold or silver nanoparticles with biorecognition elements.

5.2 OPTICAL PROPERTIES OF METALLIC NANOPARTICLES

5.2.1 Studying Light-Matter Interactions

In the mid-1850s, Michael Faraday spent most of his time exploring light-matter interactions by an "optical mode of investigation." He prepared hundreds of thin gold leaf slides and observed them under light. To make the gold leaf thin enough, he had to apply chemical means, rather than mechanical. One of the steps of this chemical treatment involved washing of the gold leaves. Faraday observed that the by-product of the washing was a ruby fluid, which darkened and changed color to a blue or gray. He reported that the incident light shone through the fluid led to a form of "opalescent" scattering (now known as the Faraday-Tyndall effect) (Figure 5.1), correctly theorizing that the ruby solutions were in fact suspensions of metallic gold particles, too small to be visible to the naked eye. Michael Faraday is therefore regarded as the discoverer of metallic gold "colloids" (a term coined by Thomas Graham).

The fascinating optical properties of gold nanoparticles originally observed by Faraday are a result of the interaction of the electromagnetic light wave and the conduction electrons in

FIGURE 5.1 (a) One of Faraday's gold colloid solutions preserved at the Royal Institution of Great Britain, London. (b) Scattering of light by Faraday's gold colloid, when exposed to a red laser demonstrating the Faraday-Tyndall effect [2]

the metal. Plasmonics is the study of these light-matter interactions, wherein materials with a negative real and small positive imaginary dielectric constant support coherent oscillations of their surface conduction electrons under light irradiation, a phenomenon known as surface plasmon resonance. When the interaction of light occurs with a particle of size comparable to that of the wavelength of light (e.g., a nanoparticle), a collective oscillation of the free electrons of the nanoparticle results in a localized surface plasmon. Figure 5.2 illustrates the difference between a propagating and localized surface plasmon.

In the early 20th century, Gustav Mie developed a solution to Maxwell's equations, which describes the absorption and scattering of light by spherical particles. On light irradiation of a spherical nanoparticle (such that $R/\lambda < 0.1$, where R is the radius of the nanoparticle and λ is the wavelength of incident light), the apparently static electromagnetic field outside the nanoparticle (E_{out}) is given by

$$E_{out}(x,y,z) = E_0 \hat{z} - \left[\frac{\varepsilon_{in} - \varepsilon_{out}}{(\varepsilon_{in} + 2\varepsilon_{out})}\right] R^3 E_0 \left[\frac{\hat{z}}{r^3} - \frac{3z(x\hat{x} + y\hat{y} + z\hat{z})}{r^5}\right] \quad (5.1)$$

where r is the radius of the nanoparticle, which is irradiated by z-polarized light, ε_{in} and ε_{out} are the dielectric constants of the metal nanoparticle and surrounding environment, respectively. For resonance condition to be met, ε_{in} should be equal to $-2\varepsilon_{out}$. It is of merit to note that this relation holds valid only for dilute colloidal solutions in nonabsorbing media. If we were to consider concentrated nanoparticle solutions, where particle-particle interactions come in to play, effective medium theories such as those developed by Maxwell Garnett [4] and Bruggemann [5] should be applied.

For metals such as Pb, Hg, In, Sn, and Cd, the plasmon band lies in the UV region of the EM spectrum; as a result, nanoparticles of these metals do not display strong color effects. In contrast, gold and silver nanoparticles, due to d-d band transition, exhibit this resonant condition in the visible region of the light spectrum, making them versatile optical probes for various analytical applications. The extinction spectrum of the metal sphere $E(\lambda)$ is given by the equation

$$E(\lambda) = \frac{24\pi^2 N R^3 \varepsilon_{out}^{3/2}}{\lambda \ln(10)} \left[\frac{\varepsilon_i(\lambda)}{\varepsilon_i(\lambda) + \chi \varepsilon_{out}^2 + \varepsilon_i(\lambda)^2}\right] \quad (5.2)$$

where ε_r and ε_i are the real and imaginary components of metal dielectric function, respectively. This equation emphasizes the wavelength dependence of the metal dielectric function. Moreover, the factor χ, which appears next to ε_{out}, is dependent on particle geometry, varying between 2 and 20 for perfect spheres to particles with higher aspect ratios. Analytical estimations of χ for unconventional particle geometries are possible only by methods such as the discrete dipole approximation (DDA) and finite-difference time-domain methods. The DDA is a useful numerical method for investigating the effect of shape on LSPR. Here, a particle is considered as a cubic array of N polarizable points (each at $\vec{r_i}$, $i = 1, ..., N$), with each point representing the polarizability of a discrete volume of material. On light irradiation, each point develops a dipole moment $\vec{P_i}$ that can be related to the local electric field $\vec{E_i}$ through a quantity called polarizability α_i, through the relation $\vec{P_i} = \alpha_i E_i$. The local field $\vec{E_i}$ in turn comprises two parts: the incident field E_{0i} and the secondary radiation field $\vec{E_i}$ (which arises from all other dipole moments). Estimation on the discrete dipole for each point ($\vec{P_i}$) is thus an iterative procedure. After solving for ($\vec{P_i}$), the LSPR of a given nanoparticle can be estimated as a function of wavelength [6].

LSPR has been applied as an excellent tool to monitor changes in local dielectric properties that may occur, for instance, by interaction or absorption of an analyte to the sensor surface. The change in LSPR wavelength maxima upon change in the dielectric environment is given by

Scheme

Propagating surface plasmon	Localized surface plasmon
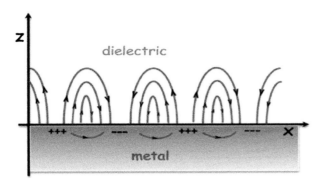	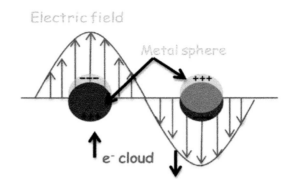

Salient features

Plasmons propagate laterally the metal-dielectric interface for hundreds of microns and decay along z-axis upto a decay length of 1/e	Localized oscillation of the electron cloud around the nanoparticle with a frequency known as the LSPR
Changes in local dielectric environment reflected as a change in SPR condition, measurable as (a) angle resolved (b) wavelength shift (c) imaging	Sensitive to changes in local dielectric environment measurable as (a) wavelength shift (b) angle resolved
Real–time kinetic data for binding reactions Yes	Yes
Sensitivity for bulk RI change High	Lower than SPR
Sensitivity for local RI change Comparable	Comparable

FIGURE 5.2 Scheme highlighting difference between propagating and localized surface plasmon. Part of the scheme reproduced from [3] with permission from the Royal Society of Chemistry.

$$\Delta\lambda_{max} = m\Delta n[1 - \exp(-2d / l_d)] \quad (5.3)$$

where m is the bulk refractive index response of the nanoparticle, Δn is the change in the local dielectric environment brought about by the analyte, d is the thickness of the analyte adsorbed, and l_d is the exponentially decaying electromagnetic field length.

5.2.2 Nanoparticle Geometry and LSPR

The frequency and intensity of the plasmon band depend upon three important properties of a system: (i) the intrinsic dielectric property of a metal, (ii) the dielectric constant of the surrounding medium, and (iii) the pattern of surface polarization [6]. By making changes to the nanoparticle morphology, composition, size, and interparticle distance, we can tailor their optical properties to meet the application needs. Interestingly, deviations from a spherical geometry of a nanoparticle result in more drastic changes in the SPR band in comparison to an increase in nanoparticle size. For instance, it was reported that an increase in size of spherical particle from 10 to 100 nm led to a shift of 47 nm in the SPR band; whereas, a change in aspect ratio from 2.5 to 3.5 for prolate ellipsoids (a considerably smaller change in overall size) led to a shift of 92 nm in the longitudinal SPR band [7].

To understand the variation of optical properties with the shape of nanoparticle, the LSPR of different Ag nanoparticles were studied (Figure 5.3) [8]. The extinction, absorption, and scattering spectra (extinction = absorption + scattering) of the Ag nanosphere were obtained by applying Mie theory, while the DDA method was used for other anisotropic shapes. From Figure 5.3a it is evident that Ag nanosphere (40 nm) primarily absorbs blue light. This gives Ag colloids their signature yellow color. Two resonant peaks, that is, a strong peak at 410 nm and a shoulder peak at 370 nm are observed. In the presence of electromagnetic (EM) waves, one side of the sphere becomes positively charged, developing a negative charge on the opposite side. This gives the nanosphere an individual particle dipole, which changes sign with the frequency of incident light. The charge separation (i.e., surface polarization) acts as the "restoring force" for electron oscillation and therefore determines the frequency and intensity of LSPR [9]. It is due to this reason that factors influencing the surface polarization, for instance, particle size, shape, and dielectric properties, also have a profound effect on the LSPR. Figure 5.3b shows

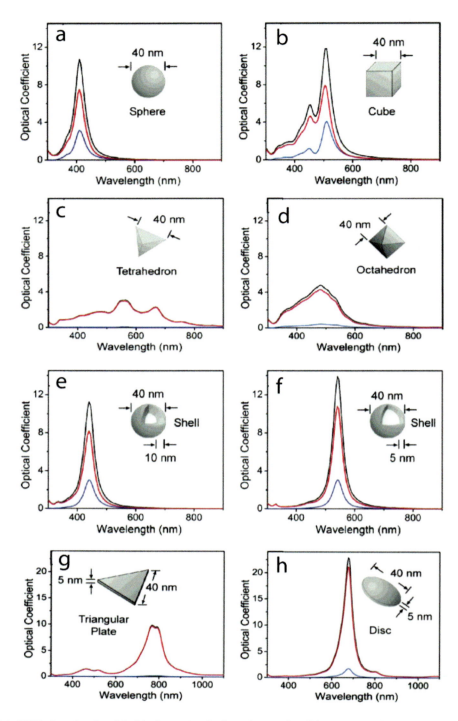

FIGURE 5.3 Calculated UV-vis extinction (black), absorption (red), and scattering (blue) spectra of silver nanostructures: isotropic nanosphere (a), anisotropic cube (b), tetrahedra (c), and octahedra (d), hollow shell (e), hollow shell with thin walls (f), 2D structures: triangular nanoplate (g), circular disc (h). Reprinted with permission from [8], copyright American Chemical Society.

the DDA calculated spectra for a 40-nm cube. Due to the presence of a larger number of symmetries for dipole resonance and sharp corners, the spectra exhibits more peaks [10], with the primary peak red shifted [11], as compared to the sphere. For shapes with sharp corners, charge accumulation at the corners leads to increase in charge separation, lowering the restoring force and hence a red shift of the resonance peak [12, 13]. The spectra for a tetrahedron and octahedron can be similarly explained (Figure 5.3c and d). The tetrahedron peak shows the maximum red shift, as its edges are the sharpest among the three shapes (i.e., cube, tetrahedron, octahedron).

For modified spherical shapes such as a hollow sphere or shell (Figure 5.3e), the resonance peak is red shifted compared to that of a solid sphere. Interestingly, in the case of shells, surface charges are induced on both the inner and outer shell, such that they are of the same sign for a particular pole of the hollow shell [14, 15], increasing the charge separation, and thus lowering the frequency of oscillation. A further red shift is seen as the thickness of shell reduces (Figure 5.3f), due to charge coupling between the charges on the inner and outer shells. DDA calculations on 2D anisotropic shapes also reveal similar concepts. For a triangular nanoplate (Figure 5.3g) snipping of the sharp corners led to a blue shift, again underlining the fact that sharp corners in a shape led to a red-shifted resonance profile [16]. Similarly, the resonance peak of a circular disc (Figure 5.3h) shows a blue shift compared to the triangular nanoplate, as the former lacks sharp corners. Additionally, the stronger light absorption and scattering by the circular disc as compared to the triangular nanoplate is a result of the larger effective dipole moment due to the circular symmetry. Therefore, as a rule of thumb, sharp corners lead to a larger charge separation and thus red-shifted resonance peaks, while symmetric nanostructures contribute to an increased intensity of the resonance peak.

5.3 METALLIC NANOPARTICLES

Noble metal nanoparticles, particularly gold and silver nanoparticles, have been extensively utilized in biomedical and many other analytical applications in the past decade. Their properties markedly differ from that of bulk metals due to the excitation of localized surface plasmons. The optoelectronic properties of the metallic nanoparticles can be tailored to different applications by changing their size, shape, or dielectric properties. The LSPR property of NPs is accompanied by distance-dependent effects that also find applications in surface enhanced spectroscopic techniques such as surface enhanced Raman spectroscopy (SERS), surface enhanced infrared absorption (SEIRA), and surface enhanced fluorescence (SEF), which have been applied for the detection of target molecules at low concentrations. Some recent reviews expertly deal with many aspects of the most popular metallic nanoparticles (i.e., gold, and silver) and may be referred to by the reader [17, 18]. Here, we intend to summarize the most salient properties of these indispensable nanomaterials for the basic understanding of the reader.

5.3.1 Gold Nanoparticles

Gold nanoparticles (AuNP), widely used in sensors, Raman scattering and electronic devices, catalysis, display devices, etc., have been the focus of much interest, both from fundamental and application points of view. Several methods for the successful synthesis of AuNPs are known, such as chemical reduction of metal salts, photolysis or radiolysis of metal salts, ultrasonic reduction of metal salts, and displacement of ligands from organometallic compounds [19]. By far the most popular route for AuNP synthesis is the chemical reduction of tetrachloroauric acid ($HAuCl_4$) using reducing agents such as sodium or potassium borohydride [20], hydrazine [21], ascorbic acid [22], nitric/hydrochloric acid [20], and dimethyl formamide [23]. Addition of a reducing agent leads to the reduction of metal ions in solution and the subsequent nucleation of the metal solid. To prepare stable nanoparticles that do not aggregate or precipitate over time, it is a common strategy to functionalize the particles with a capping agent. Cysteine [24], ionic, and nonionic surfactants [25], CS_2 [26], sodium citrate [27], nitrilotriacetate [28], 3-aminopropyltrimethoxysilane [29], 2-mercaptobenzimidazole [30], and dendrimers [31] are some capping agents that have been used in the past to prepare stable colloidal dispersions.

The distinct optical and photothermal properties of AuNPs can be fine-tuned by fabricating nanoparticles with defined morphologies and symmetries. For instance, in addition to solid spheres, hollow spheres, shells, rods, stars, cages, and clusters have been prepared (Table 5.1). Hollow or solid nanoshells act as excellent SERS substrates at the single particle level, and the synthesis process allows higher control over monodispersity from the visible to IR region, giving particles that are well suited for biological studies [34, 37].

Nanorods possess two LSPR peaks, originating from light absorption transversely (along the short axis) and longitudinally (along the long axis). Changes in rod aspect ratio lead to a change in plasmon frequency, which opens up opportunities for tailoring SPR peaks spanning a wide range of wavelengths [45]. The activity in IR range and sensitivity of the plasmon absorbance to changes in refractive index of the surroundings makes nanorods ideal for biological sensing applications [46]. Gold nanostars exhibit enhanced field strength in the vicinity of their sharp tips, which are also tunable in the NIR region, making them promising candidates for SERS, imaging, nanomedicine, etc. [40]. Like many other anisotropic particles, gold nanocages also possess tunable LSPR peaks in the NIR region. Additionally, their hollow interior and porous walls with adjustable thickness make them suitable for optical tomography, drug delivery allowing variations in drug loading efficiency, and delivery across walls of epithelial tissues [41, 47]. Luminescent gold nanoclusters with high water solubility, biocompatibility, and nontoxicity find use as *in vitro* and *in vivo* imaging probes and sensors [43].

5.3.2 Silver Nanoparticles

Silver, with its highest electrical (6.3×10^7 S/m) conductivity among all metals and a thermal conductivity of 429 W/m.K, is particularly interesting as a nanomaterial [48, 49]. With the advent of nanotechnology, silver nanoparticles (AgNPs) have become one of the most widely used materials in consumer products, being applied in the fields of medical devices, pharmacology, engineering, energy, electronics, environment, etc. [50]. Additionally, their antibacterial properties have led them to be used in the food industry, paints, cosmetics, wound dressing, women's hygiene products, contraceptives, medical textiles, sunscreen, healthcare products, etc. [51]. Over the years concerns regarding the safety of using AgNPs have risen, as the AgNPs have been found to negatively interact with the body and the ecosystem. The strong oxidative property of AgNPs releases silver ions, which may cause cytotoxicity, genotoxicity, and immune responses. Also, the AgNPs can pass through the blood-brain barrier by transcytosis of capillary endothelial cells and accumulate in other critical

TABLE 5.1
Gold nanoparticles: Geometries, Synthesis Methods, and Properties

Nanospheres

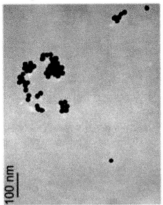

Source: Ref [32]

Morphology: Solid spheres

Synthesis: Reduction of gold salts.

Synthesis methods : (1) Turkevich method, (2) Frens method (3), Brust method, (4) Microemulsion method. Surfactant capping for particle dispersion and stabilization done using Cysteine, CS_2, sodium citrate, ionic and nonionic detergents, nitrilotriacetate, silanes, dendrimers [32]

Properties: SPR absorption in the visible range. Ratio of SPR absorption to scattering is attributed to particle size i.e. larger particles (80 nm) exhibit higher scattering and used for imaging. Smaller particles used for Photothermal therapy [33]

Hollow Nanospheres

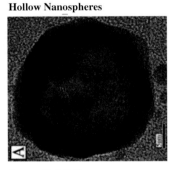

Source: Ref [34]

Morphology: Spheres with hollow interiors

Synthesis: Galvanic replacement of sacrificial templates of Ag or Co seeds [34, 35]

Properties: Size-dependent SPR response tunable from visible to IR region. SERS response at the single particle level yields greater enhancement than solid spheres [36]

Solid shells

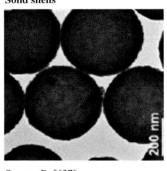

Source: Ref [37]

Morphology: Gold shells with dielectric core (usually silica, sodium sulfide, polystyrene)

Synthesis: (1) Adherence and coalescing of AuNP on silica core using silane linkers [38]

(2) In situ seed formation onto mercapto silica cores – Brust-like process [37]

Properties: Tunability in the visible to IR range, SERS, bioimaging applications.

Nanorods

Source: Ref [39]

Morphology: Cylindrical, anisotropic

Synthesis: Reduction of gold salts in the presence of soft templates (micelles of ionic surfactants, such as CTAB) [39]

Properties: Tunable longitudinal SPR peak

(Continued)

TABLE 5.1
Gold nanoparticles: Geometries, Synthesis Methods, and Properties *(Continued)*

Nanostars

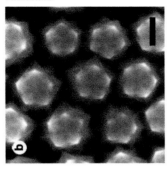

Source: Ref [40]

Morphology: Branched particles with many protruding sharp tips

Synthesis: Traditional methods produce irregular populations of particles. A new method using Au icosahedral seeds and a robust solution phase method yielded monodisperse particles [40]

Properties: Enhanced SPR as sharp tips act as hot spots, tunable to the NIR region.

Nanocages

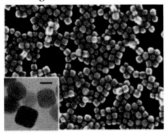

Source: Ref [41]

Morphology: Cuboidal, with hollow interiors and porous walls

Synthesis: Galvanic replacement of silver seeds, used as sacrificial template [41]

Properties: Tunability in the NIR region, applicable for bioimaging techniques such as tomography, and photothermal treatment.

Nanoclusters

Source: Ref [42]

Morphology: Around 2 nm in size, typically containing less than 100 Au atoms

Synthesis: Methods include microwave assisted, sonochemical, photoreductive, etching based, kinetically controlled, electrostatically controlled reversible phase transfer, template assisted synthesis [43]

Properties: Size dependent fluorescence, electrochemiluminescence when coupled with coreactants such as trimethylamine, [44] $K_2S_2O_8$ [42]

---

parts of the body [52, 53]. The size of AgNPs has been found to strongly dictate its cytotoxic properties, with smaller particles being more toxic [54, 55]. Besides shape, coating material properties also have an adverse effect on cellular interaction and hence the degree of toxicity [56, 57]. Similar to AuNPs, AgNPs exhibit distinct LSPR, giving rise to either radiative decay or nonradiative decay. The radiative decay property has been exploited in diagnostics and imaging, while nonradiative decay with photon energy being converted into thermal energy finds application in therapeutics [58]. The approaches to synthesize AgNPs can be classified into three major categories namely, physical, chemical, and biological. These are discussed next.

5.3.2.1 Physical Methods

The most common physical methods for AgNP preparation are laser ablation and evaporation/condensation. The evaporation/condensation method involves the synthesis of AgNPs in a tube furnace at atmospheric pressure. Although this method yields a uniform nanoparticle population and prevents solvent contamination of the prepared nanoparticle film, conventional furnaces have a high-energy consumption rate and require a long time to reach thermal stability. Laser ablation of metals in the absence of chemical reagents generates pure nano-silver colloids. The concentration and morphology of the produced nanoparticles are influenced by a number of factors, such as the laser fluence, the number of laser shots, the ablation light wavelength, pulse duration, ablation time, and the effective liquid medium. Greater laser fluence and longer ablation times lead to larger particle size and concentrations [59–62]. Inspired from laser ablation is the DC-arc discharge technique, which has been used to generate AgNPs in deionized water without any surfactants or stabilizers. In this method, silver wire electrodes are etched using electrical pulses. During the discharge, the evaporated surface layers condense in water, yielding stable and well-dispersed AgNPs [63].

5.3.2.2 Chemical Methods

Chemical reduction of a silver salt solution in the presence of a reducing agent and a capping agent, or stabilizer, constitutes the most popular chemical method of AgNP synthesis. The commonly used silver salts include nitrate, perchlorate, citrate, and acetate, among which silver nitrate is most commonly used due to its chemical stability and low cost. Many reducing agents have been used for the preparation of AgNPs, which include sodium borohydride, trisodium citrate, and organic reductants with hydroxyl or carboxylic group such as carbohydrates, alcohols, aldehydes, etc. Trisodium citrate is a weaker reducing agent, which requires elevated temperatures for the reaction. In contrast, sodium borohydride is a strong reducing agent and stabilizer, which forms smaller AgNPs and prevents their aggregation. Due to the aggregative instability of AgNPs, using a stability is a common strategy. The choice of stabilizer depends on the control of size and shape that needs to be exercised. Commonly used stabilizers include surfactants or polymers with desired functional groups such as polyvinylpyrrolidone, poly(ethylene glycol), poly(methacrylic acid), poly(methyl methacrylate), etc. [62].

Several other chemical approaches for the synthesis of AgNPs have been reported. These include the water-organic two-phase synthesis, photochemical, and electrochemical reduction methods. In the two-phase system, the metal precursor and reducing agent are present in two phases, such that their rate of interaction depends on the interphase transport between the aqueous and organic phase, which is mediated by a quaternary ammonium salt [64]. However, the large amounts of organic solvent and surfactants used in the system need to be removed from the final product, making the method less attractive. Photochemical synthesis of AgNPs involves either direct or indirect photoreduction of Ag^+ to Ag^0. The photoreduction process offers several advantages, such as being easy to execute, no need of reducing agents, yields highly pure AgNPs, more control over AgNP size and shape, etc. [65, 66]. Traditionally, AgNPs were produced by electrochemical reduction, where a metal sheet is anodically dissolved and the intermediate metal salt formed is reduced at the cathode, giving rise to metallic particles stabilized by tetraalkylammonium salts [67, 68]. This method has several advantages over chemical reduction, such as purity of the synthesized nanoparticles, amenable to control the size and shape of the nanoparticles precisely by regulating the current density, pulse duration, immersion time and period, solvent conditions, and capping agents [69].

5.3.2.3 Biological Methods

The physical and chemical approaches of AgNP synthesis allow precise control over the synthesis process, leading to a uniform population of particles. However, some of these methods are plagued by the use of noxious chemicals and are normally not cost-effective. As a result, significant efforts have been made towards the green synthesis of AgNPs using extracts from plants, bacteria, or fungi, and small molecules such as vitamins, amino acids, etc.

Phytoconstituents in several parts of the plants, like leaves, fruit, flowers, seeds, bark, etc., act as reducing agent for silver nitrate solution, producing AgNPs. These phytoconstituents usually include alkanoids, terpenoids, phenolic compounds, flavonoids, tannins, gallic acid, ascorbic acid, etc. Bacterial cells have been used for the intracellular or extracellular synthesis of AgNPs. The commonly used strains include *P. strutzeri*, *B. megaterium*, *S. aureus*, *K. pneumonia*, *E. coli*, and *E. cloacae*, among others. Extracellular synthesis is the capture of metal ions on the surface of microbial cells and their subsequent reduction in the presence of enzymes or other biological reducing agent. This is preferred over the intracellular method, as it involves easier downstream processes for large-scale production of AgNPs. Fungi are excellent candidates for the extracellular synthesis of AgNPs. The mycelia of *Humicolasp* [70] and powdered basidiocarps from *Pleurotus cornucopiae var. citrinopileatus*, and dried aqueous extracts of *Pleurotus ostreatus* [71] have all been shown to produce AgNPs extracellularly [72]. For a detailed account of the green synthesis of AgNPs, we refer the reader to the recent review by Siddiqi et al [73].

5.4 LSPR-BASED SENSORS

5.4.1 Nucleic Acid Detection Assays

The detection of specific DNA sequences is highly important in screening for many infectious diseases, cancers, gene therapy, etc. Many approaches are available for the detection of low-abundance DNA, where most of these focus on amplification of the DNA followed by signal generation using colorimetric dye, fluorescence, enzyme reaction, etc. To improve the signal to noise ratio, nanoparticle labels can be coupled with DNA probes. By virtue of its flexibility, molecular recognition, and affinity for metal ions, ssDNA acts as a good candidate for covalent or noncovalent interactions with metal nanoparticles. In a recent work by Dharanivasan et al, AuNPs conjugated bifunctional oligo probes were used for the detection of tomato leaf curl virus DNA (Figure 5.4a). The nanoparticles were labeled with virus specific ssDNA oligo probes, which acted as primers to amplify the viral DNA. In the presence of the target, the gold labels self-assembled into complex cluster arrangements and were not susceptible to salt-mediated aggregation [74].

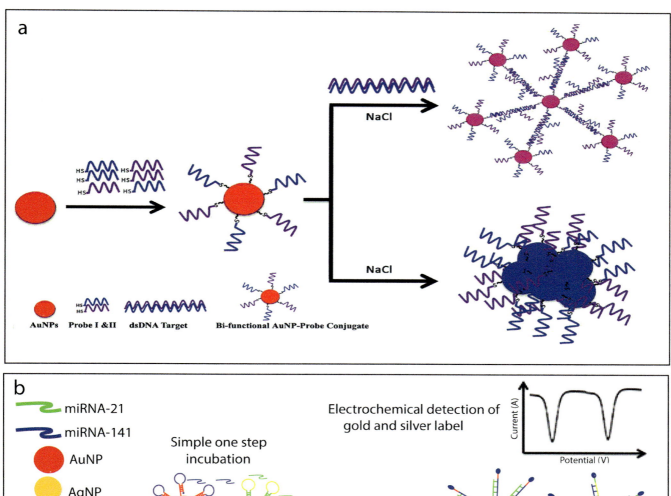

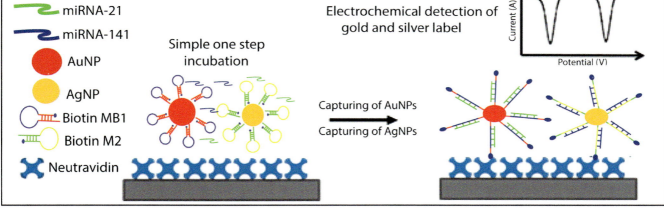

FIGURE 5.4 (a) Detection of tomato leaf curl virus DNA using nanoparticle conjugated bifunctional oligo probes Reproduced from [74], with permission from The Royal Society of Chemistry. (b) Multiplex detection of miRNA 21, and miRNA 141 using biotin conjugated molecular beacons-metal nanoparticles. Differential electrochemical signature of metal nanoparticle aids in multiplexing, Reprinted with permission from [75], copyright American Chemical Society.

MicroRNAs (miRNA) are small (17–25 nts) non-protein-coding RNAs which play critical roles in apoptosis, proliferation, differentiation, and metabolism. These microRNAs are also diagnostic and prognostic markers in the case of several diseases such as cancers, autoimmune, cardiovascular diseases, etc. Multiplexed detection of miRNA 21 and miRNA 141 has been demonstrated by using the electrochemical signature of metal nanoparticles (AgNP, and AuNP) (Figure 5.4b). Here the miRNAs causes opening of biotin-conjugated molecular beacons (MB), followed by capturing of the biotin-MB-metal-NPs on the electrode surface.

The NP labels were then detected by stripping square-wave voltammetry (SSWV). This approach yielded highly sensitive detection of miRNA with the limit of detection reaching low picomolar levels [75].

Metal nanoparticle assisted signal amplification is easily applied to the detection of DNA hybridization events. For example, sensitive detection of BRC A1 gene based on a "sandwich" detection strategy was demonstrated, where the hybridization events between an immobilized capture probe DNA (DNA-c), target DNA (DNA-t), and AuNP conjugated reporter probe DNA (DNAr. AuNP) led to change in I-V characteristics of the electrochemical quartz crystal microbalance. The method could detect as low as 1 fM of target DNA [76].

5.4.2 Immunoassays

Precise quantification of protein biomarkers is essential for disease diagnosis, as well as following response to therapy. The most commonly used approaches for biomarker quantification include enzyme-linked immunosorbent assay (ELISA) (for soluble analytes) and immunohistochemistry (IHC) (for biomarkers in tissue samples). While both techniques are practiced widely, ELISA offers only a relative concentration of the biomarker, while IHC provides no information about the expression levels. With the advance of medicine and varied treatment options available, it is often desirable to obtain digital information about the expression levels of disease markers. Nanomaterial platforms offer the sensitivity required to digitally quantify the analyte of interest and have been the center of increased scientific interest. Zhou et al have developed a AuNP probe-assisted antigen counting chip utilizing three components 1) functionalized GNPs, 2) antibody modified silicon chip, and 3) SEM for the quantification step (Figure 5.5 I). The strategy was applied for the detection of carcinoembryonic antigen to reach a limit of detection of 0.045 ng/mL, which is ~40 times more sensitive than the conventional ELISA [77].

Antigen-antibody assays are quite versatile and have been used for the detection of a myriad variety of analytes, such as pathogens, disease markers, abused drugs, hormones, pesticides, etc. Usual strategies of detection follow the sandwich or competitive immunoassay approach. While the sandwich immunoassay works well for nearly all analytes, the competitive assay is the assay of choice for smaller molecules, such as haptens, which cannot bind two antibodies because of their small size. Detection using competitive assays is challenging, as the results are the complete opposite of the sandwich assay and are often misread in the presence of small backgrounds. However, by using anti-biotin antibody-modified magnetic beads (Ab-MBs) and biotinylated-DNA gold nanoparticles (biotin-AuNPs) as competitive analyte, Lin et al, were able to demonstrate naked-eye detection of biotin (used as a model hapten) (Figure 5.5 II). Here, in the presence of biotin, the supernatant remains red because the analyte (biotin) binds the Ab-MBs, which are removed by an external magnet, leaving the biotin-AuNPs in solution [78].

Signal generation in immunoassays can be done using colorimetric dyes, fluorescence, or chemiluminescence (CL). One way to combine CL with the immunoassay is by conjugation of the CL reagent, usually luminol, with an antibody. However, CL-antibody conjugates show lowered sensitivity compared to free luminol. Additionally, the specificity of the antibody may be affected by this procedure. To overcome this limitation, a novel strategy utilizing the catalytic activity of AuNPs on luminol–$AgNO_3$ chemiluminescence (CL) reaction was developed (Figure 5.5 III). The antigen-antibody interaction induced the aggregation of antibody-functionalized AuNPs, activating the catalytic activity of AuNPs on luminol–$AgNO_3$ CL reaction. Using this reaction, Luo et al demonstrated a detection limit of 3 pg/mL for IgG [79].

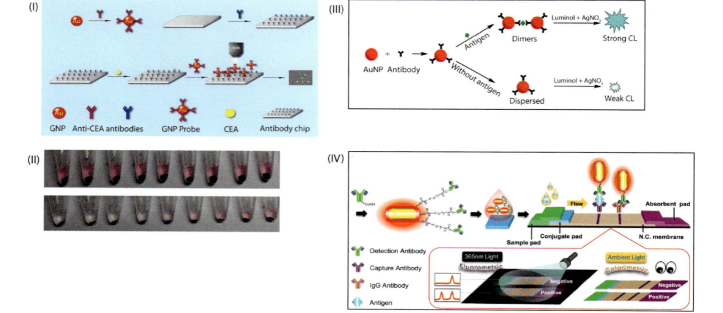

FIGURE 5.5 (I) AuNP probe assisted antigen counting chip. Reprinted with permission from [77] copyright American Chemical Society. (II) Competitive immunoassay for detection of biotin. Reprinted from [78], with permission from Elsevier. (III) Enhanced CL immunoassay for detection of IgG by catalytic activity of AuNPs on luminol–$AgNO_3$CL reaction. Reprinted from [79], with permission from Elsevier. (IV) Quantitative colorimetric and fluorescence dual mode immunoassay strip test. Reprinted with permission from [80], copyright, American Chemical Society.

AuNPs are popularly used as labels for the immunochromatographic test strips (ICTS). Although these tests have a low cost and are convenient for point-of-care (POC) analysis, efforts have been made to modify them to achieve a more precise quantitative readout. One such successful attempt was recently demonstrated by You et al, where they developed a colorimetric and fluorescence dual mode immunoassay by integrating highly fluorescent semiconducting polymer dots (Pdots) with strongly plasmonic Au nanorods to form Pdot-Au hybrid nanocomposites with dual colorimetric and fluorescent readout abilities (Figure 5.5 IV). It was shown that the fluorescence of Pdots could be enhanced by the electromagnetic coupling with Au surface plasmon, leading to high sensitivity [80].

5.4.3 Cellular Detection Assays

Contamination of food and water by pathogens is a major public health concern in developing countries. Employing metal nanoparticles, simple detection platforms can be developed for pathogen detection. Usually a biorecognition element such as an antibody or aptamer specific to the pathogen is conjugated to the nanoparticle to develop pathogen-specific labels. However, a few studies also demonstrate the application of chemical functional groups on the nanoparticle surface for detection.

Colorimetric detection assays utilizing biorecognition element coated AuNPs rely on the reduction in interparticle distance between AuNPs, leading to plasmon coupling and a change in color from red to blue. Colistin is a cationic water-soluble antibiotic that binds to lipopolysaccharides (LPS) on the bacterial cell wall in a 1:1 stoichiometry. In a recent work by Singh et al, colistin is used as a molecular recognition element in solution. In the absence of pathogens, cationic colistin aggregates anionic citrate capped AuNPs, developing a blue color. In the presence of bacteria, the colistin binds to the LPS preventing the aggregation of AuNPs and hence a red color. Using this approach, Singh et al were able to detect 10 bacterial cells/mL of solution [81].

As discussed before, metal nanoparticles can be developed into smart materials with composite properties when used together with other nanomaterials. To detection *E. coli* K12 cells at low concentrations, Zou et al used graphene for the fabrication of an electrochemical sensor. However, the chemical inertness of graphene makes it difficult to attach antibodies to its surface. Therefore polymerization of pyrrole on graphene oxide in the presence of ammonium peroxydisulfate was carried out, which generated a positively charged surface. Citrate-stabilized AuNPs could then be used to decorate the surface of the positively charged polypyrrole (PPy)-graphene composite. This allowed easy immobilization of primary antibody on the surface AuNP-PPy-graphene. A ferrocene-labeled secondary antibody was used as an electrochemical reporter probe to complete the sandwich assay [82]. Nonfunctionalized AuNPs (i.e., without antibody or aptamer conjugation) can also be applied for novel pathogen-sensing strategies. In an interesting report by Verma et al, nanostars of different sizes and degrees of branching were used for identifying clinically relevant ocular pathogens with an accuracy of 99%, namely *S. aureus, A. xylosoxidans, D. acidovorans,* and *S. maltophilia* (Figure 5.6a). The differential colorimetric response of this "chemical nose" is attributed to electrostatic aggregation of cationic gold nanostars around bacteria without the use of any biorecognion elements. Using two types of nanostars with distinct differences in color (blue and red), size, and degree of branching and mixing the original nanostar solutions in 1:1 volume to obtain a third solution of purple nanostars, the authors were able to obtain unique colorimetric outputs corresponding to the dependence of electrostatic interactions on the size and shape of nanostars and surface characteristics of bacteria [83].

Metal nanoparticles are effective reporters not only for pathogen detection but also for the detection of circulating tumor cells (CTCs) when coupled with the appropriate antibodies. CTCs are prognostic biomarkers of interest in metastasizing tumors. The major challenge in their detection lies in their extremely low counts in a complex sample (i.e., blood). In a remarkable study by Wang et al, the principle of direct plasmon enhanced electrochemistry (DPEE) was applied for the ultrasensitive and label-free detection of CTCs in blood. The electrode construction is shown in Figure 5.6b, where a glassy carbon electrode was modified with plasmonic gold nanostars (AuNSs) followed by the aptamer to selectively capture the CTCs in solution. Enhancement in the electrochemical current response occurs on LSPR excitation of AuNS, which transport hot electrons to the external circuit [84].

5.4.4 Other Biomolecular Interactions

Besides nucleic acid hybridization, antigen-antibody, and aptamer-protein interactions, nanoparticles have also been employed for probing several other biomolecular interactions, for example, protein-carbohydrate, cytochrome P450-drug, metal ion-receptor interactions, etc., which are used for the detection of small molecules.

Concanavalin A (ConA) is a lectin (carbohydrate-binding protein) that preferentially binds to the mannose residues in various glycoproteins. Utilizing the lectin-carbohydrate affinity, a study demonstrated the development of a microarray-based assay for the detection of microbial cells (Figure 5.7a). First, a monoclonal antibody was used as a capture to bind microbial cells: *E. coli* and *P. fluorescens* (gram-negative bacteria), *B. subtilis* (gram-positive bacterium), and *S. cerevisiae* (yeast). The detection of microbes was then performed by using two different kinds of lectins (ConA and RCA) conjugated gold nanoparticles. ConA binds only to gram-negative bacteria, while *Ricinus Communis Agglutini* (RCA) binds to both gram-negative bacteria and gram-positive bacteria. Using this strategy, highly selective discrimination between gram-positive, gram-negative, and yeast strains could be achieved [85].

Cytochrome P450 is a heme containing microsomal enzyme found in the liver. It metabolizes a variety of compounds, including drugs, bilirubin, xenobiotics, carcinogens, etc. The electrochemistry of cytochrome P450

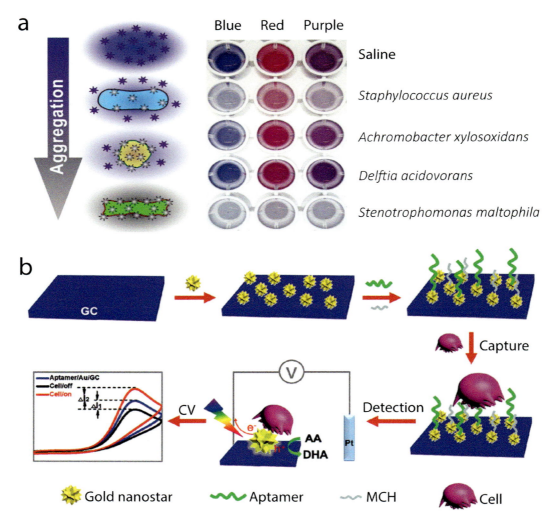

FIGURE 5.6 (a) "Chemical nose" for the detection of ocular pathogens using AuNS. Reprinted from [83], with permission from Elsevier. (b) Electrochemical platform for the ultrasensitive detection of CTCs. Reprinted with permission from [84], copyright, American Chemical Society.

enzyme and its bioelectrocatalytic reduction of three common drugs: lidocaine, cyclophosphamide (CPA), and bupropion (BUP) was electrochemically studied by depositing a biocompatible film containing AuNPs and chitosan to encapsulate the enzyme on the electrode surface. The AuNPs form a conductive layer that enables electron transfer, which was otherwise not appreciable in case of enzyme-bare electrode [86].

Heavy metal contamination of water bodies is of concern in the upkeep of public health, as well as environmental conservation. Heavy metal ions such as Hg^{+2} have been found to accumulate in the body through the food chain, leading to serious health problems of the liver, heart, and nervous system. In order to construct a miniaturized instrument for mercury detection, Yuan et al developed a fiber optic SPR sensor using 4-mercaptopyridine (4-MPY) modified AuNPs as labels (Figure 5.7b). Here, the nitrogen of the pyridine moiety in 4-MPY could coordinate with Hg^{+2} via multidentate N-bonding to form an $Hg(pyridine)_2$ complex. The plasmon coupling between the Au film on a fiber-optic probe and 4-MPY modified AuNPs labels created a strong localized SPR signal, which varies depending on the Hg^{+2} concentration in solution [87].

5.5 CONCLUSION

In this chapter we discuss the physical basis and fundamentals of LSPR, a phenomenon responsible for the remarkable absorption and scattering properties of nanoparticles. The surface plasmon band of nanoparticles is strongly dependent on their geometry, composition, and environment. It is therefore possible to engineer the plasmon absorbance by tailoring the morphology of the nanoparticles, which has led to the development of nanoparticles of various shapes (e.g., stars, rods, shells, clusters, etc.). The most applied nanoparticles in biomedical research (i.e., gold, and silver) are discussed. The morphology of AuNPs and AgNPs, which depend on their synthesis method, strongly affect their optical properties. Additionally, their stability, ease of functionalization, and biocompatibility make them the label of choice for signal amplification in biomedical, clinical, and biotechnology assays.

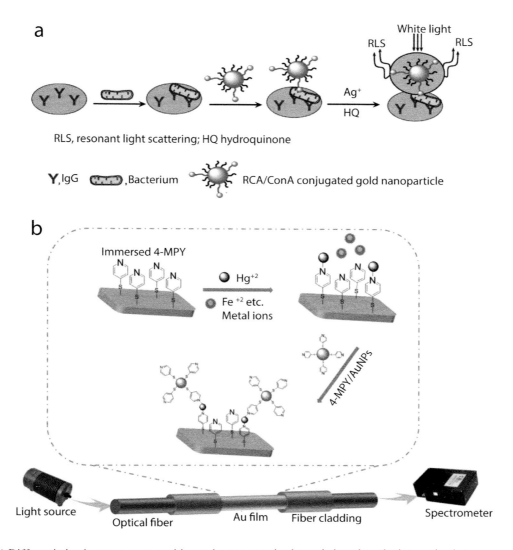

FIGURE 5.7 (a) Differentiation between gram-positive and gram-negative bacteria based on the interaction between protein (lectin) and carbohydrate. Reprinted from [85], with permission from Elsevier. (b) Fiber-optic SPR sensor for the detection of Hg^{+2} ions using 4-MPY-AuNPs as labels. Reprinted with permission from [87], copyright American Chemical Society.

ACKNOWLEDGMENT

I am thankful to Prof. Pranab Goswami for his valuable suggestions and efficient correspondence among authors.

REFERENCES

1. Mayer KM, Hafner JH. Localized Surface Plasmon Resonance Sensors. Chem Rev 2011; 111:3828–57.
2. [Page on the internet]. Available from https://www.rigb.org/our-history/iconic-objects/iconic-objects-list/faraday-goldcolloids#:~:text=These%20liquids%20are%20the%20first,properties%20of%20light%20and%20matter
3. Jana J, Ganguly M, Pal T. Enlightening surface plasmon resonance effect of metal nanoparticles for practical spectroscopic application. RSC Adv 2016; 6:86174–211.
4. (i) Maxwell Garnett JC. Colours in metal glasses and in metallic films. Philos Trans R Soc 1904; 203:385–420. (ii) Maxwell Garnett JC. Colours in metal glasses, in metallic films, and in metallic solutions- II. Philos Trans R Soc 1906; 205:237–88.
5. Bruggeman DAG. Berechnung verschiedener physikalischer Konstanten von heterogenen Substanzen. I. Dielektrizitätskonstanten und Leitfähigkeiten der Mischkörper aus isotropen Substanzen. Ann Phys 1935; 24:636.
6. Lu X, Rycenga M, Skrabalak SE, Wiley B, Xia Y. Chemical synthesis of novel plasmonic nanoparticles. Annu Rev Phys Chem 2009; 60:167–92.
7. Liz-Marzán LM. Tailoring surface plasmons through the morphology and assembly of metal nanoparticles. Langmuir 2006; 22(1):32–41.
8. Wiley BJ, Im SH, Li ZY, McLellan J, Siekkinen A, Xia Y. Maneuvering the surface plasmon resonance of silver nanostructures through shape-controlled synthesis. J Phys Chem B 2006; 110(32):15666–75.
9. Kreibig U, Vollmer M. Optical Properties of Metal Clusters. Springer: New York, 1995.
10. Fuchs R. Theory of the optical properties of ionic crystal cubes. Phys. Rev B 1975; 11:1732.
11. Haes AJ, Haynes CL, McFarland AD, Schatz GC, Van Duyne RP, Zou S. Plasmonic materials for surface-enhanced sensing and spectroscopy. MRS Bull 2005; 30:368.

12. Kottmann JP, Martin OJF, Smith DR, Schultz S. Plasmon resonances of silver nanowires with a nonregular cross section. Phys Rev B 2002; 64:235402.
13. Aizpurua J, Bryant GW, Richter LJ, García de Abajo FJ, Kelley BK, Mallouk T. Optical properties of coupled metallic nanorods for field-enhanced spectroscopy. Phys Rev B 2005; 71:235420.
14. Prodan E, Radloff C, Halas NJ, Nordlander P. A hybridization model for the plasmon response of complex nanostructures. Science 2003; 302(5644):419–22.
15. Schelm S, Smith GB. Internal electric field densities of metal nanoshells. J Phys Chem B 2005; 109: 1689.
16. Kelly KL, Coronado E, Zhao LL, Schatz GC. The optical properties of metal nanoparticles: the influence of size, shape, and dielectric environment. J Phys Chem B 2003; 107:668.
17. Abalde-Cela S, Carregal-Romero S, Coelho JP, Guerrero-Martínez A. Recent progress on colloidal metal nanoparticles as signal enhancers in nanosensing. Adv Colloid Interface Sci 2016; 233:255–70.
18. Elahi N, Kamali M, Baghersad MH. Recent biomedical applications of gold nanoparticles: a review. Talanta 2018; 184:537–56.
19. Daniel MC, Astruc D. Gold Nanoparticles: Assembly, Supramolecular Chemistry, Quantum-Size-Related Properties, and Applications toward Biology, Catalysis, and Nanotechnology. Chem Rev 2004; 104:293–346.
20. Liu Y, Male KB, Bouvrette P, Luong JHT. Control of the size and distribution of gold nanoparticles by unmodified cyclodextrins. Chem Mater 2003; 15:4172–80.
21. Chen D-H, Chen C-J. Formation and characterization of Au–Ag bimetallic nanoparticles in water-in-oil microemulsions. J Mater Chem 2002; 12:1557–62.
22. Velikov KP, Zegers GE, van Blaaderen A. Synthesis and characterization of large colloidal silver particles. Langmuir 2003; 19:1384–9.
23. Pastoriza-Santos I, Liz-Marzan LM. Formation of PVP-protected metal nanoparticles in DMF. Langmuir 2002; 18:2888–94.
24. Ma Z, Han H. One-step synthesis of cystine-coated gold nanoparticles in aqueous solution, Colloids Surf A Physicochem Eng Asp 2008; 317:229–33. doi:10.1016/j.colsurfa.2007.10.018.
25. Wang W, Chen X, Efrima S. Silver nanoparticles capped by long-chain unsaturated carboxylates. J Phys Chem B 1999; 103:7238–46.
26. Jiang X, Xie Y, Lu J, Zhu L, He W, Qian Y. Preparation, characterization, and catalytic effect of CS_2-stabilized silver nanoparticles in aqueous solution. Langmuir 2001; 17:3795–9.
27. Neiman B, Grushka E, Lev O. Use of gold nanoparticles to enhance capillary electrophoresis. Anal Chem 2001; 73:5220–7.
28. Zhu J, Liu S, Palchik O, Koltypin Y, Gedanken A. Shape-controlled synthesis of silver nanoparticles by pulse sonoelectrochemical methods. Langmuir 2000; 16:6396–9.
29. Pastoriza-Santos I, Liz-Marzan LM. Formation and stabilization of silver nanoparticles through reduction by N,N-dimethylformamide. Langmuir 1999; 15:948–51.
30. Tan Y, Jiang L, Li Y, Zhu DJ. One dimensional aggregates of silver nanoparticles induced by the stabilizer 2-mercaptobenzimidazole. Phys Chem B 2002; 106:3131–8.
31. Manna A, Imae T, Aoi K, Okada M, Yogo T. Synthesis of dendrimer-passivated noble metal nanoparticles in a polar medium: comparison of size between silver and gold particles. Chem Mater 2001; 13:1674–81.
32. Yong KT, Swihart MT, Ding H, Prasad PN. Preparation of gold nanoparticlesand their applications in anisotropic nanoparticle synthesis and bioimaging. Plasmonics 2009; 4(2):79–93.
33. Huang XH, Neretina S, El-Sayed MA. Gold nanorods: from synthesis andproperties to biological and biomedical applications. Adv Mater 2009; 21(48):4880–910.
34. Schwartzberg AM, Olson TY, Talley CE, Zhang JZ. Synthesis, characterization, and tunable optical properties of hollow gold nanospheres. J Phys Chem B 2006; 110(40):19935–44.
35. Hao E, Li S, Bailey RC, Zou S, Schatz GC, Hupp JT. Optical properties of metal nanoshells. J Phys Chem B 2004; 108(4):1224–9.
36. Talley CE, Jackson JB, Oubre C, Grady NK, Hollars CW, Lane SM, et al. Surface-enhanced Raman scattering from individual au nanoparticles and nanoparticle dimer substrates. Nano Lett 2005; 5:1569.
37. Gao Y, Gu J, Li L, Zhao W, Li Y. Synthesis of gold nanoshells through improved seed-mediated growth approach: Brust-like, in situ seed formation. Langmuir 2016; 32(9):2251–8.
38. Oldenburg SJ, Jackson JB, Westcott SL, Halas NJ. Infrared extinction propertiesof gold nanoshells. Appl Phys Lett 1999; 75(19):2897–9.
39. Jana NR, Gearheart L, Murphy CJ, Wet chemical synthesis of high aspect ratio cylindrical gold nanorods. J Phys Chem B 2001; 105(19):4065–7.
40. Niu W, Chua YAA, Zhang W, Huang H, Lu X. Highly symmetric gold nanostars:crystallographic control and surface-enhanced Raman scattering property. J Am Chem Soc 2015; 137(33):10460–3.
41. Chen J, Saeki F, Wiley BJ, Cang H, Cobb MJ, Li ZY, et al. Gold nanocages: bioconjugation and their potential use as optical imaging contrast agents. Nano Lett 2005; 5(3):473–7.
42. Li L, Liu HY, Shen YY, Zhang JR, Zhu JJ. Electrogeneratedchemiluminescence of Au nanoclusters for the detection of dopamine Anal Chem 2011; 83:661–65.
43. Cui ML, Zhao Y, Song QJ. Synthesis, optical properties and applications of ultra-small luminescent gold nanoclusters. TrAC – Trend Anal Chem 2014; 57:73–82.
44. Fang YM, Song J, Li J, Wang YW, Yang HH, Sun JJ, et al. Electrogenerated chemiluminescence from Au nanoclusters. Chem. Commun 2011; 47:2369–71.
45. Stone J, Jackson S, Wright D. Biological applications of gold nanorods Wiley Interdiscip Rev Nanomed Nanobiotechnol 2011; 3(1):100–9.
46. Pérez-Juste J, Pastoriza-Santos I, Liz-Marzán LM, Mulvaney P. Gold nanorods: synthesis, characterization and applications. Coord Chem 2005; 249(17–18):1870–1901.
47. Skrabalak SE, Chen J, Sun Y, Lu X, Au L, Cobley CM, et al. Gold nanocages: synthesis, properties, and applications. Acc Chem Res 2008; 41(12): 1587–95.
48. Rai M, Yadav A, Gade A. Silver nanoparticles as a new generation of antimicrobials. Biotechnol Adv 2009; 27:76–83.
49. Schoen DT, Schoen AP, Hu L, Kim HS, Heilshorn SC, Cui Y. High speed water sterilization using one-dimensional nanostructures. Nano Lett 2010; 10:3628–32.
50. Yu S-J, Yin Y-G, Liu J-F. Silver nanoparticles in the environment. Environ Sci Proc Impacts 2013;15:78–92.
51. Edwards-Jones V. The benefits of silver in hygiene, personal care and healthcare. Lett ApplMicrobiol 2009;49:147–52.
52. Cho J-G, Kim K-T, Ryu T-K, Lee J-W, Kim J-E, Kim J, et al. Stepwise embryonic toxicity of silver nanoparticles on Oryziaslatipes. Bio Med Res Int 2013;2013:1–7.
53. Tang J, Xiong L, Zjou G, Xi T. Silver nanoparticles crossing through and distribution in the blood-brain barrier in vitro. J Nanosci Nanotechnol 2010; 10:6313–7.
54. Johnston HJ, Hutchison G, Christensen FM, Peters S, Hankin S, Stone V. A review of the in vivo and in vitro toxicity of silver and gold particulates: particle attributes and biological mechanisms responsible for the observed toxicity. Crit Rev Toxicol 2010; 40(4):328–46.

55. Sriram MI, Kalishwaralal K, Barathmanikanth S, Gurunathani S. Size-based cytotoxicity of silver nanoparticles in bovine retinal endothelial cells. Nanosci Methods 2012; 1(1): 56–77.
56. Suresh AK, Pelletier DA, Wang W, Morrell-Falvey JL, Gu B, Doktycz MJ. Cytotoxicity induced by engineered silver nanocrystallites is dependent on surface coatings and cell types. Langmuir 2012; 28(5):2727–35.
57. Stoehr LC, Gonzalez E, Stampfl A, Casals E, Duschl A, Puntes V, et al. Shape matters: effects of silver nanospheres and wires on human alveolar epithelial cells. Fibre Toxicol 2011; 8(1): 36.
58. Wei L, Lu J, Xu H, Patel A, Chen Z-S, Chen G. Silver nanoparticles: synthesis, properties, and therapeutic applications. Drug Discov Today 2015; 20(5):595–601.
59. QuangHuy T, Van Quy N, Anh-Tuan L. Silver nanoparticles: synthesis, properties, toxicology, applications and perspectives. Adv Nat Sci Nanosci Nanotechnol 2013; 4(3):033001.
60. Tsuji T, Thang DH, Okazaki Y, Nakanishi M, Tsuboi Y, Tsuji M. Preparationof silver nanoparticles by laser ablation in polyvinylpyrrolidone solutions. Appl Surf Sci 2008; 254(16):5224–30.
61. Al-Azawi MA, Bidin N, Ali AK, Hassoon KI, Abdullah M. Effect of liquid layer thickness on the ablation efficiency and the size-control of silver colloids prepared by pulsed laser ablation. Mod Appl Sci 2015; 9(6): 20.
62. Ge L, Li Q, Wang M, Ouyang J, Li X, Xing MMQ. Nanosilver particles in medical applications: synthesis, performance, and toxicity. Int J Nanomedicine 2014; 9:2399–407.
63. (a) Tien D-C, Tseng K-H, Liao C-Y, Huang J-C, Tsung T-T. Discovery of ionic silver in silver nanoparticle suspension fabricated by arc discharge method. J Alloys Compd 2008; 463:(1–2)408–11. (b) Tien D-C, Tseng K-H, Liao C-Y, Tsung T-T. Colloidal silver fabrication using the spark discharge system and its antimicrobial effect on Staphylococcus aureus. Med Eng Phys 2008; 30(8): 948–52.
64. Krutyakov Y, Olenin A, Kudrinskii A, Dzhurik P, Lisichkin G. Aggregative stability and polydispersity of silver nanoparticles prepared using two-phase aqueous organic systems. Nanotechnol Russia 2008; 3(5–6):303.
65. Pacioni NL, Borsarelli CD, Rey V, Veglia AV. Synthetic routes for the preparation of silver nanoparticles. In: Alarcon EI, Griffith M, Udekwu KI (Eds.), Silver Nanoparticle Applications: The Fabrication Design of Medical Biosensing Devices. Springer International Publishing, Cham. 2015; pp. 13–46.
66. Haider A, Kang I-K. Preparation of silver nanoparticles and their industrial and biomedical applications: a comprehensive review. Adv Mater Sci Eng 2015; 16.
67. Rodríguez-Sánchez L, Blanco MC, López-Quintela MA. Electrochemical synthesis of silver nanoparticles. J Phys Chem B 2000; 104(41):9683–88.
68. Reetz, MT, Helbig W. Size-selective synthesis of nanostructured transition metal clusters. J Am Chem Soc 1994; 116:7401.
69. Zhang Z, Lin P-C. Chapter 7 – Noble metal nanoparticles: synthesis, and biomedical implementations. In: Emerging Applications of Nanoparticles and Architecture Nanostructures: Current Prospects and Future Trends Micro and Nano Technologies. Elsevier, Amsterdam, the Netherlands. 2018; pp. 177–233.
70. Syed A, Saraswati S, Kundu GC, Ahmad A. Biological synthesis of silver nanoparticles using the fungus Humicola sp. and evaluation of their cytoxicity using normal and cancer cell lines. SpectroActa Part A 2013; 114:144–7.
71. Owaid MN, Raman J, Lakshmanan H, Al-Saeedi SSS. Sabaratnam V, Abed IA. Mycosynthesis of silver nanoparticles by Pleurotuscornucopiae var. citrinopileatus and its inhibitory effects against Candida sp. Mater Lett 2015; 153:186–90.
72. Al-Bahrani R, Raman J, Lakshmanan H, Hassan AA, Sabaratnam V. Green synthesis of silver nanoparticles using tree oyster mushroom Pleurotusostreatus and its inhibitory activity against pathogenic bacteria. Mat Lett 2017; 186:21–5.
73. Siddiqi SK, Husen A, Rao RAK. A review on biosynthesis of silver nanoparticles and their biocidal properties. J Nanobiotechnology 2018; 16(1):14.
74. Dharanivasan G, Mohammed Riyaz SU, Michael Immanuel Jesse D, Raja Muthuramalingam T, Rajendrana G, Kathiravan K. DNA templated self-assembly of gold nanoparticle clusters in the colorimetric detection of plant viral DNA using a gold nanoparticle conjugated bifunctional oligonucleotide probe. RSC Adv 2016; 6:11773–85.
75. Azzouzi S, Fredj Z, Turner APF, Ben Ali M, Mak WC. Generic neutravidin biosensor for simultaneous multiplex detection of microRNAs via electrochemically encoded responsive nanolabels. ACS Sens 2019; 4(2):326–34.
76. Rasheed PA, Sandhyarani N. Femtomolar level detection of BRCA1 gene using a gold nanoparticle labeled sandwich type DNA sensor. Colloids Surf B Biointerfaces 2014; 117:7–13.
77. Zhou X, Yang C-T, Xu Q, Lou Z, Xu Z, Thierry B, et al. A gold nanoparticle probe-assisted antigen-counting chip using SEM. ACS Appl Mater Interfaces 2019; 11(7): 6769–76.
78. Lin WZ, Chen YH, Liang CK, Liu CC, Hou SY. A competitive immunoassay for biotin detection using magnetic beads and gold nanoparticle probes. Food Chem 2019; 271:440–4.
79. Luo J, Cui X, Liu W, Li B. Highly sensitive homogenous chemiluminescence immunoassay using gold nanoparticles as label. Spectro Chim Acta A 2014; 131:243–8.
80. You P-Y, Li F-C, Liu M-H, Chan Y-H. Colorimetric and fluorescent dual-mode immunoassay based on plasmon enhanced fluorescence of polymer dots for detection of PSA in whole-blood. ACS Appl Mater Interfaces 2019; 11:10.
81. Singh P, Gupta R, Choudhary M, Pinnaka A, Kumar R, Bhalla V. Drug and nanoparticle mediated rapid naked eye water test for pathogens detection. Sens Actuators B: Chemical 2018; 262: 603–10.
82. Zou Y, Liang J, She Z, Kraatz H-B. Gold nanoparticles-based multifunctional nanoconjugates for highly sensitive and enzyme-free detection of E.coli K12. Talanta 2019; 193:15–22.
83. Verma MS, Chen PZ, Jones L, Gu FX. "Chemical nose" for the visual identification of emerging ocular pathogens using gold nanostars. Biosens Bioelectron 2014; 61:386–90.
84. Wang S-S, Zhao X-P, Liu F-F, Younis MR, Xia X-H, Wang C. Direct plasmon-enhanced electrochemistry enables ultrasensitive and label-free detection of circulating tumor cells in blood. Anal Chem 2019; 91(7): 4413–20.
85. Gao J, Liu C, Liu D, Wang Z, Dong S. Antibody microarray-based strategies for detection of bacteria by lectin-conjugated gold nanoparticle probes. Talanta 2010; 81(4–5):1816–20.
86. Liu S, Peng L, Yang X, Wu Y, He L. Electrochemistry of cytochrome P450 enzyme on nanoparticle-containing membrane-coated electrode and its applications for drug sensing. Anal Biochem 2008; 375(2):209–16.
87. Yuan H, Ji W, Chu S, Liu Q, Qian S, Guang J, et al. Mercaptopyridine-functionalized gold nanoparticles for fiber-optic surface plasmon resonance Hg^{2+} sensing. ACS Sens 2019; 4(3):704–10.

6 Metal Nanoclusters as Signal Transducing Element

Phurpa Dema Thungon and Pranab Goswami
Indian Institute of Technology Guwahati, Assam, India

Torsha Kundu
International Management Institute, Kolkata, India

CONTENTS

6.1 Introduction 108
6.2 Principles and Theories 109
    6.2.1 Energy Levels 109
    6.2.2 The Jellium Model and "Magic" Numbers 110
6.3 Synthesis of Metal Nanoclusters 111
    6.3.1 Solid Phase Synthesis 111
    6.3.2 Gas Phase Synthesis 111
    6.3.3 Liquid Phase Synthesis 111
6.4 Gold as Metal Precursors 111
    6.4.1 Chemical Reduction Method 112
        6.4.1.1 Thiol-Groups as Stabilizing Ligand 112
        6.4.1.2 Dendrimers as Stabilizing Ligand 112
        6.4.1.3 Peptides and Proteins as Stabilizing Ligands 112
        6.4.1.4 DNA Oligonucleotides as Stabilizing Ligands 113
    6.4.2 Photoreduction Method for Synthesis of Au NCs 114
    6.4.3 Bio-reduction Method for Synthesis of Au NCs 114
    6.4.4 Electro-reduction Method for Synthesis of Au NCs 114
    6.4.5 Etching Method for Synthesis of Au NCs 114
6.5 Silver As Metal Precursors 114
    6.5.1 DNA Oligonucleotides as Stabilizing Ligand 114
    6.5.2 Dendrimers, Polymers, and Thiols as Stabilizing Ligands 116
    6.5.3 Proteins and Peptides as Stabilizing Ligands 117
    6.5.4 Inorganic Scaffolds as Stabilizer 117
6.6 Copper as Metal Precursor 117
    6.6.1 Ligand-Based Synthesis of Cu NCs 117
    6.6.2 Water-in-Oil (w/o) Microemulsion for Synthesis of Cu NCs 118
    6.6.3 Electrochemical Synthesis of Cu NCs 118
    6.6.4 Other Methods for Synthesis of Cu NCs 118
6.7 Bimetallic Nanoclusters 119
6.8 Other Metal Nanoclusters 119
6.9 Effect of Ligands on NCs 120
    6.9.1 Dendrimers as Ligands 120
    6.9.2 Thiols as Ligands 120
    6.9.3 Proteins as Ligands 121
    6.9.4 Oligonucleotides as Ligands 121
    6.9.5 Polymers as Ligands 121
6.10 Characterization Techniques for Nanoclusters 121
    6.10.1 Spectroscopic Techniques 122
        6.10.1.1 UV-Vis Spectroscopy 122
        6.10.1.2 Fluorescence Spectroscopy 122
        6.10.1.3 X-ray Photoelectron Spectroscopy 122

      6.10.1.4 Mass Spectroscopy 122
      6.10.1.5 Infrared Spectroscopy 123
  6.10.2 Microscopic Techniques 123
      6.10.2.1 Transmission Electron Microscopy (TEM) 123
  6.10.3 Chromatography 123
  6.10.4 Other Characterization Techniques 123
6.11 Optical Properties of Metal Nanoclusters 123
  6.11.1 Absorption Behavior 123
  6.11.2 Fluorescence Behavior 124
  6.11.3 Electrochemiluminescence Behavior 124
  6.11.4 Two-Photon Absorption Behavior 124
6.12 Other Properties 124
6.13 Metal Nanoclusters for Sensing Applications 124
  6.13.1 Biomolecule Detection 125
      6.13.1.1 Detection of Nucleic Acid-Based Compounds 125
      6.13.1.2 Detection of Proteins 127
      6.13.1.3 Detection of Other Organic Compounds 128
  6.13.2 Detection of Inorganic Ions 129
      6.13.2.1 Detection of Metal Ions 129
      6.13.2.2 Detection of Other Ions 131
  6.13.3 Detection of Other Substances 132
      6.13.3.1 Detection of Hydrogen Peroxide 132
      6.13.3.2 Detection of Nitroaromatic Explosives 132
6.14 Other Applications for Metal NCs 132
  6.14.1 Thermometer and pH Meter 132
  6.14.2 Biolabeling and Bioimaging 133
  6.14.3 Metal Nanoclusters as Antibacterial Agents 133
  6.14.4 Metal Nanoclusters-Based Artificial Enzymes 133
6.15 Conclusion and Future Prospects 134
References 134

6.1 INTRODUCTION

There has been extensive use of nanomaterials for developing sensors, be it biological or chemical. Among them, metal nanoparticles have received tremendous research interest for their unique characteristics found to be suitable for developing biosensors. In general, metal nanomaterials are classified into three main domains according to their size: large nanoparticles, small nanoparticles (NPs), and nanoclusters (NCs) (Figure 6.1) [1]. The size of the large nanoparticle is larger than the wavelength of the photons interacting with them. Their optical response to the external electromagnetic fields depends on their sizes, free-electron density, and surrounding medium. The small nanoparticles approach the electron mean free path (~50 nm for Au and Ag), and the dielectric function and refractive indices become strongly size dependent. In case of nanoclusters whose size is below 2–3 nm, the plasmon absorption disappears completely [2]. Since the size becomes comparable to the Fermi wavelength of an electron (i.e., ~0.5 nm for Au and Ag), the optical, electronic, and chemical properties of the clusters change dramatically from the other two size domains. These nanostructures possess discrete states and strong photoluminescence [3]. These few atom nanoclusters with molecule-like properties offer the "missing link" between metal atoms and nanoparticles [4]. The photoluminescence (PL), predicted by the free-electron model, suggests it originated from intraband transitions of the nanocluster-free electrons.

Their high-yield PL, good photostability with high emission rates, and large strokes shift, as well as low toxicity, has rendered their application in biosensing and bioimaging, as well as therapy and diagnosis. There are several metal NCs which have been synthesized consisting of gold (Au), silver (Ag), copper (Cu), and even mixed metals using either top-down or bottom-up approaches [3]. Ligands are required to stabilize these nanoclusters and help avoid their aggregation into nanoparticles. Several ligands such as thiol groups, proteins, and oligonucleotides have been explored, and interestingly, the fluorescence properties of the metal nanoclusters are dependent on these ligands. Metal NCs are considered a new class of fluorescent labels showing size-dependent fluorescence, which can be explored for developing fluorescent-based sensing systems [4]. The sensing systems depend on either the quenching of the fluorescence or the shift and change in fluorescence wavelength of metal nanoclusters and have been described in many review papers recently [5–7]. This chapter explores the principles behind the properties of metal nanoclusters, their synthesis, and their application in biosensing. A brief discussion of other applications for metal nanoclusters is also included.

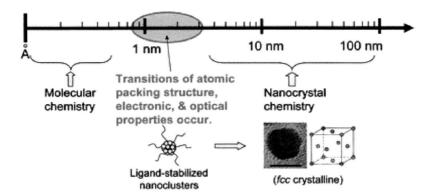

FIGURE 6.1 Classification of metal nanomaterials [1].

6.2 PRINCIPLES AND THEORIES

6.2.1 Energy Levels

The properties of metals vary significantly with their size. To understand this change, we need to know about the energy bands (valence band and conduction band) in the bulk metals and the changes they go through with their varying sizes. When excited, the electrons in the valence band can jump out and move towards the conduction band. When the electrons are in the conduction band, they have enough energy to move freely in the material. The difference between the two bands makes the band gap. In the case of bulk metals, the conduction band overlaps the valence band to some extent, is partially occupied by electrons, and creates a continuum. Hence, there is no bandgap between the valence band and the conduction band, and the electrons do not experience any barrier to populate the conduction band (Figure 6.2) [8]. This is responsible for the good electrical conductivity of bulk metals.

When the size of metals is reduced to a level that is comparable or smaller than the electron mean-free path (~50 nm for Au or Ag), the movement of electrons becomes limited by their size. The interactions of electrons are mostly on the surface, which leads to the surface plasmon resonance effects. The optical properties of the nanoparticles of such highly conductive metals depend on the collective oscillation of the conduction band electrons upon interaction with light. Usually, metal nanoparticles absorb light strongly, but they emit low to no luminescence [8]. Further reduction in the size of metal nanoparticles results in metal nanoclusters. At this stage, the size approaches the Fermi wavelength, and the continuous-band structures of metals break up into discrete energy levels [9]. The properties of particles like conductivity and surface plasmonic effect disappear, as the energy levels or bands are too far separated. However, these clusters still interact with light through electronic transitions between the energy levels and show bright luminescence, similar to organic dye molecules [6]. The Jellium model and the "magic number" rule could determine the stability and confinement of the metal NCs.

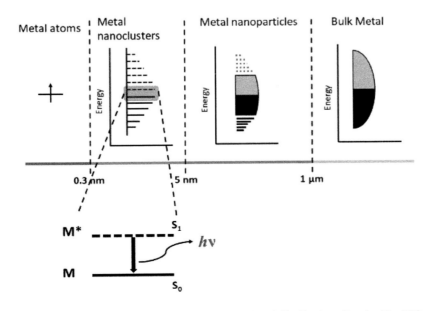

FIGURE 6.2 Illustration of the effect of the size of metals on their energy band distribution. (Inspired by [6].)

6.2.2 The Jellium Model and "Magic" Numbers

The Jellium model originated from the nuclear-shell model and was modified as a spherical Jellium model and used for determining the emission energy or the energy level spacing between adjacent levels of nanoclusters. According to the Jellium model, if there are N number of atoms in a cluster, the size of nanoclusters is related to the emission energy via the simple relation, $E_f/N^{1/3}$, where E_f is the Fermi energy of the bulk metal.

This energy scaling law was used as a model to describe the size-dependent electronic structure and relative electronic transitions of small clusters. Thus, as the number of atoms increase, the emission energy decreases [10]. Figure 6.3 shows that with the change in the size of gold (Au) nanoclusters (NCs) the energy in the band changes, leading to change in the fluorescence emission. The smallest clusters (<2 nm) show blue fluorescence, whereas the larger ones are green, and a further increase in size (~3 nm) causes them to emit red fluorescence. The metal nanoparticles and bulk metal do not produce any fluorescence.

In another study, while preparing sodium clusters, a periodic pattern of intense peaks was observed for clusters containing the number of atoms $N = 2, 8, 18, 20 \ldots$ which had greater stability than other clusters. These numbers of atoms are called "magic" numbers [9]. Similarly, during the preparation of metal NCs, according to mass spectrometry, certain masses of clusters were produced in relatively large quantities. These "magic" sizes corresponded to the closing of atomic shells or the electronic shells [11]. These clusters have a complete and regular outer geometry such as hexagonal or spherical and are described as full-shell or "magic number" clusters. Successive packing of the metal atoms into the layers or shells around a single atom provide good stability, as the structures are densely packed and provide the maximum number of metal-metal bonds [12].

The relationship between the total numbers of metal atoms in the case of a hexagonal full-shell, y, per n^{th} shell is given by:

$$y = 10n^2 + 2 \quad (6.1)$$

(where $n > 0$) [13]

Therefore, the full-shell metal clusters may contain 13 atoms (1 + 12) if there is one shell, 55 (13 + 42) if there are two shells, and 147 (55 + 92) if three shells and so on.

It may be noted that the noble metal nanocluster luminescence has been a known phenomenon for some time, though scientists paid little attention to it initially because of its low quantum yield [14, 15]. The band theory has been considered to explain the luminescence mechanism of Au NCs. According to the theory, Au NCs do not have the continuous energy band which is typical of the larger nanoparticles (causing them to have localized SPR); instead, they have discrete energy bands from which valence electrons from one atom can jump to the conduction band when excited and can emit visible-to-near infrared fluorescence. The emission can be tuned by changing the number of Au atoms composed of the nanoclusters. The study of absorption spectrum of Au NCs showed that it was mainly due to the inter-band transitions of sp→sp or d→sp bands [16]. According to the electronic band structure for Au NCs with excitation peak at 1.8 eV and the emission peak at 2.8 eV, the transitions involved were of the 5d→6sp and 6sp→5d transitions, respectively. Au NCs generally consist of core zero-valent gold atoms (Au^0) and a shell of monovalent gold ions (Au^+).

An alternative energy transfer theory, which relates to the composition of the nanoclusters, is used to explain the fluorescence of the clusters. According to this, the fluorescence is

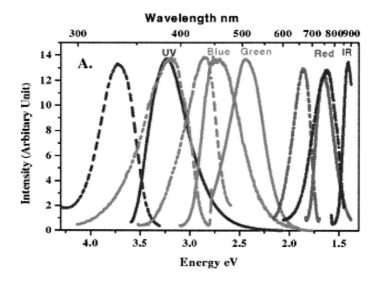
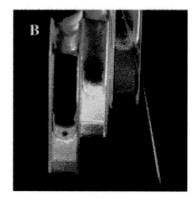

FIGURE 6.3 (a) Excitation (dashed) and emission (solid) spectra of different gold nanoclusters. Excitation and emission maxima shift to longer wavelengths with increasing initial Au concentration, suggesting that increasing nanocluster size leads to lower energy emission. (b) Emission from the three shortest wavelengths emitting gold nanocluster solutions (from left to right) under long-wavelength UV lamp irradiation (366 nm) [4].

generated by energy transfer between the core and shell [14]. Further studies also show that the fluorescence is time and temperature dependent [17, 18].

6.3 SYNTHESIS OF METAL NANOCLUSTERS

The optical properties of metal clusters were of value to early civilizations. An example is a fourth-century Roman goblet in the British museum, which is famous for its shiny colors generated by the composition of Au and Ag clusters. Even the acceptance of purple as royal colors may be due to the purple colors of colloidal Au and Cu dispersions [19]. Metal colloids were the first nano-size inorganic particles that were studied. They were prepared from metal salts in aqueous solution and have been studied for over a century. Michael Faraday performed the earliest scientific investigation of gold colloids. In the Faraday method of colloidal synthesis, the precursor metal salts undergo chemical reduction in an aqueous solution to form wine-red sols. These sols were extremely stable due to the charge stabilization via adsorbed citrate ions. The samples prepared by Faraday in the mid-1850s are displayed in the Cavendish museum in Cambridge, which still demonstrate the stability of the colloids. Current methods, though following similar concepts of reduction, vary from this method, as this cannot be used for other metals [11]. Metal nanoclusters are synthesized in many ways, but they can primarily be categorized into the solid, liquid, or gas state phase.

6.3.1 Solid Phase Synthesis

In the solid phase synthesis, all the components are in solid form. A solid metal precursor is mixed with corresponding ligand and ground well in an agate mortar, resulting in a change in color that confirms the formation of a metal-ligand complex. Then a solid reducing agent is added and ground for a few minutes. Ethanol or another solvent is added to this mixture to help reduce the excess metal and for washing unused ligands away. Many silver nanoclusters have been prepared using the solid phase method [20].

6.3.2 Gas Phase Synthesis

As the name suggests, the metal ions are reduced in the gas phase. In this method, however, aggregates of individual clusters are produced due to strong sintering between clusters. This is because in most cases no ligands are present during the cluster growth phase and individual metal clusters fuse with other clusters to form metal aggregates [11].

6.3.3 Liquid Phase Synthesis

The third and most widely studied method for synthesizing metal clusters is the liquid phase method. The cluster size in this phase is of 1–10 nm size, where cluster surface properties play a dominant role [11].

Irrespective of the type of chemical method used, metal nanocluster synthesis usually follows either of the two strategies, namely top-down of bottom-up technique. The top-down method is like a trimming process where metal nanoparticles are etched using chemicals to reduce the size from micrometer to nanometer. In the bottom-up method, individual metal atoms are clustered using various reducing agents or surface modifiers [21].

As discussed, the stability of metal NCs is important, so that they do not aggregate to form particles and lose their exceptional optical and chemical properties. However, nanoclusters are only kinetically stable and tend to aggregate into larger particles and eventually bulk metal to achieve their minimum thermodynamic energy. Thus, to obtain high-quality metal NCs, the following factors should be taken into consideration:

1. Metal precursor. Many metals have been used for the synthesis of clusters, but among them gold (Au), silver (Ag), and copper (Cu) have been mostly studied.
2. The ligand should have strong interaction with metal NCs to stabilize them. These ligands can be inorganic, like ions, or organic, like acids or thiols. The stabilization could be accomplished by either electrostatic charge, or "inorganic" stabilization and steric "organic" stabilization. In case of electrostatic stabilization, the ions adsorb onto the electrophilic metal surface, which creates an electrical layer around the cluster. This causes a repulsion force between individual particles. For steric stabilization, the metal center is surrounded by layers of a bulky material, create a steric barrier, and prevent close contact of the clusters [12].
3. The reducing condition should be strict. Methods to improve the quantum yield (QY) of the NCs such as strong reducing agents or light irradiation or sonication should be employed.
4. The reaction or aging time should be controlled. If the time is too long, then the clusters may get aggregated [22].

6.4 GOLD AS METAL PRECURSORS

Among metal nanoclusters, the gold nanoclusters (Au NCs) and their fluorescence properties have been the most widely investigated [14]. This could be because of their easy preparation, high photostability, good biocompatibility, and easy functionalization with other biomolecules. Au NCs can possess fluorescence emission from the near infrared (NIR) to visible region. The fluorescence properties can be affected by the surface ligands, structure, and charge, as well as environmental factors such as ionic strength, pH level, and temperature. These excellent properties make Au NCs a great fluorescent candidate for bio-labeling, biosensing, bio-imaging, and targeted cancer therapy applications [21]. The synthesis of Au NCs were first reported in 1966, in which phosphine was used as the ligand [23]. Since then, a wide range of methods have been used to synthesize Au nanoclusters of size less than 2 nm, using both bottom-up and top-down strategies.

6.4.1 CHEMICAL REDUCTION METHOD

In general, Au NCs are prepared by reduction of gold precursor like chloroauric acid (HAuCl4) and chloro(triphenylphosphine) gold (I) salt using reducing reagents like sodium borohydride (NaBH4), citrate, hydrazine hydrate, tetrakis (hydroxymethyl) phosphonium chloride (THPC), and ascorbic acid in the presence of a stabilizing ligand such as thiol-groups, proteins, peptides, DNA oligonucleotides, dendrimers, and polymers [21, 24]. The ligands used for stabilizing are important for the synthesis and the characteristics of Au NCs.

6.4.1.1 Thiol-Groups as Stabilizing Ligand

Thiol compounds are often used as a stabilizing ligand for the synthesis of Au NCs because they can form strong Au–S covalent bonding with Au atoms/ions [24]. The common reducing agents used to prepare Au NCs are sodium borohydride (NaBH$_4$) and THPC. In 1998, glutathione (GSH) (a tripeptide composed of glycine, cysteine, and glutamic acid) was first used as a ligand to synthesize Au NCs via chemical reduction. A polymeric gold glutathione derivative (Au(I)SG) solution was prepared, followed by reduction in an aqueous methanol medium using NaBH$_4$ as the reducing agent. After purification and separation, various sizes of GSH–Au NCs possessing fluorescence emission from blue to NIR region with the QYs of <0.1% were obtained. Using electrospray ionization mass spectrometry (ESI MS) analysis, these different nanoclusters were isolated and assigned as Au$_{10}$(GSH)$_{10}$, Au$_{15}$(GSH)$_{13}$, Au$_{18}$(GSH)$_{14}$, Au$_{22}$(GSH)$_{16}$, Au$_{22}$(GSH)$_{17}$, Au$_{25}$(GSH)$_{18}$, Au$_{29}$(GSH)$_{20}$, Au$_{33}$(GSH)$_{22}$, and Au$_{39}$(GSH)$_{24}$, depending on the number of Au atoms and GSH groups [25]. The size of Au$_{25}$(GSH)$_{18}$ was comparable to that of Au$_{28}$(GSH)$_{16}$ which was isolated earlier by Schaaff [25, 26]. This isolation and assignment of Au$_{25}$(GSH)$_{18}$ led to the revolution around this cluster, which was stable and showed red fluorescence. Since then, GSH is often used as the stabilizing ligand for studying the structure and size of Au NCs [27, 28]. By adapting a similar approach, many other thiols such as tiopronin [29], phenylethylthiolate [30], polyethylene glycol appended lipoic acid [31], and thiolate cyclodextrin [32] have been used to prepare Au NCs.

The particle size and the QY of the thiol-stabilized Au NCs usually decrease upon increasing the molar ratio of thiol- to-Au ions, and clusters stabilized by thiols using this harsh chemical reduction method usually have low QYs [24]. A more effective approach for the production of highly fluorescent thiol-stabilized Au NCs is to use GSH as a ligand as well as a reducing agent, under neutral conditions. GSH acts as a weak reducing agent but a strong capping agent. Hence, it incompletely reduces Au^{3+} to Au^0, which forms the core of the Au NCs, and reduces it to Au$^+$, which forms a complex with the thiol groups. Each core of the clusters consists of Au^0 atoms, which is stabilized, with a monolayer of thiolate-Au$^+$ complexes. With a relatively low thiol-to-Au molar ratio (1.5:1), the Au NCs formed had QY of about 15% [33]. This structure of Au NCs has been explored, and used to understand Au NCs stabilized by other ligands with sulfur groups in them, such as some amino acids and polymers.

6.4.1.2 Dendrimers as Stabilizing Ligand

In addition to thiol compounds, dendrimers have been used to prepare Au NCs with high QY. Dendrimers are repetitively branched molecules with small cavities, whose terminal groups can also be tailored to increase bioconjugation with ligands [34]. Polymers such as poly(amidoamine) (PAMAM) and poly(propylene imine) (PPI) are common dendrimer templates used to prepare Au NCs along with a reducing agent like NaBH$_4$. The stabilization of the clusters is due to the complex coordination between Au$^+$ ions and the amino or carboxylic groups of the dendrimers.

PAMAM is the common dendrimer used to prepare Au NCs with QY as high as 41% [34]. Furthermore, by changing the molar ratio of PAMAM to gold precursor from 1:1 to 1:15 different sizes of fluorescent Au NCs, Au$_5$, Au$_8$, Au$_{13}$, Au$_{23}$, and Au$_{31}$ were prepared [4]. The fluorescent Au NCs prepared with PAMAM exhibited bright ultraviolet (UV) to NIR fluorescence emission with QYs ranging from 10% to 70% [21]. More about this dendrimer will be discussed in the ligand section.

6.4.1.3 Peptides and Proteins as Stabilizing Ligands

Biomolecules such as peptides and proteins have been used as structure-defined scaffolds to induce the nucleation and growth of clusters through "green" synthesis to produce biocompatible fluorescent Au NCs [21]. By varying the molar ratio of protein/Au^{3+} (gold precursor), different sizes of Au NCs with various energy levels (emission spectrum) and QYs can be prepared. Bovine serum albumin (BSA) is the most common protein used for the preparation of Au NCs. It was first used by Xie et al. to prepare fluorescent Au NCs at 37°C under highly basic pH [35]. The reduction and stability mechanism were reported to be due to cysteine and histidine residues in BSA that can coordinate with Au^{3+} ions, while the tyrosine residues reduce Au^{3+} ions to form Au NCs that are stabilized by BSA. The pH of the solution needs to be above the pKa value of tyrosine (~10.0) to obtain a strong reducing strength of the tyrosine residues. The as-prepared BSA-Au NCs consist of 25 Au atoms (Au$_{25}$), which exhibit high QY (~6%), good biocompatibility, and excellent photostability [35]. This synthesis not only contributed to a novel alternative method for fluorescent Au NCs but also provided a new protocol for studying other metals and proteins [14].

To further explore the mechanism of Au NCs in proteins, matrix-assisted laser desorption ionization mass spectrometry (MALDI-MS) and x-ray photoelectron spectroscopy (XPS) were employed to characterize the behavior of BSA and another protein, lactoferrin (NLf). When Au^{3+} solution is added to the protein (NLf) solution, an Au$^+$-protein complex is formed, which consist of 13–14 gold atoms per protein. As the pH is adjusted to 12 or above, the bound Au$^+$ begin to further reduce to Au^0 while free protein subsequently emerges, and finally after incubation Au$_{25}$ clusters with Au$^0_{13}$ nuclei and Au$^{+1}_{12}$ shells are formed (Figure 6.4) [36].

One drawback of using protein is that to achieve strong reducing strength and capping capability, a high concentration

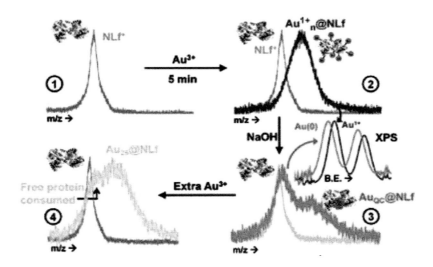

FIGURE 6.4 MALDI-MS data of time-dependent biomineralization of Au^{3+} by native lactoferrin (NLf) to form NLf-Au NCs at pH ~12.4 showing the presence of Au_{13} and Au_{25} cores (Reprinted (adapted) with permission from [36]. Copyright (2020) American Chemical Society).

(usually 0.62 mM) is required [14]. Early synthesis involved long incubation at room temperature or 37°C; however heating-assisted and microwave-assisted approaches have been used to accelerate this process [37, 38]. For example, the preparation time of BSA-Au NCs was shortened from 12 hours to less than 20 min when the preparation was conducted at 70°C instead of 30°C [24, 38].

Another widely used protein to synthesize Au NCs is human serum albumin (HSA) [39–41]. This protein is similar to BSA, and fluorescent Au NCs have stabilized using HSA. Many other proteins such as pepsin [42], insulin [43], and horseradish peroxidase (HRP) [44] have been used to prepare Au NCs, with QYs ranging from 4.3% to 12.0%.

The synthesis of protein-stabilized Au NCs can be optimized by controlling solution pH and ionic strength and reaction temperature. Solution pH is important in controlling the reducing strength, protein conformation, and capping capability of the protein, and thus various sizes of Au NCs can be prepared at various pH values. Pepsin-stabilized Au NCs with different sizes of Au_5 (Au_8), Au_{13}, and Au_{25} were prepared at pH values of 9.0, 1.0, and 12.0, respectively. They emitted blue, green, and red fluorescence, respectively [42]. The reaction temperature can be used to control the reaction rate and protein conformation and thus the size of Au NCs.

Peptides as a template for preparing Au NCs can allow precise control of both the size and surface functional properties of the Au NCs, as the amino acid composition and length of the peptide influence the physico-chemical properties of synthetic Au NCs [14]. Tyrosine and tryptophan play an important role in redox reactions during the formation of Au NCs, whilst cysteine-containing thiol groups participate in reducing and stabilizing the metal ions [45]. Based on this principle, a functional peptide was designed with a CCY terminal sequence to reduce Au^{3+} ions to form a thiol intermediate, which when further reduced to Au^0 gave peptide-Au clusters with a separate biomolecule targeting domain. Targeted imaging of different organelles or cancer-specific proteins was achieved by changing the amino acid sequence in the functional region of the peptide. For example, the CCYTAT amino acid sequence-based Au NCs can target imaging of the nucleus [46].

6.4.1.4 DNA Oligonucleotides as Stabilizing Ligands

Another excellent scaffolds for the synthesis of fluorescent Au NCs are DNA oligonucleotides and single nucleotides. Poly-cytosine (C) and poly-adenine (A) DNA oligonucleotides were used to prepare blue fluorescent Au NCs with citrate as the reducing agent under low and neutral pH conditions, respectively. The interaction between gold and cytosine was weaker than that between silver and cytosine, about which we will learn more in the next section. However, the emission intensity of these Au NCs can be enhanced by receiving energy from nucleobases [47].

Different types of DNA sequences such as hairpin DNAs (HP-DNAs), single-stranded DNAs (ssDNA), and fully matched DNAs as ligands affect the characteristics of the Au NCs [48] as follows. (1) The emission behavior of the HP-DNA stabilized Au NCs is dependent on the loop sequences and loop length of the strand. The cytosine loop was the most efficient host to produce fluorescent Au NCs followed by guanine loops. However, the loop composed of thymine and adenine produces Au NCs with a much weaker emission. (2) The emission behavior of Au NCs hosted by the ss-DNAs with an identical base composition as the corresponding HP-DNAs also exhibits a cytosine-rich dependence. (3) The fully matched DNAs seem to be least efficient of all structures.

The sequence-dependent formation of fluorescent Au NCs could be caused by differences in binding nucleophilicity of the DNA heterocyclic nitrogen and exocyclic keto groups to the hydrolyzed Au^{3+} species. The pH conditions and incubation time are also important parameters for the emission behavior of DNA stabilized Au NCs [49]. Blue, green, and yellow emission Au NCs could be simultaneously prepared using single cytidine units as the ligand by using different reaction times and/or pH environments [21].

6.4.2 Photoreduction Method for Synthesis of Au NCs

Compared to the chemical reduction synthesis of Au NCs, photoreduction avoids the use of hazardous reducing agents such as sodium borohydride and provides a nontoxic and eco-friendly approach. The metal precursor such as Au^{3+} in presence of a polymer (ligand) is irradiated with a UV light source to produce Au NCs with QY from 3.8% to 20.1% [50, 51]. The fluorescence properties and the size of the Au NCs formed depended on the polymer (ligand) and can be controlled by varying the molar ratio of polymer to Au ions [24].

6.4.3 Bio-reduction Method for Synthesis of Au NCs

Fluorescent Au NCs can be biosynthesized in a cancer cell line such as human hepatocarcinoma (HepG2) and leukemia (K562) by the reduction of Au^{3+} inside the cell's cytoplasm [52]. This interesting biosynthesis does not occur in normal cells. Thus, this mechanism may provide a promising chance for *in vivo* bioimaging of tumors cells as well as sensing cancer cells. However, the exact metabolic pathway of the biosynthesis of these Au NCs is not yet clear.

6.4.4 Electro-reduction Method for Synthesis of Au NCs

Fluorescent Au NCs can be prepared by electro-reduction of a gold electrode as the precursor with poly(N-vinylpyrrolidone) (PVP) as a ligand. One of the smallest reported Au NCs was prepared with this method [53]. The as-prepared Au NCs consist of only two and three atoms and still show very stable photoluminescent properties.

6.4.5 Etching Method for Synthesis of Au NCs

Besides reduction, Au NCs of fluorescent properties can be prepared through a top-down approach, by etching large AuNPs through the ligand-induced etching method. A large quantity of ligands is required in this process. Green fluorescent Au NCs were prepared using polyethylenimine (PEI) containing multivalent imine groups to replace the original capping agent dodecylamine and to etch the AuNPs. These clusters have high QY of 10–20% [54]. Thiol ligands such as 11-mercaptoundecanoic acid (11-MUA) were also used as an etching agent to prepare Au NCs from AuNPs. This reaction was performed under high pH conditions, during which 11-MUA shows strong etching ability to etch the surface Au atoms of the NPs and simultaneously provide strong coordination to eventually form stable and fluorescent 11-MUA-capped Au NCs [55]. The advantage of this approach is that the size and the optical properties of Au NCs can be controlled by using different thiol compounds. Using core etching, GSH-stabilized Au NCs ($Au_{25}GSH_{18}$) can be further etched with octanethiol, leading to the formation of brightly red fluorescent $Au_{23}GSH_{18}$ [56]. Moreover, a prepared nonfluorescent Au NC (stabilized with didodecyldi-methyl-ammonium bromide [DDAB]) can be converted to another fluorescent Au NC (AuNC@DHLA) (dihydrolipoic acid [DHLA]) via a simple ligand exchange reaction, as shown in Figure 6.5 [57].

6.5 SILVER AS METAL PRECURSORS

Gold and silver belong to the same group in the periodic table. However, they show different properties in many cases. Silver (Ag) in the zero valent state is much more reactive and can be easily oxidized as compared to gold; thus, it is difficult to study their clusters in detail [58]. However, synthesis of silver nanoclusters (Ag NCs) is easier than Au NCs, with minute variations in parameters of the synthesis, including the temperature, the reducing method, the stabilizers, and the initial ratio of silver: the stabilizer may lead to the formation of large nonfluorescent Ag NPs. Thus, defining rules for Ag NCs synthesis remains difficult because similar reagents and reaction conditions may lead to Ag NPs instead of NCs. In general, silver clusters in solution are prepared by the reduction of silver ions. The chemical reduction and the photo-reduction are the common methods used for preparing Ag NCs [59]. Though the true mechanism of fluorescence in Ag NCs is not fully known, the specific properties of Ag NCs, such as the stability and QY, depend largely on the ligand used during reduction.

The first precise composition of monolayer-protected Ag NCs using a chemical reduction method was determined using ESI MS analysis [60]. The clusters were synthesized using $AgNO_3$ dissolved in a suitable solvent, and the ligand was added to the solution. Ag-ligand intermediates are formed and then $NaBH_4$ is added as a reducing agent, which led to further reduction to Ag NCs [7]. The color of the final solution changes as the cluster formation occurs, which can be monitored using UV-vis spectroscopy. Further purification by the precipitation method results in a purified cluster solution. As discussed, the ligand as well as its ratio with respect to silver is important for synthesis of photoluminescent Ag NCs.

Among the several advantages of Ag NCs as a fluorescence sensor, they can also be functionalized to attach to proteins or cells, and due to their small size, they do not create much perturbance to the protein or cell to which they are attached [59]. Furthermore, according to the tests using HeLa cells, Ag NCs exhibit no significant toxicity [61]. In this section, we will go through the different types of ligands used for the Ag NCs synthesis.

6.5.1 DNA Oligonucleotides as Stabilizing Ligand

The strong interaction between silver ions and DNA bases has led to the use of them as templates for the synthesis of Ag NCs. Silver ions are more attracted to the heterocyclic bases over the phosphates [62, 63] and prefer the single-stranded DNA (ssDNA) over the double-stranded DNA [64, 65].

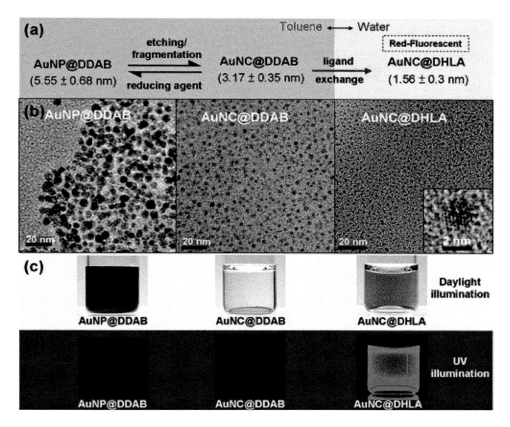

FIGURE 6.5 (a) General strategy to fabricate water-soluble fluorescent AuNCs. DDAB-stabilized AuNPs (AuNP@DDAB) are etched by the addition of Au precursors (HAuCl or AuCl) to smaller nanoclusters (AuNC@DDAB). By the addition of reducing agent (TBAB) the AuNCs grow again reversibly into bigger Au NPs. The hydrophobic AuNC@DDAB become water soluble upon ligand exchange with dihydrolipoic acid (AuNC@DHLA); (b) TEM images of AuNP@DDAB, AuNC@DDAB, and AuNC@DHLA; 100 particles were randomly selected for measuring the size distribution, resulting in 5.55 ± 0.68 nm, 3.17 ± 0.35 nm, and 1.56 ± 0.3 nm in diameter, respectively. (c) Pictures of particle solutions under daylight. Contrary to AuNP@DDAB solution, which features the red color of surface plasmon absorption, AuNC@DDAB and AuNC@DHLA display a colorless and brown translucent solution without plasmon absorption, respectively. (d) Pictures of the same particle solutions under UV excitation. The AuNC@DHLA solution shows red photoluminescence [57].

However, for the synthesis of highly luminescent and photostable Ag NCs, which do not aggregate into nanoparticles, the proper oligonucleotides, the silver to DNA ratio, and pH and temperature are crucial.

In 2004, DNA was first used as a template for synthesis for Ag NCs. A 12 base (5′-AGGTCGCCGCCC-3′) was employed for the time-dependent synthesis [66]. The clusters have intense absorption in the region 400–550 nm and emission at around 630 nm. Circular dichroism associated with the silver cluster electronic transitions is evidence that the clusters are bound to DNA (Figure 6.6). The Ag^+ ions first strongly interacted with the DNA strands followed by reduction. Further reduction to Ag NCs occurred with the addition of $NaBH_4^-$. According to the ESI-MS study, 1–4 Ag atoms interacted with the DNA strand, thus forming small clusters.

The mechanism of the cluster formation includes Ag^+ ions first strongly interacting with the DNA strands, followed by reduction into Ag NCs with the addition of $NaBH_4^-$ or other reducing agents. According to ESI-MS study, 1–4 Ag atoms interacted with a DNA strand, thus forming small clusters [66].

The cluster properties are highly dependent on the sequence, and different sequences of DNA strands lead to different well-defined clusters with distinct emission. Twelve-base DNA strands with different sequences were used to prepare Ag NCs [67]. Ag NCs with blue, green, yellow, and red emission were prepared depending on the oligonucleotides sequence, with QY as high as 30%.

In terms of bases, an MS-based study revealed that both cytosine (C) and guanine (G) have equally good chemical affinity towards silver [65]. While another ^{1}H NMR and theoretical study revealed that only cytosine had the best affinity with silver [68]. This difference in reports may be due to the conformation of oligonucleotides, as well as the different relative concentration used in their experiments [8]. According to the MS study 0.29 silver per nucleotide was used and the theoretical study reported 0.5 silver per nucleotide was used. These discoveries led the further use of cytosine-rich oligonucleotides as templates for preparation of Ag NCs.

A 12-mer cytosine (5′-CCCCCCCCCCCC-3′; also denoted as dC) was used in the stoichiometry ratio of 2:1:1 with respect to oligonucleotides: Ag^+: BH_4^- [69]. The nanoclusters formed were excited at different wavelengths and various emission peaks, mainly at 485 nm (blue), 525 nm (green), and 665 nm (red) were observed. This report revealed the presence of

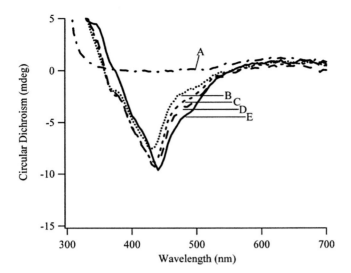

FIGURE 6.6 Induced CD spectra for the electronic transitions associated with the nanoclusters. For these spectra, [oligonucleotide] = 10 µM, [Ag$^+$] = 60 µM, and [BH$_4^-$ = 60 µM, and the cell path length was 5 cm. The spectra were collected 2 min (A, dashed-dotted line), 20 min (B, dotted line), 40 min (C, fine dashed line), 60 min (D, coarse dotted line), and 150 min (E, solid line) after adding the BH$_4^-$ [66].

multiple electronic transitions and different emitters in the clusters formed. Interestingly, chemical conversion between the emitters occurs with time, the peak 665 nm, and increase in the emission peak 525 nm.

The oxidation of reduced NCs with red emission led to formation of oxidized NCs with green emission. Reversely, when the reductant was added, a chemical reduction of green emitters to red emitters occurs [69, 70].

As discussed, different bases have a different affinity towards silver and their interactions produce different emissions. The influence of thymine (T) and cytosine (C) bases and their combination in oligonucleotides revealed that the Ag NCs synthesized by dT$_{12}$ and dT$_4$C$_4$T$_4$ have similar fluorescence properties [70]. The emission intensity in both cases increases up to pH 9.5 and then decreases. This pH is close to the pK$_a$ value of N3 of the thymine base, which indicates that after this pH, the deprotonated thymine forms a complex with the fluorescent Ag NCs leading to the quenching. The length and the conformation of the DNA also influences the properties of Ag NCs formed. Longer oligonucleotides can produce different emission clusters with high QY of 64% [71]. Among the conformations, Ag NCs stabilized with hairpin structure with C-loops showed the highest fluorescence [65]. The fluorescence properties and stability of silver cluster in hairpins depend on the number of cytosines in the loop [72]. This is because cytosine-rich oligonucleotides can form red emitting Ag NCs and red emitters are brighter than green emitters [8].

Lastly, DNA stabilized Ag NCs can be easily functionalized to enable their attachment to biological molecules, like proteins. Ag NCs stabilized by thiolated 24-mer oligocytosine was linked to proteins via chemical coupling [73]. These functionalized DNA-Ag NCs can serve as alternatives to organic dyes and semiconductor quantum dots.

6.5.2 DENDRIMERS, POLYMERS, AND THIOLS AS STABILIZING LIGANDS

There have been several reports of polymers and their dendrimers for the synthesis of Ag NCs via chemical reduction and photoreduction. The end functional groups in the dendrimers play an important role in NCs as well their fluorescence properties. Highly photostable fluorescent Ag NCs were synthesized in aqueous solution using an OH-terminated poly(amidoamine) (PAMAM) dendrimer as the ligand [74]. With a specific dendrimer to silver ions ratio (OH: Ag ~ 1:0.2), Ag NCs with emission from 533 to 648 nm, were prepared using photoreduction with a stabilization of Ag blue light. In another study, polymer poly(N-isopropylacrylamide-acrylic acid-2-hydroxyethyl acrylate) (poly(NIPAM-AA-HEA)) microgel was used as the ligand [75]. The NCs were prepared in a molar ratio COOH: Ag$^+$ of 1: 1 with UV irradiation for a defined time interval. However, the microgel contained both OH-groups and COOH-groups, and the latter were reported to be critical for the formation of Ag NCs. Interestingly, the ligand is sensitive to external factors like temperature and pH, which facilitate these Ag NCs for various applications.

An acrylates-based polymer was used to synthesize Ag NCs by γ-radiation, but no fluorescence was detected [76]. However, when the density of COOH-groups was increased by using a star-shaped polymer with 36 arms, fluorescent Ag NCs were formed under similar conditions [77]. Poly(methacrylic acid) (PMAA) is an excellent ligand to produce NCs with high QY, high photostability, and is itself very sensitive to the environment, especially pH [78]. However, most of these clusters were prepared using UV radiation [59], and they cause many health issues. Linear PMAA was used to prepare Ag NCs in water solution by photoreduction with visible light [79]. This was a breakthrough because a more environmentally friendly procedure was demonstrated. A sono-chemical method that exploits the chemical effects of high-intensity ultrasound waves with PMAA as a ligand is another easy and hassle-free method for synthesis of Ag NCs [80]. The stoichiometry ratio of the ligand groups to Ag$^+$ play a significant role, as the silver clusters are capable of dynamic transfer between molecules. Silver clusters were prepared by photoactivation using PMAA as a scaffold, and the change in the Ag$^+$: PMAA ratio results in distinct spectral bands [79]. A blue shift was achieved by the addition of pure PMAA, and if the amount of polymer decreases to the initial ratio, say Ag$^+$:PMAA ~ 0.5:1, the new optical band would move back to the solution with the same ratio (0.5:1). This blue shift was reported to be due to the redistribution of the formed silver clusters in the newly available PMAA chains.

Other than polymers and dendrimers, small molecules with carboxylic groups or thiols such as captopril, glutathione [81], and dihydrolipoic acid [82] were used to stabilize Ag NCs in solution. The molar ratio of thiol groups or COOH groups was important to produce a large strokes shift. Etching of large Ag NPs was used to produce Ag7 and Ag8 nanoclusters with mercaptosuccinic acid (MSA) as the ligand [83]. A decrease in temperature led to up to a 9% increase in the fluorescence QY of Ag8 nanoclusters. MSA was also used to prepare Ag9 nanoclusters via a solid-state route. For this preparation, silver nitrate powder was mixed with MSA powder. Sodium borohydride powder was added to this mixer and grinded to produce clusters in solid state [20]. The clusters were purified by PAGE and were stable in solvent mixtures with water.

Fluophore thioflavin T (ThT) was used to prepare fluorescent Ag NCs by photoreduction which were used to image proteins [84]. The emission was at ~450 nm and was proposed to have originated by the fluorescent silver clusters and by metal-enhanced fluorescence of ThT. Using these highly fluorescent clusters, a single fibril could be detected. The ligand that is used for the synthesis impacts the application of the Ag NCs that are formed. For example, the clusters formed using thiols were sensitive to Hg(II) ions [82], while those prepared in PMAA were sensitive to cysteine [85] and Cu(II) ions [86] and that with ThT lead to imaging of proteins [84].

6.5.3 Proteins and Peptides as Stabilizing Ligands

The interactions between proteins and silver have been known for a long time and were explored to develop a common staining method for the cell nucleolus [87, 88]. Nucleolin and other proteins of the nucleolus show high affinity to silver ions due to their amino-terminal domain. Addition of silver ions to the cells leads to the formation of an Ag NPs stain [89]. By tuning the staining conditions, cells can be stained with fluorescent Ag NCs instead of Ag NPs [90]. Fixed cells were stained with a low concentration of silver nitrate solution and allowed to reduce by photoactivation under ambient conditions, which resulted in the formation of Ag NCs exhibiting emission of broad spectra between 500 and 700 nm under blue excitation. This discovery leads to using silver salts for various biological applications.

General and definitive conclusions regarding the attraction between silver ions and peptides cannot be easily extracted. However, some authors have demonstrated a higher degree of silver binding in peptides rich in proline and hydroxyl residues [91], whereas others showed a preferential affinity of silver for methionine-containing peptides as compared to their non-methionine-containing counterparts [92].

Besides the formation of luminescent Ag NCs in fixed cells with the presence of proteins, Ag NCs have been synthesized using just proteins as ligands. Enzymes turn out to be a useful protein to work with because their intrinsic properties effect the applications as well. Bovine pancreatic α-chymotrypsin (CHT), a digestive enzyme that performs proteolysis of proteins, was used as a stabilizing ligand to produce photostable fluorescent Ag NCs via chemical reduction by $NaBH_4$ [93]. During the synthesis, the enzymatic activity of CHT reduced by 2.8 times as compared to the native enzyme. This is because the high amount of $NaBH_4$ (molar ratio of CHT: Ag^+:BH_4^-) was 1:10:100, which led to denaturation of the protein by cleaving disulfide bonds. The increase in the pH of the reaction solution also aggregated the protein.

As mentioned earlier, the high affinity of silver ions to nucleolin prompted the use of synthetic peptides derived from nucleolin as a stabilizing ligand to produce fluorescent Ag NCs [90]. Oligopeptides with amino acids like glutamic acid (E), lysine (K), and aspartic acid (D), which are present in nucleolin as well as cysteine (C), which is known to bind silver, were used for the Ag NCs preparation via chemical reduction [94]. However, the clusters formed had very low photostability (3 days). A histidine-rich peptide (AHHAHHAAD) was used for the formation of fluorescent Ag NCs [95]. Another oligopeptide with D, C, and K but also histidine (H) and asparagine (N) or leucine (L) "HDCNKDKHDCNKDKHDCN" was used to produce Ag NCs exhibiting fluorescence at 630 nm when excited at 400 nm [59]. Furthermore, sunlight was used for photoreduction of Ag NCs with a dipeptide-based supramolecular hydrogel as ligand. There were free COOH-groups in each peptide that formed a complex with silver ions [96].

The major disadvantage of protein- and peptide-encapsulated Ag NCs is the relatively low fluorescence QY (about 3%) [73]. Though these nanoclusters could be loaded in living cells for staining, the relation between their structure and the Ag NCs' optical properties are still not clear [90, 95].

6.5.4 Inorganic Scaffolds as Stabilizer

Inorganic glasses provide a solid matrix that stabilizes Ag NCs by immobilization and prevent their tendency to aggregate to large nanoparticles. Activation of the clusters is often achieved by laser irradiation [97] or synchrotron irradiation [98]. Zeolites are highly porous material with cage-like structures. Ag NCs were stabilized within the confinement of these zeolite cages. The clusters were prepared using UV light [99] or heat treatment [100]. These clusters possess high photostability and could be used for various applications such as wavelength converters for fluorescent lamps and biocompatible labels [59].

6.6 COPPER AS METAL PRECURSOR

As compared to Au NCs and Ag NCs, there have been fewer reports of the synthesis of copper nanoclusters (Cu NCs). This is because they are very small and therefore more difficult to prepare and are easily oxidized [6]. There have been several methods that were used to successfully synthesize Cu NCs. Like the earlier NCs, the ligand or template that is used to stabilize the NCs plays an important role in the optical and chemical properties of the Cu NCs.

6.6.1 Ligand-Based Synthesis of Cu NCs

Template- or ligand-based synthesis has the advantage of controlling the core size of the nanoparticles or clusters according to the ligand's state, functional groups, and the cavity they provide [101]. Copper nanoclusters (Cu NCs) are often stabilized by phosphate or thiolate ligands, but there are reports of carbenes, DNA, and dendrimers being used as well. Electronically neutral ligands like phosphines, amines, and carbenes will lead to neutral clusters with a Cu^0 state [102]. However, when anionic ligands like hydride or thiolate is used, the Cu atoms are in the oxidation state I. In such nanoclusters with Cu^+, there are not many Cu-Cu interactions, only Van der Waal interaction between the atoms occurs. The stability in this case is provided by the interaction of the anionic ligands with the surface Cu^+ [103]. In larger nanoclusters, the inner atoms are at Cu^0 and the surface atoms are at Cu^+ (very similar to Au NCs) [36, 104].

PAMAM dendrimers with an OH terminal is a good scaffold to synthesize fluorescent Cu NCs via chemical reduction of the Cu ions by $NaBH_4$ [105, 106]. DNA is another effective ligand for Cu NCs synthesis. Both double-stranded (ds) [107] and single stranded (ss) DNA [108] are employed to produce fluorescent clusters. In ds-DNA, not only the length of the stand but also the type of base located in the major groove is important for the size of the Cu NCs and their properties [109]. Furthermore, it was found that thymine plays a dominant role in producing red-emissive fluorescent CuNPs on ss-DNA templates [108].

6.6.2 WATER-IN-OIL (W/O) MICROEMULSION FOR SYNTHESIS OF CU NCS

Microemulsion is a liquid mixture of aqueous and organic phases. It acts as an ideal reactor for the preparation of metallic NPs with precise control of the NP shape, size, and distributions [110]. A series of small atomic Cu NCs was successfully synthesized following this strategy [111]. The core size and photoluminescence properties of these clusters depend on the percentages of the reducing agent added during the synthesis procedure. The microemulsion system was created by mixing high-purity sodium-dodecyl sulfate (SDS, as a surfactant), isopentanol (as a cosurfactant), cyclohexane (as an oily phase), and Cu (II) sulfate solution (as an aqueous phase). Then, a suitable amount of $NaBH_4$ (reducing agent, aqueous solution) was slowly added dropwise into the formed w/o microemulsion. Photoluminescent Cu NCs with blue emission (Cu_n, n (number of atoms) ≤ 13) can be obtained using very low percentages of the reducing agent (<10%), while as the percentage of reducing agent is increased, the core size increased and photoluminescent clusters disappeared and surface plasmon would appeared (Figure 6.7).

6.6.3 ELECTROCHEMICAL SYNTHESIS OF CU NCS

Fluorescent Cu NCs could be prepared using a simple electrochemical approach. A Cu anode is used as the metal precursor with a ligand like tetrabutylammonium nitrate. During the electrolysis, fluorescent Cu_n ($n \leq 14$) NCs are synthesized on the cathode [112]. Cu NCs prepared by this method are very stable with unusual high QY of 13% in the visible range. These clusters also exhibit amphiphilic characteristic (disperse in both polar and nonpolar solvents), which holds great advantage for biological applications.

6.6.4 OTHER METHODS FOR SYNTHESIS OF CU NCS

The Brust-Schiffrin method is a two-phase method where a phase transfer agent is used for the preparation of nanomaterials. It has been modified to prepare stable Cu NCs with 2-mercapto-5-n-propylpyrimidine (MPP) as a protecting ligand [113]. The clusters exhibited emission at 425 and 593 nm with QY of 3.5% and 0.9%, respectively. MPP was also used as a protecting ligand for synthesizing Cu NCs via a one pot microwave-assisted method. The clusters show dual emission at 423 and 593 nm [114]. In another microwave-assisted method, no ligand or reducing agent is used to produce blue fluorescent Cu NCs [115]. $CuCl_2$ (precursor), and NaOH in ethylene glycol is microwaved, after which HCl is added. The synthesized clusters are finally extracted with diethyl ether. These Cu NCs were highly resistant to oxidation

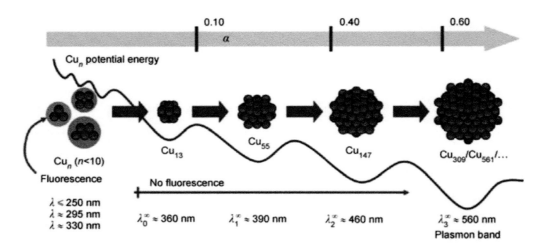

FIGURE 6.7 Schematic of the evolution of copper cluster size with an increase in the percentage of $NaBH_4$ (α). Copper clusters, Cu_n, with n (number of atoms) ≤ 13, show fluorescence. Estimated wavelengths of the plasmon band for different sized copper clusters are shown at the bottom [111].

and exhibited strong fluorescence. Cu NCs have been used for several applications such as H_2O_2 detection and Hg^{2+}, which will be explored in a later section.

6.7 BIMETALLIC NANOCLUSTERS

As mentioned earlier, the photoluminescence properties of metal nanoclusters are directly correlated to their structures and composition; thus, there is the possibility to manipulate such properties through precise variations in composition [116]. This led to the exploration of luminescent bimetallic nanoclusters. Bimetallic clusters or alloy clusters consist of two or more metals and are completely distinct in structural and photoluminescence properties as compared to their monometallic counterparts [117]. One metal is doped during the synthesis of the other metal nanostructure, and this doping could improve the cluster stability.

Palladium (Pd) doped into gold protected by thiols (SR) ($PdAu_{24}(SR)_{18}$) was the very first example of an atomically precise bimetallic nanocluster [118]. These clusters differ from their monometallic analogue $Au_{25}(SR)_{18}$ and show enhanced photostability. Further analysis showed that Pd atom was in the central position of the icosahedral core with Au atoms around.

A 13-atom Au-Ag alloy cluster was prepared to manipulate the composition and hence the luminescence [119]. The preparation of these clusters involves three steps. First Ag NPs are protected by mercaptosuccinic acid (H_2MSA) to form Ag-H_2MSA NPs. Then via interfacial etching, an $Ag_{7,8}$ cluster (mixture of seven and eight atoms of Ag clusters) was synthesized. Finally, addition of gold precursor ($HAuCl_4$) to the as-prepared $Ag_{7,8}$ clusters would lead to the bimetallic cluster, which emitted red fluorescence.

A protein such as BSA was used to synthesize another Au/Ag cluster [116]. AuNCs and Ag NCs were used as the starting materials, and either through NCs-NCs interaction or galvanic exchange reaction the alloy clusters were formed. Mass spectroscopy and other elemental analysis confirmed the cluster formation, and it further suggested that clusters were formed across the entire compositional window (Figure 6.8).

Copper (Cu) was doped with Au NCs to provide not only low-cost materials but also to understand the formation mechanism of bimetallic clusters in general. The bimetallic Au/Cu NCs were of 2–3 nm in size and displayed a photoluminescence that could be tuned by changing the composition of the alloy itself [120]. When the molar ratio of Cu in the alloy was varied from 0% to 100%, the photoluminescence shifts from 947 to 1067 nm (excitation at 360 nm). These bimetallic NCs are promising, as their emission can extend to NIR spectral regions and can be used to prepare tunable NIR nanoprobes for various application. Furthermore, they show excellent photostability and low cytotoxicity [3]. However, the exact mechanism of the photoluminescence of these alloy metal clusters are still unknown and requires more studies in terms of their coordination with metals in the alloy.

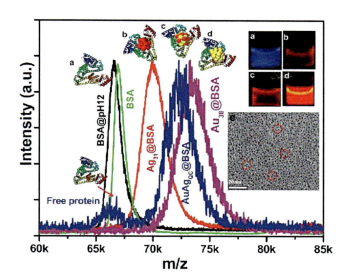

FIGURE 6.8 MALDI-MS data shows the tunability of the composition of the Au-Ag alloy clusters in protein templates. MS of native BSA (violet) and BSA at pH 12 (black). The products of 900 mL Ag NC@BSA + 100 μL Au NC@BSA(90:10) peak at m/z 71 100 (orange trace), 500 mL Ag NC@BSA+ 500 μL Au NC@BSA(50:50) peak at m/z 72 300 (blue trace), and 100 mL Ag NC@BSA + 900 mL Au NC@BSA(10:90) peak at m/z 72 800 (olive trace) lie between Ag NC (red trace) and Au NC@BSA (magenta trace), suggesting compositional variation in the alloys. Inset (a) Ag NC@BSA; (b) Au NC: Ag NC@BSA(10:90); (c) Au NC: Ag NC (50:50); (d) Au NC: Ag NC(90:10) and (e) TEM image of Au NC: Ag NC(90:10). The ratio mentioned here is v/v, which can also be considered as a molar ratio since Au and Ag solutions of the same molarity were used [116].

6.8 OTHER METAL NANOCLUSTERS

Platinum (Pt) and palladium (Pd) were also used to prepare luminescent NCs. In the absence of any stabilizing ligand, Pt NCs [121] and Pd NCs [122] were synthesized in a one-pot synthesis with N,N-dimethylformamide (DMF) solution. The precursor for Pt NCs was chloroplatinic acid (H_2PtCl_6) and $PdCl_2$ for Pd clusters, and the modified reaction time was 8 hours for Pt and 6 hours for Pd.

The as-prepared Pt NCs were stable for at least 6 months when stored in the dark. Under different excitations, different emissions were observed for the PtNCs, which suggests that there is more than one emitter in the solution, or the emission originates from the electronic transitions from multiple excited states to the ground state. Blue emitting Pt NCs were prepared using fourth-generation polyamidoamine dendrimer (PAMAM (G_4-OH)) as the stabilizing ligand by reducing chloroplatinic acid with $NaBH_4$ [123]. Later this ligand was replaced with mercaptoacetic acid (MAA) because PAMAM (G4-OH) underwent simple oxidation during the process. So, in order to avoid the possible fluorescence interference from oxidation species of the dendrimer, the ligands were exchanged. The as-prepared $Pt_5(MAA)_8$ NCs shows blue photoluminescence with QY of 18% and was used for HeLa cell detection [123]. A solid-state route was used to synthesize atomically precise blue emitting platinum clusters [124].

Protein-PtNCs were synthesized which showed oxidase-enzyme activity [125]. The clusters had blue emission but very poor QY (3×10^{-3}). Most recently, water-soluble PdNCs with blue-green emission were synthesized with methionine as the ligand [126]. A water bath heating method was used, and ascorbic acid was used for reducing the metal ions. These clusters have high QY of 5.47% and can be used to detect hemoglobin with a detection limit of 50 nM.

6.9 EFFECT OF LIGANDS ON NCs

As mentioned in the previous section, the role of the ligand in the synthesis of metal nanoclusters is critical, as they not only sterically stabilize the metal nanoclusters but are important for determining the eventual structure and stable size of the cluster [11]. The emission wavelength of the clusters not only correlates with its size but also its protected molecules [10]. Figure 6.9 shows different ligands forming different types of clusters. Without stabilization, metal nanoclusters tend to strongly interact with each other and lead to aggregation into large particles. Thiols, dendrimers, polymers, oligonucleotides, peptides, and proteins have been commonly used as ligands. Many of these ligands have already been discussed in regard to synthesis of different metal NCs. In this section we will briefly discuss how they affect the nanoclusters' structural and photoluminescence properties.

6.9.1 DENDRIMERS AS LIGANDS

Dendrimers can stabilize metal ions in solution and have been used to synthesize Au [34], Ag [74], Cu [105, 106], and even Pt NCs [123]. By just changing the ratio of the dendrimer to metal concentration, as well as the dendrimer generation, metal nanoclusters with a range of emission wavelength from the UV to NIR region can be obtained [4]. A poly(amidoamine) dendrimer (PAMAM) with repeatedly branched molecules with different generations is the commonly used dendrimer. The OH groups in the PAMAM dendrimers act as good scaffold and as the generation increases, the number of this group increases and thus provides more stability. Both photoreduction and chemical reduction method of synthesis have been used with QY of 42% [34].

6.9.2 THIOLS AS LIGANDS

Brust et al. first used thiols to prepare AuNPs in 1994, and ever since, these are used as the most common ligand to stabilize metal NCs [285]. This mainly is due to the strong interaction between thiols and Au or Ag metal. These as-prepared NCs have large Stokes shifts, and their fluorescence emission is tunable via selection of appropriate length of thiolated ligands [55]. Glutathione (GSH) is the most common thiol to be used for synthesizing MNCs [26, 127–129]. Some of the other thiols used are tiopronin [29, 30, 130], meso-2,3-dimercaptosuccunic acid [131], phenylethylthiolate [132], dodecanethiol [133], and mercaptoundecanol [55, 134] and dihydrolipoic acid (DHLA) [82]. The emission wavelength of these clusters ranges from red to infrared [10]. While using thiols, the luminescence efficiency of the MNCs depends heavily on the thiol to metal concentration ratio. The higher this ratio, the smaller the NCs and a blue shift of the emission wavelength [82]. Other than the common chemical reduction, the etching method has also been used for synthesis of MNCs. In this method, thiols play a significant role [55], and different

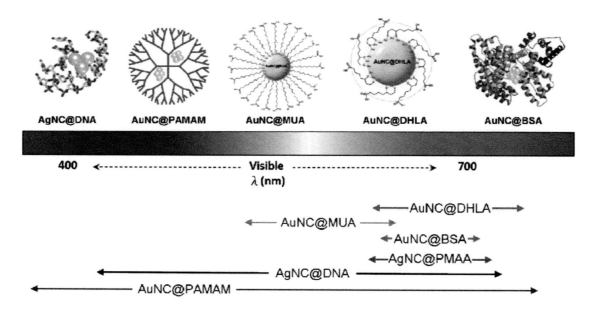

FIGURE 6.9 Representative fluorescent noble-metal nanoclusters scaled as a function of their emission wavelength superimposed over the spectrum. Different ligands show different capabilities to tune the emission wavelength of metallic NCs from current reports [10]. DHLA, dihydrolipoic acid; PAMAM, poly(amidoamine); MUA, mercapto-undecanoic acid; PMAA, poly(methacrylic acid); BSA, bovine serum albumin; DNA, deoxyribonucleic acid.

kinds of MNCs can be obtained by changing the pH of the reaction solution [135].

Two possible mechanisms are described to explain the etching-based strategy: (1) The metal atoms are removed from the nanoparticle surface by the thiols which is in excess and form Au(I)-thiolate complexes, which then undergo strong aurophilic interactions and form the metal NCs. (2) Thiols etch the metal atoms of the nanoparticles directly to form the MNCs. In both the cases, we observe that thiol is explained to be responsible for the etching. Interestingly, Au_{23}NCs can also be etched from Au_{25} NCs [56].

6.9.3 Proteins as Ligands

In nature, organisms undergo biomineralization process in which inorganic ions or minerals are sequestered by proteins to decrease their harmful effects. Keeping this process in mind, proteins and peptides have been used to synthesize metal NCs. BSA is the first protein to be used as a template to prepare 25 gold atoms Au NCs [35]. Proteins as a ligand for synthesizing NCs function as both a stabilizing agent and a reducing agent. This "green" synthesis of MNCs paved the way for a plethora of studies around the use of other proteins as a ligand for synthesizing fluorescent metal NCs [5]. Proteins such as lysozyme [136] and transferrin [137, 138], insulin [43, 139] and pepsin [42] have been used as ligands to synthesize a variety of metal and alloy NCs. Proteins are also functional as etching agents [140], and even denatured protein can be used as a ligand [141]. Enzymes have been used to prepare MNCs. Multifunctional nanoprobes can be constructed using the integration of the catalytic function of the enzyme integrated with the luminescence of the MNCs in a single cluster. HRP [44] and bovine pancreatic α-chymotrypsin [93] were employed as templates for the synthesis of MNCs. The former showed enzymatic activity; however, due to excess reducing agent ($NaBH_4$), the latter lost its enzyme activity.

Protein-stabilized NCs, because of their biocompatibility, have been used in several applications, and many of them have been discussed in the application section [7, 37, 142–147]. Another attractive feature of these protein-stabilized MNCs is that most of the proteins retain their intrinsic biological activity, leading to the preparation of MNCs with interesting biofunction. Insulin-stabilized fluorescent Au NCs have been proved to retain bioactivity and biocompatibility in blood glucose regulation as the natural insulin in the *in vivo* study [43]. In addition, HRP retained its activity in HRP–Au NCs, which enables catalytic reduction/oxidation of H_2O_2 [44].

6.9.4 Oligonucleotides as Ligands

Metal cations known to interact with DNA have led to the use of DNA as ligands to stabilize MNCs [148]. As discussed in the previous section, silver ions in particular possess high affinity to the cytosine bases of the single-stranded DNA [64, 149]. This interaction has been exploited to prepare several DNA-stabilized Ag NCs [65, 66, 69–71, 150, 151]. Not only the bases of the DNA [70] but also the sequence and the secondary structure of the oligonucleotide strand [65] influence the cluster formation and the fluorescence properties of the cluster. Thus, variation in these parameters can produce Ag NCs with distinct emissions ranging from blue to NIR [67]. The cytosine loops in the DNA hairpins also affect the fluorescence and photostability of the NCs [72]. DNA-oligonucleotides have been used to etch AuNPs to prepare Au NCs [152].

6.9.5 Polymers as Ligands

Polymers in solution and gels have been used to stabilize metal NCs. Among them, polymers with abundant carboxylic acid groups (COOH groups) were preferred as a ligand [77]. Different polymers include poly(N-isopropylacrylamide-acrylic acid-2-hydroxyethyl acrylate) [75], poly(methacrylic acid) (PMAA) [78-80, 153], and polyethylenimine (PEI) [54]. In this case as well, the polymer to metal concentration ratio is important, and variation in this ratio leads to formation of nonfluorescent to fluorescent NCs with particle size between 1.1 and 1.7 nm [154]. Extremely small metal NCs consisting of only two or three atoms were also reported, which were synthesized using poly(N-vinylpyrrolidone) (PVP) via a simple electrochemical technique [53].

6.10 CHARACTERIZATION TECHNIQUES FOR NANOCLUSTERS

The characterization of the synthesized metal NCs is an essential step as different metals, treated with different physical and chemical processes, give rise to distinct NCs. The first key goal of NC characterization is to establish the particle size and overall composition. Characterization techniques often used include TEM, UV-vis spectroscopy, fluorescence spectroscopy, mass spectroscopy, elemental analysis, and many more (Figure 6.10). X-ray crystallography is an ideal tool for complete characterization, but NCs generally do not crystallize. It has been postulated that their often-spherical shapes hinder long-range ordering [155].

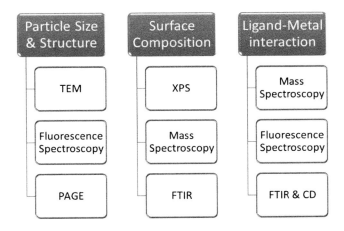

FIGURE 6.10 Different characterization techniques for metal NCs.

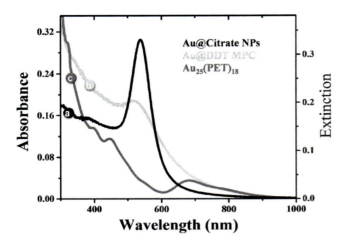

FIGURE 6.11 UV-vis spectra of three different kinds of particles: (a) 25 nm Au@citrate particles, (b) 3 nm Au@DDT clusters, and (c) 2 nm $Au_{25}(PET)_{18}$ clusters, respectively. Several differences are seen between clusters and nanoparticles and among clusters with difference size. Extinction (label on the right) is applicable for (a) and (b) [7].

6.10.1 Spectroscopic Techniques

6.10.1.1 UV-Vis Spectroscopy

The simplest way to differentiate the large nanoparticles from nanoclusters is by their optical absorbance spectra [156]. We considered here three different NCs for the comparison (Figure 6.11): (a) the citrate protected Au NPs exhibit a sharp SPR absorption; (b) large monolayer protected clusters (MPCs) (3 nm) show broadening of the same peak; and (c) $Au_{25}(PET)_{18}$ clusters, which are fluorescent NCs, do not show any plasmon effect on the spectra [7]. These small clusters show completely different absorption patterns compared to nanoparticles and quite similar to a dye molecule.

As the particle size decreases, the λmax shifts to shorter wavelengths due to the corresponding increase in band gap [157]. The UV-vis spectra can also be used to determine particle size and the degree of cluster aggregation. So, when metal NCs aggregate to their counter NPs, they can be characterized by UV-Vis spectroscopy.

6.10.1.2 Fluorescence Spectroscopy

Photoluminescence, another distinct property of NCs, arises because of the transition between the HOMO (highest-occupied molecular orbital) and LUMO (lowest-unoccupied molecular orbital) of the clusters [128]. Fluorescence spectroscopy has provided a new direction to the investigations of these NCs, which can emit from the visible to the NIR region, depending on their core size [127]. Fluorescence spectroscopy could be used to measure not just the emission and excitation spectra but also calculate the QY of the metal NCs (Figure 6.3). It is possible to increase the QY by decreasing the particle size, as the density of states decreases with the number of atoms in the cluster, and this results in a larger gap between the adjacent energy levels for electrons and holes [127]. The photostability of the clusters could be investigated using this technique as well. Other than these, the interactions of NCs with their ligand or analyte can be characterized by the fluorescence, especially the quenching mechanism. Most of the sensing methods using metal NCs are based on fluorescence, which make this spectroscopy one of the most widely exploited characterization tool.

6.10.1.3 X-ray Photoelectron Spectroscopy

X-ray photoelectron spectroscopy (XPS) has been used to evaluate the chemical states of the metal cores in nanomaterial studies. For instance, the oxidation state of BSA-Au NCs was investigated by XPS (Figure 6.12(a)). The Au 4f 7/2 spectrum could be deconvoluted into two distinct components (red and blue curves) centered at binding energies of 84.0 and 85.1 eV, which could be assigned to Au^0 and Au^+, respectively. Such results suggest that the Au NCs could be properly synthesized [35].

6.10.1.4 Mass Spectroscopy

A molecular system could be adequately understood using several complementary tools. Mass spectrometry is one of the ways of understanding such a system, although this has

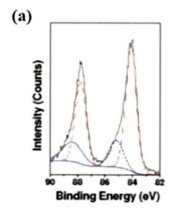
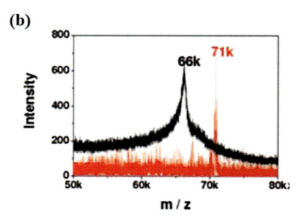

FIGURE 6.12 (a) XPS spectra of Au 4f for BSA-Au NCs (black); (b) MALDI-TOF mass spectra of BSA (black) and BSA-Au NCs (red) [35].

been the single most important technique in this category of materials. The ligand shell required molecular techniques to characterize the functional groups, inter cluster interactions, solvation shells, and changes in the media. These studies have revealed a wealth of information. Researchers have demonstrated the effectiveness of mass spectrometry to characterize clusters. It can give information about the core size as well as the ligand composition. Various advanced mass-spectrometric techniques (e.g., MALDI-TOF-MS and ESI MS) have been used to understand the molecular formula of purified clusters, as in Figure 6.4 [36]. MALDI-MS analysis of an $Au_{25}(SR)_{18}$ cluster was investigated, and a molecular peak around 10.4 kDa was assigned to an isolated glutathione protected gold cluster [25]. The exact composition was later investigated via ESI MS measurements. They identified all nine isolated clusters based on their peak positions [26]. Mass spectroscopic investigation of BSA-Au NCs confirmed the synthesis of the cluster with respect to the mass (Figure 6.12(b))[35].

6.10.1.5 Infrared Spectroscopy

Fourier-transform infrared spectroscopy (FTIR) is widely used to study the formation mechanism and the surface chemical environment of metal clusters. Even the interactions of the analytes, especially biomolecules, can be investigated using FTIR. The interaction of β-d-glucose with L-cysteine-capped Au cluster colloids has been confirmed from their FTIR measurements [158].

6.10.2 MICROSCOPIC TECHNIQUES

6.10.2.1 Transmission Electron Microscopy (TEM)

The most widely used technique for characterizing NCs is TEM and high-resolution TEM (HR-TEM), which provides direct visual information on the size, shape, disparity, structure, and morphology of nanoclusters. TEM is capable of routine magnifications of ≥400,000, to, typically, a ± 2 A° resolution. As we can see in Figure 6.8 (inset e), high resolution of up to the scale of 2 nm can be observed. Potential drawbacks of this technique include electron-beam-induced nanocluster structural rearrangement, aggregation, or decomposition, and samples must be dried and examined under high-vacuum conditions, meaning that no direct information could be gained on how the nanoclusters exist in solution. Despite these potential limitations, TEM is considered the technique of choice for the initial characterization of NCs due to the atomic-level resolution possible, the speed of analysis, and the powerful visual images that are obtained [12].

6.10.3 CHROMATOGRAPHY

For organic soluble clusters, chromatographic techniques such as HPLC (high performance liquid chromatography) and TLC (thin layer chromatography) are suitable for their characterization [159]. Reverse-phase HPLC was an extremely effective means of separating ligand combinations when working with metal clusters protected by two different types of thiolates [160]. TLC could separate even different metal clusters, which is simple yet surprisingly efficient [161].

6.10.4 OTHER CHARACTERIZATION TECHNIQUES

Some unique techniques are used in special cases to give extremely important information regarding the metal NCs as well as their interaction with ligands and analytes. Polyacrylamide gel electrophoresis (PAGE) is one such effective technique for separating clusters of different sizes, especially in the case of protein-stabilized NCs [20, 26, 162]. The clusters are separated based on their size and aqueous soluble clusters; in particular, ligands with carboxylate functionality work well for PAGE. Separation can easily be observed by looking at the gels with different colors, which are then cut and extracted in appropriate solvents to get the purified cluster.

Circular dichroism (CD) spectroscopy is an extremely helpful tool to understand the interaction of the NCs with DNA and proteins, which may be the target or the ligand [137]. We discussed earlier that the induced change in the CD spectra associated with the Ag NCs electronic transitions is further evidence that the clusters are bound to the DNA (Figure 6.6) [66].

Cyclic voltammetry, including differential pulse voltammetry (DPV), can be used to investigate the energy gap between the HOMO and LUMO and other features of the electronic structure [163, 164].

6.11 OPTICAL PROPERTIES OF METAL NANOCLUSTERS

Understanding the different properties of metal NCs is not only helpful to elucidate their structure and composition but also may help to design novel strategies to generate signals for sensing applications. The most important property when it comes to sensing application is the photoluminescence (fluorescence mostly) of the NCs. Other properties such as electro-chemiluminescence have also been used. This section discusses the common and unique characteristics of the metal NCs in general.

6.11.1 ABSORPTION BEHAVIOR

The optical properties of metal NPs are generally described by the collective oscillation of electrons at their surfaces, known as SPR. In case of NCs, as the size reduces to the sub-nanometer scale, they become too small to support plasmons. Then they transit from metal to a molecule type entity. Meaning that instead of forming continuous densities of states and behaving as conductors, these few-atom NCs possess discrete energy levels. Well-defined metal NCs are expected to possess characteristic absorption features and can be distinguished from each other by their absorption profiles [140]. GSH-Au NCs exhibit several characteristic absorption features in the range 400–1000 nm, which are believed to arise from intraband (sp←sp) or interband transitions (sp←d) of the bulk gold

[16]. Discrete electronic transitions clearly indicate that these few atom metal NCs exhibit molecular behavior. The intraband transition tends to be blue-shifted with decreasing core size of GSH-Au NCs [26]. As the size decreases, the spacing between the discrete states in each band increases, leading to a blue shift in the absorption peaks. Absorption spectra of a few-atom Ag NCs thus exhibit discrete electronic transitions instead of collective plasmon excitations [66].

6.11.2 Fluorescence Behavior

Photoluminescence, especially fluorescence, is another distinct property of NCs, which arises because of the transition between the HOMO and LUMO of the clusters [33, 128]. They vary from UV to NIR wavelengths. The bulk or metal NPs do not exhibit any luminescence because of the fast nonradiative relaxation in their continuous band structure. Fluorescence of metal MNs highly depend on local environments such as the protecting ligands and solvents. Surface ligands can influence the fluorescence of the clusters in two different ways [165]. First can be due to the ligands to the metal cluster core charge transfer (LMCCCT) through the Au–S bonds, and second due to the direct donation of delocalized electrons of electron-rich atoms or groups of the ligands to the metal core. One study on GSH-Au NCs showed that the high QY of the clusters could be attributed by the presence of more electron-rich atoms (O, N, etc.) or groups ($-COOH$, $-NH_2$) which facilitate the LMCCCT. Another report also supports the role of ligands in the origin of fluorescence of thiol-stabilized clusters [166]. A solution phase temperature-dependent study of thiol-Au NCs was performed [167]. At low temperature, multiple bands exist as fine structures, which overlap at higher temperatures and form a broad peak. At higher temperature, the energy difference between the emission bands decreases, which in turn increases the overlapping possibility to form a broad peak. Such broad emission features are normally seen for metal quantum clusters [168–170]. This fluorescence property of metal NCs has been exploited in the development of sensing system which we will discuss more in section 6: 13.

6.11.3 Electrochemiluminescence Behavior

Electrochemiluminescence (ECL) refers to the emission of light due to a high-energy electron transfer reaction between electrogenerated species, which possesses several advantages over photoluminescence in analytical chemistry, such as low cost and high sensitivity. In the ECL system of metal NCs, the NCs usually show strong ECL in the presence of co-reactant. A co-reactant is a species that produces a highly reactive intermediate upon oxidation or reduction, which reacts with a luminescent species to either produce an excited state or commence an excitation pathway by a one-electron oxidation or reduction [171]. The most common excitation pathway in ECL is the oxidation-reduction excitation pathway, in which the luminescent species is first oxidized by a cathodically produced oxidizing radical and then reduced by one energetic electron to the excited form of its original oxidation state. PMAA-protected Ag NCs exhibiting ECL were first reported in 2009 [79]. BSA-stabilized Au NCs also exhibited ECL using triethylamine as the co-reactant on the Pt electrode surface [172]. In another report of BSA-Au NCs also exhibited ECL on a modified indium tin oxide electrode surface, where $K_2S_2O_8$ was used as the co-reactant [173]. An organic-soluble fluorescent Au_8NCs was synthesized through a unique heterophase ligand-exchange-induced etching of AuNPs, and the fluorescent Au_8NCs exhibited both the annihilation ECL in organic solution and the co-reactant ECL in aqueous solution [174]. A detailed account on this subject is given in Chapter 11.

6.11.4 Two-Photon Absorption Behavior

Two-photon absorption (TPA) is the simultaneous absorption of two photons of identical or different frequencies in order to excite a molecule from one state (usually the ground state) to a higher energy electronic state [175]. Goodson group reported that two-photon emission of Au_{25} NCs is observed at 830 nm by exciting at 1290 nm, and the two-photon absorption cross-section was measured to be 42,700 GM (Göpert-Mayer unit, $10-50$ cm^4 s), which is superior to the TPA cross-sections of many organic chromophores [176]. Recently, GSH-Au NCs also exhibited strong two-photon emission, and the TPA cross-section was determined [177]. In addition, DNA-protected Ag NCs show two-photon emission [178, 179].

6.12 OTHER PROPERTIES

The solvatochromic effect is the way the spectrum of a substance (the solute) varies when the substance is dissolved in a variety of solvents. Metal NPs have been known to exhibit obvious solvatochromic effects due to the SPR phenomenon. Metal NCs were also reported to display similar properties. The solvatochromic effect of Au NCs has been attributed to the rich surface properties of the clusters [152]. Those studies indicate that the change in chemical environment may induce electronic energy splitting and electron redistribution on the cluster surface, leading to a variation of the optical properties of the metal NCs.

The fluorescence lifetime in metal NCs is the time an electron spends in the excited state before returning to the ground state by emitting a photon. Most metal NCs exhibit monoexponential nanosecond fluorescence lifetimes. Some polymer and peptide-protected metal NCs show biexponential lifetime decays, with a hundreds of picoseconds component and the other a 2–3 nanoseconds component [180]. However, thiol-protected metal NCs present extremely short picoseconds fluorescence lifetimes [20]. Microsecond lifetimes were also detected from some reported metal NCs [181].

6.13 METAL NANOCLUSTERS FOR SENSING APPLICATIONS

The ultra-small size and unique optical (both UV and fluorescence) properties, as well as magnetic and catalytic properties, and rapid reactiveness of the metal NCs make them a new class of reporters in various detection systems. Metal NCs can be used to detect a huge range of organic and inorganic molecules such as proteins, DNAs, or metals, including hazardous metals (such as Hg^{2+})[22]. Most of the sensing is based on the fluorescence mechanism, either quenching or enhancing, of the MNCs when they interact with the targets. In this section, we will discuss the different sensing of molecules reported as well as look at the mechanism that has been described for these sensing.

6.13.1 Biomolecule Detection

Biomolecules include a large list of organic materials such as nucleic acids (DNA), proteins, carbohydrates, and many other molecules found in biological sources. Detection of these molecules is important, as they serve as biomarkers for diseases and contamination by pathogens, as well as for forensic applications. These markers may be found in food, water, or biological fluids (serum, sweat, blood). Compared to NPs, ultra-small NCs have enhanced fluorescence, exhibit high QY, and have lower cytotoxicity, which make them better detectors in biological applications [182]. With good photostability, lower toxicity, and large stokes shift and ease to functionalization, metal NCs outclass other fluorescence materials such as traditional fluorescent dyes and semiconductor quantum dots. A quick look into the different sensors for biomolecules is provided in Table 6.1.

6.13.1.1 Detection of Nucleic Acid-Based Compounds

Malfunction or mutation of nucleic acids in animals may lead to many disorders, including cancer. DNA is also a unique way to recognize any organism or pathogen. Among the metal NCs, Ag NCs are mostly stabilized by DNA templates and show high affinity towards DNA in itself (Ag metal) [65]. Along with excellent physicochemical properties such as high biocompatibility, strong luminescence, and excellent photostability, Ag NCs are suitable to be used as a probe in intracellular detection techniques.

6.13.1.1.1 Detection of DNA

DNA probes are designed for detection of their complementary DNA strands. A nanocluster beacon (NCB) was designed with two DNA probes: one probe strand was for the stabilization of the Ag NCs and the other probe was complementary to the target G-rich strand. This beacon, when brought close to the target DNA, would "light-up" (Figure 6.13(a)), the reason being that emission of DNA-Ag NCs can be enhanced when placed close to G-rich DNA sequences. A sequence from influenza A virus was detected using this method [183]. Another sensor to detect G-rich DNA sequences was developed with exonuclease III-catalyzed amplification, which could enhance the fluorescence in proximity to the target DNA [184]. This method could be used to develop better a fluorescent sensing platform for the detection of target DNA.

TABLE 6.1
Sensing Properties of Metal Clusters for Biomolecules

Cluster Core	Ligand	Analyte	Sensing Mechanism	LOD	References
Gold (Au)	MUA	PDGF AA	Turn on	0.5 nM	[134]
	BSA	Proteinase K	Turn off	1 ng mL^{-1}	[205]
	BSA	Cys C	Turn on	4.0 ng mL^{-1}	[206]
	Peptide	Histone deacetylase	Turn off	5:00 PM	[208]
	Glutathione	Tyrosinase	Turn off		[210]
	BSA	Cysteine	Turn on	1.2 nM	[286]
	BSA	Glutathione	Turn on	7 nM	[287]
	N-acetyl-L-cysteine	Glutathione	Turn off	0.32 nM	[288]
	Glutathione	Glutathione transferase	Turn off	0.25 μM	[289]
	PAMAM	Immunoglobulin G	Turn off	1 nM	[290]
	Peptide	Protein kinase	Turn on	0.02 U mL^{-1}	[291]
Silver (Ag)	DNA	miRNA	Turn off	0.5 μM	[194]
	DNA	human BRAF oncogene	Turn on	10 nM	[183]
	DNA	G rich DNA	Turn on	0.1 nM	[184]
	DNA	MiRNA	Electro-chemical	67 fM	[185]
		DNA	Turn on		[190]
	DNA-apt	Thrombin	Turn off	1 nM	[199]
	DNA-apt	α-Thrombin	Turn on	1 nM	[200]
	DNA	DNA	Turn on	<1 μM	[292]
Copper (Cu)	BSA	Glycoprotein	Turn off	2.6 nM	[220]
	4-methylthiophenol	β-galactosidase	Turn off	0.9 Unit L^{-1}	[221]

Turn off: Fluorescence quenching; turn on: fluorescence enhancement.

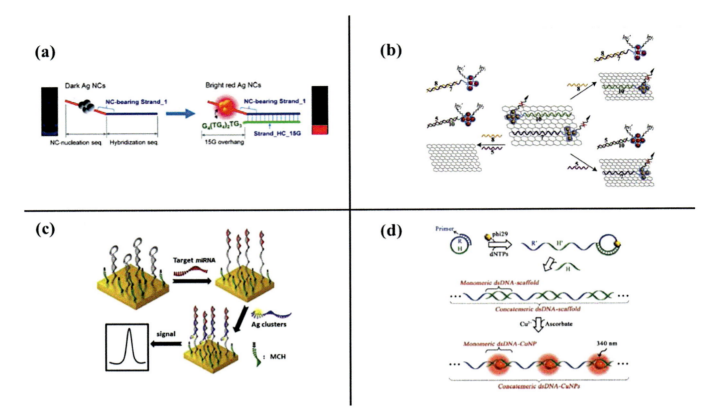

FIGURE 6.13 (a) The "light-on" mechanism of the NCB [183]; (b) multiplexed analysis of the HBV gene (5) and of the HIV gene (8) using the near-infrared- and red-emitting Ag NCs/GO hybrid [187]; (c) Illustration of electrochemical detection of MiRNA using oligonucleotide encapsulated Ag-NCs [185]; (d) Schematic showing microRNA detection based on the principle of the RCR-mediated concatemeric dsDNA-CuNC strategy [186].

The fluorescence of DNA-Ag NCs is quenched in the presence of G-quadruplex/hemin complexes due to photo-induced electron transfer (PET) between the cluster and the complexes [6]. This mechanism was used to develop an NCB for detecting target DNA with high sensitivity and selectivity [184]. Integration of graphene oxide (GO) with DNA-Ag NCs creates interesting hybrid materials for DNA sensing [187]. The DNA sequence stabilizing the Ag NCs had the probe sequence as well the complementary sequence to the target DNA. When DNA-Ag NCs is adsorbed into the GO, the fluorescence of the clusters is completely quenched, but in the presence of target DNA, the formation of the duplex between DNA-Ag NCs and target DNA leads to removal of the clusters from the GO sheet and the fluorescence is regenerated. The hepatitis B virus (HBV) gene and the human immunodeficiency virus (HIV) gene were detected using this beacon (Figure 6.13(b)).

6.13.1.1.2 Detection of a Single Nucleotide Polymorphism (SNP)

SNPs is the substitution or mutation of a single nucleotide that occurs at a specific position in the genome and is a promising disease marker for detection of polymorphism-related diseases like sickle cell anemia. DNA duplexes with an inserted cytosine loop were used as a template for fluorescent Ag NCs, which were used to detect SNPs associated with sickle cell anemia [188]. There would be quenching of the clusters in the presence of the polymorphism. The formation of the clusters was so highly sequence dependent that even a single-nucleotide mismatch located two bases away from the NC would forbid the generation of fluorescence.

Based on site-specific enhancement of fluorescence, Ag NCs protected by a mismatched dsDNA was used to detect other mismatched DNA. The clusters would localize near a mismatch and the fluorescence would be retained [189]. A chameleon-type NCB was fabricated, which lights up into different colors upon binding to different SNP targets [190]. The color change in the emission depended on the alignment between the Ag NCs and the DNA enhancer sequence, and SNPs could be detected by the naked eye under UV excitation.

In addition, Cu NCs stabilized by dsDNA were employed for the detection of SNPs. The fluorescence intensity of Cu NCs was highly sensitive to the base type located in the major groove, and a sensitive fluorescence sensor to detect mismatch in DNA was developed [109].

6.13.1.1.3 Detection of MicroRNA (miRNA)

miRNAs are small, highly conserved, noncoding RNA molecules involved in the regulation of gene expression [191]. They are used in the diagnostics of several human diseases such as cancer (lung cancer, breast cancer, and liver cancer are most common), viral diseases (HCV, HIV-I, influenza), and neurodegenerative diseases (Alzheimer's disease and

Parkinson's disease) [192]. Fluorescent DNA-Ag NCs are commonly used for detecting miRNA. In the DNA-Ag NCs conjugate, the DNA is responsible for the specificity towards the target RNA, and the fluorescence comes from Ag NCs. The principle behind the detection system is that when the fluorescent DNA-Ag NCs encounters the specific target miRNA, the fluorescence intensity of the conjugate decreases due to quenching [193, 194]. The influence of the secondary structure of the nucleic acid on the DNA-Ag NCs was investigated and found that the reduction in secondary structures led to the quenching of the emission. Based on this mechanism, a low emissive DNA-Ag NCs probe was developed whose fluorescence increased in the presence of the formation of a secondary structure. The redesigned DNA-Ag NCs showed a dramatic increase in red emission in the presence of target RNA-miR172 [195].

An electrochemical probe for miRNA detection was fabricated with functional DNA as molecular beacon (MB) probe with both recognition sequence for hybridization and template sequence for stabilization of DNA-Ag NCs [185]. The developed clusters possess metal mimic enzyme properties for catalyzing H_2O_2 reduction. The MB was immobilized on gold electrodes. After the MB probe subsequently hybridizes with the target miRNA and functional probe, DNA-Ag NCs come close to the electrode surface and a selective and sensitive detection signal with a detection limit of 67 fM was produced in response to H_2O_2 reduction (Figure 6.13(c)).

Other than Ag NCs, Cu NCs with dsDNA as a template was used to develop a concatemeric-based miRNA detection system. A rolling circle replication (RCR) was introduced into Cu NCs synthesis (Figure 6.13(d)). The target miRNA as the primer could trigger RCR and be further converted to a long concatemeric dsDNA scaffold through hybridization, which was capable of being used as the template to synthesize Cu NCs, with a detection limit of 10 pM [186]. Cancer cells and miRNA were detected by target-triggered formation of dsDNA polymers and *in situ* formation of Cu NCs with strong fluorescence intensity, with a detection limit of 0.25 nM for microRNA [196].

6.13.1.2 Detection of Proteins

Proteins serve as an efficient biomarker, as their presence, overproduction, or irregularity could be an indication for many diseases such as cancer and Alzheimer's [197]. Thus, biosensors for such proteins is an important strategy for early diagnosis as well as treatment. There are several challenges to develop these sensors, though, which are mainly due to the complex structural diversity of these biomolecules. A highly sensitive recognition molecule for the detection of proteins is the aptamers. A protein aptamer is a functional DNA or RNA structure that can recognize and align to certain proteins with high affinity – a more detailed description is mentioned in another chapter of this book. Based on an aptamer as the stabilizing ligand for NCs, a detection method for thrombin was presented [198]. Thrombin is a coagulation protein in the bloodstream and is responsible for catalyzing many other coagulation-related reactions. An irregularity in this protein could imply many diseases. A DNA-Ag NCs probe was designed with two active domains: one strand was the aptamer specific to thrombin and the other would stabilize Ag NCs. In presence of the target thrombin, DNA-Ag NCs showed significant quenching, as the thrombin bound to its aptamer sequence, and using this method, nano-molar concentrations of thrombin could be detected [199]. A turn-on assay was designed for the detection of thrombin. In this approach, aptamer binding-induced DNA hybridization and fluorescent enhancement of Ag NCs were employed. The DNA for preparing Ag NCs had two aptamer sites, and when thrombin binds with its two aptamers, a hybridization process would be initiated between the complementary sequences attached to each aptamer. Thereby, the G-rich site will overhang in proximity with Ag NCs and resulting in a significant fluorescence enhancement [200]. This was an extremely sensitive sensor with limit of detection (LOD) 1 nM. In another report, a duplex DNA system with a G rich DNA and thrombin aptamer was combined with a DNA-Ag NCs. The aptamer strand inhibited the G-rich DNA to enhance the fluorescence. However, in addition of thrombin, the aptamer would interact with thrombin to form G-quadruplex structure and be released from the duplex structure. The free G-rich DNA then would enhance the fluorescence intensity of DNA/Ag NCs indication thrombin detection [184]. C-rich DNA was used to prepare a bimetallic Ag-PtNCs, which exhibited high peroxidase-mimicking activity. A thrombin aptamer was sandwiched in these clusters to design a label-free colorimetric aptasensor, which showed high sensitivity and selectivity [201]. DNA-Ag NCs were also absorbed into GO to create a hybrid system with thrombin aptamer as the functional probe for thrombin detection [187]. The DNA-Ag NCs-GO did not show fluorescence, but desorption of the cluster in the presence of thrombin due to interaction with its aptamer led to recovery of fluorescence, a mechanism similar to the detection of the HBV gene [187].

Protein-stabilized Au NCs are also employed to detect protein. Different protein analytes could interact with the protein ligand of the Au NCs and influence the fluorescence process differently so that distinct fluorescence image patterns could be obtained. Ten kinds of protein could be detected using protein-Au NCs [202]. A series of dual-ligand, co-functionalized fluorescent Au NCs were prepared with similar fluorescence and diverse surface properties to build a protein-sensing array. Using the "chemical nose/tongue" strategy, eight proteins have been well distinguished at low concentration [203].

Any irregularity or dysfunctional of the enzymes could be associated with many diseases. Proteases are one such enzyme where the alteration in its level is directly linked to a number of diseases such as cancer, arthritis, and neurodegenerative and cardiovascular diseases [204]. A nanoscale platform was developed using protein-Au NCs for sensitive detection of proteases. The enzyme activity unto the protein ligand was exploited here. In the presence of protease, the protein shell of the Au NCs is degraded and the clusters are exposed to O_2 which leads to quenching of the fluorescence, thus leading to a one-step fluorescence protease detection system [205]. Cystatin C (Cys C) is a significant cysteine protease inhibitor in human bodies that is related to kidney injury detection. BSA-Au NCs, along with papain, was used a system to detect Cys C. The method relies on degradation of the BSA scaffold by papain and no fluorescence. However, in presence of Cys C, which inhibits papain's activity, the BSA structure is retained and the fluorescence is turned on [206].

Enzymes related to protein post-translational modifications (PTMs) play key roles in functional proteomics, and its quantitative detection would make an excellent therapeutic tool [207]. Two PTM enzymes, histone deacetylase 1 and protein kinase A, were detected using different peptide Au NCs. The enzymes could make chemical modifications to the peptides of the Au NCs and quench its fluorescence [208]. A "top-down" etching process was used to detect hydrolytic enzymes, like esterase and alkaline phosphatase, for the generation of fluorescent Au NCs followed by the turning on of the fluorescence [209]. In addition, GSH-Au NCs were used to detect tyrosinase activity along with dopamine. Tyrosinase is a polyphenol oxidase and can catalyze the oxidization of DA to o-quinone, and the fluorescence of the clusters is quenched by quinones [210].

Enzymes have an impact on the structure of DNA. This mechanism was exploited to develop a detection system for enzymes. Adenosine deaminase (ADA), a crucial element in the differentiation and maturation of the lymphoid system, has been exploited for developing sensors for its detection [211, 212]. An aptamer specific to adenosine was introduced to the DNA-Ag NCs. Initially, the adenosine formed a complex with its aptamer, but in the presence of ADA it was converted to inosine. The free aptamer parts and the DNA template with a six-base cytosine loop could then generate red-emitting fluorescence due to the formation of stable Ag NCs. This way the concentration of the enzyme and its inhibitor could be detected [212]. Polycytosine oligonucleotide-based Ag NCs were used to monitor the activity and inhibition of protein kinase (PKA) [213]. A detection system for other enzymes like endonucleases [214] and acetylcholinesterase (AChE) [215, 216] have also been prepared using DNA-stabilized Ag NC.

Along with Au NCs and Ag NCs, Cu NCs have also been used as probe for enzyme detection. A dsDNA-specific fluorescent Cu NCs was used as a "green" nano-dye for polymerization-mediated biochemical analysis. Polymerase was detected using this system [217]. Another novel label-free turn-on fluorescent strategy using dsDNA-Cu NCs was used to detect alkaline phosphatase (ALP) under physiological conditions [218]. Apart from the dsDNA template, the poly(thymine) (poly T) sequence was used to form Cu NCs. These clusters were used to prepare a sensitive nuclease assay. In the presence of nuclease, poly T was digested to mono- or oligonucleotide fragments with a decrease of fluorescence, which could be recognized as a signal for the detection of nuclease activity [219]. BSA-Cu NCs were further functionalized by 3-aminophenylboronic acid and consumed in the sensitive detection of glycoproteins (as low as 2.6 nM) via a fluorescence quenching response [220].

An aggregation-induced emission (AIE) probe using Cu NCs was used to detect the β-galactosidase activity by fluorescence "turn-off" response [221]. Cu NCs with low emission were bound to high luminescent Cu NCs via hydrophobic interaction. 4-Nitrophenyl-β-D-galactopyranoside (NPGal) was used as the substrate to detect β-galactosidase. In the presence of the enzyme, the substrate was hydrolyzed into galactose and 4-nitrophenol. The 4-nitrophenol adsorbed on the Cu NCs and disrupted the hydrophobic interactions and quenched the luminescence (Figure 6.14).

6.13.1.3 Detection of Other Organic Compounds

Other than nucleotides and proteins there are many small biomolecules like urea, glucose, and dopamine that can be detected by metal NCs.

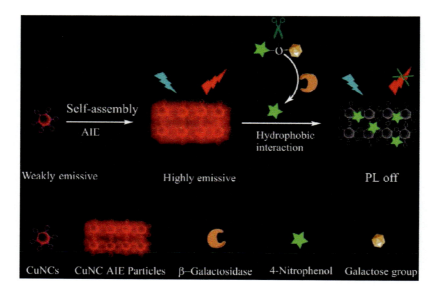

FIGURE 6.14 Schematic illustration of self-assembly of AIE particles of Cu NCs mediated by hydrophobic interaction and application in detection of β-galactosidase activity using 4-nitrophenol-releasing substrate [221].

6.13.1.3.1 Detection of Glucose

Cu NCs are found to possess intrinsic peroxidase-like activity, which was utilized to detect glucose. Cu NCs, along with TMB (tri-methyl benzoate) and GOx (glucose oxidase), detect the glucose concentration in the reaction solution. The reaction takes place as follows: glucose reacts with GOx and produces gluconic acid and H_2O_2. Cu NCs then oxidizes the colorless TMB into blue-colored TMB with the help of H_2O_2. The intensity of the blue-colored oxidized TMB is directly proportional to the concentration of the glucose in the solution [222].

6.13.1.3.2 Detection of Urea

Urea is a critical biomarker for evaluating uremic toxin levels and kidney and hepatocellular functions. It also acts as the indicator of nonprotein nitrogen in food products such as milk, since it is known that urea adulteration is utilized as an indicator of protein feeding efficiency. Au NCs were stabilized by urease AuC@Urease (here AuC = Au NCs) which exhibited NIR emission. Urea was selectively detected based on the enzyme-specific conversion of urea to ammonium ions, which facilitates pH-induced aggregation of Au NCs, leading to fluorescence quenching [223].

6.13.1.3.3 Detection of Dopamine

Dopamine (DA) is an important neurotransmitter and plays a vital role in the function of the central nervous system and renal, hormonal, and cardiovascular systems. Both reduced and increased levels of DA may result in various neural diseases such as Parkinson's disease (reduced DA level) and schizophrenia (increased DA level). Several biosensors for DA detection based on metal NCs have been reported [210, 224, 225]. Highly fluorescent BSA-Au NCs exhibiting high peroxidase-like activity were used for dual detection of DA. In the presence of dopamine, the fluorescence intensity of the Au NCs decreases significantly through a PET process, as well as the colorimetric response of the clusters was observed [224]. ECL-based biosensors for DA detection have also been reported [173].

6.13.1.3.4 Detection of Thiols

Abnormal levels of biological thiols (e.g., such as cysteine [Cys]), homocysteine [Hcy], and glutathione [GSH]) act as markers to several diseases, such as leukocyte loss, psoriasis, hair depigmentation, liver damage, cancer, and AIDS [226]. So quantitative assessment of thiols in the human body is very important. Huang et al. found that the fluorescence response pattern of Ag NCs to a specific analyte was highly dependent on the nature of DNA templates. The group developed a novel fluorescence turn-on assay for thiol compounds with high sensitivity by modulating DNA-templated silver nanoclusters (DNA-Ag NCs). The technique showed high selectivity for the determination of biothiols among amino acids found in proteins, as well as detection in serum samples [227].

6.13.1.3.5 Detection of Reactive Oxygen Species (ROS)

ROS is a type of unstable molecule that easily reacts with other molecules in a cell. Accumulation of ROS in cells may cause damage to DNA, RNA, and proteins, and may induce cell death. ROSs are free radicals that play a very important role in a variety of pathogenic processes, including inflammation, carcinogenesis, ischemia-reperfusion injury, and signal transduction [228]. A nano-complex displaying single-excitation and dual-emission fluorescent properties has been developed by Chen et al., through a crown-like assembly of dye-encapsulated silica particles decorated with satellite Au NCs for live cell imaging of highly ROS (hROS), including -OH, ClO$^-$, and ONOO$^-$. The strong fluorescence of Au NCs can be sensitively and selectively quenched by these hROS. The nanocomplex is biocompatible and stable for long-term observations. The results show that the nanocomplex provides a sensitive sensor for rapid imaging of hROS signaling with high selectivity and contrast [229].

6.13.1.3.6 Detection of Bilirubin

Free bilirubin levels in blood serum has an enormous clinical importance in probing hyper-bilirubinemia conditions, such as bilirubin encephalopathy in newborn babies. Hence, quantitative detection of free bilirubin is essential to assess the risk, diagnosis, and treatment [40]. A dual (fluorometric and colorimetric) probe for detecting free bilirubin was developed using has-stabilized Au NCs (HSA Au NCs). In the presence of bilirubin, the fluorescence of nanoclusters was quenched. This behavior was linked to the inherent specific interaction between bilirubin and HSA [40]. The same authors further developed an amperometric biosensor using HSA Au NCs for bilirubin detection. With the interaction between HSA and bilirubin, the Au NCs acted as an electron transfer bridge between them and the electrode to produce the signals. This biosensor had high sensitivity and an LOD of 86.32 nM. This system was used to measure bilirubin spikes in serum samples [40].

6.13.2 Detection of Inorganic Ions

6.13.2.1 Detection of Metal Ions

Metal ions play fundamental roles in biology by serving as essential cofactors in processes such as energy metabolism and storage, signal transduction, and nucleic acid processing. In excess (even small amounts), they may become toxic; hence, tight regulation of metal import is necessary at a cellular level [230]. Some heavy metal ions, such as Hg^{2+}, Cd^{2+}, Pb^{2+}, and Cu^{2+}, are acutely toxic to humans and aquatic species even at low concentrations, and bioaccumulation of these metal ions can cause serious health problems and diseases [231]. Developing selective and sensitive qualitative and quantitative methods for metal ions are of utmost importance. A quick look into the different sensors for inorganic ions are listed in Table 6.2.

6.13.2.1.1 Detection of Mercury Ions (Hg^{2+})

Hg^{2+} is a highly toxic and widespread pollutant ion and is responsible for causing damaging effects to the brain, nervous system, and kidney even at very low concentrations [232]. Au NCs were used to prepare a detection system for Hg^{2+} sensing. The sensing mechanism is based on the high-affinity metallophilic 5d10–5d10 interaction between Hg^{2+}

TABLE 6.2
Sensing Properties of MNCs for Inorganic Ions

Cluster Core	Ligand	Analyte	Sensing Mechanism	LOD	References
Gold (Au)	Pepsin	Hg^{2+}	Turn off	50 ± 10 nM	[42]
	Lysozyme	Hg^{2+}	Turn off	10 nM	[136]
	DHLA	Hg^{2+}	Turn off	0.5 nM	[168]
	Glutathione	Cu^{2+}	Turn off	3.6 nM	[235]
	Glutathione	Pb^{2+}	Turn off	2 nM	[240]
	BSA	Ag^+	Disappearance of color	0.204 µM	[248]
	GSH	Cr^{3+} & Cr^{4+}	Turn off	2.5 µg L^{-1} & 0.5 µg L^{-1}	[250]
	BSA	CN^-	Turn off	20 µM	[251]
	Lysozyme	CN^-	Turn off	190 nM	[252]
	BSA	NO_2^-	Turn off	1 nM	[254]
	Peptide	As^{3+}	Turn off	53.7 nM	[259]
	BSA	Hg^{2+}	Turn off	0.5 nM	[293]
	BSA	Cu^{2+}	Turn off	0.33 nM	[294]
	Histidine	Cu^{2+}	Turn off	0.1 nM	[295]
	L-proline	Fe^{3+}	Turn off	0.2 µM	[296]
	BSA	Ag^+	Turn on	0.1 µM	[297]
	GSH	I^-	Turn on	400 nM	[298]
Copper (Cu)	BSA	Pb^{2+}	Turn off	>20 ppm	[144]
Bimetallic (Ag/Au)	DNA	S^{2-}	Turn off	0.83 nM	[257]
	BSA	Al^{3+}	Turn off	0.8 µM	[299]

Turn off: Fluorescence quenching; Turn on: fluorescence enhancement.

and Au^+. This interaction alters the electronic structures of Au NCs, which efficiently quenches the fluorescence of Au NCs [233]. Furthermore, BSA-stabilized Au NCs displayed high selectivity towards Hg^{2+} over other environmentally relevant metal ions. The estimated LOD for Hg^{2+} ions was 0.5 nM, which is much lower than the maximum level of mercury in drinking water (10 nM) permitted by the U.S. Environmental Protection Agency (EPA). By taking advantage of this interaction between Hg^{2+} and Au^+ many sensitive and selective sensing systems have been developed for the detection of Hg^{2+} using Au NCs stabilized by BSA [38, 145, 233], lysozyme [136], dihydrolipoic acid (DHLA) [168], and l-amino acid oxidase [234].

6.13.2.1.2 Detection of Copper Ions (Cu^{2+})

Cu^{2+} is another significant environmental pollutant and an essential trace element in biological systems. To detect Cu^{2+}, a simple and sensitive fluorescent sensor based on fluorescent PMAA-Ag NCs (poly (methacrylic acid)-templated Ag NCs) was designed. When Cu^{2+} interacts with the carboxylic group of PMAA, the fluorescence quenches [86]. Similarly, glutathione-protected Au NCs (GSH-Au NCs) were highly sensitive to Cu^{2+} based on aggregation-induced fluorescence quenching [235]. This assay had good selectivity toward Cu^{2+} over other metal ions, with a LOD of 3.6 nM. Moreover, the Cu^{2+}-quenched fluorescence could be efficiently recovered upon addition of a metal ion chelator, EDTA. However, these turn-off assays may compromise specificity, since other quenchers or environmental stimulus might also lead to fluorescence quenching and report "false-positive" results. To remove this disadvantage, Chang and coworkers recently demonstrated a turn-on fluorescent assay for Cu^{2+} using DNA-templated Ag NCs [236]. The introduction of Cu^{2+} resulted in the formation of DNA-Cu/Ag bimetallic NCs with more stabilization from DNA templates and thus enhanced the fluorescence. Furthermore, by using a combination of DNA-Cu/Ag NCs and 3-mercaptopropionic acid (MPA), they observed that MPA-induced fluorescence quenching of DNA-Cu/Ag NCs was suppressed in the presence of Cu^{2+} ions, allowing turn-on detection of Cu^{2+} [237]. An NIR fluorescent assay for Cu^{2+} was prepared based on the quenching of luminescent Au NCs prepared via heat-assisted reduction of a gold(I)-thiol complex [238]. This NIR fluorescence-based method enabled the sensitive detection of Cu^{2+} with a LOD down to 1.6 nM.

6.13.2.1.3 Detection of Lead Ions (Pb^{2+})

Pb^{2+} ions are a highly toxic environmental pollutant that can cause damage to the kidney, the liver, and the nervous system and pose severe effects on human health [239]. The U.S. EPA sets the maximum contamination level for lead in drinking water at 75 nM. A method was developed using GSH-Au NCs for ultrasensitive and selective Pb^{2+} sensing [240]. The LOD was determined to be 2 nM with a signal-to-noise ratio of 3. GSH-Cu NCs have also been used for Pb^{2+} detection. GSH-Cu NCs prepared via sono-chemical synthesis and utilized for a fluorescent "turn-off" sensing of Pb^{2+} was as low as 1.0 nM [241]. In another case, a "turn-on" system for detecting Pb^{2+} was developed [242]. An aggregation-induced quenching system using BSA-Cu NCs for selective fluorescent "on-off" Pb^{2+} detection system has been reported [144].

6.13.2.1.4 Detection of Silver Ions (Ag^+)

Ag ions are toxic to a lot of bacteria, viruses, algae, and fungi. Therefore, the antibacterial activity of Ag^+ has been employed into many application fields such as toiletry, timbering, and clinical material. However, once the concentration of Ag^+ is high enough, they can still bring harmful side effects to the environments and human health [243]. There are several systems using fluorescent Au NCs in the sensing of Ag^+. A one-pot method for the synthesis of green fluorescence Au NCs using THPC and 11-MUA as reduction and stabilizer agents has been developed. These Au NCs had high sensitivity and selectivity for the Ag^+ detection except Hg^{2+}, which could be masked using EDTA [244]. A fluorescence enhancement of GSH-Au NCs was applied to the detection of Ag^+ [245]. The fluorescence signal of Au NCs was greatly enhanced by Ag^+ deposition on the surface of NCs metal core. The method revealed good selectivity for the detection of Ag^+, and the LOD was 200 nM, which is lower than the maximum permissive concentration of Ag^+ (460 nM) in drinking water proposed by the EPA. A similar strategy was used for Ag^+ detection using BSA-Au NCs [246]. It is known that the BSA-Au NCs reduced Ag^+ to Ag^0 to form a bimetallic Au-Ag NCs. The mechanism also provided a simple and mild method for the synthesis of Au-Ag NCs [247]. A colorimetric sensor for Ag^+ was developed using BSA-Au NCs which possess the peroxidase-like activity that could catalyze the oxidation of 3,3′,5,5′-tetramethylbenzidine (TMB) by H_2O_2. Ag^+ selectively reacts with Au^0 through redox reaction, which induces an apparent inhibition of the peroxidase-like activity of BSA-Au NCs and the color in the solution disappears [248].

6.13.2.1.5 Detection of Chromium Ions

Chromium mainly exists in two valence states in aqueous solutions: Cr^{3+} and Cr^{4+}. Cr^{3+} in trace amounts is required for sugar and fat metabolism. Its high concentration is believed to induce oxidative DNA damage, causing cancer. While Cr^{4+} is ~100 times more toxic and is considered a human carcinogen with adverse impact on human skin, stomach, lung, liver, and kidneys [249]. Ag NCs were prepared via microwave-assistant synthesis and were used for the sensitive and selective determination of Cr^{3+} ions based on a fluorescence quenching mechanism [153]. The detection limit was found to be 28 nM. Another method using GSH Au NCs was reported for selective determination of Cr^{3+} and Cr^{4+} in environmental water samples based on target-induced fluorescence quenching [250].

6.13.2.2 Detection of Other Ions

6.13.2.2.1 Cyanide (CN^-)

Cyanide is a highly toxic ion, because it inhibits the activity of cytochrome C oxidase in mitochondria and hinders cell respiration. Au NCs have been used for the detection of CN^-. In the presence of oxygen, CN^- can transform Au atoms into water-soluble Au $(CN)^{2-}$ ion leading to the quenching of Au NCs fluorescence [247]. Based on this principle, BSA-Au NCs were used to achieve the selective detection of CN^- [251]. The LOD reported was 200 nM, which is far lower than that of the permissive CN^- concentration (2.7 μM) of the World Health Organization (WHO) in drinking water. Based on a similar principle, Lys-Au NCs have also been used in the highly selective detection of CN^-. The LOD was found to be 190 nM [252]. *In situ* synthesis of Au NCs in a macroporous polymer film was applied for CN^- detection [253]. The macroporous structures of the polymer film are considered to enhance the interaction between Au and CN^- and visual detection was possible.

6.13.2.2.2 Nitrite (NO_2^-)

Nitrite ions are generally scarce in nature, but a high amount is found in some wastewater. Under acid conditions, it can react with secondary amine to generate nitrosamines with carcinogenicity. NO_2^- are highly reactive toward proteins like BSA and cause aggregation of Au NCs stabilized by BSA, thus serving as a sensitive detection system for NO_2^- [254]. As low as one nM detection was achieved using this system. Further study into the mechanism behind the quenching found that the electron binding energy of Au 4f7/2 in BSA-Au NCs increased after the addition of BSA-Au, indicating that the addition of NO_2 could lead to the increase of the oxidation state of Au NCs [255]. A nanocomposite was developed with dual-wavelength emission by means of the electrostatic interaction and hydrogen bonding between positively charged GO and BSA-Au NCs [256]. The red fluorescence of BSA-Au NCs revealed a specific response to NO^{2-}, while the blue fluorescence intensity of GO was kept constant. The color would change from red to blue with the addition of NO^{2-}. Based on this, a fluorescent ratio sensor was constructed for the detection of NO^{2-} with a LOD of 46 nM [247].

6.13.2.2.3 Sulfide (S^{2-})

Sulfide ions are a common environmental contaminant and an important gas signal emitter. Similar with the toxicity of CN^-, S^{2-} could chelate with iron in mitochondrial cytochrome oxidase and hinder cell respiration. DNA-Au/Ag NCs were used for the detection of S^{2-} in hot spring water and seawater [257]. The S^{2-} quenching fluorescence signal was caused by the strong interaction between S^{2-} and Au or Ag, which led to the configuration change of DNA. Based on a similar principle, BSA-Au NCs were applied to the high selective and sensitive detection of S^{2-} in river samples. The LOD was 29 nM [258]. GSH-Ag NCs were also used to construct a fluorescent sensor for the detection of S^{2-} [170]. In this case, the GSH-Ag NCs solution gradually become yellow with the increase in concentration of S^{2-}, indicating the generation of Ag_2S. Simultaneously, a broad absorption band in the UV-visible absorption spectrum with no SPR absorption peak of Ag NPs was observed, proving the formation of Ag_2S after the addition of S^{2-}.

6.13.2.2.4 Arsenic Ions (As^{3+})

Arsenic ions are currently one of the most important environmental global contaminants and toxicants, particularly in the developing countries. Dipeptide L-cysteinyl-L-cysteine

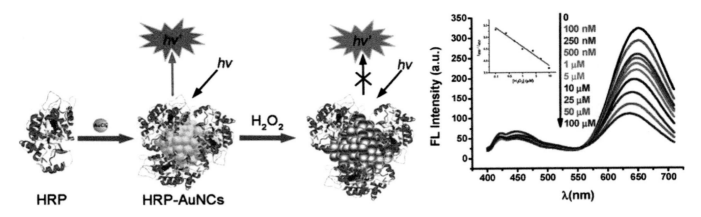

FIGURE 6.15 Schematic of the formation and the H_2O_2 directed quenching of HRP-Au NCs [264].

was used to prepare Au NCs for a "turn-on" detection of As^{3+} [259]. The addition of As^{3+} induced a quite significant fluorescence enhancement of Au NCs, which should be attributed to the fact that positively charged As^{3+} ions interact with the negatively charged gold clusters and the electrons can flow from the electron rich gold clusters system to the electron deficient As^{3+}, resulting in an increase in the radiative decay rate Au NCs. The LOD reached as low as 53.7 nM, which is far below the permissible limit (133 nM) of As^{3+} in drinking water permitted by U.S. EPA.

6.13.3 Detection of Other Substances

6.13.3.1 Detection of Hydrogen Peroxide

H_2O_2 is an important analyte, which is involved in several chemical, biological, and environmental processes. In particular, H_2O_2 is one of the products of enzymatic reactions by almost all oxidases, thus enabling quantitative assays of the activity of the enzyme as well as various enzyme substrates such as glucose [260, 261]. A quantitative sensor was developed using 11-MUA-Au NCs for determination of H_2O_2 [262]. In the presence of H_2O_2, 11-MUA units, which are bound to the cluster surface through Au-S bonds, get easily oxidized to form an organic disulfide product, resulting in reduced luminescence. Further combination of the luminescent BSA-Au NCs with glucose oxidase enabled the sensitive determination of glucose [263]. The sensing mechanism is based on the enzymatically generated H_2O_2-induced degradation of Au NCs, which resulted in the fluorescence quenching.

A new strategy was reported to construct enzyme-functionalized fluorescent Au NCs for the detection of H_2O_2 (Figure 6.15). Bifunctional fluorescent Au NCs could be formed *in situ* using HRP as a scaffold [264]. The enzyme remains active in the clusters and enables the catalytic reaction of HRP-Au NCs and H_2O_2, resulting in the fluorescence quenching that can be applied to H_2O_2 detection with high sensitivity (LOD: 30 nM).

6.13.3.2 Detection of Nitroaromatic Explosives

Nitrobenzene explosives are a class of chemicals that pose a serious threat to security, particularly 2,4,6-trinitrotoluene (TNT), which is commonly used in terrorist attacks, military production, and industrial blasting. A double fluorescence emission and ultrasensitive probe for the TNT detection was reported [265]. The fluorescent probe was prepared using Au nanoflower bonded fluorescein isothiocyanate (FITC) and then BSA-Ag NCs was modified on the surface of Au nanoflower. BSA-Ag NCs with red fluorescence were used as the TNT sensitive probes, while FITC with green fluorescence was used as a reference fluorophore. In presence of TNT, the red fluorescence of Ag NCs was completely quenched because of the formation of the Meisenheimer complex, while the green fluorescence intensity of FITC remained constant. The probe provided a highly selective visualization method for the determination of TNT. Using BSA-Au NCs a test paper based, simultaneous determination of TNT and 4-nitrophenol were reported [266]. The detection limits were 10 pM for TNT and 1 pM for 4-nitrophenol. BSA-Au NCs nanofibers combined with polyethylene glycol were prepared by means of electrospinning and were used for real-time, visual inspection of TNT [267]. After the addition of TNT, the membrane color produced a marked change. The excellent properties were attributed to the specific surface area of the nanofiber membrane and the strong adsorption capacity of TNT. The different sensors for other analytes are listed in Table 6.3.

6.14 OTHER APPLICATIONS FOR METAL NCs

Metal NCs have been used a great deal in bioimaging due to their fluorescence property. They are also employed in physical sensors (pH and temperature) and gene therapy. Some of these applications are discussed briefly here to show the applications of metal nanoclusters other than sensing.

6.14.1 Thermometer and pH Meter

Lipoic acid-protected Au NCs were used to prepare a nanothermometer, which can be used for precise temperature measurements in biological systems [181]. In addition to this, BSA-Au NCs are used as a highly promising nanoscale thermometer in which at high temperatures the fluorescence was quenched [268], contributed by the thermal denaturation of the

TABLE 6.3
Sensing Properties of MNCs for Other Analytes

Cluster Core	Ligand	Analyte	Sensing Mechanism	LOD	References
Gold (Au)	BSA	Ascorbic acid	Turn off	0.2 μM	[300]
	HAS	Bilirubin	Absorbance and Turn off	248 ±12 nM	[39]
	BSA	Dopamine	Turn off	10 nM	[224]
	11-MUA	H_2O_2	Turn off	30 nM	[262]
	BSA	Glucose	Turn off	5 μM	[263]
	BSA	Glucose	Turn off	5 μM	[263]
	BSA	H_2O_2	absorbance/colorimetric	20 nM	[282]
	BSA	Cholesterol	Turn off	12 μM	[301]
	GSH	SDS	Turn off	20 ng mL^{-1}	[302]
	MUA	H_2S	FRET	1.83 μM	[303]
Silver (Ag)	DNA	NAD$^+$	Turn off	22.3 μM	[162]
	BSA	TNT	Fluorescence change	10 pM	[265]

Turn off: Fluorescence quenching; Turn on: fluorescence enhancement.

biomolecular template. Au NCs have also been used as fluorescence pH sensors. BSA-Ce/Au NCs were prepared and exhibited, with two fluorescence intensities, 410 and 650 nm, which were pH dependent and independent, respectively. The stability and biocompatibility of the clusters, as well as the linear relationship of the intensity ratio against pH (pH 6.0–9.0), enable it to be used as a probe for monitoring local pH values inside HeLa cells [269]. Cu NCs and Au/Cu bimetallic NCs were also effective as pH sensing fluorescence indicators [270, 271].

6.14.2 Biolabeling and Bioimaging

Biolabeling is the process of labeling individual cells, often by the use of metal dyes/nanomaterials, and the subsequent imaging process is called bioimaging. Bioimaging helps to capture real-time visualization of cells and many cellular processes that play a very important role in disease analysis and disease progress. Due to the fluorescence properties and biocompatibility, nanoclusters have been used for fluorescence bioimaging [6]. This imaging technique has high sensitivity, multiplex detection capabilities, and low analysis cost. Metal NCs can be inserted into the cells via microinjection, which is conjugate with a cell-penetrating peptide, using a transfection agent, or simple endocytosis. Ag NCs in combination with a fluorophore, thioflavin T (ThT), were used for the labeling and imaging of amyloid fibrils [84]. Living cells could be stained by intracellularly generated fluorescent peptide-encapsulated Ag NCs [90]. Ag NCs stabilized by DNA and avidin were used to label cell surfaces [73]. A peptide was designed that had two functional sides, one that stabilized Au NCs and another that could conjugate with specific peptides in the nucleus. This Au NCs exhibited red emission and was used to stain the nucleus of three types of cells, namely HeLa, GES-1, and MRC-5 [46]. Self-bioimaging is also possible by reduction of Au^{3+} inside cancer cells to biosynthesize nanoclusters producing fluorescence [52].

The application of nanomaterials for the treatment of cancer is mostly based on early tumor detection and diagnosis by nano-devices capable of selective targeting and delivery of chemotherapeutic drugs to the specific tumor site [272]. Doxorubicin immobilized Au NCs conjugated with methionine (Met) were prepared and hydrophilic indocyanine green (ICG) derivative (MPA), an NIR fluorescent dye, was used for tumor imaging and treatment [273]. In another report Au NCs and doxorubicin dual-loaded photosensitive liposome used a supercritical CO_2 method to produce a liposome-based drug carrier. Upon application of light irradiation, the carrier could release the drug in a short time, exhibiting a light-triggered nano-switch for controlled drug release and tumor therapy [274]. Metal nanocluster-based hybrid nanocomposites were also successfully applied in gene therapy. Poly-ethyleneimine (PEI) stabilized Au NCs has become a novel gene carrier. The interaction has been contributed by the Au$^+$ on the surface of the Au NCs and the negatively charged DNA [275, 276].

6.14.3 Metal Nanoclusters as Antibacterial Agents

Nanoclusters, especially Au NCs and Ag NCs, show antibacterial activity, as these can act against both gram-positive and gram-negative bacteria. Lysozyme stabilized Au NCs with the bioactive lysozyme were used as broadband labeling agent for pathogenic bacteria and inhibited the growth of the bacteria [277]. Ag NCs inhibit bacterial growth by destroying the outer cellular membrane and permeating into the cells, followed by the antibacterial effect of the internalized Ag NCs and released silver ions [278].

6.14.4 Metal Nanoclusters-Based Artificial Enzymes

Nanomaterials such as metal NPs and NCs that mimic the catalytic activity of a natural enzyme are called artificial enzymes or nanozymes. Apart from catalytic activity, enzyme mimics have certain advantages over natural enzymes such as tunable structures, excellent tolerance to experimental conditions, lower cost, and comparatively easy synthesis and storage

[279]. These nanoclusters have promising potential in environmental detection and biomedical applications [280, 281]. For the first time in 2011 BSA-Au NCs were reported to exhibit intrinsic peroxidase-like activity. Due to the ultra-small size, good photostability, and biocompatibility these NCs were better candidates for bioanalysis as compared to metal nanoparticles. A colorimetric assay was prepared using the BSA-Au NCs, which could detect H_2O_2 as low as 2.0×10^{-8} M [282]. This detection method was extended to a sensitive xanthine detection method using xanthine oxidase and the BSA-Au NCs in urine and human serum samples. BSA-Au NCs also showed peroxidase mimic activity. A luminol-hydrogen peroxide system was employed for the detection of H_2O_2 and glucose using these clusters. In the presence of the analytes, the chemiluminescence of the clusters was enhanced [283].

The enzyme activity of the clusters could be regulated with the addition of graphene oxide (GO). A hybrid GO-Au NCs were synthesized using lysozyme as the ligand, which unlike other NCs, showed peroxidase like activity over a broad pH range [284]. Lysozyme stabilized PtNCs also exhibited oxidase-like activity [125].

6.15 CONCLUSION AND FUTURE PROSPECTS

Metal nanocluster behaves differently from the metal nanoparticles in terms of the disappearance of the SPR effect and the emergence of fluorescence. The electron intraband gap theory can explain the fluorescence characteristics of metal nanoclusters. The Jellium model describes reasonably the size-dependent electronic structure and relative electronic transitions of these small clusters. Several methods emerged for the synthesis of metal NCs that belong to the "bottom-up" and "top-down" approach. In the former process, metal precursors and a stabilizing ligand are required for the synthesis of the clusters. In case of the "top-down" method, nanoclusters are etched out of their counter metal nanoparticles by either radiations or ligands. Many metals are used to prepare luminescent clusters, among which Au followed by Ag are the most widely used metals. The method of NCs synthesis varies from metal to metal due to the obvious reason of different electronic structures of the metals. Ligands play an important role in the synthesis process, as they not only stabilize the clusters from aggregating into their metal NPs but also influence their fluorescence properties and their interactions with analytes. Among the ligands, thiols, proteins, and DNA have been widely used. The functional groups present in these ligands support to stabilize the nanoclusters and in their sensing mechanisms. For the characterization of the NCs, fluorescence-based spectroscopy is extensively used as the surrounding chemical and physical environments influence this optical behavior of the NCs. MNCs have been used to detect a wide array of targets ranging from ions to macromolecules from both organic and inorganic classes. Both "turn-off" and "turn-on" strategies for the fluorescence signals are exploited for developing the sensors.

There has been intensive basic research on different fronts of MNCs such as optical, electronic, functional, and chemical. However, the underlying principles of many of these phenomena have yet to be explained clearly. The elucidation of the precise structure and size is a challenging task for a routine study of these extremely small nanomaterials. There is also difficulty in the crystallization of the metal NCs with their ligands, particularly with the most commonly used ligands DNA and protein for better characterization. In addition, the preparation methods of highly stable metal NCs still need to be standardized, because these optical reporting materials are highly sensitive to the contacted chemical and physical environments. With regard to the QY (and hence the intensity of the signal), we find it is considerably lower than that of many organic fluorescent dyes. Hence, a new process to prepare highly fluorescent clusters needs to be explored. Moreover, a more general mechanism for NC-based detection of target analytes needs to be investigated for better understanding, as well as applications of these interesting class of nanomaterials.

REFERENCES

1. Jin R. Quantum sized, thiolate-protected gold nanoclusters. Nanoscale. 2010;2:343–62.
2. Hostetler MJ, Wingate JE, Zhong C, et al. Alkanethiolate gold cluster molecules with core diameters from 1.5 to 5.2 nm: Core and monolayer properties as a function of core size. Langmuir. 1998;14:17–30.
3. Sun H, Sakka Y. Luminescent metal nanoclusters: Controlled synthesis and functional applications. Sci Technol Adv Mater. 2014;15:1–13.
4. Zheng J, Zhang C, Dickson RM. Highly fluorescent, water-soluble, size-tunable gold quantum dots. Phys Rev Lett. 2004;93(7):5–8.
5. Shang L, Dong S, Nienhaus GU. Ultra-small fluorescent metal nanoclusters: Synthesis and biological applications. Nano Today. 2011;6:401–18.
6. Zhang L, Wang E. Metal nanoclusters: New fluorescent probes for sensors and bioimaging. Nano Today . 2014;9(1):132–57.
7. Chakraborty I, Pradeep T. Atomically precise clusters of noble metals: Emerging link between atoms and nanoparticles. Chem Rev. 2017;117:8208–71.
8. Díez I, Ras RHA. Few-atom silver clusters as fluorescent reporters. In: Advanced Fluorescence Reporters in Chemistry and Biology II Springer Series on Fluorescence (Methods and Applications), 2010, pp. 307–32.
9. Zheng J, Nicovich PR, Dickson RM. Highly fluorescent noble-metal quantum dots. Annu Rev Phys Chem. 2007;58:409–31.
10. Lin CJ, Lee C-H, Hsieh J, et al. Review: Synthesis of fluorescent metallic nanoclusters toward biomedical application: Recent progress and present challenges. J Med Biol Eng. 2009;29(6):276–83.
11. Wilcoxon JP, Abrams BL. Synthesis, structure and properties of metal nanoclusters. Chem Soc Rev. 2006;35:1162–94.
12. Aiken III JD, Finke RG. A review of modern transition-metal nanoclusters : their synthesis, characterization, and applications in catalysis. J Mol Catal A Chem. 1999;145:1–44.
13. Schmid G. Clusters and colloids: Bridges between molecular and condensed material. Endeavour. 1990;14:172–8.
14. Zhang Y, Zhang C, Xu C, et al. Ultrasmall Au nanoclusters for biomedical and biosensing applications: A mini-review. Talanta. 2019;200:432–42.

15. Mooradian A. Photoluminescence of metals. Phys Rev Lett. 1969;22:185–187.
16. Zhu M, Aikens CM, Hollander FJ, Schatz GC, Jin R. Correlating the crystal structure of a thiol-protected Au 25 cluster and optical properties. J Am Chem Soc. 2008;130(18):5883–5.
17. Sakanaga I, Inada M, Saitoh T, et al. Photoluminescence from excited energy bands in Au25 nanoclusters. Appl Phys Express. 2011;4:544–548.
18. Wen X, Yu P, Toh Y, Tang J. Structure-correlated dual fluorescent bands in BSA-protected Au25 nanoclusters. J Phys Chem C. 2012;116(21):11830–6.
19. Kreibig U, Vollmer M. Optical properties of metal clusters. Springer Series in Material Science 25, 1995.
20. Rao TUB, Nataraju B, Pradeep T. Ag9 quantum cluster through a solid-state route. J Am Chem Soc. 2010;132:16304–7.
21. Zheng Y, Lai L, Liu W, Jiang H, Wang X. Recent advances in biomedical applications of fluorescent gold nanoclusters. Adv Colloid Interface Sci. 2017;242:1–16.
22. Guo S, Wang E. Noble metal nanomaterials: Controllable synthesis and application in fuel cells and analytical sensors. Nano Today. 2011;6:240–64.
23. Naldini L, Cariati F, Simonetta G, Malatesta L. Gold-tertiary phosphine derivatives with intermetallic bonds. Chem Commun. 1966;(18):647–648.
24. Chen LY, Wang CW, Yuan Z, Chang HT. Fluorescent gold nanoclusters: Recent advances in sensing and imaging. Anal Chem. 2015;87(1):216–29.
25. Schaaff TG, Knight G, Sha MN, Borkman RF, Whetten RL. Isolation and selected properties of a 10.4 kDa gold: Glutathione cluster compound. J Phys Chem B. 1998;102(52):10643–6.
26. Negishi Y, Nobusada K, Tsukuda T. Glutathione-protected gold clusters revisited: Bridging the gap between gold(I)−thiolate complexes and thiolate-protected gold nanocrystals. J Am Chem Soc. 2005;127(14):5261–70.
27. Negishi Y, Chaki N, Shichibu Y, Whetten R, Tsukuda T. Origin of magic stability of thiolated gold clusters: A case study on Au25(SC6H13)18. J Am Chem Soc. 2007;129(37):11322–3.
28. Shichibu Y, Negishi Y, Tsunoyama H, Kanehara M, Teranishi T, Tsukuda T. Extremely high stability of glutathionate-protected Au25 clusters against core etching. Small. 2007;3(5):835–9.
29. Huang T, Murray RW. Visible luminescence of water-soluble monolayer-protected gold clusters. J Phys Chem B. 2001;105(50):12498–502.
30. Wang G, Huang T, Murray RW, Menard L, Nuzzo R. Near-IR luminescence of monolayer-protected metal clusters. J Am Chem Soc. 2005;127(3):812–3.
31. Aldee F, Muhammed MAH, Palui G, Zhan N, Mattoussi H. Growth of highly fluorescent polyethylene glycol- and zwitterion-functionalized gold nanoclusters. ACS Nano. 2013;7(3):2509–21.
32. Rabara L, Aranyosiova M, Velic D. Thiolated cyclodextrin self-assembled monolayer-like characterized with secondary ion mass spectrometry. Appl Surf Sci. 2010;257(6):1886–92.
33. Lou Z, Yuan X, Yu Y, et al. From aggregation-induced emission of Au (I)-thiolate complexes to ultrabright Au(0)@Au (I)-thiolate core-shell nanoclusters. J Am Chem Soc. 2012;134(40):16662–70.
34. Zheng J, Petty JT, Dickson RM. High quantum yield blue emission from water-soluble Au 8 nanodots. J Am Chem Soc. 2003;125:7780–1.
35. Xie J, Zheng Y, Ying JY. Protein-directed synthesis of highly fluorescent gold nanoclusters. J Am Chem Soc. 2009;131(3):888–9.
36. Chaudhari K, Xavier PL, Pradeep T. Understanding the evolution of luminescent gold quantum clusters in protein templates. ACS Nano. 2011;5(11):8816–27.
37. Yue Y, Liu T-Y, Li H-W, Liu Z, Wu Y. Microwave-assisted synthesis of BSA-protected small gold nanoclusters and their fluorescence-enhanced sensing of silver(I) ions. Nanoscale. 2012;4:2251–4.
38. Chen P-C, Chiang C-K, Chang H-T. Synthesis of fluorescent BSA – Au NCs for the detection of Hg^{2+} ions. J Nanoparticle Res. 2013;15:1336.
39. Santhosh M, Chinnadayyala SR, Kakoti A, Goswami P. Selective and sensitive detection of free bilirubin in blood serum using human serum albumin stabilized gold nanoclusters as fluorometric and colorimetric probe. Biosens Bioelectron. 2014;59:370–6.
40. Santhosh M, Chinnadayyala SR, Singh NK, Goswami P. Human serum albumin-stabilized gold nanoclusters act as an electron transfer bridge supporting specific electrocatalysis of bilirubin useful for biosensing applications. Bioelectrochemistry. 2016;111:7–14.
41. Yan L, Cai Y, Zheng B, et al. Microwave-assisted synthesis of BSA-stabilized and HSA-protected gold nanoclusters with red emission. J Mater Chem. 2012;22:1000–5.
42. Kawasaki H, Hamaguchi K, Osaka I, Arakawa R. Ph-dependent synthesis of pepsin-mediated gold nanoclusters with blue green and red fluorescent emission. Adv Funct Mater. 2011;21(18):3508–15.
43. Liu C, Wu H, Hsiao Y, et al. Insulin-directed synthesis of fluorescent gold nanoclusters: preservation of insulin bioactivity and versatility in cell imaging. Angew Chemie - Int Ed. 2011;50(31):7056–60.
44. Wen F, Dong Y, Feng L, Wang S, Zhang S, Zhang X. Horseradish peroxidase functionalized fluorescent gold nanoclusters for hydrogen peroxide sensing. Anal Chem. 2011;83:1193–6.
45. Liu R, Wang Y-L, Cui Y-Y, Sun Z-P, Wei Y-T, Gao X-Y. Recent development of noble metal clusters for bioimaging and in vitro detection. Prog Biochem Biophys. 2013;40(10):977–89.
46. Wang Y, Cui Y, Zhao Y, et al. Bifunctional peptides that precisely biomineralize Au clusters and specifically stain cell nuclei. Chem Commun. 2012;48(6):871–873.
47. Kennedy TAC, Maclean JL, Liu J. Blue emitting gold nanoclusters templated by poly-cytosine DNA at low pH and poly-adenine DNA at neutral pH. Chem Commun. 2012;48(54):6845–7.
48. Liu G, Shao Y, Wu F, Xu S, Peng J, Liu L. DNA-hosted fluorescent gold nanoclusters: Sequence-dependent formation. Nanotechnology. 2013;24(1):015503.
49. Jiang H, Zhang Y, Wang X. Single cytidine units-templated syntheses of multi-colored water-soluble Au nanoclusters. Nanoscale. 2014;6(17):10355–62.
50. Zhang H, Huang X, Li L, et al. Photoreductive synthesis of water-soluble fluorescent metal nanoclusters. Chem Commun. 2012;48(4):567–9.
51. Li L, Li Z, Zhang H, Zhang S, Majeed I, Tan B. Effect of polymer ligand structures on fluorescence of gold clusters prepared by photoreduction. Nanoscale. 2013;5(5):1986–92.
52. Wang J, Zhang G, Li Q, et al. In vivo self-bio-imaging of tumors through in situ biosynthesized fluorescent gold nanoclusters. Sci Rep. 2013;3:1157.
53. González BS, Rodríguez MJ, Blanco C, Rivas J, Lopez-Quintela MA, Martinho JMG. One step synthesis of the smallest photoluminescent and paramagnetic PVP-protected gold atomic clusters. Nano Lett. 2010;10(10):4217–21.
54. Duan H, Nie S. Etching colloidal gold nanocrystals with hyperbranched and multivalent polymers: A new route to

54. fluorescent and water-soluble atomic clusters. J Am Chem Soc. 2007;129(9):2412–3.
55. Huang C, Yang Z, Lee K, Chang H. Synthesis of highly fluorescent gold nanoparticles for sensing mercury(II). Angew Chemie - Int Ed. 2007;46(36):6948–6952.
56. Muhammed MAH, Verma PK, Kumar RCA, Paul S, Omkumar RV, Pradeep T. Bright, NIR-emitting Au23 from Au25: Characterization and applications including biolabeling. Chem A Eur J. 2009;15(39):10110–20.
57. Lin CJ, Yang T, Lee C, et al. Synthesis, characterization, and bioconjugation of fluorescent gold nanoclusters toward biological labeling applications. ACS Nano. 2009;3(2):395–401.
58. Udayabhaskararao T, Pradeep T. New protocols for the synthesis of stable Ag and Au nanocluster molecules. J Phys Chem Lett. 2013;4:1553–64.
59. Diez I, Ras HAR. Fluorescent silver nanoclusters. Nanoscale. 2011;3:1963–70.
60. Wu Z, Lanni E, Chen W, Bier ME, Ly D, Jin R. High yield, large scale synthesis of thiolate-protected Ag7 clusters. J Am Chem Soc. 2009;131:16672–4.
61. Antoku Y, Hotta J, Mizuno H, Dickson RM, Vosch T. Transfection of living HeLa cells with fluorescent poly-cytosine encapsulated Ag nanoclusters. Photochem Photobiol Sci. 2010;9:716–21.
62. Marzilli LG. Metal–ion interactions with nucleic acids and nucleic acid derivatives. In: Progress in Inorganic Chemistry, 1977, pp. 255–377.
63. Izatt RM, Christensen JJ, Ryttingí H. Sites and thermodynamic quantities associated with proton and metal ion interaction with ribonucleic acid, deoxyribonucleic acid, and their constituent bases, nucleosides and nucleotides. Chem Rev. 1971;71(5):439–81.
64. Luk KFS, Maki AH, Hoover RJ. Studies of heavy metal binding with polynucleotides using optical detection of magnetic resonance. Silver(I) binding. J Am Chem Soc. 1975;97(I):1241–2.
65. Gwinn EG, Neill PO, Guerrero AJ, Bouwmeester D, Fygenson DK. Sequence-dependent fluorescence of DNA-hosted silver nanoclusters. Adv Mater. 2008;20:279–83.
66. Petty JT, Zheng J, Hud N V, Dickson RM. DNA-templated ag nanocluster formation. J Am Chem Soc. 2004;126:5207–12.
67. Richards CI, Choi S, Hsiang J, et al. Oligonucleotide-stabilized Ag nanocluster fluorophores. J Am Chem Soc. 2008;130:5038–9.
68. Soto-Verdugo Í, Metiu H, Gwinn E. The properties of small Ag clusters bound to DNA bases. J Chem Phys. 2010;132:195102.
69. Ritchie CM, Johnsen KR, Kiser JR, Antoku Y, Robert M, Petty JT. Ag nanocluster formation using a cytosine oligonucleotide template. J Phys Chem C Nanomater Internfaces. 2007;111:175–81.
70. Sengupta B, Ritchie CM, Buckman JG, Johnsen KR, Goodwin PM, Petty JT. Base-directed formation of fluorescent silver clusters. J Phys Chem C. 2008;112:18776–82.
71. Sharma J, Yeh H-C, Yoo H, Werner JH, Martinez JS. A complementary palette of fluorescent silver nanoclusters. Chem Commun. 2010;46:3280–3282.
72. Neill PRO, Velazquez LR, Dunn DG, Gwinn EG, Fygenson DK. Hairpins with poly-C loops stabilize four types of fluorescent Ag:DNA. J Phys Chem C. 2009;113(11):4229–33.
73. Yu J, Choi S, Richards CI, Antoku Y, Dickson RM. Live cell surface labeling with fluorescent Ag nanocluster. Photochem Photobiol Sci. 2009;84(6):1435–1439.
74. Zheng J, Dickson RM. Individual water-soluble dendrimer-encapsulated silver nanodot fluorescence. J Am Chem Soc. 2002;124:13982–3.
75. Zhang I, Xu S, Kumacheva E. Photogeneration of fluorescent silver nanoclusters in polymer microgels. Adv Mater. 2005;17:2336–40.
76. Ershov BG, Henglein A. Time-resolved investigation of early processes in the reduction of Ag+ on polyacrylate in aqueous solution. J Phys Chem B. 1998;102:10667–71.
77. Shen BZ, Duan H, Frey H. Water-soluble fluorescent Ag nanoclusters obtained from multiarm star poly (acrylic acid) as "molecular hydrogel" templates. Adv Mater. 2007;19:349–52.
78. Shang L, Dong S. Facile preparation of water-soluble fluorescent silver nanoclusters using a polyelectrolyte template. Chem Commun. 2008;9:1088–90.
79. Díez I, Pusa M, Kulmala S, et al. Color tunability and electrochemiluminescence of silver nanoclusters. Angew Chemie - Int Ed. 2009;48:2122–5.
80. Xu H, Suslick KS. Sonochemical synthesis of highly fluorescent Ag nanoclusters. ACS Nano. 2010;4(6):3209–14.
81. Cathcart N, Kitaev V. Silver nanoclusters: Single-stage scaleable synthesis of monodisperse species and their Chirooptical properties. J Phys Chem C. 2010;16010–7.
82. Adhikari B, Banerjee A. Facile synthesis of water-soluble fluorescent silver nanoclusters and HgII sensing. Chem Mater. 2010;22:4364–71.
83. Rao TUB, Pradeep T. Luminescent Ag7 and Ag8 clusters by interfacial synthesis. Angew Chemie - Int Ed. 2010;49:3925–9.
84. Makarava N, Parfenov A, Baskakov IV. Water-soluble hybrid nanoclusters with extra bright and photostable emissions: A new tool for biological imaging. Biophys J. 2005;89(1):572–80.
85. Shang L, Dong S. Sensitive detection of cysteine based on fluorescent silver clusters. Biosens Bioelectron. 2009;24:1569–73.
86. Shang L, Dong S. Silver nanocluster-based fluorescent sensors for sensitive detection of Cu(II). J Mater Chem. 2008;18:4636–40.
87. Ayres JG, Crocker JG, Skilbeck NQ. Differentiation of malignant from normal and reactive mesothelial cells by the argyrophil technique for nucleolar organiser region associated proteins. Thorax. 1988;43:366–70.
88. Giuffre G, Mormandi F, Barresi V, Bordi C, Tuccari G, Barresi G. Quantity of AgNOR in gastric endocrine carcinoid tumours as a potential prognostic tool. Eur J Histochem. 2006;50(1):45–50.
89. Kerenyi L, Gallyas F. A highly sensitive method for demonstrating proteins in electrophoretic, immunoelectrophoretic and immunodiffusion preparations. Clin Chim Acta. 1972;38:465–7.
90. Yu J, Patel SA, Dickson RM. In vitro and intracellular production of peptide-encapsulated fluorescent silver nanoclusters. Angew Chemie - Int Ed. 2007;46:2028–30.
91. Naik RR, Stringer SJ, Agarwal G, Jones SE, Stone MO. Biomimetic synthesis and patterning of silver nanoparticles. Nat Mater. 2002;1:169–72.
92. Li H, Siu KWM, Guevremont R, Blanc JCY Le. Complexes of silver(I) with peptides and proteins as produced in electrospray mass spectrometry. J Am Soc Mass Spectr. 1997;8:781–92.
93. Narayanan SS, Pal SK. Structural and functional characterization of luminescent silver-protein nanobioconjugates. J Phys Chem. 2008;112:4874–9.
94. Andersson L-O. Study of some silver-thiol complexes and polymers: Stoichiometry and optical effects. J Polym Sci. 1972;10:1963–73.

95. Peyser-Capadona L, Zheng J, Gonzalez JI, Lee T, Patel SA, Dickson RM. Nanoparticle-free single molecule anti-stokes Raman spectroscopy. Phys Rev Lett. 2005;94:058301.
96. Adhikari B, Benerjee A. Short-peptide-based hydrogel: A template for the in situ synthesis of fluorescent silver nanoclusters by using sunlight. Chem A Eur J. 2010;16:13698–13705.
97. Bellec M, Royon A, Bourhis K, Choi J, Bousquet B. 3D patterning at the nanoscale of fluorescent emitters in glass. J Phys Chem C. 2010;114(37):15584–8.
98. Eichelbaum M, Rademann K, Hoell A, et al. Photoluminescence of atomic gold and silver particles in soda-lime silicate glasses. Nanotechnology. 2008;19(13):135701.
99. Cremer G De, Antoku Y, Roeffaers MBJ, et al. Photoactivation of silver-exchanged zeolite A. Angew Chemie - Int Ed. 2008;47:2813–2816.
100. Cremer G De, Coutino-Gonzalez E, Roeffaers MBJ, et al. Characterization of fluorescence in heat-treated silver-exchanged zeolites. J Am Chem Soc. 2009;131(8):3049–56.
101. Yizhong L, Wentao W, Wei C. Copper nanoclusters: Synthesis, characterization and properties. Chinese Sci Bull. 2012;57(1):41–7.
102. Fuhr O, Dehnen S, Fenske D. Chalcogenide clusters of copper and silver from silylated chalcogenide sources. Chem Rev. 2013;4:1871–906.
103. Dhayal RS, van Zyl WE, Liu CW. Polyhydrido copper clusters: Synthetic advances, structural diversity, and nanocluster-to-nanoparticle conversion. Acc Chem Res. 2016;49(1):86–95.
104. Deutsch C, Krause N, Lipshutz BH. CuH-catalyzed reactions. Chem Rev. 2008;108(8):2916–27.
105. Zhao M, Sun L, Crooks RM. Preparation of Cu nanoclusters within dendrimer templates. J Am Chem Soc. 1998;120(19):4877–8.
106. Balogh L, Tomalia DA. Poly(amidoamine) dendrimer-templated nanocomposites. 1. Synthesis of zerovalent copper nanoclusters. J Am Chem Soc. 1998;120(29):7355–6.
107. Rotaru A, Dutta S, Elmar Jentzsch, Gothelf K, Mokhir A. Selective dsDNA – Templated formation of copper nanoparticles in solution. Angew Chemie Int Ed. 2010;49(33):5665–7.
108. Liu G, Shao Y, Peng J, et al. Highly thymine-dependent formation of fluorescent copper nanoparticles templated by ss-DNA. Nanotechnology. 2013;24(34):345502.
109. Jia X, Li J, Han L, Ren J, Yang X, Wang E. DNA-hosted copper nanoclusters for Fluorescent identication of single nucleotide polymorphisms. ACS Nano. 2012;6:3311–7.
110. López-Quintela MA, Tojo C, Blanco MC, Rio LG, Leis JR. Microemulsion dynamics and reactions in microemulsions. Curr Opin Colloid Interface Sci. 2004;9:264–78.
111. Vázquez-vázquez C, Bañobre-lópez M, Mitra A, López-quintela MA, Rivas J. Synthesis of small atomic copper clusters in microemulsions. Langmuir. 2009;25(14):8208–16.
112. Vilar-Vidal N, Blanco MC, López-Quintela MA, Rivas J, Serra C. Electrochemical synthesis of very stable photoluminescent copper clusters. J Phys Chem C. 2010;114(38):15924–30.
113. Nguyen TD, Cook AW, Wu G, Hayton TW. Subnanometer-sized copper clusters: A critical re-evaluation of the synthesis and characterization of Cu8(MPP)4 (HMPP=2-Mercapto-5-n-propylpyrimidine). Inorg Chem. 2017;56(14):8390–6.
114. Wei W, Lu Y, Chen W, Chen S. One-pot synthesis, photoluminescence, and electrocatalytic properties of subnanometer-sized copper clusters. J Am Chem Soc. 2011;133(7):2060–3.
115. Kawasaki H, Kosaka Y, Myoujin Y, Narushima T, Yonezawa T, Arakawa R. Microwave-assisted polyol synthesis of copper nanocrystals without using additional protective agents. Chem Commun. 2011;47:7740—7742.
116. Mohanty JS, Xavier PL, Chaudhari K, et al. Luminescent, bimetallic AuAg alloy quantum clusters in protein templates. Nanoscale. 2012;4:4255–62.
117. Jin R, Zhao S, Xing Y, Jin R. All-thiolate-protected silver and silver-rich alloy nanoclusters with atomic precision: Stable sizes, structural characterization and optical properties. Cryst Eng Comm. 2016;18:3996–4005.
118. Fields-Zinna CA, Crowe MC, Dass A, Weaver JEF, Murray RW. Mass spectrometry of small bimetal monolayer-protected clusters. Langmuir. 2009;25(13):7704–10.
119. Udayabhaskararao T, Sun Y, Goswami N, Pal SK, Balasubramanian K, Pradeep T. Ag7Au6: A 13-atom alloy quantum cluster. Angew Chemie - Int Ed. 2012;51(9):2155–9.
120. Andolina CM, Dewar AC, Smith AM, Marbella LE, Hartmann MJ, Millstone JE. Photoluminescent gold-copper nanoparticle alloys with composition-tunable near-infrared emission. J Am Chem Soc. 2013;135(14):5266–9.
121. Kawasaki H, Yamamoto H, Fujimori H, Arakawa R, Inada M, Iwasaki Y. Surfactant-free solution synthesis of fluorescent platinum subnanoclusters. Chem Commun. 2010;46:3759–61.
122. Hyotanishi M, Isomura Y, Yamamoto H, Kawasaki H, Obora Y. Surfactant-free synthesis of palladium nanoclusters for their use in catalytic cross-coupling reactions. Chem Commun. 2011;47:5750–2.
123. Tanaka S, Miyazaki J, Tiwari DK, Jin T, Inouye Y. Fluorescent platinum nanoclusters: Synthesis, purification, characterization, and application to bioimaging. Angew Chemie - Int Ed. 2011;50(2):431–5.
124. Chakraborty I, Bhuin RG, Bhat S, Pradeep T. Blue emitting undecaplatinum clusters Indranath. Nanoscale. 2014;6:8561–4.
125. Yu C-J, Chen T-H, Jianga J-Y, Tseng W-L. Lysozyme-directed synthesis of platinum nanoclusters as a mimic. Nanoscale. 2014;6(16):9618–24.
126. Peng Y, Wang P, Luo L, Liu L, Wang F. Green synthesis of fluorescent palladium nanoclusters. Mater. 2018;11(2):E191.
127. Bigioni TP, Whetten RL. Near-infrared luminescence from small gold nanocrystals. J Phys Chem B. 2000;104:6983–6.
128. Link S, Beeby A, Fitzgerald S, El-sayed MA, Schaaff TG, Whetten RL. Visible to infrared luminescence from a 28-atom gold cluster. J Phys Chem B. 2002;106(13):3410–5.
129. Negishi Y, Takasugi Y, Sato S, Yao H, Kimura K, Tsukuda T. Magic-numbered Au-n clusters protected by glutathione monolayers (n=18, 21, 25, 28, 32, 39): Isolation and spectroscopic characterization. J Am Chem Soc. 2004;126:6518–9.
130. Wang G, Guo R, Kalyuzhny G, Choi J, Murray RW. NIR luminescence intensities increase linearly with proportion of polar thiolate ligands in protecting monolayers of Au 38 and Au 140 quantum dots. J Phys Chem B. 2006;110:20282–9.
131. Negishi Y, Tsukuda T. Visible photoluminescence from nearly monodispersed Au12 clusters protected by meso-2,3-dimercaptosuccinic acid. Chem Phys Lett. 2004;383:161–5.
132. Lee D, Donkers RL, Wang G, Harper AS, Murray RW. Electrochemistry and optical absorbance and luminescence of molecule-like Au38 nanoparticles. J Am Chem Soc. 2004;126:6193–9.
133. Yang Y, Chen S. Surface manipulation of the electronic energy of subnanometer-sized gold clusters: An electrochemical and spectroscopic investigation. Nano Lett. 2003;3(1):75–9.
134. Huang C, Chiang C, Lin Z, Lee K, Chang H. Bioconjugated gold nanodots and nanoparticles for protein assays based on photoluminescence quenching. Anal Chem. 2008;80(5):1497–504.
135. Muhammed MAH, Ramesh S, Sinha SS, Pal SK, Pradeep T. Two distinct fluorescent quantum clusters of gold starting

from metallic nanoparticles by ph-dependent ligand Etching. Nano Res. 2008;1:333–40.
136. Wei H, Wang Z, Yang L, Tian S, Hou C, Lu Y. Lysozyme-stabilized gold fluorescent cluster: Synthesis and application as Hg^{2+} sensor. Analyst. 2010;135(6):1406.
137. Xavier PL, Chaudhari K, Verma PK, Pal SK, Pradeep T. Luminescent quantum clusters of gold in transferrin family protein, lactoferrin exhibiting FRET. Nanoscale . 2010;2(12): 2769.
138. Guevel X Le, Daum N, Schneider M. Synthesis and characterization of human transferrin-stabilized gold nanoclusters. Nanotechnology. 2011;22(27).
139. Garcia AR, Rahn I, Johnson S, et al. Biointerfaces human insulin fibril-assisted synthesis of fluorescent gold nanoclusters in alkaline media under physiological temperature. Colloids Surfaces B Biointerfaces. 2013;105:167–72.
140. Muhammed MAH, Verma PK, Pal SK, et al. Luminescent quantum clusters of gold in bulk by albumin-induced core etching of nanoparticles: Metal ion sensing, metal-enhanced luminescence, and biolabeling. Chem A Eur J. 2010;16:10103–12.
141. Guo C, Irudayaraj J. Fluorescent Ag clusters via a protein-directed approach as a Hg(II) ion sensor. Anal Chem. 2011;(Ii):2883–9.
142. Liu X, Fu C, Ren X, Liu H, Li L, Meng X. Fluorescence switching method for cascade detection of salicylalde- hyde and zinc (II) ion using protein protected gold nanoclusters. Biosens Bioelectron. 2015;74:322–8.
143. Li H, Guo Y, Xiao L, Chen B. Selective and sensitive detection of acetyl- cholinesterase activity using denatured protein- protected gold nanoclusters as a online, label-free probe. Analyst. 2014;139:285–9.
144. Goswami N, Giri A, Bootharaju MS, Xavier PL, Pradeep T, Pal SK. Copper quantum clusters in protein matrix: Potential sensor of Pb^{2+} Ion. Anal Chem. 2011;83:9676–80.
145. Hu D, Sheng Z, Gong P, Cai L. Highly selective fluorescent sensors for Hg^{2+} based on bovine serum albumin-capped gold nanoclusters. Analyst. 2010;135:1411–6.
146. Durgadas CV, Sharma CP, Sreenivasan K. Fluorescent gold clusters as nanosensors for copper ions in live cells. Analyst. 2011;136:933–40.
147. Zhou W, Cao Y, Sui D, Guan W, Lu C, Xie J. Ultrastable BSA-capped gold nanoclusters with a polymer-like shielding layer against reactive oxygen species in living cells. Nanoscale. 2016;8:9614–20.
148. Pitchiaya S, Krishnan Y. First blueprint, now bricks: DNA as construction material on the nanoscale. Chem Soc Rev. 2006;35:1111–21.
149. Dattagupta N, Crothers DM. Solution structural studies of the Ag(I)-DNA complex. Nucleic Acids Res. 1981;9(12):2971–2985.
150. Vosch T, Antoku Y, Hsiang J, Richards CI, Gonzalez JI, Dickson RM. Strongly emissive individual DNA-encapsulated Ag nanoclusters as single-molecule fluorophores. PNAS. 2007;104(31).
151. Sengupta B, Springer K, Buckman JG, et al. DNA templates for fluorescent silver clusters and I-motif folding. J Phys Chem C. 2009;113:19518–24.
152. Zhou R, Shi M, Chen X, Wang M, Chen H. Atomically monodispersed and fluorescent sub-nanometer gold clusters created by biomolecule-assisted etching of nanometer-sized gold particles and rods. Chem A Eur J. 2009;15:4944–51.
153. Liu S, Lu F, Zhu J-J. Highly fluorescent Ag nanoclusters: Microwave-assisted green synthesis and Cr^{3+} sensing. Chem Commun. 2011;47:2661–3.
154. Schaeffer N, Tan B, Dickinson C, et al. Fluorescent or not? Size-dependent fluorescence switching for polymer-stabilized gold clusters in the 1.1-1.7 nm size range. Chem Commun. 2008;3986–8.
155. Schmid G. Clusters and colloids: From theory to applications. Vol. VCH Publis. 1994;183.
156. Li B, Li J, Zhao J. Silver nanoclusters emitting weak NIR fluorescence biomineralized by BSA. Spectrochim Acta Part A Mol Biomol Spectrosc. 2015;134:40–7.
157. Wilcoxon JP, Newcomer PP, Samara GA. Synthesis and optical properties of MoS_2 and isomorphous nanoclusters in the quantum confinement regime. J Appl Phys. 1998;81:7934.
158. Hussain AMP, Sarangi SN, Kesarwani JA, Sahu SN. Au-nanocluster emission based glucose sensing. Biosens Bioelectron. 2011;29(1):60–5.
159. Negishi Y, Nakazaki T, Malola S, et al. A critical size for emergence of nonbulk electronic and geometric structures in dodecanethiolate-protected Au. J Am Chem Soc. 2015;137(3):1206–12.
160. Niihori Y, Matsuzaki M, Uchida C, Negishi Y. Advanced use of high-performance liquid chromatography for synthesis of controlled metal clusters. Nanoscale. 2014;6:7889–96.
161. Ghosh A, Hassinen J, Pulkkinen P, Tenhu H, Ras RHA, Thalappil P. Simple and efficient separation of atomically precise noble. Anal Chem. 2014;86(24):12185–90.
162. Jain P, Chakma B, Patra S, Goswami P. Hairpin stabilized fluorescent silver nanoclusters for quantitative detection of NAD^+ and monitoring $NAD^+/NADH$ based enzymatic reactions. Anal Chim Acta . 2017;956:48–56.
163. Templeton AC, Cliffel DE, Murray RW. Redox and fluorophore functionalization of water-soluble, tiopronin-protected gold clusters. J Am Chem Soc. 1999;121(30):7081–9.
164. Hicks JF, Templeton AC, Chen S, et al. The monolayer thickness dependence of quantized double-layer capacitances of monolayer-protected gold clusters. Anal Chem. 1999;71(17):3703–11.
165. Wu Z, Jin R. On the ligand 's role in the fluorescence of gold nanoclusters. Nano Lett. 2010;25:2568–73.
166. Wang S, Zhu X, Cao T, Zhu M. A simple model for understanding the fluorescence behavior of Au25 nanoclusters. Nanoscale. 2014;6:5777–81.
167. van Wijngaarden JT, Toikkanen O, Liljeroth P, Quinn BM, Meijerink A. Temperature-dependent emission of monolayer-protected Au38 clusters. J Phys Chem C. 2010; 114(38):16025–8.
168. Shang L, Yang L, Stockmar F, et al. Microwave-assisted rapid synthesis of luminescent gold nanoclusters for sensing Hg^{2+} in living cells using fluorescence imaging. Nanoscale. 2012;4:4155–60.
169. Sharma J, Rocha RC, Phipps ML, et al. A DNA-templated fluorescent silver nanocluster with enhanced stability. Nanoscale Commun. 2012;4:4107–10.
170. Zhou T, Rong M, Cai Z, Yang CJ, Chen X. Sonochemical synthesis of highly fluorescent glutathione-stabilized Ag nanoclusters and S^{2-} sensing. Nanoscale. 2012;4:4103–6.
171. Zhang L, Li J, Xu Y, Zhai Y, Li Y, Wang E. Solid-state electrochemiluminescence sensor based on the Nafion/poly (sodium 4-styrene sulfonate) composite film. Talanta. 2009;79(2):454–9.
172. Fang Y, Song J, Li J, et al. Electrogenerated chemiluminescence from Au nanoclusters. Chem Commun. 2011;47:2369–71.
173. Li L, Liu H, Shen Y, Zhang J, Zhu J. Electrogenerated chemiluminescence of Au nanoclusters for the detection of dopamine. Anal Chem. 2011;83(3):661–5.

174. Guo W, Yuan J, Wang E. Organic-soluble fluorescent Au8 clusters generated from heterophase ligand-exchange induced etching of gold nanoparticles and their electrochemiluminescence. Chem Commun. 2012;48:3076–8.
175. So PTC, Dong CY, Masters BR, Berland KM. Two-photon excitation fluorescence microscopy. Annu Rev Biomed Eng. 2000;2:399–429.
176. Ramakrishna G, Varnavski O, Kim J, Lee D, Goodson T. Quantum-sized gold clusters as efficient two-photon. J Am Chem Soc. 2008;130(15):5032–3.
177. Polavarapu L, Manna M, Xu Q. Biocompatible glutathione capped gold clusters as one- and two-photon excitation fluorescence contrast agents for live cells imaging. Nanoscale. 2011;3:429–34.
178. Patel SA, Richards CI, Hsiang J-C, Dickson RM. Water-soluble Ag nanoclusters exhibit strong two-photon-induced fluorescence. J Am Chem Soc. 2008;130(35):11602–3.
179. Yau SH, Abeyasinghe N, Orr M, et al. Bright two-photon emission and ultra-fast relaxation dynamics in a DNA-templated nanocluster investigated by ultra-fast spectroscopy. Nanoscale. 2012;4:4247–54.
180. Choi S, Dickson RM, Yu J. Developing luminescent silver nanodots for biological applications. Chem Soc Rev. 2012;41:1867–91.
181. Shang L, Stockmar F, Azadfar N, Nienhaus GU. Intracellular thermometry by using fluorescent gold nanoclusters. Angew Chemie - Int Ed. 2013;52(42):11154–7.
182. Mathew A, Pradeep T. Noble metal clusters: Applications in energy, Environment, and Biology. Part Part Syst Charact. 2014;31:1017–1053.
183. Yeh H, Sharma J, Han JJ, Martinez JS, Werner JH. A DNA-silver nanocluster probe that fluoresces upon hybridization. Nano Lett. 2010;10(8):3106–10.
184. Zhang L, Zhu J, Zhou Z, et al. A new approach to light up DNA/Ag nanocluster-based beacons for bioanalysis. Chem Sci. 2013;4:4007–10.
185. Dong H, Jin S, Ju H, et al. Trace and label-free microRNA detection using oligonucleotide encapsulated silver nanoclusters as probes. Anal Chem. 2012;84(20):8670–4.
186. Xu F, Shi H, He X, et al. Concatemeric dsDNA-templated copper nanoparticles strategy with improved sensitivity and stability based on rolling circle replication and its application in microRNA. Anal Chem. 2014;86(14):6976–82.
187. Liu X, Wang F, Aizen R, Yehezkeli O, Willner I. Graphene oxide/nucleic-acid-stabilized silver nanoclusters: Functional hybrid materials for optical aptamer sensing and multiplexed analysis of pathogenic DNAs. J Am Chem Soc. 2013;135(32):11832–9.
188. Guo W, Yuan J, Dong Q, Wang E. Highly sequence-dependent formation of fluorescent silver nanoclusters in hybridized DNA duplexes for single nucleotide mutation identification. J Am Chem Soc. 2010;132(3):932–4.
189. Huang Z, Pu F, Hu D, Wang C, Ren J, Qu X. Site-specific DNA-programmed growth of fluorescent and functional silver nanoclusters. Chem A Eur J. 2011;17:3774–80.
190. Yeh H, Sharma J, Shih I, Vu DM, Martinez JS, Werner JH. A fluorescence light-up Ag nanocluster probe that discriminates single-nucleotide variants by emission color. J Am Chem Soc. 2012;134(28):11550–8.
191. Macfarlane L, Murphy PR. MicroRNA: Biogenesis, function and role in cancer. Curr Genomics. 2010;11(7):537–61.
192. Li Y, Kowdley KV. MicroRNAs in common human diseases. Genomics Proteomics Bioinforma. 2012;10(5):246–53.
193. Shah P, Cho SK, Thulstrup PW, et al. Effect of salts, solvents and buffer on miRNA detection using DNA silver nanocluster (DNA/Ag NCs) probes. Nanotechnology. 2014;25(4):045101.
194. Yang SW, Vosch T. Rapid detection of MicroRNA by a silver nanocluster DNA. Anal Chem. 2011;83(18):6935–9.
195. Shah P, Rørvig-lund A, Chaabane S Ben, et al. Design Aspects of Bright Red Emissive Silver Nanoclusters/DNA Probes for MicroRNA Detection. ACS Nano. 2012;6(10):8803–14.
196. Zhang Y, Chen Z, Tao Y, Wang Z, Ren J, Qu X. Hybridization chain reaction engineered dsDNA for Cu metallization: An enzyme-free platform for amplified detection of cancer cells and microRNAs. Chem Commun. 2015;51:11496–9.
197. Ross JS, Fletcher JA. The HER-2/neu oncogene in breast cancer: Prognostic factor, predictive factor, and target for therapy. Stem Cells. 1998;16(6):413–28.
198. Wei H, Li B, Li J, Wang E, Dong S. Simple and sensitive aptamer-based colorimetric sensing of protein using unmodified gold nanoparticle probes. Chem Commun (Camb). 2007;36:3735–7.
199. Sharma J, Yeh H-C, Yoo H, Wernera JH, Martinez JS. Silver nanocluster aptamers: In situ generation of intrinsically fluorescent recognition ligands for protein detection. Chem Commun. 2011;47:2294–6.
200. Li J, Zhong X, Zhang H, Le XC, Zhu J-J. Binding-induced fluorescence turn-on assay using aptamer-functionalized silver nanocluster DNA probes. Anal Chem. 2012;84(12):5170–4.
201. Zheng C, Zheng AX, Liu B, et al. One-pot synthesized DNA-templated Ag/Pt bimetallic nanoclusters as peroxidase mimics for colorimetric detection of thrombin. Chem Commun (Camb). 2014;50(86):13103–6.
202. Kong H, Lu Y, Wang H, Wen F, Zhang S, Zhang X. Protein discrimination using fluorescent gold nanoparticles on plasmonic substrates. Anal Chem. 2012;84(10):4258–61.
203. Yuan Z, Du Y, Tseng YT, et al. Fluorescent gold nanodots based sensor array for proteins discrimination. Anal Chem. 2015;87(8):4253–9.
204. Drag M, Salvesen GS. Emerging principles in protease-based drug discovery. Nat Rev Drug Discov. 2010;9(9):690–701.
205. Wang Y, Wang Y, Zhou F, Kim P, Xia Y. Protein-Protected Au clusters as a new class of nanoscale biosensor for label-free fluorescence detection of proteases. Small. 2012;(24):3769–73.
206. Lin H, Li LL, Lei C, et al. Immune-independent and label-free fluorescent assay for cystatin C detection based on protein-stabilized Au nanoclusters. Biosens Bioelectron. 2013;41:256–61.
207. Walsh CT, Garneau-Tsodikova S, Gatto JG. Protein posttranslational modifications: The chemistry of proteome diversifications. Angew Chemie - Int Ed. 2005;44(45):7342–72.
208. Wen Q, Gu Y, Tang L, Yu R, Jiang J. Peptide-templated gold nanocluster beacon as a sensitive, label-free sensor for protein post-translational modification enzymes. Anal Chem. 2013;85:11681–5.
209. Chen Y, Zhou H, Wang Y, et al. Substrate hydrolysis triggered formation of fluorescent gold nanoclusters–a new platform for the sensing of enzyme activity. Chem Commun. 2013;49(84):9821–3.
210. Teng Y, Jia X, Li J, Wang E. Ratiometric fluorescence detection of tyrosinase activity and dopamine using thiolate-protected gold nanoclusters. Anal Chem. 2015;87(9):4897–902.
211. Zhang M, Guo SM, Li YR, Zuo P, Ye BC. A label-free fluorescent molecular beacon based on DNA-templated silver nanoclusters for detection of adenosine and adenosine deaminase. Chem Commun (Camb). 2012;48(44):5488–90.
212. Zhang K, Wang K, Xie M, et al. DNA-templated silver nanoclusters based label-free fluorescent molecular beacon for

212. the detection of adenosine deaminase. Biosens Bioelectron. 2014;52:124–8.
213. Shen C, Xia X, Hu S, Yang M, Wang J. Silver nanoclusters-based fluorescence assay of protein kinase activity and inhibition. Anal Chem. 2015;87(1):693–8.
214. Qian Y, Zhang Y, Lu L, Cai Y. A label-free DNA-templated silver nanocluster probe for fluorescence on – off detection of endonuclease activity and inhibition. Biosens Bioelectron. 2014;51:408–12.
215. Coggan JS, Bartol TM, Esquenazi E, Stiles JR, Lamont S, Martone ME. Evidence for ectopic neurotransmission at a neuronal synapse. Science (80). 2005;309(5733):446–51.
216. Zhang Y, Cai Y, Qi Z, Lu L, Qian Y. DNA-templated silver nanoclusters for fluorescence turn-on assay of acetylcholinesterase activity. Anal Chem. 2013;85(17):8455–61.
217. Qing Z, Qing T, Mao Z, et al. dsDNA-specific fluorescent copper nanoparticles as a "green" nano-dye for polymerization-mediated biochemical analysis. Chem Commun. 2014;50(84):12746–8.
218. Zhang L, Zhao J, Duan M, Zhang H, Jiang J, Yu R. Inhibition of dsDNA-templated copper nanoparticles by pyrophosphate as a label-free fluorescent strategy for alkaline phosphatase assay. Anal Chem. 2013;85(8):3797–801.
219. Qing Z, He X, Qing T, et al. Poly(thymine)-templated fluorescent copper nanoparticles for ultrasensitive label-free nuclease assay and its inhibitors screening. Anal Chem. 2013;85(24):12138–43.
220. Li X-G, Zhang F, Gao Y, et al. Facile synthesis of red emitting 3-aminophenylboronic acid functionalized copper nanoclusters for rapid, selective and highly sensitive detection of glycoproteins. Biosens Bioelectron. 2016;86:270–6.
221. Zhao M, Qian Z, Zhong M, Chen Z, Ao H, Feng H. Fabrication of stable and luminescent copper nanocluster-based AIE particles and their application in β-galactosidase. ACS Appl Mater Interfaces. 2017;9(38):32887–95.
222. Hu L, Yuan Y, Zhang L, Zhao J, Majeed S, Xu G. Copper nanoclusters as peroxidase mimetics and their applications to H_2O_2 and glucose detection. Anal Chim Acta . 2013;762: 83–6.
223. Nair L V., Philips DS, Jayasree RS, Ajayaghosh A. A near-infrared fluorescent nanosensor (AuC@Urease) for the selective detection of blood urea. Small. 2013;9(16):2673–7.
224. Tao Y, Lin Y, Ren J, Qu X. A dual fluorometric and colorimetric sensor for dopamine based on BSA-stabilized Au nanoclusters. Biosens Bioelectron. 2013;42:41–6.
225. Govindaraju S, Ankireddy SR, Viswanath B, Kim J. Fluorescent Gold Nanoclusters for Selective Detection of Dopamine in Cerebrospinal fluid. Sci Rep. 2017; 7: 40298. doi: 10.1038/srep40298
226. Zhu J, Song X, Gao L, et al. A highly selective sensor of cysteine with tunable sensitivity and detection window based on dual-emission Ag nanoclusters. Biosens Bioelectron. 2014;53:71–5.
227. Huang Z, Pu F, Lin Y, Ren J, Qu X. Modulating DNA-templated silver nanoclusters for fluorescence turn-on detection of thiol compounds. Chem Commun. 2011;47(12):3487–9.
228. Chen T, Hu Y, Cen Y, Chu X, Lu Y. Interfering with ROS metabolism in cancer cells: The potential role of quercetin. Cancers (Basel). 2010;2(2):1288–311.
229. Chen T, Hu Y, Cen Y, Chu X, Lu Y. A dual-emission fluorescent nanocomplex of gold-cluster-decorated silica particles for live cell imaging of highly reactive oxygen species. J Am Chem Soc. 2013;135(31):11595–602.
230. Yang Y, Han A, Li R, Fang G, Liu J, Wang S. Synthesis of highly fluorescent gold nanoclusters and its use in sensitive analysis of metal ions. Analyst. 2017;142:4486–93.
231. Lu Y, Chen W. Sub-nanometre sized metal clusters: From synthetic challenges to the unique property discoveries. Chem Soc Rev. 2012;41:3594–623.
232. Holmes P, JAmes KAF, Levy LS. Is low-level environmental mercury exposure of concern to human health? Sci Total Environ. 2009;408(2):171–82.
233. Xie J, Zheng Y, Ying JY. Highly selective and ultrasensitive detection of Hg^{2+} based on fluorescence quenching of Au nanoclusters by Hg^{2+}–Au^+ interactions. Chem Commun. 2010;46(6):961–3.
234. Qiao Y, Zhang Y, Zhang C, et al. Water-soluble gold nanoclusters-based fluorescence probe for highly selective and sensitive detection of Hg^{2+}. Sensors Actuators, B Chem. 2016;224:458–64.
235. Chen W, Tu X, Guo X. Fluorescent gold nanoparticles-based fluorescence sensor for Cu^{2+} ions. Chem Commun. 2009;13:1736–8.
236. Lan G-Y, Huang C-C, Chang H-T. Silver nanoclusters as fluorescent probes for selective and sensitive detection of copper ions. Chem Commun. 2010;46:1257–9.
237. Su YT, Lan GY, Chen WY, Chang HT. Detection of copper ions through recovery of the fluorescence of DNA-templated copper/silver nanoclusters in the presence of mercaptopropionic acid. Anal Chem. 2010;82(20):8566–72.
238. Tu X, Chen W, Guo X. Facile one-pot synthesis of near-infrared luminescent gold nanoparticles for sensing copper (II). Nanotechnology. 2011;22(9):095701.
239. Needleman H. Lead Poisoning. Annu Rev Med. 2004; 55:209–22.
240. Yuan Z, Peng M, He Y, Yeung ES. Functionalized fluorescent gold nanodots: Synthesis and application for Pb^{2+} sensing. Chem Commun. 2011;47:11981–3.
241. Wang C, Cheng H, Huang Y, Xu Z, Lin H, Zhang C. Facile sonochemical synthesis of pH-responsive copper nanoclusters for selective and sensitive detection of Pb^{2+} in living cells. Analyst. 2015;140:5634–9.
242. Han B, Hou X, Xiang R, et al. Detection of lead ion based on aggregation-induced emission of copper nanoclusters. Chinese J Anal Chem. 2017;45(1):23–7.
243. Li B, Du Y, Dong S. DNA based gold nanoparticles colorimetric sensors for sensitive and selective detection of Ag (I) ions. Anal Chim Acta. 2009;644(1–2):78–82.
244. Sun J, Yue Y, Wang P, He H, Jin Y. Facile and rapid synthesis of water-soluble fluorescent gold nanoclusters for sensitive and selective detection of Ag^+. J Mater Chem C. 2013;1:908–13.
245. Wu Z, Wang M, Yang J, et al. Well-defined nanoclusters as fluorescent nanosensors: A case study on Au 25(SG)18. Small. 2012;8(13):2028–35.
246. Li H, Yue Y, Liu T, Li D, Wu Y. Fluorescence-enhanced sensing mechanism of BSA-protected small gold-nanoclusters to silver (I) ions in aqueous solutions. J Phys Chem C. 2013;117(31):16159–65.
247. Xia-hong Z, Ting-yao Z, Xi C. Applications of metal nanoclusters in environmental monitoring. Chinese J Anal Chem. 2015;43(9):1296–305.
248. Chang Y, Zhang Z, Hao J, Yang W, Tang J. BSA-stabilized Au clusters as peroxidase mimetic for colorimetric detection of Ag^+. Sensors Actuators B Chem. 2016;232:692–7.
249. Li D, Li J, Jia X, Xia Y, Zhang X, Wang E. A novel Au-Ag-Pt three-electrode microchip sensing platform for chromium (VI) determination. Anal Chim Acta. 2013;804(4):98–103.

250. Zhang H, Liu Q, Wang T, et al. Facile preparation of glutathione-stabilized gold nanoclusters for selective determination of chromium (III) and chromium (VI) in environmental water samples. Anal Chim Acta. 2013;770:140–6.

251. Liu Y, Ai K, Cheng X, Huo L, Lu L. Gold-nanocluster-based fluorescent sensors for highly sensitive and selective detection of cyanide in water. Adv Funct Mater. 2010;20(6):951–6.

252. Lu D, Liu L, Li F, et al. Lysozyme-stabilized gold nanoclusters as a novel fluorescence probe for cyanide recognition. Spectrochim Acta Part A Mol Biomol Spectrosc. 2014;121:77–80.

253. Zong C, Zheng LR, He W, Ren X, Jiang C, Lu L. In situ formation of phosphorescent molecular gold (I) cluster in a macroporous polymer film to achieve colorimetric cyanide sensing. Anal Chem. 2014;86(3):1687–1692.

254. Liu H, Yang G, Abdel-Halim ES, Zhu J-J. Highly selective and ultrasensitive detection of nitrite based on fluorescent gold nanoclusters. Talanta. 2013;104:135–9.

255. Unnikrishnan B, Wei S-C, Chiu W-J, Cang J, Hsu P-H, Huang C-C. Nitrite ion-induced fluorescence quenching of luminescent BSA-Au25 nanoclusters: Mechanism and application. Analyst. 2014;139:2221–8.

256. Xu H, Zhu H, Sun M, et al. Graphene oxide supported gold nanoclusters for the sensitive and selective detection of nitrite ions. Analyst. 2015;140:1678–85.

257. Chen W, Lan G, Chang H. Use of fluorescent DNA-templated gold/silver nanoclusters for the detection of sulfide ions. Anal Chem. 2011;83(24):9450–5.

258. Cui M-L, Liu J-M, Wang X-X, et al. A promising gold nanocluster fluorescent sensor for the highly sensitive and selective detection of S^{2-}. Sensors Actuators B Chem Actu. 2013;188:53–8.

259. Roy S, Palui G, Banerjee A. The as-prepared gold cluster-based fluorescent sensor for the selective detection of AsIII ions in aqueous solution. Nanoscale. 2012;4:2734–40.

260. Shang L, Chen H, Deng L, Dong S. Enhanced resonance light scattering based on biocatalytic growth of gold nanoparticles for biosensors design. Biosens Bioelectron. 2008;23(7):1180–4.

261. Shang L, Dong S. Design of fluorescent assays for cyanide and hydrogen peroxide based on the inner filter effect of metal nanoparticles. Anal Chem. 2009;81(4):1465–70.

262. Shiang Y-C, Huang C-C, Chang H-T. Gold nanodot-based luminescent sensor for the detection of hydrogen peroxide and glucose. Chem Commun. 2009;3437–9.

263. Jin L, Shang L, Guo S, et al. Biomolecule-stabilized Au nanoclusters as a fluorescence probe for sensitive detection of glucose. Biosens Bioelectron. 2011;26(5):1965–9.

264. Wen F, Dong Y, Feng L, Wang S, Zhang S, Zhang X. Horseradish peroxidase functionalized fluorescent gold nanoclusters for hydrogen peroxide sensing. Anal Chem. 2011;1193–6.

265. Mathew A, Sajanlal PR, Pradeep T. Selective visual detection of TNT at the sub-zeptomole level. Angew Chemie - Int Ed. 2012;51(38):9596–600.

266. Yang X, Wang J, Su D, et al. Fluorescent detection of TNT and 4-nitrophenol by BSA Au nanoclusters. Dalt Trans. 2014;43:10057–63.

267. Senthamizhan A, Celebioglu A, Uyar T. Ultrafast on-site selective visual detection of TNT at sub-ppt level using fluorescent gold cluster incorporated single nanofiber. Chem Commun. 2015;51:5590–3.

268. Chen X, Essner JB, Baker GA. Exploring luminescence-based temperature sensing using protein-passivated gold nanoclusters. Nanoscale. 2014;6:9594–8.

269. Chen Y-N, Chen P-C, Wang C-W, et al. One-pot synthesis of fluorescent BSA–Ce/Au nanoclusters as ratiometric pH probes. Chem Commun. 2014;50:8571–4.

270. Wang C, Wang C, Xu L, Cheng H, Lin Q, Zhang C. Protein-directed synthesis of pH-responsive red fluorescent copper nanoclusters and their applications in cellular imaging and catalysis. Nanoscale. 2014;6(3):1775–81.

271. Chen P-C, Ma J-Y, Chen L-Y, et al. Photoluminescent AuCu bimetallic nanoclusters as pH sensors and catalysts. Nanoscale. 2014;6:3503–7.

272. Singh P, Pandit S, Mokkapati VRSS, Garg A, Ravikumar V, Mijakovic I. Gold nanoparticles in diagnostics and therapeutics for human. Int J Mol Sci. 2018;19(7):E1979.

273. Chen H, Li B, Ren X, et al. Multifunctional near-infrared-emitting nano-conjugates based on gold clusters for tumor imaging and therapy. Biomaterials. 2012;33(33):8461–76.

274. Gui R, Wan A, Liu X, Jin H. Intracellular fluorescent thermometry and photothermal-triggered drug release developed from gold nanoclusters and doxorubicin dual-loaded liposomes. Chem Commun. 2014;50(13):1546–8.

275. Tian H, Guo Z, Chen J, et al. PEI conjugated gold nanoparticles: Efficient gene carriers with visible fluorescence. Adv Healthc Mater. 2012;1(3):337–41.

276. Tao Y, Li Z, Ju E, Ren J, Qu X. Polycations-functionalized water-soluble gold nanoclusters: A potential platform for simultaneous enhanced gene delivery and cell imaging. Nanoscale. 2013;5:6154–60.

277. Chen, W-Y, Lin, J-Y, Chen WJ, Luo L, Diau, E. W-G, Chen, Y-C. Functional gold nanoclusters as antimicrobial agents for antibiotic-resistant bacteria. Nanomedicine. 2010;5(5): 755–64.

278. Jin J-C, Wu X-J, Xu J, Wang B-B, Jiang F-L, Liu Y. Ultrasmall silver nanoclusters: Highly efficient antibacterial activity and their mechanisms. Biomater Sci. 2017;5:247–57.

279. Kuah E, Toh S, Yee J, Ma Q, Gao Z. Enzyme Mimics: Advances and applications. Chemistry (Easton). 2016;22(25):8404–30.

280. Dong Z, Luo Q, Liu J. Artificial enzymes based on supramolecular scaffolds. Chem Soc Rev. 2012;41:7890–908.

281. Lin Y, Ren J, Qu X. Nano-gold as artificial enzymes: Hidden talents. Adv Mater. 2014;26(25):4200–17.

282. Wang X, Wu Q, Shan Z, Huang Q. BSA-stabilized Au clusters as peroxidase mimetics for use in xanthine detection. Biosens Bioelectron. 2011;26:3614–9.

283. Deng M, Xu S, Chen F. Enhanced chemiluminescence of the luminol-hydrogen peroxide system by BSA-stabilized Au nanoclusters as a peroxidase mimic and its application. Anal Methods. 2014;16:3117–23.

284. Tao Y, Lin Y, Huang Z, Ren J, Qu X. Incorporating graphene oxide and gold nanoclusters: A synergistic catalyst with surprisingly high peroxidase-like activity over a broad pH range and its application for cancer cell detection. Adv Mater. 2013;25(18):2594–9.

285. Brust M, Walker M, Bethell D, Schiffrin D J and Whyman R. Synthesis of thiol-derivatised gold nanoparticles in a two-phase Liquid–Liquid system. J. Chem. Soc., Chem. Commun. 1994;7:801-802.

286. Cui ML, Liu J-M, Wang X-X, et al. Selective determination of cysteine using BSA-stabilized gold nanoclusters with red emission. Analyst. 2012;137:5346–5351.

287. Tian D, Qian Z, Xia Y, Zhu C. Gold nanocluster-based fluorescent probes for near-infrared and turn-on sensing of glutathione in living cells. Langmuir. 2012;28:3945–3951.
288. Peng H, Jian M, Huang Z, et al. Facile electrochemiluminescence sensing platform based on high-quantum- yield gold nanocluster probe for ultrasensitive glutathione detection. Biosens. Bioelectron. 2018;105:71–76.
289. Chen C, Chen W, Liu C, Chang L, Chen Y. Glutathione-bound gold nanoclusters for selective-binding and detection of glutathione S-transferase-fusion proteins from cell lysates. Chem. Commun. 2009;48:7515–7517.
290. Triulzi RC, Micic M, Giordani S, Serry M, Chiou W-A, Leblanc RM. Immunoasssay based on the antibody-conjugated PAMAM-dendrimer-gold quantum dot complex. Chem. Commun. 2006;5068–5070.
291. Liu Q, Na W, Wang L, Su X. Gold nanocluster-based fluorescent assay for label-free detection of protein kinase and its inhibitors. Microchim. Acta. 2017;2017:3381–3387.
292. Petty JT, Sengupta B, Story SP, Degtyareva NN. DNA sensing by amplifying the number of near-infrared emitting, oligonucleotide-encapsulated silver clusters. Anal. Chem. 2011;83:5957–5964.
293. Xie J, Zheng Y, Ying JY. Highly selective and ultrasensitive detection of Hg^{2+} based on fluorescence quenching of Au nanoclusters by Hg^{2+}-Au^+ interactions. Chem. Commun. 2010;46: 961–963.
294. Luo M, Di J, Li L, Tu Y, Yan J. Copper ion detection with improved sensitivity through catalytic quenching of gold nanocluster fluorescence. Talanta 2018;187:231–236.
295. Liu Y, Ding D, Zhen Y, Guo R. Amino acid-mediated 'turn-off/turn-on' nanozyme activity of gold nanoclusters for sensitive and selective detection of copper ions and histidine. Biosens. Bioelectron. 2017;92:140–146.
296. Mu X, Qi L, Dong P, Qiao J, Hou J, Nie Z, Ma H. Facile one-pot synthesis of L-proline-stabilized fluorescent gold nanoclusters and its application as sensing probes for serum iron. Biosens. Bioelectron. 2013;49:249–255.
297. Yue Y, Liu T-Y, Li H-W, Liu Z, Wu Y. Microwave-assisted synthesis of BSA-protected small gold nanoclusters and their fluorescence-enhanced sensing of silver (I). Nanoscale 2012;4:2251–2254.
298. Wang M, Wu Z, Yang J, Wang G, Wang H, Cai W. $Au25(SG)18$ as a fluorescent iodide sensor. Nanoscale 2012;4:4087–4090.
299. Zhou T, Lin L, Rong M, Jiang Y, Chen X. Silver-gold alloy nanoclusters as a fluorescence-enhanced probe for aluminum ion sensing. Anal. Chem. 2013;85:9839–9844.
300. Wang X, Wu P, Hou X, Lv Y. An ascorbic acid sensor based on protein-modified Au nanoclusters. Analyst 2013;138:229–233.
301. Chen X, Baker GA. Cholesterol determination using protein-templated fluorescent gold nanocluster probes. Analyst 2013;138:7299–7302.
302. Zheng C, Ji Z, Zhang J, Ding S. A fluorescent sensor to detect sodium dodecyl sulfate based on the glutathione-stabilized gold nanoclusters/poly diallyldimethylammonium chloride system. Analyst 2014;139:3476–3480.
303. Yu Q, Gao P, Zhang KY, et al. Luminescent gold nanocluster-based sensing platform for accurate H2S detection in vitro and in vivo with improved anti-interference. Light Sci. Appl. 2017;6: e17107.

7 Nanozymes as Potential Catalysts for Sensing and Analytical Applications

Smita Das and Pranab Goswami
Indian Institute of Technology Guwahati, Assam, India

CONTENTS

- 7.1 Introduction 144
- 7.2 How do Nanozymes Work? 144
  - 7.2.1 Types of Enzyme-Like Activity 144
    - 7.2.1.1 Oxidase 144
    - 7.2.1.2 Superoxide Dismutase (SOD) 145
    - 7.2.1.3 Catalase 145
    - 7.2.1.4 Peroxidase 146
  - 7.2.2 Reactive Oxygen Species, Types and Their Interactions 146
  - 7.2.3 An Insight into the Mechanism of Action of Nanozymes 147
    - 7.2.3.1 Ping-Pong Mechanism 148
    - 7.2.3.2 Eley-Rideal Mechanism 148
- 7.3 Characterization of the Enzyme-Mimicking Activity of Nanozymes 148
  - 7.3.1 Colorimetry 149
  - 7.3.2 Fluorescence 149
  - 7.3.3 Electron Spin Resonance (ESR) 149
  - 7.3.4 Oxygen Meter 152
  - 7.3.5 UV-Vis Absorption Spectroscopy 152
- 7.4 Classification of Nanozymes 152
  - 7.4.1 Metal-Based Nanozymes 152
    - 7.4.1.1 Gold Nanoparticles (AuNPs) 152
  - 7.4.2 Metal Oxide-Based Nanozymes 153
    - 7.4.2.1 Iron Oxide (Fe_3O_4) Nanoparticles 153
    - 7.4.2.2 Cobalt Oxide (Co_3O_4) Nanoparticles 154
    - 7.4.2.3 Cerium Oxide (CeO_2) Nanoparticles 154
  - 7.4.3 Carbon-Based Nanozymes 155
    - 7.4.3.1 Fullerene as a SOD Mimic 155
    - 7.4.3.2 Graphene as a Peroxidase Mimic 155
    - 7.4.3.3 Carbon Nanotube as a Peroxidase Mimic 155
    - 7.4.3.4 Carbon Dot as a Peroxidase and Catalase Mimic 155
  - 7.4.4 Nanoalloy-Based Nanozymes 156
- 7.5 Applications of Nanozymes 156
  - 7.5.1 H_2O_2 Sensing 156
  - 7.5.2 Diagnosis of Cancer and Detection of Microorganisms 157
  - 7.5.3 Nucleic Acid Sensing 157
  - 7.5.4 Detection of Metal Ions 158
  - 7.5.5 Detection of Bilirubin 158
  - 7.5.6 For Therapeutics Applications 158
- 7.6 Challenges and Future Prospects 159
- References 160

7.1 INTRODUCTION

Enzymes have been extensively studied since their discovery and have found widespread applications, with the principal sectors being medicine, food processing, cosmetics, brewing, cleaning (detergents), and pulp and paper. However, despite the high catalytic activity and substrate specificity, natural protein-based enzymes possess inherent drawbacks, which are obstacles for commercial applications. The stability as well as the catalytic activity of enzymes may be drastically affected by a variety of environmental factors such as pH and temperature. Also, the large-scale production and purification of enzymes are often challenging and expensive. All these bottlenecks led to the discovery of artificial enzymes or enzyme mimetics, which intend to imitate the properties of natural enzymes. In 1970, Professor Ronald Breslow at Columbia University coined the term "artificial enzymes" for natural enzyme alternatives as reviewed in [1]. In his book, entitled *Artificial Enzymes*, he mentioned how cyclodextrins (cyclic oligosaccharide with a hydrophilic outer surface and lipophilic cavity) could bind to hydrophobic substrates imitating an enzyme-substrate complex. However, some of the artificial enzymes, such as DNAzymes, hemin, and porphyrin, are biomolecules that are susceptible to activity loss due to fluctuations in temperature and pH. These led researchers to explore better alternatives for enzyme mimics.

Over the past few years, it has been established that nanomaterials can be a potential candidate for enzyme mimics (nanozymes) due to their various advantageous properties. In contrast to natural enzymes, nanozymes are highly stable, easily synthesized, robust, low cost, and can be stored for an extended period. The presence of high surface area to volume ratio and size-dependent tunable catalytic properties further enhance the enzymatic efficiency. These nanomaterial-based artificial enzymes have found profound applications in the field of biosensing, bioimaging, environmental protection, and therapeutics [2]. The performance of many nanozyme actions is dependent on its ability to generate or consume reactive oxygen species (ROS) [3].

The concept of the nanozyme first came into view in the year 1996 with the introduction of two hydroxyl derivatives of buckminsterfullerene (C_{60}) in cortical neuron culture for its free radical scavenging property in neuronal injury and apoptotic death [4]. Laura L. Dugan observed that both the fullerene derivatives could effectively scavenge hydroxyl radicals generated from the decomposition of hydrogen peroxide, which resulted in reduced neuronal apoptotic death. Eight years later, Samesh S. Ali introduced another fullerene (C_{60}) derivative (*tris*-malonic acid (C3)) with potential superoxide dismutase (SOD)-like activity [5]. Superoxide, an anionic radical, and a common toxic by-product in biochemical reactions, is dismutated by SOD in our body. However, the high-level activity of SOD can cause tissue damage, neural injury, and inflammation. To overcome this, the introduction of a SOD mimic *in vivo* was explored. *Tris*-malonic acid (C_3) was found to be capable of eliminating superoxide radicals. The reaction between superoxide and C_3 released the same product H_2O_2 and regenerated oxygen, as is found in the case of SOD-based catalysis. Also, structurally there was no modification of C_3 after the reaction, which further indicated that C_3 has an enzymatic property. However, compared to SOD, the rate of reaction of C_3 was slow.

The discovery of gold nanoparticles with glucose oxidase-like activity was a breakthrough in the field of nanozymes. In 2004, Comotti et al. synthesized naked gold nanoparticles with an average size of 3.6 nm, which mimicked glucose oxidase, catalyzing glucose in the presence of oxygen to gluconic acid and H_2O_2 [6]. Another milestone in the history of nanozymes came in 2007 with the discovery of peroxidase-like activity of iron oxide magnetic nanoparticles by Lizeng Gao and his co-workers. Fe_3O_4 magnetic nanoparticles of different sizes were able to catalyze the oxidation of the peroxidase substrates (i.e., TMB, DAB, and OPD) in the presence of H_2O_2 [7]. The Fe_3O_4 magnetic nanoparticle with peroxidase-like activity was later used in the immunoassay for the detection of antigen.

In this chapter, we will focus on the different enzyme-like activity shown by nanozymes along with their reaction mechanism, characterization techniques, classification, applications, future direction, and the challenges faced in nanozyme applications.

7.2 HOW DO NANOZYMES WORK?

7.2.1 Types of Enzyme-Like Activity

As the concept of nanozyme emerges, understanding its mechanism of action is considered prerequisite to move for their practical applications. So far, nanozymes have been most commonly found to exhibit oxidase-like, catalase-like, SOD-like, and peroxidase-like activity. The properties, as well as the reaction mechanism of each of these enzymes, are discussed in the following sections.

7.2.1.1 Oxidase

Oxidases are a subclass of the oxidoreductase class of enzymes, with oxygen as the electron acceptor. The reaction mechanism involves the transfer of two electrons from oxygen, which results in the formation of either H_2O or H_2O_2 and an oxidized substrate.

$$SH + O_2 \rightarrow S + H_2O$$

$$SH + O_2 \rightarrow S + H_2O_2$$

Different types of oxidase enzymes are known, and in most cases, these enzymes are named depending upon the substrate that is being oxidized. For example, the glucose oxidase (GOx) enzyme that catalyzes the oxidation of the glucose molecule. However, due to the limitations of less stability and high cost, researchers have attempted to develop nanomaterial-based oxidase mimics with promising activity. Table 7.1 lists the utilization of some of these mimics in various applications.

TABLE 7.1
Nanomaterials with Oxidase-Like Activity

Nanozyme	Target Enzyme	Application	Remarks	References
Naked AuNP	GOx-like	Glucose detection	Glucose was converted to gluconic acid and H_2O_2	[6]
AuNP	GOx-like(glucose oxidase-like)	Glucose detection	The size of AuNP increased after conversion of glucose to gluconic acid and H_2O_2.	[8]
CeO_2 NP	Oxidase-like	Detection of cancer biomarker	pH tunable catalytic activity. Converts Ampliflu to resazurin.	[9]
Au@Pt	Oxidase-like	Detection of mouse Interleukin-2	Oxidize OPD and TMB in the presence of dissolved oxygen.	[10]
$MnFe_2O_4$	Oxidase-like	-	Activity is dependent on the morphology of MFe_2O_4.	[11]
PtNP-MnO_2 3D scaffold nanocomposite	Oxidase-like	Detection of glutathione and dopamine	Stability and catalytic activity of Pt NP is enhanced in the presence of MnO_2 nanosheets.	[12]
CeO_2 metal organic framework (MVC-MOF)	Oxidase-like	Detection of Hg^{2+} ion	ssDNA inhibits the oxidase activity of MVC-MOF. Hg^{2+} binding to the thymine-rich ssDNA retains the activity of MVC-MOF.	[13]
NiO NP	Oxidase-like	Bioimaging	Amplex red was converted to resorufin at physiological pH.	[14]

7.2.1.2 Superoxide Dismutase (SOD)

SOD, with cofactors Cu-Zn or Mn, catalyzes the conversion of superoxide (an anionic form of oxygen, $O_2^{.-}$) to molecular oxygen and hydrogen peroxide. The chemical reaction involves an electron transfer from the superoxide anion to two hydrogen molecules, generating hydrogen peroxide and converting the anion to molecular oxygen. Since there is a simultaneous oxidation and reduction of superoxide anion, this type of reaction is known as dismutation. The overall reaction is shown here:

$$2O_2^- + 2H^+ \rightarrow H_2O_2 + O_2$$

There are three different types of SOD enzymes (SOD1, SOD2, and SOD3) in the living system, each differing in their gene sequence, protein structure, and location of expression. SOD1 is expressed primarily in the cytoplasm, SOD2 in mitochondria, and SOD3 is secreted out of the cell (i.e., extracellular).

Due to the disadvantages associated with native SOD enzyme, the development of SOD mimics was significant. They have been found to exhibit neuroprotective action as well as anti-inflammatory and antioxidant activity. The SOD mimics have been further found to improve the growth of stem cells (Table 7.2).

7.2.1.3 Catalase

Catalase, a ubiquitous enzyme found mostly in peroxisomes, is present in both aerobic prokaryotes and eukaryotes. It catalyzes the decomposition of H_2O_2 into molecular oxygen and water. The catalase-based enzymatic reaction occurs in two steps. In the first stage of the reaction, the catalase enzyme interacts with a molecule of H_2O_2, causing oxidation of the heme iron, which results in the formation of an intermediate form of the enzyme

TABLE 7.2
Nanomaterials with Superoxide Dismutase-Like Activity

Nanozyme	Target Enzyme	Application	Remarks	Reference
Buckminsterfullerene (C_{60})	SOD-like	Neuroprotection	Scavenged hydroxyl radicals generated from the decomposition of H_2O_2.	[4]
Tris- malonic acid	SOD-like	Neuroprotection	Scavenged hydroxyl radicals generated from the decomposition of H_2O_2.	[5]
Nanoceria (CeO_2)	SOD-like	Antioxidant	Scavenge superoxide anion. Ratio of $Ce^{3+/4+}$ surface oxidation states influence enzyme-like activity	[15]
CeO_2 NP in PLGA scaffold	SOD-like	Stem cell growth	Improved mechanical, topographical and biological properties of stem cell.	[16]
Apoferritin encapsulated CeO_2NP	SOD-like	Antioxidant	Encapsulation enhanced the scavenging activity of CeO_2NP	[17]
CeO_2 NP	SOD-like	Oxidative stress	Protected cardiac progenitor cells from H_2O_2-induced cytotoxicity	[18]

TABLE 7.3
Nanomaterials with Catalase-Like Activity

Nanozyme	Target Enzyme	Application	Remarks	References
CeO_2 NP	Catalase-like	H_2O_2 detection	CeO_2 NP with +4 oxidation state had better activity compared to +3 oxidation state. H_2O_2 decomposition is dependent on +3/+4 oxidation state.	[19]
Fe_3O_4 NP	Catalase-like	Cytotoxicity study	In the presence of H_2O_2, gas bubbles were generated in the presence of Fe_3O_4 confirming catalase activity.	[20]
Co_3O_4 NP	Catalase-like	Ca^{2+} sensor	Different morphologies of Co_3O_4 NP exhibited different intensity of catalytic activity.	[21]
Co_3O_4 NP	Catalase-like	-	Oxygen generation as a result of decomposition of H_2O_2 by Co_3O_4 NP	[22]
Au@Pt NP	Catalase-like	Cancer cell lines	H_2O_2 was decomposed to O_2 in cancer cells	[23]

(compound I). The oxygen-oxygen bond in peroxide (R-O-O-H) is cleaved heterolytically, such that one oxygen remains with the heme and the other leaves with a molecule of water. The reaction between compound I with another molecule of H_2O_2 in the second stage causes it to reduce back to the ferric enzyme, releasing a molecule of oxygen and water.

$$E - Fe(III) + H_2O_2 \rightarrow E-^{\cdot+}Fe(IV) = O + H_2O_2$$

$$E-^{\cdot+}Fe(IV) = O + H_2O_2 \rightarrow E - Fe(III) + O_2 + H_2O$$

Table 7.3 illustrates some of the reported nanoparticles possessing catalase-like activity.

7.2.1.4 Peroxidase

A peroxidase enzyme catalyzes the reduction of peroxides such as hydrogen peroxide (H_2O_2) or lipid peroxide, and at the same time, causes oxidation of a redox substrate. The first stage of the reaction mechanism of peroxidase is the same as catalase, which involves the binding of an H_2O_2 molecule to the heme (Fe (III)) cofactor resulting in the heterolytic cleavage of the oxygen-oxygen bond in peroxide. Two-electron oxidation of the heme results in the formation of an intermediate form of the enzyme (compound I), which can be converted back to the resting enzyme via two successive single-electron oxidation of a redox substrate.

$$SH_2 + H_2O_2 \rightarrow S + 2H_2O$$

Horseradish peroxidase (HRP)-based colorimetric analysis is most commonly used in diagnostics. However, its inherent disadvantages, including denaturation at high temperature, expensive purification steps, and storage complications, led to the development of peroxidase nanozymes, which are highly stable compared to the natural counterpart. Some of the peroxidase mimics are listed in Table 7.4

7.2.2 Reactive Oxygen Species, Types and Their Interactions

In all of the reaction mechanisms noted earlier, ROS is either being generated or consumed during the reaction. Figure 7.1 shows the different types of ROS. The ROS evolving from oxygen constitutes the critical by-products of aerobic metabolism. At high concentrations, they cause an imbalance of the oxidation state of the cell, causing "oxidative stress," which can have a deleterious effect on the biomolecules. Various defensive enzymes such as SOD, catalase, and peroxidase accomplish detoxification of these ROS. ROS includes both free radicals such as superoxide anion ($O_2^{\cdot-}$), hydroxyl radical (OH·), and nonradicals such H_2O_2 and singlet oxygen ($^1O_2$) [29]. The lifetime of these oxygen derivatives is shown in Table 7.5.

Molecular oxygen in its ground state, also known as triplet oxygen, has two unpaired electrons located in separate orbits, making it a bi-radical molecule. These two unpaired electrons have the same spin, which is responsible for the paramagnetic

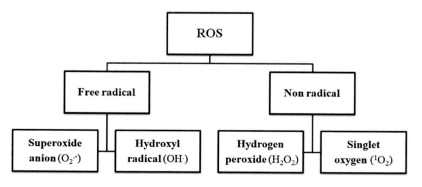

FIGURE 7.1 Types of reactive oxygen species (ROS).

TABLE 7.4
Nanomaterials with Peroxidase-Like Activity

Nanozyme	Target Enzyme	Application	Remarks	References
Fe_3O_4 MNP	Peroxidase-like	Immunoassay	Oxidized different peroxidase substrate in the presence of H_2O_2. pH and temperature stability is more compared to natural HRP.	[7]
Fe_3O_4 MNP	Peroxidase-like	H_2O_2 and glucose detection	Oxidized ABTS in the presence of H_2O_2. Limit of detection (LOD) H_2O_2: 3×10^{-6} M Glucose: 3×10^{-5} M	[24]
Fe_3O_4 MNP	Peroxidase-like	Glucose detection	Peroxidase-like activity is dependent on the surface characteristics (coatings) of the NP.	[25]
BSA-AuNC	Peroxidase-like	H_2O_2 and xanthine detection	Biocompatible. Highly stable over a wide range of pH and temperature. LOD H_2O_2: 2×10^{-8} M Xanthine: 5×10^{-7} M	[26]
HSA-AuNC	Peroxidase-like	Bilirubin detection	Excellent stability. Oxidized free bilirubin in the presence of H_2O_2.	[27]
2D Ni metal organic framework	Peroxidase-like	H_2O_2 detection	Oxidized TMB in the presence of H_2O_2	[28]

TABLE 7.5
Lifetimes of Different Reactive Oxygen Species (ROS)

ROS Types	Lifetime
Singlet oxygen ($^1O_2$)	3 μs
Superoxide anion (O_2^-)	1 μs
Hydrogen peroxide (H_2O_2)	1 ms
Hydroxyl radical ($OH^.$)	1 ns

behavior of O_2. Absorption of energy results in the activation of oxygen, which causes a reversal of the spin of one of the unpaired electrons that further leads to the formation of singlet oxygen in which the electrons are paired and have opposite spins. In another approach, the triplet state of oxygen sequentially reduces by gaining electrons one at a time, resulting in the generation of reactive oxygen intermediates leading to the formation of H_2O. The primary ROS formed from the one-electron reduction of molecular oxygen is the superoxide anion. Transition metal ions present in the active site of SOD generates the secondary ROS (i.e., H_2O_2) from the dismutation of superoxide anion. Eventually, the produced H_2O_2 reacts with catalase in a second dismutation reaction to form H_2O and O_2. Also, both H_2O_2 and superoxide can react to form hydroxyl radical, the most reactive form of ROS, by the Haber-Weiss reaction mechanism that utilizes the Fenton chemistry of Fe (II) (Figure 7.2). The reaction consists of two steps: one is the reduction of Fe^{3+} by singlet oxygen, followed by the oxidation of hydrogen peroxide [29].

$$Fe^{3+} + O_2^- \rightarrow Fe^{2+} + O_2$$

$$Fe^{2+} + H_2O_2 \rightarrow Fe^{3+} + OH^- + OH^.$$

Using the Fenton chemistry, hydroxyl radicals can also be directly produced from H_2O_2, and this chemistry is mainly responsible for the peroxidase-like action of nanomaterials.

$$Fe^{2+} + H_2O_2 \rightarrow OH^- + OH^. + Fe^{3+}$$

7.2.3 An Insight into the Mechanism of Action of Nanozymes

Under different environmental conditions, nanoparticles, especially metal nanoparticles such as Au, Ag, Pt, Cu, Co, and Fe, can generate ROS. Typically, the ability of nanoparticles to generate and scavenge ROS forms the primary foundation behind their enzyme-mimicking property. Until now, many nanomaterials with different types of enzyme-like activity have been discovered with applications in diverse fields. For instance, AuNPs can behave like GOx enzymes, catalyzing glucose in the presence of oxygen to gluconic acid and H_2O_2. The property of cerium oxide nanoparticles to scavenge

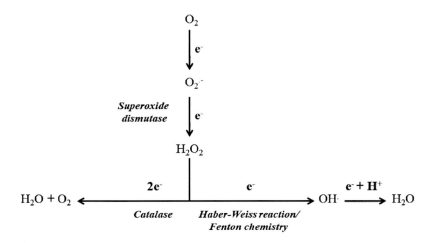

FIGURE 7.2 Schematic representation of generation of different types of reactive oxygen species.

superoxide anion has shown promising application as an antioxidant and anti-inflammatory agent.

With different types of enzyme-like activity, the underlying mechanism of action also varies. The presence of two different redox states and oxygen vacancies on the surface of cerium oxide nanozyme is proposed to be responsible for its SOD-like activity. The oxygen vacancies act as hot spots for the catalytic activity, as they allow the binding of the superoxide ion followed by its dismutation. Additionally, the ratio of redox states has a massive impact on the catalytic activity. For example, with a higher Ce^{3+}/Ce^{4+} ratio, the cerium oxide nanoparticles possess superior activity [15].

Peroxidase mimics, in most cases, have been found to follow the Fenton reaction mechanism that generates hydroxyl radicals. Metal nanoparticles with redox potential lower than H_2O_2/H_2O (1.77 V) are mostly favorable for this type of mechanism [30]. In particular, iron-oxide magnetic nanoparticles, the first reported nanozyme with peroxidase-like activity, with standard redox potential value of 0.77 V (Fe^{3+}/Fe^{2+}), can cause the reduction of H_2O_2 into a hydroxyl radical. As mentioned earlier, the peroxidase-based reaction involves H_2O_2, peroxidase substrate, and the enzyme. Nanozymes following the Fenton reaction mechanism first react with H_2O_2, resulting in its decomposition to hydroxyl radical, which further brings about the oxidation of the peroxidase substrate. Oxidation results in the generation of a colored end product that is visible to the naked eye. In contrast to that, some nanozymes do not follow the Fenton reaction mechanism. For example, the standard redox potential of cobalt oxide nanoparticles is 1.808 V (Co^{3+}/Co^{2+}), which is higher than the redox potential value of H_2O_2 (1.77 V) [22]. A detailed mechanism study showed that in this case, the route of electron transfer is from peroxidase substrate to nanozyme and finally to H_2O_2. In the case of carbon nanomaterials, however, the oxygen-containing functional groups on the surface of the nanozyme play an essential role in catalytic activity, especially in the proton transfer reaction. Although the electron transfer route varies, the peroxidase mimic nanozymes undergo a ping-pong reaction mechanism (described in the next section).

The catalase-like activity has also been observed in a wide range of nanozymes. The reaction mechanism is closer to the natural catalyzed reaction. However, in most instances, its activity is found to be higher at neutral pH compared to acidic pH [22].

7.2.3.1 Ping-Pong Mechanism

Most of the enzymatic reactions are bisubstrate based, wherein two substrate molecules bind to the enzyme and participate in the reaction. These types of reaction proceed in different pathways. Like the natural counterpart, nanomaterials with peroxidase-like activity have been found to exhibit a *ping-pong mechanism* where one substrate first binds to the enzyme mimic, is converted to product, and then dissociates before the binding of the second substrate (Figure 7.3).

7.2.3.2 Eley-Rideal Mechanism

In 1938, Dan Eley and Eric Rideal put forward a mechanism depicting one of the possible ways how molecules can react on the surface of the catalyst. Herein, one of the reactant molecules adsorbs on the catalyst surface, whereas the second reactant molecule, which is in the gas phase, does not adsorb. The reaction takes place when the gas molecule (second reactant species) reacts with the adsorbed reactant molecule (Figure 7.4). AuNPs with GOx activity have been found to exhibit an Eley-Rideal mechanism wherein the glucose molecule first adsorbs on the surface of the nanoparticles, followed by interaction of the oxygen molecule with the adsorbed glucose. This reaction results in the formation of gluconic acid and H_2O_2 [8].

7.3 CHARACTERIZATION OF THE ENZYME-MIMICKING ACTIVITY OF NANOZYMES

Many techniques have been used to characterize the type of enzyme-like activity as well as to observe the type of ROS generated during the time of reaction. Briefly, oxidase-like activity is monitored for the decrease in the level of oxygen concentration and the formation of the oxidized product. The

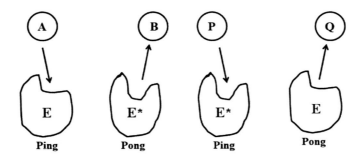

FIGURE 7.3 Ping-pong reaction pathway of enzymes.

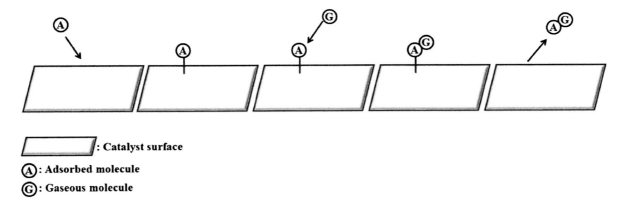

FIGURE 7.4 Eley-Rideal mechanism.

catalase-like activity investigates the expenditure of H_2O_2 and the generation of molecular oxygen. In the case of SOD-like enzyme mimicking activity, the utilization of superoxide anion and the formation of H_2O_2 are monitored. The peroxidase-like activity detects the ability of the nanozyme to decompose H_2O_2 and the formation of an oxidized substrate. The frequently used techniques for the characterization of nanozyme are discussed next.

7.3.1 Colorimetry

It is the most commonly employed characterization method to detect the enzyme-like activity of nanomaterials. The peroxidase-like catalytic activity of nanomaterials is investigated by the catalytic oxidation of the peroxidase substrate in the presence of H_2O_2. In contrast, the oxidase-like activity of nanozyme is characterized by the reaction of the peroxidase substrate with the nanomaterial in the absence of H_2O_2. The formation of a colored product corresponding to the oxidation of the peroxidase substrate confirms the presence of oxidase-mimicking activity. Table 7.6 lists the most commonly used peroxidase substrate, the maximum absorbance of the oxidized substrate, and the type of color produced after the reaction.

7.3.2 Fluorescence

Other than colorimetry, ROS can also be detected using fluorescent probes. Some of the prominent probes are shown in Table 7.7. Furthermore, compared to colorimetry, the fluorescence-based detection of ROS is more sensitive.

7.3.3 Electron Spin Resonance (ESR)

The short lifetime of ROS sometimes poses a hindrance in their detection by the earlier two mentioned methods. Electron paramagnetic resonance (EPR) is currently the most popular and robust technique to detect the presence of free radicals, especially ROS, both qualitatively and quantitatively. Therefore, it is the first approach for analyzing the ROS for studying the enzyme-mimicking efficiency of nanomaterials.

There are two main methods of ESR for detecting ROS: *spin trapping* and *spin labeling*. As mentioned earlier, ROS is short-lived, and so the *spin trapping technique* makes use of chemical entities, called spin traps, that can bind with the free radicals or ROS present in the sample, forming a stable adduct with a half-life feasible for ESR analysis. Different types of spin traps having nitrone, nitroso, and piperidine/pyrrolidine groups are available. Nitrone spin

TABLE 7.6
List of Different Peroxidase Substrates

Peroxidase Substrate	Maximum Absorbance	Color of Oxidized Substrate
2,2′-azino-bis-(3-ethyl-benzothiazoline-6-sulfonic acid) (ABTS)	414 nm and lesser maxima at 395 nm, 640 nm, 725 nm	Green
3,3′,5,5′-Tetramethylbenzidine (TMB)	655 nm (Continuous reaction)	Blue
	450 nm (Reaction stopped by acid)	Yellow
o-dianisidine (ODA)	405 nm	Yellow-orange
3,3′-diaminobenzidine (DAB)	465 nm	Brown
o-phenylenediamine (OPD)	450 nm	Orange-brown

TABLE 7.7
List of Different Fluorescent Probes

Fluorescent Probe	Excitation/Emission Wavelengths (nm)	Type of ROS Detected	End Product
Amplex red	563/587	H_2O_2	Highly fluorescent resorufin
Fluorescein	495/515	OOR·	Fluorescence decay upon oxidation
Dihydrorhodamine 123 (DHR)	505/529	H_2O_2	Highly fluorescent rhodamine 123

traps such as DMPO (5, 5,-dimethypyrroline N-oxide), BMPO (5-tertbutoxycarbonyl-5-methyl-1-pyrroline-N-oxide), POBN (α-(4-pyridyl-1-oxide)-N-tert-butylnitrone), and DEPMPO (5-diethoxyphosphoryl-5-methyl-1-pyrroline-N-oxide) are most commonly used to detect OH· and OOH· radicals. However, the DMPO/OOH· adduct is unstable and therefore often gets decomposed to the DMPO/OH· adduct, leading to false-positive results. In contrast, the BMPO/OOH· adduct is the most stable and does not decompose into the corresponding hydroxyl adduct. The adducts resulting from the interaction of nitroso spin traps with ROS are both photochemically and thermally unstable and therefore not suitable for applications detecting ROS. Piperidine spin traps such as TEMP (2, 2, 6, 6-tetramethylpiperidine) and 4-oxo-TEMP (4-oxo-2, 2, 6, 6-tetramethyl piperidine) are mostly used to trap singlet oxygen [30].

DMPO, BMPO, and POBN are the most commonly used spin traps to capture hydroxyl radicals generated by metal nanoparticles such as AuNP, AgNP, TiO_2, ZnO, and other nanoparticles. The hydroxyl radical captured by different spin traps as a result of the interaction of AuNP with H_2O_2 is shown in Figure 7.5 [31]. The sharp peaks in ESR spectrum also demonstrate that only in the presence of the enzyme mimic the hydroxyl radical/spin trap adduct is formed. The ESR spectrum of POBN produces six lines, whereas both BMPO and DMPO spin traps show four distinct lines with intensities in the ratio of 1:2:2:1.

The ESR experiment can validate the intrinsic peroxidase-like activity of nanomaterials. As shown in Figure 7.6, the hydroxyl radical/spin trap adduct is not so evident compared to hydroxyl radical/spin trap/Fe_2O_3 or Fe_3O_4, demonstrating that both Fe_2O_3 and Fe_3O_4 converts H_2O_2 to hydroxyl radical. Additionally, the hydroxyl radical/spin trap adduct peak intensity increases with increasing concentration of both Fe_2O_3 and Fe_3O_4, indicating that the peroxidase-like activity of both Fe_2O_3 and Fe_3O_4 may arise from its ability to convert H_2O_2 to hydroxyl radical (OH·) [20].

A superoxide is another important free radical that is produced during enzymatic reactions. AuNPs are known to efficiently scavenge superoxide anion, mimicking the property of SOD. Figure 7.7 shows the ESR spectrum of BMPO/OOH·adduct with the relative intensity of 1:1:1:1. Similar to the control experiment (a and b), the addition of AuNP decreased the intensity of BMPO/OOH· spin adduct peak (c and d) [31].

The catalase-mimicking activity of nanomaterials can be investigated using both ESR spin trap and spin label oximetry methods. In the case of the ESR spin trap technique, the typical hydroxyl radical/DMPO adduct signal intensity decreases in the presence of the nanozyme, demonstrating the catalase-like behavior of the nanomaterial, which lowered the production of OH· from H_2O_2 [32].

ESR oximetry uses stable nitroxide free radical as a spin label that can be used to monitor the dissolved oxygen generated as a result of H_2O_2 decomposition by either catalase or its mimic. The spin label consists of unpaired electrons, and therefore the spectrum that is obtained, is because of its collision with oxygen. Figure 7.8 shows the ESR oximetry spectra of catalase and AuNPs with different surface coatings. Here, CTPO (3-carbamoyl-2, 2, 5, 5-tetramethy-3-pyrroline-1-yloxyl) is used as a spin label. Because of the hyperfine interaction of the unpaired electron with the nucleus, the ESR spectra of CTPO display several lines (control). Compared with the control with no catalyst, broader line width and low peak intensity are observed both in the case of catalase and in the case of different AuNPs indicating the generation of oxygen [31].

Nanozymes as a Potential Catalyst

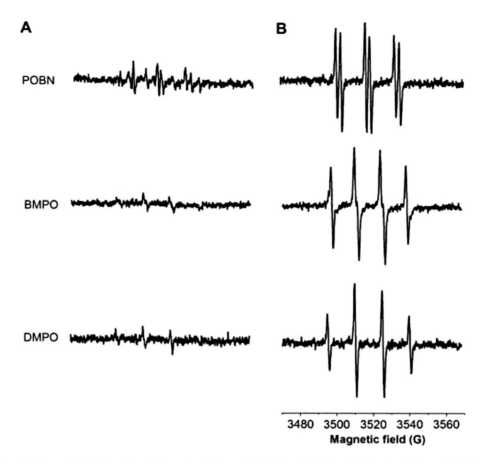

FIGURE 7.5 Captured hydroxyl radicals using different spin traps. (A) Control (H_2O_2 + different spin traps), (B) Test (AuNP + H_2O_2 + different spin traps) [31].

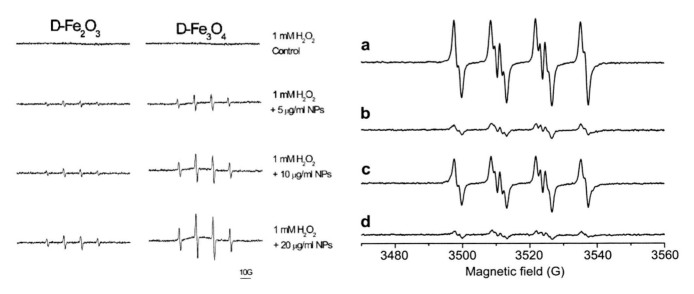

FIGURE 7.6 ESR spectra of OH·/DMPO adduct with different concentrations of the NP. Reprinted (adapted) with permission from reference [20]. Copyright (2012) American Chemical Society.

FIGURE 7.7 ESR spectra of OOH· BMPO/adduct. (a) Control, (b) Superoxide dismutase (SOD), (c) AuNP coated with polyvinylpyrrolidone (PVP), (d) AuNP coated with tannic acid) [31].

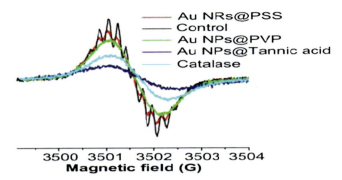

FIGURE 7.8 ESR oximetry spectra of CTPO spin label in the presence or absence of catalyst [31].

7.3.4 OXYGEN METER

The generation of oxygen by the catalase-mediated reaction can also be quantified using a dissolved oxygen meter. Oxygen generated because of the enzymatic reaction is visible in the form of bubbles.

7.3.5 UV-VIS ABSORPTION SPECTROSCOPY

The catalase-like activity can also be monitored using the UV-VIS absorption spectroscopy technique. H_2O_2 has an absorbance maximum at 240 nm. Because of the catalase-like activity of the nanomaterial, H_2O_2 is decomposed to O_2, resulting in a decrease in the absorbance at 240 nm.

7.4 CLASSIFICATION OF NANOZYMES

Since the discovery of nanozymes, different nanomaterials are being routinely explored for an enzyme-mimicking property. The vast majority of the reported nanozymes can be broadly classified into metal, metal-oxide, carbon, and nano-alloy based nanozymes.

7.4.1 METAL-BASED NANOZYMES

Metal nanoparticles have always been the subject of interest for researchers because of their vast applications in the field of biomedical science as well as nanotechnology. One of the reasons for this increased importance is mainly because of its easy synthesis procedure and modification with functional groups, which allows its conjugation to biomolecules, drugs, and ligands. The utilization of metal nanoparticles for drug delivery and bioimaging over traditional pharmaceutical agents has further led to its increased popularity. The enzyme-like activity of metal nanoparticles is another asset added to the list. The efficiency of using them as an alternative to natural enzymes has led to new opportunities, especially in the development of the sensor platform. Among the reported metal nanoparticles, the usage of gold, silver, and platinum nanoparticles as enzyme mimics is important. To enable their use in bioimaging, therapeutics, and diagnostics, they have continuously been modified. Thus, in this section, some of the metallic nanoparticles will be discussed along with their already established applications.

7.4.1.1 Gold Nanoparticles (AuNPs)

AuNPs are the most widely used metal nanoparticles because of their attractive optical and electronic properties. Its high surface area to volume ratio, biocompatibility, low toxicity, stability, and facile functionalization have increased its versatility. Already several metallic catalysts have been employed for a range of chemical reactions such as hydrogenation, oxidation, and reduction. However, due to the high ionization potential, metallic gold cannot quickly lose electrons from its outermost shell that makes it catalytically inactive. The discovery of AuNPs with intrinsic enzyme-like activity was a significant turning point in the field of nanotechnology, implying that it is active in the nanomolar scale.

7.4.1.1.1 AuNP as a Glucose Oxidase (GOx) Mimic

With a dimension in the range of 3–6 nm, citrate-coated "naked" AuNPs can convert glucose to gluconic acid and H_2O_2 in the presence of oxygen, thus mimicking the activity of GOx [6]. However, the catalytic activity of the nanoparticle is inversely proportional to its size. This property can be correlated with the exposed number of atoms and the size of the nanoparticle. With a smaller diameter, the number of exposed atoms is larger, which ultimately increases the catalytic activity. Although GOx enzymes and "naked" AuNPs have similar stoichiometry, the mechanism of enzymatic activity differs. That is because the AuNPs-based catalysis follows the Eley-Rideal mechanism wherein the glucose molecule adsorbs onto the surface of AuNP, followed by the reaction of oxygen with the adsorbed glucose molecule, resulting in the formation of gluconic acid and H_2O_2 [33]. AuNPs with GOx activity also possesses self-catalytic and self-limiting properties [8]. The H_2O_2 produced as a result of oxidation of glucose can reduce chloroauric acid (HAuCl4) into Au^0, which is deposited on the surface of AuNP. This event ultimately increases the size of the AuNPs. However, the enlargement of AuNPs decreases its activity. Additionally, the deposition of gluconic acid on the surface of AuNP reduces the number of exposed surface-active atoms, which further hinders the catalytic activity (Figure 7.9).

7.4.1.1.2 AuNP as a Peroxidase Mimic

Apart from GOx-like activity, AuNPs also display peroxidase-like behavior. The peroxidase-based catalytic reaction is mainly because of electron transfer reactions between the peroxidase substrate, peroxidase enzyme, and H_2O_2. The surface charge characteristic of AuNPs plays a vital role in this electron transfer-mediated reaction [34]. Depending on the different precursors used for the synthesis of AuNPs, they acquire different surface charges. Compared to negatively charged AuNPs, the catalytic efficiency of positively charged AuNPs to oxidize 3,3′,5,5′-tetramethylbenzidine (TMB) in the presence of H_2O_2 is maximum. Modification of the surface properties of AuNPs leads to differences in the binding

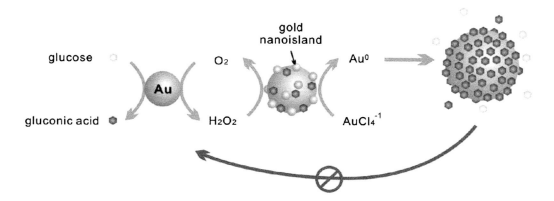

FIGURE 7.9 AuNPs with GOx-like activity. Reprinted (adapted) with permission from reference [8]. Copyright (2010) American Chemical Society.

affinity of the peroxidase substrate. This property again influences the performance of the peroxidase-mimicking activity of the nanoparticles. TMB contains two amino groups, whereas ABTS consists of sulfate groups, which results in their affinity for negatively and positively charged groups, respectively. Therefore, amino-modified (positively charged) and citrate-capped (negatively charged) AuNPs react differently with both the peroxidase substrates. Unmodified AuNPs have the same affinity to oxidize TMB and ABTS because of the absence of any functional groups [35].

The peroxidase-mimicking property of AuNPs is enhanced in the presence of different ions. Hg^{2+} is one such ion that can accelerate the peroxidase-like catalytic activity of AuNPs, resulting in the intense blue color of the oxidized TMB [36]. Citrate sodium used in the preparation of AuNP can reduce Hg^{2+} to Hg^0. Later these Hg^0 are distributed on the surface of AuNP, which modifies the surface characteristics and thus results in the amplification of the catalytic activity of AuNP [36].

7.4.2 Metal Oxide-Based Nanozymes

With improved physical, chemical, and electrical attributes over the bulk counterpart, metal oxide nanoparticles constitute one of the emerging nanomaterials that are routinely used in numerous applications. The discovery of the enzyme-like activity of metal oxide nanoparticles is another essential feature added to the list. Several metal oxide nanozymes with different enzyme-mimicking characteristics have been established.

7.4.2.1 Iron Oxide (Fe_3O_4) Nanoparticles

One of the most studied metal oxide nanoparticles with promising applications, especially in the field of biomedicine, is Fe_3O_4. Predominantly it exists in two forms: magnetite (Fe_3O_4) and maghemite (γ-Fe_2O_3) that possess excellent superparamagnetic behavior at the nanoscale level followed by high surface to volume ratio. With the discovery of peroxidase-like and catalase-like activity, the potentiality of the nanomaterial has further increased.

7.4.2.1.1 Fe_3O_4 Nanoparticle as a Peroxidase Mimic

Iron oxide magnetic nanoparticles (Fe_3O_4) were the first metal oxide nanoparticles found to possess intrinsic peroxidase-like activity [7]. Three different dimensions of the Fe_3O_4 nanoparticle (30, 150, and 300 nm) catalyzed the oxidation of the peroxidase substrates viz. TMB, OPD, and DAB, with higher activity found in the case of smaller nanoparticles. Similar to HRP, the catalytic activity of Fe_3O_4 nanoparticles is dependent on pH, temperature, and H_2O_2 concentration and follows the ping-pong reaction mechanism. In addition, being an inorganic nanoparticle, the Fe_3O_4 nanoparticle is more stable at different temperatures and pH compared to HRP. Its magnetic property allows easy recovery or recycling in various applications. The combination of intrinsic peroxidase-like activity and magnetic property of Fe_3O_4 nanoparticles were utilized in a capture probe to develop a novel immunoassay [7]. The peroxidase-like activity the nanozyme can be further employed for colorimetric detection of H_2O_2, as well as for other targets that generates H_2O_2 as one of its by-products.

Differences in structural properties affect the catalytic activity of Fe_3O_4 NPs. Comparison of three different Fe_3O_4 nanostructures: cluster sphere, octahedral, and triangular plates shows that the peroxidase-like activity is dependent on the number of catalytically active iron atoms exposed on the surface. The cluster sphere having more substantial surface area had the maximum number of exposed atoms on its surface, resulting in the highest activity followed by triangular plates and octahedral. The size and surface area of both triangular plates and octahedral Fe_3O_4 nanostructures were comparable, but the arrangement of surface atoms differed, leading to differences in their catalytic efficiency [37].

Like AuNPs, the modification of surface charge of Fe_3O_4 nanoparticles also plays an essential role in increasing the affinity of the Fe_3O_4 nanoparticles for different peroxidase substrates. With different surface coatings, the electrostatic interaction between Fe_3O_4 nanoparticle and peroxidase substrate differs that ultimately causes variations in their peroxidase-like activity. Positively charged Fe_3O_4 showed a maximum affinity for the negatively charged peroxidase substrate, ABTS carrying the SO_4^{2-} and vice versa [25].

7.4.2.1.2 *Fe_3O_4 Nanoparticle as a Catalase Mimic*

Both Fe_3O_4 and γ-Fe_2O_3 can decompose H_2O_2 into H_2O and oxygen at neutral pH. This attribute shows that iron oxide nanoparticles can also mimic catalase enzymes [20].

7.4.2.2 Cobalt Oxide (Co_3O_4) Nanoparticles

Another metal oxide that has drawn attention in the last few years is the cobalt oxide (Co_3O_4) nanoparticle. It is a p-type semiconductor with diverse applications in the field of catalysis, electrochemistry, gas sensors, and lithium-ion batteries. Like Fe_3O_4 nanoparticle, it also possesses multiple valence states and therefore was expected to exhibit enzyme-like activity. Nanostructured cobalt oxide has been found to display peroxidase-like, catalase-like, and SOD-like activity.

7.4.2.2.1 *Co_3O_4 Nanoparticle as a Peroxidase Mimic*

Like HRP, Co_3O_4 nanoparticles can also catalyze the oxidation of TMB in the presence of H_2O_2. The activity is further dependent on the concentration of Co_3O_4 nanoparticle, H_2O_2 concentration, pH, and temperature. While a surplus amount of H_2O_2 inhibits HRP activity, the same principle does not hold for Co_3O_4 nanoparticles because of its high stability. Compared to neutral and alkaline conditions, the nanoparticle is catalytically more active in acidic conditions. However, the peroxidase-like activity of the Co_3O_4 nanoparticle does not originate from the generation of OH. The mechanism of catalytic activity has been speculated as the Co_3O_4 nanoparticle transferring electrons between TMB and H_2O_2. It is assumed that during the reaction, the TMB molecule gets adsorbed on the surface of the nanoparticle, followed by the donation of the lone pair of electrons from the amino group to the Co_3O_4 nanoparticle. As a result, there is an increase in electron mobility and density in the nanoparticle, which further expedites the electron transfer process from Co_3O_4 to H_2O_2. This feature altogether increases the rate of oxidation of TMB [38].

7.4.2.2.2 *Co_3O_4 Nanoparticle as a Catalase Mimic*

In the presence of only H_2O_2, the Co_3O_4 nanoparticles can generate oxygen, which is a product of H_2O_2 decomposition. When analyzed with a dissolved oxygen meter, it can be seen that the concentration of dissolved oxygen increases with an increasing concentration of Co_3O_4 nanoparticle. This behavior corroborates the catalase-like activity of the nanoparticle. In addition, compared to Fe_3O_4 nanoparticles, the Co_3O_4 nanoparticles have many-fold higher catalase-like activity. Presumably, this is because of higher redox potential of Co^{3+}/Co^{2+} (standard redox potential 1.808 V) than Fe^{3+}/Fe^{2+} (standard redox potential 0.771 V) [22].

7.4.2.2.3 *Co_3O_4 Nanoparticle as a Superoxide Dismutase Mimic*

In addition to decomposing H_2O_2, the Co_3O_4 nanoparticles can act as potent superoxide radical scavenger. Co_3O_4 nanoparticles, when mixed in a system that generates superoxide, can significantly reduce its production. This property can be validated by using ESR spectroscopy with BMPO as a spin trap. With the increasing concentration of Co_3O_4 nanoparticles, the intensity of the ESR signal corresponding to the spin adduct decreases [22].

7.4.2.3 Cerium Oxide (CeO_2) Nanoparticles

Cerium with atomic number 58 is one of the rare earth metals belonging to the lanthanide series. It usually exists in either metallic form or oxide form; the latter form is commonly used for polishing glass. Additionally, it can also switch from Ce^{4+} to Ce^{3+} oxidation states. In the past few years, the utilization of nanoceria in oxygen sensors, in the automobile exhaust system as catalytic converters, in fuel cells as an electrolyte, and as an ultraviolet absorbent has become popular [39]. Although the physical and chemical properties are comparable to the bulk form, nanostructured cerium-oxide nanoparticles with high surface area and oxygen vacancies result in increased catalytic activity [15]. The variations in oxidation states of cerium-oxide nanoparticles are responsible for the formation of oxygen vacancies or defects in the lattice structure as a result of the loss of oxygen or electrons. However, differences in physical parameters such as temperature, ions, and partial pressure of oxygen can cause variation in the valence state, as well as in the lattice structure defects. The increase of oxygen vacancies in nanoceria also causes a sudden increase in the Ce^{3+}/Ce^{4+} ratio.

7.4.2.3.1 *CeO_2 Nanoparticle as a Superoxide Dismutase Mimic*

The presence of Ce^{3+}/Ce^{4+} oxidation states on the surface of nanostructured cerium-oxide nanoparticles is shown to be responsible for its ROS scavenging activity. This switching of oxidation states of the nanoparticle can be related to the cofactors present in SOD enzyme that undergo similar kind of redox reaction mechanism and so have been called as SOD mimics. A high Ce^{3+}/Ce^{4+} ratio of the nanoceria shows the higher catalytic activity, as only Ce^{3+} can oxidize, resulting in the generation of H_2O_2 from superoxide anions. When tested in neuron culture, the presence of CeO_2 nanoparticles has shown to expand the lifetime of the neuronal cell. According to a previously presented model, H_2O_2 first binds to two Ce^{4+} sites present in oxygen vacancy on the surface of nanoceria. This event results in the release of two protons and two electrons that reduce the Ce^{4+}, followed by the liberation of molecular oxygen, which allows the binding of superoxide anion to the reduced oxygen vacancy sites. The release of an electron and the addition of two protons causes the formation of H_2O_2 that is subsequently released. The reaction of a second superoxide molecule brings back the Ce^{4+} oxidation state on the surface of CeO_2 nanoparticle [40].

7.4.2.3.2 *CeO_2 Nanoparticle as a Catalase Mimic*

While the high Ce^{3+}/Ce^{4+} ratio is related to SOD-like activity, the low Ce^{3+}/Ce^{4+} ratio results in catalase-like activity. The mechanism involves H_2O_2 reacting with Ce^{4+} resulting in its reduction to Ce^{3+} and oxidation of H_2O_2 to O_2. According to a previously proposed model, the first four steps are the same as the SOD-like activity of the nanoceria. In the fifth step, instead of superoxide anion, a molecule of H_2O_2 binds to the two Ce^{3+} sites on the surface. As a result of the uptake of two protons followed by the homolysis of the O-O bond, the H_2O molecule is generated and the Ce^{4+} site is reverted [40].

7.4.2.3.3 CeO$_2$ Nanoparticle as an Oxidase Mimic

Nanoceria also causes the oxidation of peroxidase substrates such as ABTS and TMB to a colored product even in the absence of H_2O_2. This action shows that apart from SOD-like and catalase-like activity, CeO_2 nanoparticles also exhibit oxidase-like activity. Furthermore, the activity is dependent on pH, temperature, and coating on its surface [41].

7.4.3 CARBON-BASED NANOZYMES

While the vast majority of the nanozymes are metallic, a few carbon nanoparticles also exhibit excellent enzyme-like activity. The most predominant being the fullerene, graphene, carbon nanotube, and carbon dots that have equal potential to serve as a replacement for natural enzymes. As mentioned in the first section, fullerene and its derivatives were the first nanozymes to be discovered to have SOD-like activity, and since then, ample research has been undertaken to uncover enzyme-like activities of other existing nanomaterials. Herein, we will focus more on the carbon nanomaterial's enzyme mimicking mechanism.

7.4.3.1 Fullerene as a SOD Mimic

The interaction of the fullerene derivative C_3 with a superoxide anion showed a catalytic dismutation reaction instead of the stoichiometric scavenging route. By using semi-empirical quantum-mechanical calculations, a model was put forward to anticipate the electron distribution on the surface of C_3. The study showed that the electron-deficient region on the surface of C_3 could electrostatically attract the superoxide anion, followed by its dismutation. The protons of the carboxyl groups and water molecules play a very significant role during the entire reaction mechanism [5] (Figure 7.10).

7.4.3.2 Graphene as a Peroxidase Mimic

Graphene is one of the emerging nanomaterials that have attracted attention due to its amazing optical, electronic, and mechanical properties. With a thickness of one atom and sp^2-bonded carbon atoms, the graphene is a tightly packed two-dimensional honeycomb crystal lattice. Graphene oxide (GO), a derivative of graphene, has been found to exhibit peroxidase-like activity. Due to the low cost, high stability against biodegradation and denaturation, easy dispersion in solvents, and simple synthesis procedure compared to HRP, this nanomaterial has a significant benefit for practical applications. Like the HRP enzyme-catalyzed reaction, the GO-based peroxidase-like activity also follows the ping-pong mechanism. However, the K_m value of GO with H_2O_2 as a substrate is remarkably higher than HRP, which illustrates its requirement for high H_2O_2 concentrations for catalytic activity.

Interactions of H_2O_2 with GO results in a bathochromic shift of the GO spectra, suggesting the transfer of electrons occurring from the top of the valence band of graphene to the lowest unoccupied molecular orbital of H_2O_2. The route of electron transfer has been proposed as TMB donating a lone pair of electrons to GO that increases its Fermi level, and this ultimately increases the charge transfer from GO to H_2O_2 [42].

7.4.3.3 Carbon Nanotube as a Peroxidase Mimic

Both single-walled carbon nanotubes (SWNTs) and helical carbon nanotubes are found to have intrinsic peroxidase-like activity. However, the catalytic activity of the latter is dependent on the presence of metal residues, such as Fe [43]. One of the significant limitations of the peroxidase-like activity of the carbon nanotube is its activity is restricted only to a narrow pH range with deficient activity in the neutral pH. Modification of the surface properties of the carbon nanotube with oxygen-rich functional groups results in the enhancement of peroxidase-like activity acting over a broad pH range. In contrast to the electrostatic repulsion and van der Waals force of interaction between the pristine carbon nanotube and H_2O_2, the existence of hydrogen bonding between H_2O_2 and the oxygenated functional groups in modified carbon nanotube results in improved binding capacity between the two. The role of different oxygen functional groups has a different impact on the activity of the nanoparticle. Phenyl hydrazine (PH), benzoic anhydride (BA), and 2-bromo-1-phenylethanone (BrPE) are known to deactivate the carbonyl group, hydroxyl group, and carboxylic acid group, respectively. The carbon nanotube, when modified with PH, results in an 85% decrease of the peroxidase-like activity in comparison to the other two-carbon nanotubes. This event illustrates the active role of the carbonyl group in catalytic activity. In addition, the binding capacity of these functional groups with H_2O_2 plays a critical role in the catalytic process. A carbonyl group with negligible binding capacity with H_2O_2 shows the maximum ability to decompose H_2O_2 compared to carboxyl and hydroxyl groups, with superior binding capacity and, therefore, lesser activity [44].

7.4.3.4 Carbon Dot as a Peroxidase and Catalase Mimic

One of the newcomers to the carbon nanomaterial family, carbon dots, has attracted the attention due to its excellent photoluminescence property. Due to its high aqueous solubility, stability, low toxicity, robustness, facile functionalization, tunable photoluminescence, and biocompatibility, the carbon dots have proved to be one of the promising candidates for diverse applications. Other than these attributes, carbon dots have also been found to have enzyme-mimicking behavior such as peroxidase-like and catalase-like. Since the starting material for carbon dot preparation does not contain metal ions, the catalytic activity can be ascribed to its intrinsic property. In addition, being an inorganic nanomaterial, the activity is stable over a broad pH and temperature range. The mechanism of peroxidase-like activity of carbon dots is closer r to graphene but yet to be elucidated thoroughly.

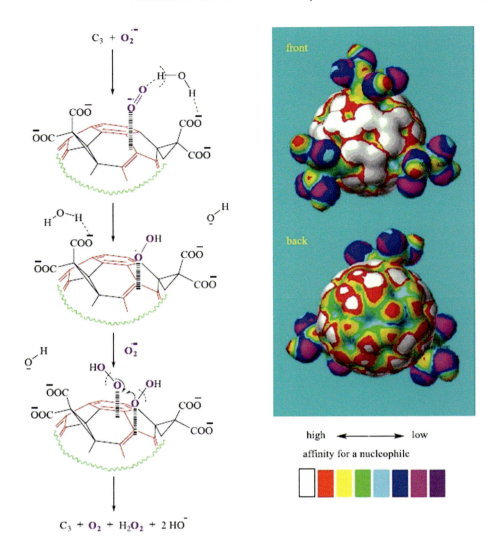

FIGURE 7.10 Proposed SOD-like activity of fullerene derivate [5].

7.4.4 NANOALLOY-BASED NANOZYMES

Instead of single metal-based nanoparticles, bimetallic nanoparticles with the synergistic effect of two different metallic elements results in the improvement of the properties. As a result, it also leads to the enhancement of catalytic activity in nanoalloy-based nanozymes. However, the difference between the electrochemical redox potential values of the two metals in a nanoalloy should be large. The rate of oxidation of peroxidase substrates in the presence of H_2O_2 is further dependent on the alloy composition, which is considered to be due to differences in electronic structure as a result of mixing.

7.5 APPLICATIONS OF NANOZYMES

7.5.1 H_2O_2 SENSING

Detection of H_2O_2 in the field of healthcare, environment safety, and the food industry plays a vital role. H_2O_2 can promote the oxidation of several substrates (e.g., ABTS, TMB, ODA) in the presence of a peroxidase enzyme. The oxidized product of ABTS, TMB, and ODA has a distinct color that can be visualized with the naked eye. The intensity of color change of the oxidized products at their corresponding wavelengths is directly proportional to the concentration of H_2O_2 present in the sample. The nanomaterials bearing peroxidase-mimicking activity can be utilized for the detection of H_2O_2 in diverse applications, as H_2O_2 is one of the by-products of several oxidase-based enzymes. Therefore, by detecting the concentration of H_2O_2, one can indirectly determine the corresponding substrate of the oxidase enzyme. Most of these substrates are biologically essential molecules whose increased or decreased concentration plays a crucial role in the clinical diagnosis. Glucose is one such analyte that is routinely tested in the medical field. GOx, belonging to the oxidoreductase class of enzyme, acts as the biorecognition element that reacts with glucose in the presence of oxygen to form gluconic acid and H_2O_2. In the presence of a peroxidase mimic, this H_2O_2 can cause the oxidation of the peroxidase substrate resulting in a visually observable color change. Likewise, other oxidase enzymes such as cholesterol oxidase or xanthine oxidase, in combination with peroxidase nanozymes, have also been employed for the development of a simple and sensitive colorimetric sensor for the detection of their corresponding substrates.

7.5.2 Diagnosis of Cancer and Detection of Microorganisms

The role of cancer-specific biomarkers in understanding and monitoring the diseased condition is predominant. Enzyme-linked immunosorbent assay (ELISA) tests that make use of antibody labeled with HRP are usually performed for the detection of these biomarkers. However, due to the low sensitivity of colorimetric dyes, as well as the instability of HRP, the need for another alternative arises. Although fluorescent dyes are highly sensitive, their quenching in the presence of H_2O_2 might result in an incorrect result. A nanoceria-based ELISA test has been developed wherein the oxidase-like activity of cerium oxide nanoparticles was tuned by regulating the pH. At acidic pH, the cerium oxide nanoparticles undergo complete oxidation that leads to the conversion of the nonfluorescent substrate Ampliflu to a nonfluorescent product resazurin, whereas at neutral pH due to partial oxidation of Ampliflu, a fluorescent and stable resorufin is formed. Antibodies conjugated to the surface of these cerium oxide nanoparticles was employed for the detection of cancer biomarkers such as folate receptor and epithelial cell adhesion molecule (EpCAM) at pH 7 with a readout time of 3 hours compared to 15 hours for traditional ELISA test [9] (Figure 7.11).

Apart from its application in the diagnosis of cancer, the nanozymes have also proved to be equally robust in detecting microorganisms. Platinum (Pt) and palladium (Pd) nanoparticles are known to have peroxidase-like activity. In an application of dual lateral flow immunoassay for simultaneous detection of *Salmonella enteritidis*, and *Escherichia coli* O157:H7 pathogen, a Pt-Pd nanocomposite was used. The symbiotic interaction of the two nanomaterials increased the detection signal. Later, antiantibodies specific for the two microorganisms was coated onto the nanocomposite, and the hybrid system could oxidize TMB in the presence of H_2O_2. Further, to make the system cost-effective and portable, the immunoassay was integrated with a smartphone for recording the developed color and interpreting the result [45].

7.5.3 Nucleic Acid Sensing

Nanozymes have also been used in the detection of nucleic acid. Hemin on graphene exhibits peroxidase-like activity that can convert TMB to ox-TMB in the presence of H_2O_2. The hybrid nanomaterial possesses different affinities for ssDNA and dsDNA in 0.6 M Na^+. With ssDNA, the hybrid is well dispersed in the solution, whereas with dsDNA, it forms a precipitate. This attribute is because of the π-π stacking interaction between the nucleotides of ssDNA and hexagonal rings of graphene oxide. The ssDNA/hemin graphene hybrid without any precipitation can efficiently oxidize TMB in the presence of H_2O_2 as compared to dsDNA. Due to the presence of peroxidase-like activity as well as nucleic acid discriminating property, a label-free colorimetric method was developed for detecting single nucleotide polymorphism (SNP) [46].

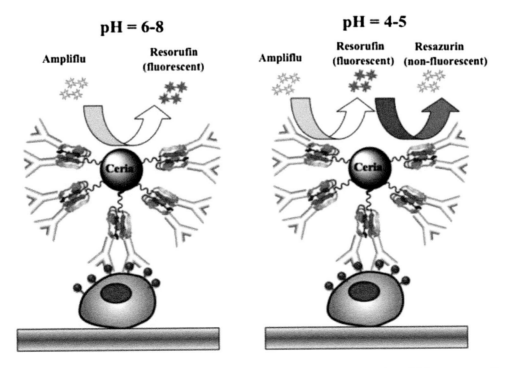

FIGURE 7.11 pH dependent oxidase-like activity of cerium oxide nanoparticles for the detection of folate receptor. Reprinted (adapted) with permission from reference [9]. Copyright (2011) American Chemical Society.

In most cases, nucleic acid detection uses labels that require chemical modification leading to a tedious process and complicated systems. Hence, the need for a label-free detection system arises that can knock out the intricacies. Mesoporous silica (mSiO$_2$), known for its porous structure, is generally used as a carrier to deliver drug molecules. In a recent approach, these mSiO$_2$ nanoparticles were coupled to Pt nanoparticles and used as "smart" reporters for label-free detection of nucleic acid. The smart reporter was synthesized by implanting Pt nanoparticles into mSiO$_2$, followed by immobilization of a probe DNA on the surface of the nanocomposite. As a result, the peroxidase-like activity of the Pt nanoparticles to oxidize TMB in the presence of H$_2$O$_2$ is halted. In the presence of the complementary target DNA, the probe DNA hybridizes with it and dissociates itself from the mSiO$_2$-Pt nanocomposite, which then fastens the oxidation of TMB. The detection method can be further applied for the discrimination of single base pair mismatches. When four targets with different mismatches were allowed to hybridize to the complementary probe adsorbed onto the surface of mSiO$_2$, the one with the lowest mismatches gave the maximum absorbance value of ox-TMB [47].

In another study, cerium oxide nanoparticles (CeO$_2$ NP), one of the popular oxidase-mimicking nanozymes, was utilized for label-free detection of nucleic acid because of its ability to oxidize peroxidase substrates (ABTS, TMB) even in the absence of H$_2$O$_2$. Binding of the target DNA to the CeO$_2$ NP inhibited the oxidase-like catalytic activity. The assay took less than 1 s for detection and was successfully validated by diagnosing *C. trachomatis* in human urine samples [48].

7.5.4 Detection of Metal Ions

Efforts have also been made to allocate nanozymes in the sensing of metal ions. The Hg^{2+} ion is known to interact actively with two thymine (T) bases resulting in the formation of T-Hg^{2+}-T base pairing. This property was used to design a label-free colorimetric sensor to detect both the Hg^{2+} ion and target DNA molecules. The hairpin structured capture probe with T-Hg^{2+}-T base pair, when hybridized with the target DNA, mediated the release of the Hg^{2+} ion and subsequently resulted in the oxidation of TMB in the presence of BSA stabilized AgNC [49] (Figure 7.12).

Due to the property of Hg^{2+} to form an Au-Hg amalgam by metallophilic interaction that has been found to have positive implications in enhancing the peroxidase-like activity of the AuNPs, a gold-nanozyme paper-based platform was developed for the detection of Hg^{2+} ion. The presence of Hg^{2+} increased the speed of decomposition of H$_2$O$_2$, leading to the generation of a blue-colored stain corresponding to the oxidized TMB on the paper chip [50].

7.5.5 Detection of Bilirubin

Hyperbilirubinemia(i.e., increased concentration of bilirubin in the blood) is a prevalent disorder in neonates. Enzyme-based detection of bilirubin with bilirubin oxidase as the biorecognition element experiences poor stability and has a high cost. A nonenzymatic detection method was developed for the sensitive detection of bilirubin by employing a human serum albumin-stabilized gold nanocluster (HSA-AuNC) as a peroxidase mimic (Figure 7.13) [27]. With a core of Au$_{18}$ and a dimension of ~1.0 nm diameter, the ability of HSA-AuNC to oxidize bilirubin in the presence of H$_2$O$_2$ was examined. Dueto the inherent property of bilirubin to bind with stronger affinity with HSA, only the unbound bilirubin was oxidized to a colorless compound by H$_2$O$_2$, and hence the free bilirubin present in the blood serum could be tested. The peroxidase-like activity of the HSA-AuNC was monitored by observing the absorbance change at 440 nm and could detect as low as 200±19 mM free bilirubin. Additionally, the synthesized HSA-AuNC exhibited an electrochemical response with redox behavior in the presence of bilirubin [51]. However, since the electrochemical behavior of the nanozyme is not yet established in comparison to the colorimetric behavior, it has not been further focused on the chapter.

7.5.6 For Therapeutics Applications

Nanozymes have also been employed in the field of therapeutics, mostly due to their ability to scavenge ROS. In most instances, these chemically reactive species are generated during a diseased condition that keeps on accumulating in the body. As mentioned in the first section, the notion of the nanozyme came into the picture when two hydroxyl derivatives of buckminsterfullerene (C$_{60}$) were found to scavenge free radicals in cortical neuron culture.

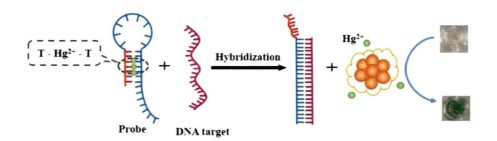

FIGURE 7.12 Detection of Hg^{2+} ion by BSA stabilized AgNC [49].

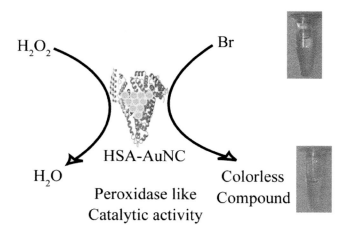

FIGURE 7.13 Detection of bilirubin using HSA-AuNC [51].

Cerium oxide nanoparticles, with mixed valence states of Ce^{3+} and Ce^{4+} and oxygen vacancies, can act as an SOD mimic and also possess attractive anti-inflammatory, antioxidant, and neuroprotective activity. Free radical, nitric oxide produced from nitric oxide synthase is responsible for inflammation in the cellular system. In an approach, the free radical scavenging property of cerium oxide nanoparticles and its ability to obstruct inflammation were examined in the J774A.1 murine macrophage. The results showed that the nanoparticles decreased the production of ROS in the J774A.1 cell. Remarkably, the cerium oxide nanoparticles were also found to scavenge elevated levels of toxic reactive oxygen intermediates within the retinal neuron. This feature signifies that the nanoparticle can be employed for recovering cell death promoted by reactive oxygen intermediates in a variety of different diseases [52].

The ability of apoferritin-encapsulated platinum nanoparticles to scavenge superoxide and peroxide was tested in the mammalian cell line Caco-2. These cell lines constitutively express ferritin receptors that allow internalization of the nanocomposite inside the cell. As a result, when H_2O_2 was stimulated externally into the cell, the cell viability was found to increase and there was a decrease in the generation of ROS [53].

There has been considerable research done to develop ROS generating systems that can bring about oxidative stress in cancerous cells. In an approach, a self-assembled biomimetic nanoflower nanozyme MnO_2@PtCo was developed that generated a set of biochemical reactions. These resulted in the production of ROS in both hypoxic and normoxic conditions. The PtCo nanoparticle with oxidase enzyme-mimicking property initiated a cascade of oxidation reactions that resulted in oxidative damage inside the cell, whereas the MnO_2 nanoparticle acted as catalase that rapidly converted the H_2O_2 into oxygen. This event had two consequences (1) the generated oxygen alleviated the hypoxic condition in a tumor, and (2) the developed nanozyme promoted apoptosis through the ROS-mediated mechanism [54].

7.6 CHALLENGES AND FUTURE PROSPECTS

The replacement of natural enzyme with nanozyme offers a novel and exciting addition to the property of nanomaterials. Since the discovery of fullerene derivatives with enzyme-like activity, there has been extensive growth in the field of nanozymes followed by their utility in various applications. Despite the numerous advantages over natural enzymes, the downside of nanozymes cannot be neglected and need to be resolved in order to have better activity in future research and development.

Although experimental findings have shown nanozymes to have remarkable stability, recyclability, and low cost in comparison to natural enzymes, there is still a need for improving the activity for high performance as well as for application in other allied fields. The functional groups on the surface of nanozymes play a vital role in the catalytic activity, especially in the electron transfer process. Modification of its surface properties using different strategies can be used as a target for enhancing the enzyme-like activity. Furthermore, the inside core of the nanozyme might also influence the activity of nanozyme, as coating and conjugation results in a decrease in its activity. Therefore, refinement of the type of coating and conjugation strategy can boost the enzyme-like catalytic activity of nanomaterials. Another essential feature of nanozyme is a better understanding of the underlying mechanism of catalytic activity. The reported nanozymes to date are mostly redox-based that involve electron transfer reactions. Although the electron transfer reaction is shown to be the same as the natural enzyme, many other nanomaterials have been found to transfer electrons by a different route. The available techniques that are used for nanomaterial-based catalysis reactions are either time consuming or not proper for studying such electron transfer reactions. Hence, results in conclusions often are probability-based. The computational approach makes it is easier to predict the electron distribution on the surface of a nanomaterial, as well to visualize its interaction with target molecules. So, by integrating both experimental findings and computational study, the catalytic mechanism can be elucidated more effectively.

Nanozymes that are reported to date have mainly focused on redox-based reactions. Bringing versatility to conduct other types of reactions will be another addition to the property of nanozymes. This attribute can be achieved by judiciously selecting the source material to fabricate the nanozyme followed by novel surface modification strategies.

The toxicity of nanozymes is one of the main challenges that needs to be addressed, especially for biomedical applications, as well as for health, environment, and safety concerns. Despite the utilization of nanozymes with different enzymatic activity in animal models, its ability to react in human cells should also be tested. Resovist, an iron oxide magnetic nanoparticle with peroxidase-like activity, has been found to cause stem cell growth and has been further approved by the Food and Drug Administration (FDA) for clinical use [1].

Another critical concern is the low selectivity of nanozymes. Although nanozymes have been used as a replacement for HRP in the detection of glucose and cholesterol, the selectivity of the reactions is mainly because of the presence of the respective oxidase enzymes such as GOx and cholesterol oxidase coupled to the system rather than the nanozyme. Therefore, further research is needed to improve the selectivity of the nanozymes for the target of interest.

REFERENCES

1. Wei H, Wang E. Nanomaterials with enzyme-like characteristics (nanozymes): next generation artificial enzymes. Chem Soc Rev. 2013; 42(14): 6060–6093.
2. Wu J, Wang X, Wang Q, et al. Nanomaterials with enzyme-like characteristics (nanozymes): next generation artificial enzymes (II). Chem Soc Rev. 2019; 48(4): 1004–1076.
3. He W, Liu Y, Wamer WG, Yin JJ. Electron spin resonance spectroscopy for the study of nanomaterial-mediated generation of reactive oxygen species. J Food Drug Anal. 2014; 22(1): 49–63.
4. Dugan LL, Gabrielsen JK, Yu SP, et al. Buckminsterfullerenol free radical scavengers reduce excitotoxic and apoptotic death of cultured cortical neurons. NeurobiolDis. 1996; 3(2): 129–135.
5. Ali SS, Hardt JI, Quick KL, et al. A biologically effective fullerene (C_{60}) derivative with superoxide dismutase mimetic properties. Free Radic Biol Med. 2004; 37(8): 1191–1202.
6. Comotti M, Pina CD, Matarrese R, Rossi M. The catalytic activity of "naked" gold particles. Angew. Chem. Int. Ed. 2004; 43(43): 5812–5815.
7. Gao L, Zhuang J, Nie L, et al. Intrinsic peroxidase-like activity of ferromagnetic nanoparticles. Nature Nanotechnol. 2007; 2: 577–583.
8. Luo W, Zhu C, Su S, et al. Self-catalyzed, self-limiting growth of GOx mimicking AuNPs. ACS Nano. 2010; 4(12): 7451–7458.
9. Asati A, Kaittanis C, Santra S, Perez JM. pH-tunable oxidase-like activity of cerium oxide nanoparticles achieving sensitive fluorigenic detection of cancer biomarkers at neutral pH. Anal Chem. 2011; 83(7): 2547–2553.
10. He W, Liu Y, Yuan J, et al. Au@Pt nanostructures as oxidase and peroxidase mimetics for use in immunoassays. Biomaterials. 2011; 32(4): 1139–1147.
11. Vernekar AA, Das T, Ghosh S, Mugesh G. A remarkable efficient $MnFe_2O_4$-based oxidase nanozyme. Chem Asian J. 2016; 11(1): 72–76.
12. Liu J, Meng L, Fei Z, et al. On the origin of the synergy between the Pt nanoparticles and MnO_2 nanosheets in Wonton-like 3D nanozyme oxidase mimic. Biosens Bioelectron. 2018;121: 159–165.
13. Wang C, Tang G, Tan H. Colorimetric determination of mercury(II) via the inhibition by ssDNA of the oxidase-like activity of a mixed valence state cerium-based metal organic framework. Microchim Acta. 2018;185:1–8.
14. Li D, Liu B, Huang PJJ, et al. Highly active fluorigenic oxidase-mimicking NiO nanozymes. Chem Commun. 2018; 54(88): 12519–12522.
15. Heckert EG, Karakoti AS, Seal S, Self WT. The role of cerium redox state in the SOD mimetic activity of nanoceria. Biomaterials. 2008; 29(18): 2705–2709.
16. Mandoli C, Pagliari F, Pagliari S, et al. Stem cell aligned growth induced by CeO_2 nanoparticles in PLGA scaffolds with improved bioactivity for regenerative medicine. Adv Funct Mater. 2010; 20(10): 1617–1624.
17. Liu X, Wei W, Yuan Q, et al. Apoferritin-CeO_2 nano-truffle that has excellent artificial redox enzyme activity. Chem. Comm. 2012; 48(26): 3155–3157.
18. Pagliari F, Mandoli C, Forte G, et al. Cerium oxide nanoparticles protect cardiac progenitor cells from oxidative stress. ACS Nano. 2012; 6(5): 3767–3775.
19. Pirmohamed T, Dowding JM, Singh S, et al. Nanoceria exhibit redox state –dependent catalase mimetic activity. Chem Commun. 2010; 46(16): 2736–2738.
20. Chen Z, Yin JJ, Zhou YT, et al. Dual enzyme-like activities of iron oxide nanoparticles and their implication for diminishing cytotoxicity. ACS Nano. 2012; 6(5): 4001–4012.
21. Mu J, Zhang L, Zhao M, Wang Y. Catalase mimic property of Co_3O_4 nanomaterials with different morphology and its application as a calcium sensor. ACS Appl Mater Interfaces. 2014; 6(10): 7090–7098.
22. Dong J, Song L, Yin JJ, et al. Co_3O_4 nanoparticles with multi-enzyme activities and their application in immunohistochemical assay. ACS Appl Mater Interfaces. 2014; 6(3): 1959–1970.
23. Liang H, Wu Y, Ou XY, Li JY, Li J. Au@Pt nanoparticles as catalase mimics to attenuate tumor hypoxia and enhance immune cell-mediated cytotoxicity. Nanotechnology. 2017; 28(46): 465702.
24. Wei H, Wang E. Fe_3O_4 magnetic nanoparticles as peroxidase mimetics and their applications in H_2O_2 and glucose detection. Anal Chem. 2008; 80(6): 2250–2254.
25. Yu F, Huang Y, Cole AJ, Yang VC. The artificial peroxidase activity of magnetic iron oxide nanoparticles and its application to glucose detection. Biomaterials. 2009; 30(27): 4716–4722
26. Wang XX, Wu Q, Shan Z, Huang QM. BSA-stabilized Au clusters as peroxidase mimetics for use in xanthine detection. Biosens Bioelectron. 2011; 26(8): 3614–3619.
27. Santhosh M, Chinnadayyala SR, Kakoti A, Goswami P. Selective and sensitive detection of free bilirubin in blood serum using human serum albumin stabilized gold nanoclusters as fluorometric and colorimetric probe. Biosens Bioelectron. 2014; 59: 370–376.
28. Chen J, Shu Y, Li H, et al. Nickel metal-organic framework 2D nanosheets with enhanced peroxidase nanozyme activity for colorimetric detection of H_2O_2. Talanta. 2018; 189: 254–261.
29. Sharma P, Jha AB, Dubey RS, Pessarakli M. Reactive oxygen species, oxidative damage, and antioxidative defense mechanism in plants under stressful conditions. JBot. 2012; 21737. doi:10.1155/2012/217037.
30. He W, Wamer W, Xia Q, et al. Enzyme-like activity of nanomaterials. J Environ Sci Health C Environ Carcinog Ecotoxicol Rev. 2014; 32(2): 186–211.
31. He W, Zhou YT, Wamer WG, et al. Intrinsic catalytic activity of Au nanoparticles with respect to hydrogen peroxide decomposition and superoxide scavenging. Biomaterials. 2013; 34(3): 765–773.
32. Shi W, Wang Q, Long Y, et al. Carbon nanodots as peroxidase mimetics and their applications to glucose detection. Chem Commun. 2011; 47(23): 6695–6697.
33. Beltrame P, Comotti M, Pina CD, Rossi M. Aerobic oxidation of glucose: II. Catalysis by colloidal gold. Appl Catal A-Gen. 2006; 297(1): 1–7
34. Jv Y, Li B, Cao R. Positively-charged AuNPs as peroxidase mimic and their application in hydrogen peroxide and glucose detection. Chem Commun. 2010; 46(42): 8017–8019.
35. Wang S, Chen W, Liu AL, et al. Comparison of the peroxidase-like activity of unmodified, amino-modified, and citrate-capped AuNPs. Chem Phys Chem. 2012; 13(5): 1199–1204.
36. Long YJ, Li YF, Liu Y, et al. Visual observation of the mercury-stimulated peroxidase mimetic activity of gold nanoparticles. Chem Commun. 2011; 47(43): 11939–11941.

37. Liu S, Lu F, Xing R, Zhu JJ. Structural effects of Fe_3O_4 nanocrystals on peroxidase-like activity. Chem Eur J. 2011; 17(2): 620–625.
38. Mu J, Wang Y, Zhao M, Zhang L. Intrinsic peroxidase-like activity and catalase-like activity of Co_3O_4 nanoparticles. Chem Commun. 2012; 48(19): 2540–2542.
39. Korsvik C, Patil S, Seal S, Self WT. Superoxide dismutase mimetic properties exhibited by vacancy engineered ceria nanoparticles. Chem Commun. 2007; (10): 1056–1058.
40. Celardo I, Pederson JZ, Traversa E, Ghibelli L. Pharmacological potential of cerium oxide nanoparticles. Nanoscale 2011; 3(4): 1411–1420.
41. Asati A, Santra S, Kaittanis C, et al. Oxidase-like activity of polymer-coated cerium oxide nanoparticles. Angew. Chem. Int. Ed. 2009; 48(13): 2308–2312.
42. Song Y, Qu K, Zhao C, et al. Graphene oxide: Intrinsic peroxidase catalytic activity and its application to glucose detection. Adv Mater. 2010; 22(19): 2206–2210.
43. Cui R, Han Z, Zhu JJ. Helical carbon nanotubes: intrinsic peroxidase catalytic activity and its application for biocatalysis and biosensing. Chem Eur J. 2011; 17(**34**): 9377–9384.
44. Wang H, Li P, Yu D, et al. Unraveling the enzymatic activity of oxygenated carbon nanotubes and their application in the treatment of bacterial infections. Nano Lett. 2018; 18(6): 3344–3351.
45. Cheng N, Song Y, Zeinhom MMA, et al. Nanozyme-mediated dual immunoassay integrated with smartphone for use in simultaneous detection of pathogens. ACS Appl. Mater. Interfaces. 2017; 9(46): 40671–40680.
46. Guo Y, Deng L, Li J, et al. Hemin- graphene hybrid nanosheets with intrinsic peroxidase-like activity for label-free colorimetric detection of single-nucleotide polymorphism. ACS Nano.2011; 5(2): 1282–1290.
47. Wang Z, Yang X, Feng J, et al. Label-free detection of DNA by combining gated mesoporous silica and catalytic signal amplification of platinum nanoparticles. Analyst. 2014;139(23): 6088–6091.
48. Kim MI, Park KS, Park HG. Ultrafast colorimetric detection of nucleic acids based on the inhibition of the oxidase activity of cerium oxide nanoparticles. Chem Commun. 2014; 50(67): 9577–9580.
49. Wang GL, Jin LY, Wu XM, et al. Label-free colorimetric sensor for mercury (II) and DNA on the basis of mercury (II) switched-on the oxidase mimicking activity of silver nanoclusters. Analytica Chimica Acta.2015; 871: 1–8.
50. Han KN, Choi JS, Kwon J. Gold nanozyme-based paper chip for colorimetric detection of mercury ions. Scientific Reports. 2017; 7(1): 2806.
51. Santhosh M, Chinnadayyala SR, Singh NK, Goswami P. Human serum albumin-stabilized gold nanoclusters act as an electron transfer bridge supporting specific electrocatalysis of bilirubin useful for biosensing applications. Bioelectrochemistry. 2016; 111:7–14.
52. Hirst SM, Karakoti AS, Tyler RD, et al. Anti-inflammatory properties of cerium oxide nanoparticles. Small. 2009; 5 (24): 2848–2856.
53. Zhang L, Laug L, Münchgesang W, et al. Reducing stress on cells with apoferritin –encapsulated platinum nanoparticles. Nano Lett. 2010; 10(1): 219–223.
54. Wang Z, Zhang Y, Ju E, et al. Biomimetic nanoflowers by self-assembly of nanozymes to induce intracellular oxidative damage against hypoxic tumors. Nat Commun. 2018; 9(1):3334.

8 Carbon-Based Nanomaterials for Sensing Applications

Naveen K Singh
University of California San Diego, California, USA

Manoharan Sanjay
Universidad de Buenos Aires, Argentina

Pranab Goswami
Indian Institute of Technology Guwahati, Assam, India

CONTENTS

- 8.1 Introduction ... 164
- 8.2 Carbon Dots ... 164
  - 8.2.1 Synthesis of Carbon Dots ... 165
    - 8.2.1.1 Bottom-Up Approach ... 165
    - 8.2.1.2 Top-Down Approach ... 165
  - 8.2.2 Properties of Carbon Dots ... 165
    - 8.2.2.1 Optical Properties ... 165
  - 8.2.3 CDs-Based Sensor Design ... 167
    - 8.2.3.1 CDs-Based Cation (M+) Sensing ... 168
    - 8.2.3.2 CDs-Based Anions (M−) Sensing ... 168
    - 8.2.3.3 CDs-Based Microorganism Sensing ... 168
    - 8.2.3.4 CDs-Based Protein Sensing ... 169
    - 8.2.3.5 CDs-Based DNA Sensing ... 169
    - 8.2.3.6 CDs for Bioimaging and Biolabeling Applications ... 169
    - 8.2.3.7 CDs for Disease Detection ... 170
    - 8.2.3.8 CDs-Based Detection of Explosives ... 171
- 8.3 Carbon Nanotube (CNT) ... 171
  - 8.3.1 Synthesis of CNTs ... 171
    - 8.3.1.1 Arc Discharge Method ... 171
    - 8.3.1.2 Laser Ablation Method ... 171
    - 8.3.1.3 Chemical Vapor Deposition ... 172
    - 8.3.1.4 Liquid Electrolysis Method ... 172
    - 8.3.1.5 Natural Occurrence ... 172
  - 8.3.2 Types of Carbon Nanotubes ... 172
    - 8.3.2.1 Single-Walled Carbon Nanotube (SWCNT) ... 172
    - 8.3.2.2 Double-Walled Carbon Nanotube (DWCNT) ... 172
    - 8.3.2.3 Multiwalled Carbon Nanotube (MWCNT) ... 172
  - 8.3.3 Carbon Nanotube Properties ... 175
    - 8.3.3.1 Mechanical Properties ... 175
    - 8.3.3.2 Electronic Properties ... 175
    - 8.3.3.3 Chemical Properties ... 175
    - 8.3.3.4 Optical Properties ... 175
  - 8.3.4 Carbon Nanotube-Based Sensor ... 176
    - 8.3.4.1 CNT-Based Detection of DNA ... 176
    - 8.3.4.2 CNT-Based Detection of Protein Biomarkers ... 176
    - 8.3.4.3 CNT-Based Microorganism Detection ... 177
    - 8.3.4.4 CNT-Based Small Molecule Detection ... 177

8.4 Graphene 178
   8.4.1 Synthesis of Graphene 179
   8.4.2 Application of Graphene 179
       8.4.2.1 Graphene-Based Chemical and Electrochemical Sensors 179
       8.4.2.2 Graphene-Based Optical Sensor 180
8.5 Conclusion and Future Prospects 181
References 181

8.1 INTRODUCTION

Carbon is the 15th most abundant element present on earth, and based on its chemical properties, it is unique because it can form strong bonds with itself and other elements in different ways and resist chemical forces under ambient conditions. This unique property has led to the formation of compounds with long chains and rings of carbon atoms. The field of synthetic organic chemistry supports the development of a series of carbon nanoscale allotropes and nanomaterials with tailored functional modifications like fullerenes, diamondoids, carbon onions, carbynes, nanodiamonds, carbon nanotubes, carbon dots, and graphene. Carbon nanomaterials have numerous physical, chemical, and electronic properties, which differ from their corresponding bulk materials. This provides it an inimitable place in the field of sensor development. The sp^2 hybridized carbon nanomaterials are classified into different category-based dimension such as zero dimension (all the dimensions in the nanoscale, like carbon dots), one dimension (one in the microscale and two dimensions in the nanoscale, like carbon nanotubes), and two dimension (two dimensions in the microscale and one dimension in the nanoscale, like graphene) (Figure 8.1). The unique properties of the carbon nanomaterials inspired their wide range of applications through interdisciplinary approaches in different areas including the emerging fields of sensors, biosensors, and allied analytical sciences.

8.2 CARBON DOTS

Carbon dots (CDs) have a very broad application among various nano-sized carbon-based materials. The average size of a CD can vary between 2 nm and 10 nm and come under zero dimension category (0D) due to the nano-size range in all dimensions. Xu and coworkers [1] accidentally discovered CDs during the purification of a carbon nanotube, and, through further characterization with microscopy, they established it as a nanomaterial. Based on composition and biocompatibility, CDs appear to be a suitable alternative to metal-based quantum dots (QDs). The structure of the CD mainly consists of sp^2/sp^3 carbon atoms along with oxygen/nitrogen elements or with polymeric aggregation. The building block of this type of CD is mainly composed of a carbon skeleton accompanied by some doped heteroatoms that enhances its properties positively. CDs exhibit better biocompatibility characteristics than QDs, which consist of a heavy metal core that negatively impacts the biocompatibility. The CDs can be divided into graphene quantum dots (GQDs), carbon nanodots (CNDs), and polymeric dots (PDs). The GQDs possess high charge carrier density and exhibit very fast electron mobility because of several layers of graphene with various functional groups attached to the edge and are anisotropic in shape. The CNDs are spherical and can be categorized into carbon nanoparticles and carbon quantum dots based on the absence or presence of a crystal lattice, which governs different photoluminescent regions for different types of CNDs. PDs are grouped or crosslinked polymers arranged from a linear polymer or monomer. In recent years, CDs have received remarkable scientific attention from various fields. It is an ideal alternative to various existing metal-based nanomaterials. These CDs exhibit low toxicity, biocompatibility, high solubility, and stability. Moreover, they can be produced following simple synthesis protocols and easily functionalized. Besides, some distinct

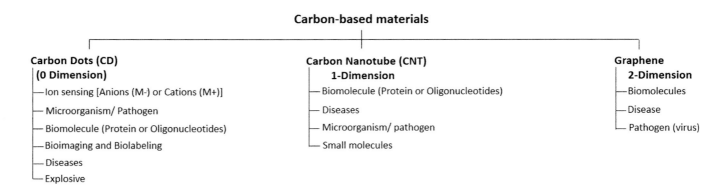

FIGURE 8.1 Types of carbon-based materials and their potential applications (the list of applications is provided here for materials based on examples used in the chapter; otherwise, all materials could be used based on sensor design needs).

features of CDs make them more useful than other existing fluorometric materials, which are discussed in the following sections.

8.2.1 Synthesis of Carbon Dots

Ideally, CDs can be synthesized from many carbon-containing compounds. Even milk, amino acids, carbohydrates, carbon materials, plant parts or extracts, etc., can be used as substrates. The synthesis of CDs can be categorized into bottom-up and top-down approaches. The bottom-up method uses a carbon source as the precursor, while the top-down method generally uses graphite materials; CDs are also denoted as graphene nanodots.

8.2.1.1 Bottom-Up Approach

The "bottom-up" approach is an effective way to synthesize fluorescent CDs. Small molecules may undergo dehydration and carbonization reactions to form CDs. The CDs produced through these methods retain hydroxyl, carboxylic, keto, and amine function groups on their surface depending upon the precursor molecules being used for the synthesis. The dehydration and carbonization reaction can be achieved through an array of techniques, such as pyrolysis, microwave, ultrasonic and hydrothermal treatments. However, it is tough to regulate these synthesis methods to prevent polydispersity in CDs. The production of CDs with uniform surface functionality, fewer surface defects, and identical molecular weight and size may be possible by using designed precursors and standard optimized conditions. Interestingly CDs synthesized from different methods have distinctive structural, physical, and chemical properties.

8.2.1.2 Top-Down Approach

In this process, nanoparticles are produced by breaking down larger materials like graphite into desirable nanostructures. However, to generate fluorescent CDs from these materials, their size and surface chemistry need to be cautiously modulated by various methods like laser ablation, arc discharge, chemical or electrochemical oxidation, and solvothermal approaches. The CDs prepared through oxidative passivation method from poly ethyl glycol (PEG) exhibit better fluorescence. Hence, passivation emerges as an effective method to improve emission efficiency. The monolayer and multilayer CDs with high electrochemical luminescence (ECL) can be prepared from CX-72 carbon black [2]. CDs with fluorescence of different wavelengths have been prepared by altering the potential in the electrochemical approach [3]. Fluorescent carbon dots with tunable light emission have also been synthesized by using laser irradiation in carbon materials suspended in an organic solvent [4].

8.2.2 Properties of Carbon Dots

The chemical and physical properties of CDs have been intensively studied for their applications in diverse fields. CDs exhibit good dispersibility, catalytic functions, biocompatibility, and photoluminescence behaviors suitable for sensing and imaging applications. The studies on electrocatalytic properties of the CDs are still in its infancy. Currently, a great effort has been seen on the utilization of CDs for various applications exploring their fluorescence, photocatalytic, and electronic characteristics.

8.2.2.1 Optical Properties

The CDs possess several interesting optical properties that have greatly encouraged their application in cutting-edge scientific fields such as photocatalysis, photovoltaics, sensing, and bioimaging. Among the various optical properties of CDs, the excitation-dependent emission with high quantum yield, and better photostability have been significantly explored for biosensing applications. Moreover, the optical activity of CDs in NIR (near infra-red) region makes it suitable for bioimaging applications, since NIR light has higher tissue penetration without any interference from biological moieties. Various optical properties of CDs and their applications have been discussed in the following sections.

8.2.2.1.1 Absorption

CDs show absorbance in the UV-visible region. The sharp absorption peak that appears in the UV region is attributed to the π-π* transition from C=C bands, and the apparent broad peak in the visible region appears due to n-π* transitions of C=O groups or a designated size distribution of the sp^2 cluster in CDs that were doped inside the sp^3 cluster matrix. In addition to this, functionalized groups may contribute to absorption in the UV-visible region.

8.2.2.1.2 Fluorescence

It is one of the most fundamental and application-oriented properties in the biosensing and bioimaging fields. Some of the critical phenomena, such as excitation wavelength-dependent emission, intensity, and up-conversion fluorescence properties, make CDs stand out from the conventional fluorescence agents. The mechanism of luminance of CDs is well explained with the help of the quantum confinement effect, surface state, and hole-electron pair recombinations.

8.2.2.1.3 Quantum Effect or Quantum Size Confinement

The particle size of the material when reduced below a particular nanometer range causes the quantum confinement to occur. The Fermi energy level near the quasi-continuous electronic level moves into discrete energy levels. Here, the word "confinement" implies limiting the random motion of freely moving electrons to defined energy levels (discreteness). Therefore, synthesized CDs exhibit different optical properties from their precursor counterparts. The CDs exhibit size-dependent fluorescence phenomena [5]. For example, CDs with 1.2 nm, 1.5–3.0 nm, and 3.8 nm sizes emit fluorescence lights in UV, visible, and near-infrared regions, respectively, which indicates that the gap distance is inversely proportional to particle size. Thus, the wavelength of fluorescence emission is directly proportional to the size of CDs.

8.2.2.1.4 Surface Factor

A diverse range of functional group may be present on the surface of CDs. These functional groups may influence the surface energy level and energy gap of the CDs. Passivating the CDs with some agents (e.g., PEG-1500N) enhances their fluorescence quantum yield to a great extent, say up to 50%. The nitrogen-containing functional groups successfully passivate the surface of CDs, leading to enhancement of the fluorescence quantum yield. Theoretical calculations have shown that surface coverage of CDs with $-NH_2$ can cause a bathochromic shift (redshift) of emission wavelength. A redshift in emission was also observed for GQDs following their functionalization [6]. Thus the fluorescence properties can be tuned by the addition of functional groups that change the bandgap size of the CDs. Also, the shielding of CDs in a hydrophobic environment leads to fluorescence enhancement. Usually, after excitation, electrons move to the first singlet state (S1), and surplus vibrational energy is rapidly lost to the solvent. CDs have more dipoles in an excited state than ground state conditions. As a result of this, the solvent dipoles reorient or ease themselves around the excited state of the fluorophore, which decreases the excited state energy. Hence, the emission energy of CDs has an inverse relationship with solvent polarity.

8.2.2.1.5 Electron Hole and Recombination of Radiation Theory

In semiconducting materials, a hole is formed when an excited electron from the valence band moves to the conductance band. Later the electron returns to the vacant space in the valence band by energy relaxation (Figure 8.2). The excited electron quickly (~10–12 s) loses energy. Here, the excited electron first moves close to the bottom of the conduction band, and the holes rise close to the top of the valence band in a time frame of 10^{-9} to 10^{-6} s. Eventually, the electron dropped across the energy gap into the empty state is represented by the hole, and the process is termed electron-hole

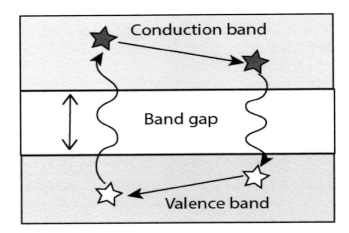

FIGURE 8.2 Recombination of electrons and holes generated by light absorption. The white spot depicts the hole and the red the electron.

recombination. During this process, an emission wavelength (radiation energy) is released that is nearly equal to the bandgap energy. The nanoscale dimension of CDs facilitates the movement of electron-hole recombination between the valence and conductance band. In nitrogen-doped heterogeneous CDs, a radiation rearrangement is a common phenomenon, and it improves the fluorescence intensity. The surface structure of CDs also influences radiation rearrangement.

8.2.2.1.6 Phosphorescence

It is related to fluorescence, but it emits radiation for a longer duration of time from a forbidden energy state or metastable state (Figure 8.3). Many CDs exhibit phosphorescence. The phosphorescence property of CDs came late in the discovery timeline as it occurs at a low temperature, low concentration of oxygen or other impurities, and in the complete absence of a quencher. CDs, when dispersed in polyvinyl alcohol (PVA) followed by excitation with UV light, generate

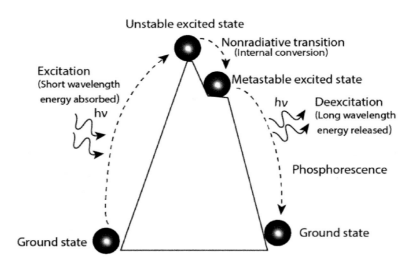

FIGURE 8.3 Phosphorescence mechanism.

phosphorescence. It was originated from an excited triplet state from an aromatic carbonyl functional group present at the surface of the CDs. The PVA molecule, along with hydrogen bonds, protects excited triplet state energy from rotational or vibrational energy loss.

8.2.2.1.7 Chemiluminescence

It is a physicochemical phenomenon. Here, two compounds react with each other leading to the formation of a higher energy state intermediate complex. The complex is then broken down and releases some of the energy in the form of light to reach the ground state. When CDs are incubated with some oxidants like $KMnO_4$ and Ce, holes are formed in the dots leading to an increase in the rate of electron-hole recombination/annihilation. Subsequently, the nanomaterials release energy in the form of chemiluminescence (CL). Positive effects on CL have been observed by increasing the temperature and concentration of CDs. Further, the CL properties can also be controlled with surface modification [7]. The CL of CDs brings up new prospects in the determination of reductive elements.

8.2.2.1.8 Electrochemical Luminescence

This is also called electrogenerated chemiluminescence. The CDs are semiconducting, and on applying an electric potential over CDs, intermediate forms are electrochemically generated. These undergo a very exergonic reaction to create an electronically excited state that releases radiation on returning to the ground state. The emission of a particular wavelength of light happens without photoexcitation. For example, synthesized CDs via carbonization oxidation strategy or carbonization extraction strategy can differentiate oxidized CDs (oCDs) and reduced CDs (rCDs) based on a high or low oxidation state, respectively. The ECL can be controlled with the diffusion of oCDs (i.e., related to direct oxidation of oCDs). The presence of an oxygen-containing group on the surface of oCDs supports generation of oCD radicals. The reduction of these radicals (S_2O_8) releases a strong oxidizing agent (SO_4C), which absorbs an electron from anionic oCDs and produces ECL emission [8].

8.2.2.1.9 Upconversion Fluorescence

A common fluorophore emits a longer wavelength of light following the excitation with a shorter wavelength of light; this phenomenon is called downconversion fluorescence (DCF). On the other hand, upconversion fluorescence (UCF) involves the emission of a shorter wavelength of light than the wavelength of the exciting radiation [9]. The consecutive or sequential absorption of two or more photons during excitation has been attributed to UCF. Leaky constituents from the second diffraction in a monochromator are ascribed as the reason for the occurrence of such absorptions [10]. The UCF primarily occurs in the near-infrared region (NIR) (700–1000 nm), which makes this phenomenon suitable to study in biomedical (e.g., biosensing, bioimaging) applications. Of note, the biological samples have low absorbance in the NIR region; hence, the technique may offer low background fluorescence and autofluorescence in biological samples.

8.2.2.1.10 Photoinduced Electron Transfer (PET) Property

Many CDs exhibit this phenomenon. Photosynthesis and fluorescence quenched by some fluorophores are simple examples of PET. This phenomenon occurs when a molecule is first excited by a specific wavelength of light. The excited electron is then transferred from the molecule (donor) to an acceptor molecule, completing the process of PET. CD-based PET is widely used for sensing applications. There is a stark difference between PET and foster resonance energy transfer (FRET). In PET, the fluorescence of the acceptor is quenched and not perceived, since the acceptor is transformed into a radical anion. The charge transfer state can be probed by various time-resolved techniques, like transient absorption. In FRET, conversely, a spectral overlap between the emission of the donor and the absorption of the acceptor is essential. These fascinating PET properties of CDs as an electron donor/acceptor offer new prospects for catalysis, light energy conversion, and related applications [8].

8.2.3 CDs-Based Sensor Design

In general, three approaches are known for developing fluorescent CDs-based sensors. One is a direct approach in which the target analyte directly interacts with the CD and quenches its fluorescence. Following this approach, different types of metal ions in a broad range can be detected by monitoring the fluorescence quenching of pristine CDs. The functional group and surface charge present over the CDs interact with the target metal ions. Various oxidative analytes could also be detected following this approach, probably through the oxidation of CDs surface. The second approach involves post-modification or functionalization of CDs. Specific receptors (aptamer, antibody, and chemical reagents) are conjugated over CDs, which facilitates the binding-specific target analyte in a different population of analytes. For example, surface modification of CD with boronic acid facilitates the binding and detection of glucose with high selectivity. The third approach is an integration of CDs with other sensory materials such as suitable fluorophores to initiate FRET and some substrates to quench fluorescence. For instance, the fluorescence of CD is quenched first using a quenching element, following which the fluorescence activity is recovered back in the presence of the analyte of interest. As mentioned earlier, the addition of metal ions quenches the fluorescence of CDs. This quenched CD-metal complex could be applied for sensing a range of analytes such as anions. Some dual-mode sensors, through which more than one analyte can be detected by integrating fluorescence quenching and recovery of CD fluorescence, are known. Furthermore, a ratiometric sensor has also been developed in which the ratio of fluorescence intensities at two or more wavelengths is measured. This ratiometric approach is more reliable and is less likely to interfere with the background signal [11].

Due to the interesting optoelectronic properties, CDs can be used for sensing a wide array of analytes, such as macromolecules, small molecules, cells, ions, and bacteria, following the approaches discussed earlier. Some of the CD-based sensing of analytes is presented in the following sections.

8.2.3.1 CDs-Based Cation (M+) Sensing

A plethora of research works indicates the application potential of CDs to detect various metal ions such as Fe^{3+}, Cu^{2+}, Hg^{2+}, Ag^+, Cr^{6+}, and Hg^{2+} from a complex solution by using the fluorescent and quenching phenomenon of CDs.

8.2.3.1.1 CDs-Based Hg^{2+} Sensing

Mercury is known for its toxicity to human health. It is considered an environmental pollutant because of improper disposal by industry and agriculture. It can cause various diseases, including colitis, dizziness, dyspnea, and kidney failure, even at extremely low concentrations (ppm or ppb). Mercury is a d-block element. It can efficiently capture an extra electron from the surface of CDs into their empty d-shell, leading to the fluorescence quenching (turn-off method) of CDs. This redox activity also causes a change in the UV-visible spectra of CDs. Hence, this interaction can be used for sensitive detection of Hg^{2+} in aqueous samples. Mercury can also be detected by using CDs following the fluorescence method, as discussed earlier. In a report, CDs were conjugated with bis(dithiocarbamate) copper(II) that facilitates fluorescence quenching of CDs through electron and energy transfer. Hg^{2+} inhibits this energy and electron transfer process, resulting in the recovery of CD fluorescence. This strategy has been used for quantitative measurement of Hg^{2+} with LOD as low as 4 ppb on a paper platform [12].

8.2.3.1.2 CDs-Based Pb^{2+} Sensing

Lead (Pb^{2+}) is a toxic metal and is naturally present in the earth's crust. The continuous extraction of lead from earth's crust and its application in a wide range of industrial products, including in fuel and batteries, contaminates the environment. CD can be used to detect lead in aqueous samples. Like mercury, lead has empty d-orbitals. It can efficiently capture the electron from the excited state of fluorescent CDs better than any other metal ions. It was observed that the quenching process was mainly due to static interaction between CDs and Pb^{2+} and was confirmed by time-resolved fluorescence spectroscopy [13].

8.2.3.1.3 CDs-Based Cobalt Ions (Co^{2+}) Sensing

Co^{2+} is a vital trace metal micronutrient necessary for life. However, a high accumulation may lead to various disabilities and diseases, such as asthma, heart disease (decreased cardiac output, cardiac enlargements), dermatitis, and lung disease. Therefore, monitoring the Co^{2+} level in the environment or biological samples is important. The CD-based detection of Co^{2+} mostly relies on the CL resonance energy transfer through Fenton-like chemistry. Transition metals [e.g., Co(II), Mn(II), Fe(III), and Cu(II)] catalyze the breakdown of hydrogen peroxide in a heterogenous or homogenous environment by a process known as a Fenton-like reaction or process. The Co^{2+}-mediated breakdown of H_2O_2 releases nascent oxygen (reactive oxygen species [ROS]), which further induces CL emission of CDs. This system shows high selectivity toward Co^{2+} and results in quick *OH production in bulk by Co^{2+} compared with other transition metals [14].

8.2.3.2 CDs-Based Anions (M-) Sensing

There are fewer reports on the application of CDs for anion detection than metal ion detection. The fluorescence of CD is quenched in the presence of transition metal, as discussed previously. This quenching event may be exploited to detect anion. The addition of an anion to the system facilitates its binding with the metal cations, regenerating the fluorescence of CDs. This phenomenon of fluorescence restoration has been exploited for the detection of various anions such as PO_4^{3-}, S^{2-}, and CN^- using corresponding suitable metal ion-anion conjugates. In addition to metal ions, the gold nanoparticles-CD system can be used to detect anions such as SCN^- through the on-off strategy. Further, the up- and down-conversion fluorescence quenching phenomenon was also exploited for the detection of hypochlorite (ClO^-). The fluorescence lifetime analysis of the CDs in the absence and presence of hypochlorite ion using the time-resolved fluorescence study revealed the static quenching mechanism. The hypochlorite anion selectively abolishes the surface passivation layer of CDs leading to the static quenching [15].

8.2.3.3 CDs-Based Microorganism Sensing

There are reports on CDs as an optical probe for the detection of bacteria, particularly in the field of diagnosis and health care. However, these CD-based methods are generally limited to the detection of individual bacterial cells rather than the detection of biofilm, colonies, or bacterial metabolites. Then again, these fluorescence-based optical nanoprobes have appeared as a versatile tool for the detection of different bacteria with morphological information about the species. These methods are technically simple and inexpensive. Researchers have developed long amphiphilic chained (hydrocarbon chains) CDs that get attached to a bacterial membrane and induce a fluorescence response. The interaction is influenced by the type of bacterial strain that gets attached. This technique offers the ability to determine a specific bacterial strain in diverse bacteria populations [16]. CDs have also been successfully used to differentiate between live and dead microbial cells. Conventionally, bacterial viability assay is performed by different techniques such as cell division study and NADP/NADPH production ability, among others. However, these are mostly time-consuming and labor-intensive methods. Huan and coworkers [17] synthesized CDs from *Staphylococcus aureus* and *Escherichia coli* as carbon sources with unique surface properties. These CDs can efficiently differentiate between live and dead microbial cells based on their functional group, surface charge, polarity, and size. These properties facilitate CDs to penetrate inside damaged cell walls or membrane of microbial cells (Figure 8.4A), leading to differentiative fluorescent staining of live and dead cells.

The CD-aptamer conjugate has also been successfully used for the detection of pathogenic bacteria. In this method, an aptamer was used as a biorecognition element, and a CD was used as a reporter molecule. A specific amine-modified aptamer against *Salmonella typhimurium* was developed through SELEX (systemic evolution of ligand through

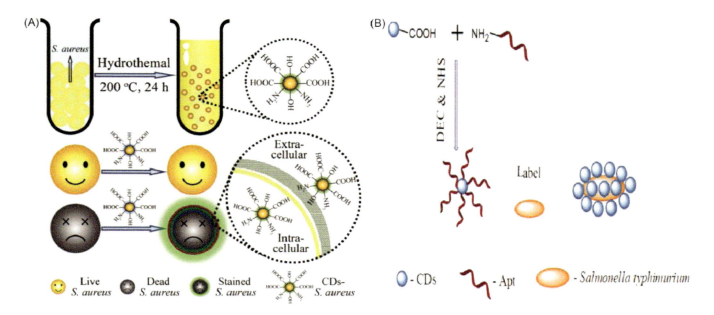

FIGURE 8.4 (A) Schematic representation of selective differentiation of live and dead microbial cells with CDs. [17]. (B) Specific detection of pathogenic bacteria with CDs aptamer conjugates [18].

exponential enrichment) and conjugated to carboxylic group modified CD through (-CO-NH-) peptide bond formation. In the presence of the target, the CD-aptamer conjugate is attached to the surface of bacteria producing intense fluorescence intensity (Figure 8.4B) due to aggregation of fluorescent CDs [18].

8.2.3.4 CDs-Based Protein Sensing

The CD may be used as a probe and a reporter molecule for the sensitive and selective detection of different proteins. For example, hemoglobin (Hb) can quench the fluorescence of CDs due to the presence of ferrous ions inside its tetra pyrrole ring. The emission (400 λ–550 λ) and absorption (400 λ–600 λ) bands of CDs and Hb, respectively, are overlapped to support FRET between them. The Hb molecule is adsorbed onto the surface of the CD through hydrophobic interaction, and the fluorescence intensity of the CD decreases linearly with the increasing concentration of protein in a range of 0.05–250 nM [16]. Another strategy was developed [19] for the detection of Hb in which CDs, along with H_2O_2, were used. In the presence of Hb, H_2O_2 is degraded to produce ROS, OH•, and superoxide, which triggers the fluorescence quenching of CDs. In another study, a fluorescent recovery strategy exploiting the metal ion-CDs interaction was developed for the detection of different proteins. For example, alkaline phosphatase hydrolyses adenosine triphosphate (ATP) into adenosine monophosphate or diphosphate and inorganic phosphate ions (iP). In the presence of cerium ions, the fluorescence of CDs is quenched due to their aggregation. On the other hand, the iP so formed has a stronger affinity toward cerium ion. Because of that, $CePO_4$ (cerium phosphate) is formed, impelling the CDs to disaggregate, and regenerating the initial fluorescence intensity of CDs corresponding to the increasing concentration of alkaline phosphatase. This method offers a quantitative assessment of alkaline phosphatase in a broad range (4.6–383.3 U/L), with a detection limit of 1.4 U/L. The advantage of the method has been ascribed as its potential application in human serum and scope for recycling of CDs [20].

8.2.3.5 CDs-Based DNA Sensing

The identification of a specific nucleic acid (DNA or RNA) sequences present in a clinical sample using the CDs-based methods has received some interest recently. The analysis of the specific DNA sequence provides much information related to genetic aberration, pathogenic microorganisms, and so on. A probe DNA sequence is usually conjugated to the CD for detection of the target gene in the sample. At an optimized condition, the probe DNA in the conjugated assembly hybridized with the target DNA in the sample, resulting in a remarkable fluorescence enhancement of CDs. The hybridization also prevents the CDs from aggregation. An increase in fluorescence lifetime by ~2 ns and quantum yield approximately ~4% was observed for the CDs-DNA array compared to solitary CDs [21]. In another work, a simple and sensitive CL method was developed for the detection of DNA over a paper surface using a simple, fast wax-screen-printing technique. The DNA sensor utilized N,N'-disuccinimidyl carbonate (DSC) to immobilize the probe DNA on μPADs and CDs-dotted nanoporous gold (CDs@NPG) for generating optical CL signal [22].

8.2.3.6 CDs for Bioimaging and Biolabeling Applications

QDs such as CdSe and CdTe are widely used for *in vivo* or *in vitro* cell imagining. Of note, CDs are emerging as a suitable biocompatible alternative to QDs for bioimaging applications (Figure 8.5A). The biocompatibility study revealed that the

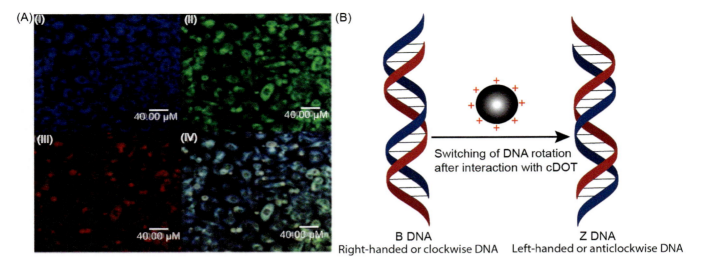

FIGURE 8.5 Laser scanning confocal microscopy images of C-dots labeled HepG-2 cells. The samples overlay was (I) excited at 405 nm (II) excited at 488 nm; (III) excited at 543 nm; (IV) [23]. (B). B to Z transition of DNA structure with interaction with CDs.

concentration of CDs at 50 mg/kg does not affect the blood structure and function of its components. Because of low cytotoxicity, wavelength-dependent excitation-emission, and high photostability, CDs are growingly used for bioimaging [23]. CDs exhibit excitation at longer wavelengths, usually at the NIR region, which is ideal for *in vivo* imaging. Moreover, it produces low background autofluorescence signals and better contrast in terms of image visualization. The CDs were mostly hydrophilic because of the hydrophilic functional group located at their surface, which facilitates their cellular internalization, which is useful for bioimaging applications. The other advantage of CDs is their potential to generate multicolor photoluminance that can be monitored by confocal microscopy or fluorescence microscopy.

Some optical imaging studies of cells revealed that CDs are mostly localized in the cytoplasm, and only a minute amount of these are present in the nucleus region. As a result, the cytoplasm is observed to be brighter than the nucleus. The application of CDs for *in vivo* imaging is, however, still at an infant stage. A recent study revealed that CDs could affect the DNA conformation under physiological salt conditions, along with conformation and sequence selectivity (Figure 8.5B). The study indicated that the positively charged surface of spermine-functionalized CDs could induce B-DNA to Z conformation change in the phosphate buffer solution at pH 7.2 [24]. This study further suggests that the nano-size dimension (2–4 nm) of CDs could bind inside the major grove in a GC-rich sequence region. Recent work on CDs indicates that it can be linked to various cells, molecules, or chemicals through modifications with different targeting groups. Besides the advantages of QDs discussed earlier, these nanomaterials encounter some risks associated with health and environmental issues due to their heavy metal constituents such as cadmium (Cd), cesium (Cs), and lead (Pb).

8.2.3.7 CDs for Disease Detection

The significance of luminescent CDs in biomedical research is well recognized. The research is boosted by the emerging tools and techniques such as fluorescence microscopy, laser technology, and allied instruments in optical spectroscopy/microscopy domains. The CDs have the potential to work as nanolanterns for studying biochemical interaction and intracellular mechanisms, which has potential theranostic applications.

Diabetes is a medical condition in which the body's homeostasis system is unable to maintain the normal glucose level. An electrochemical CD sensor was developed [25], where glucose oxidase (GOD) was self-assembled with reduced graphene oxide (rGO) and CDs (GOD/rGO-CDs). In the presence of glucose substrate, GOD catalyzes the reaction mentioned in equation (i). The reaction generates hydrogen peroxide as a by-product, which is later broken down by the CD due to its peroxidase-like activity. The free electron generated in the breakdown reaction is transferred to an electrode through rGO. The number of electrons transferred to the electrode is directly proportional to the number of glucose molecules present in human blood.

$$\text{Glucose} + H_2O + O_2 \xrightarrow{\text{Glucose oxidase}} \text{Gluconic acid} + H_2O_2$$

$$H_2O_2 \xrightarrow{\text{Carbon dots}} H_2O + 1/2\, O_2 + \text{electron (e}^-)$$

The CDs are also used for *in vitro* detection of cancer cells for diagnosis. The expression of the folic acid receptor (FR) is highly upregulated in the cancer cells as compared to the normal cell. Therefore, the overexpression of FR in cancer cells has clinical significance for therapeutics and diagnosis applications. A Trojan horse strategy was developed to specifically detect cancer cells [26]. In this strategy, the folic acid was conjugated with CDs surfaces through amidation covalent chemistry. The folic acid/CDs conjugate was then engulfed inside only the cancer cells via receptor-mediated endocytosis. The cancer cells could then be conveniently identified from the normal cells using fluorescence microscopy.

A sensitive fluorescence-based aptasensor for the detection of *Plasmodium falciparum* glutamate dehydrogenase (*Pf*GDH) biomarker for diagnosing malaria was developed recently [27]. A *Pf*GDH-specific amine-modified aptamer was covalently conjugated with CDs prepared from glutamic acid through the pyrolysis method. The CDs/aptamer conjugate specifically binds with the *Pf*GDH protein in diluted serum leading to protein-induced fluorescence enhancement (PIFE) phenomenon. The mechanism behind this was explored through surface docking study, which revealed that CDs/aptamer conjugates bind inside the hydrophobic pocket of protein, leading to protein-induced fluorescence enhancement of CDs. The PIFE sensor based on CDs displayed decent analytical performance for the detection of *Pf*GDH in the nanomolar range.

8.2.3.8 CDs-Based Detection of Explosives

The 4-chloro-2,6-dinitroaniline is widely used to prepare explosives. CDs can be used for the detection of the explosive compound in random samples. One such method involves the synthesis of fluorescent CDs from aqueous activated carbon suspension following an acid oxidation method. The synthesized CDs were 12 nm in size and had a quantum yield of 3.94%. The CDs were then functionalized with polyethyleneimine/poly (amino amine), 1,2-ethylenediamine, branched, or PAMAM-NH2 dendrimer through carboxylic and amine covalent chemistry. The functionalized CDs@PAMAM-NH2 dendrimer size and quantum yield were increased to ~5 (65 nm) and ~2 (6.33%) times, respectively. The synthesized CDs@PAMAM-NH2 conjugate was used as a ratiometric fluorescent probe for 4-chloro-2,6-dinitroaniline detection. The fluorescence of CD was decreased (at 465 nm) following the interaction. An additional fluorescence peak emerged at 507 nm, and it is related to the formation of the Meisenheimer complex. Ratiometric fluorescence intensity at 465 nm and 507 nm was used for the sensing applications to minimize the background noise. It was observed that the method could detect 4-chloro-2,6-dinitroaniline in a broad linear range with LOD of 2 µM [28].

8.3 CARBON NANOTUBE (CNT)

CNT was discovered by a Japanese scientist Iijima in 1991 [29] while studying the C60 carbon structure or Buckyball. CNTs are considered to have a one-dimensional (1D) structure as they have a length up to micrometers and a very small diameter in the nanometer range, which directly influences their properties. They have remarkable mechanical, optical, thermal, and electrical properties, which make it a material of great scientific interest. The basic properties of CNTs are described in Chapter 3.

8.3.1 Synthesis of CNTs

Over the years, several techniques have been developed for the synthesis of CNTs. The most commonly used procedures for producing CNTs are the chemical vapor deposition (CVD) technique, the laser ablation technique, and the carbon arc discharge technique. The latter two techniques involve high temperatures and are not suitable to regulate precisely the length, diameter, orientation, density, and alignment of the CNTs. Hence, there was an urge to develop a more advanced method for the synthesis of CNTs.

8.3.1.1 Arc Discharge Method

In this procedure, an arc voltage is applied between two graphite electrodes in an inert gas environment and heated up to ~4000 K. Commonly, helium is used as the inert gas, which, however, can be replaced by hydrogen or methane [30]. The presence or absence of catalysts controls the type of CNTs. When a graphite anode holding a metal catalyst like Fe or Co is used with a pure graphite cathode, SWCNTs are present in the form of soot. Different metal catalysts have been explored with an interest in increasing the yield of CNTs. Saito et al. showed that Ni-Y graphite mixtures could produce SWNTs in high yields of up to 90% [31]. Henceforth, this mixture has been used extensively for SWNT synthesis using the arc discharge technique. In the absence of a catalyst, fullerene is deposited as soot, and MWCNTs are formed on the cathode. The main advantage of this method is the ability to produce large quantities of CNTs with fewer structural defects when compared with other methods. However, this method provides limited control over the various features of the nanotubes, and due to the presence of metal catalysts, additional purification is required.

8.3.1.2 Laser Ablation Method

The principle of the laser ablation method is similar to the arc discharge technique. Guo and coworkers first developed the method. The research group produced metal molecules by blasting metals with a laser. They discovered that when metals were substituted with graphite rods, MWCNTs are formed [32]. In this technique, a high power laser directed to a graphite target at 1200± C in an argon atmosphere is used. The graphite is then vaporized and condensed as CNTs on the colder areas of the reactor [33]. Both MWNTs and SWNTs can be formed with this procedure in the presence or absence of a catalyst, similar to the arc discharge technique. In order to generate SWNTs, the graphite target must be supplemented with a metallic catalyst similar to the arc discharge technique. Studies showed that the diameter of the nanotubes could be controlled by the power of the laser used. A higher laser power produced thinner tubes [34]. This method results in low metallic impurities compared to arc discharge, since the metal catalyst atoms evaporate from the end of the tube once it is closed. On the other hand, structural deformities and branching are commonly observed in the nanotubes obtained by this method. It is also expensive due to the requirement of a high-power laser and high purity graphite rods.

8.3.1.3 Chemical Vapor Deposition

CVD is a well-accepted method for the production of CNTs because of its advantages over the other methods. It offers better control over the characteristics of the CNTs produced and is economically viable. There are variations of CVD methods, including catalytic chemical vapor deposition (CCVD), thermal, plasma-enhanced, oxygen-assisted CVD, water-assisted CVD, microwave plasma-enhanced CVD (MPECVD), radio-frequency CVD, and hot-filament (HFCVD) [33]. However, CCVD is the most commonly used process. In this process, a carbon-containing gas and a carrier gas like nitrogen, hydrogen, or ammonia is passed into a reactor in the presence of a metal catalyst at 600–1200°C to decompose the hydrocarbon. CNTs grown on the catalyst in the reactor are collected after cooling the system to room temperature. The most commonly used hydrocarbon gases are ethylene, acetylene, methane, or ethanol. Liquid hydrocarbons like benzene, alcohol, or volatile materials like camphor, naphthalene, and ferrocene can also be used as the carbon source [35]. The mechanism of CNT growth has yet to be properly elucidated; the only recognized fact is the influence of a carbon precursor, metal catalyst, particle size, temperature, and pressure. Many mechanisms have been proposed, while the widely accepted one is mentioned here. The carbon-containing gas vapor decomposes into its constituent's carbon and hydrogen at high temperature on the surface of the metal catalyst. The carrier gas removes the hydrogen generated in this process. At the same time, the carbon is dissolved in the metal until the metal reaches a saturation point. Post-saturation carbon precipitates out and crystallizes as energetically stable cylindrical grids [36].

8.3.1.4 Liquid Electrolysis Method

Recently, a team of scientists have reported a strategy for converting carbon dioxide to CNTs in liquid salts on the electrode surface. Carbon dioxide is introduced along with the molten salt at high pressure to increase the solubility of carbon. The principle of CNT synthesis here is cathodic reduction of dissolved carbon dioxide to carbon allotropes on the metal electrode. However, CNTs are not the only allotrope obtained from this method; it is found as a constituent in a mixture of amorphous carbon, crystalline graphite, and nanofibers [37].

8.3.1.5 Natural Occurrence

The occurrence of CNT in nature has surfaced lately, with researchers claiming to have isolated it from oil wells, ice caps, volcanic areas, explosions, and even as a result of wildfires. Though there is no doubt that these environments are conducive for the formation of various carbon allotropes, the control over allotrope features cannot be expected. These phenomena have inspired researchers to develop controlled explosive environments for CNT synthesis, which promises to be cost-effective and facilitate large-scale production with high yields [38].

8.3.2 Types of Carbon Nanotubes

There are differences among CNTs that have directed its classification into various categories. As discussed previously, CNT morphologically appears as a rolled graphite sheet in tubular form. The presence of single, double, or several concentric tubes attached by surface interaction forces can be differentiated into single-walled, double-walled, or multiwalled carbon nanotubes.

8.3.2.1 Single-Walled Carbon Nanotube (SWCNT)

The SWCNT is a seamless cylinder with a single concentric layer of graphene with carbon atoms arranged in a honeycomb lattice. The diameter of SWCNT varies from 0.4 nm to 1.5 nm and the length up to micrometer dimensions. The SWCNTs exhibit exceptional electronic properties and dual nature as metals and semiconductors. The decisive factor for the properties mentioned earlier depends on the chirality of the carbon atom. The chirality can be defined as the amount of graphite twist when the sheet is rolled and is explained with two vectors (m, n). When m is equal to n (m = n) then SWCNT behaves like a metallic conductor. In the case where m is not equal to n (m ≠ n), it leads to semiconducting behavior. In a particular case, where n-m is a multiple of three, it corresponds to a special type of semiconductor known as a small gap semiconductor (Table 8.1). The actual metallic or semiconductor properties depend on the bandgap of SWCNT, which further depends on the diameter and chirality of the tubes.

8.3.2.2 Double-Walled Carbon Nanotube (DWCNT)

The DWCNT is a seamless cylinder with a double concentric layer of graphite or graphene nested within each other and carbon atoms arranged in a honeycomb lattice. This double-walled system of DWCNT makes it suitable for studying the inter-wall interaction and properties of CNT. The DWCNTs have higher thermal, mechanical, and different optical and electrical properties as compared to SWCNT. The inter-wall distance of a DWCNT varies from 0.33 nm to 0.42 nm and length up to micrometers in size. A tube inside a tube of the SWCNT pattern is present in DWCNT, and each sheet looks like a graphene sheet rolled along with chiral vector $Ch = na_1 + ma_2$ into a tubular structure. The properties of DWCNTs do not merely rely on the superposition of inner and outer layer properties. Besides this, the inter-wall distance also affects the properties of DWCNTs. In some reports of DWCNTs, the outer and inner layers are attributed to corresponding metallic- and semiconductor-like properties [39]. Recent studies, however, revealed that both walls could be semiconducting. The DWCNT has several useful properties as compared to MWCNTs, such as an enhanced lifetime; high stability against chemical, mechanical, and thermal treatments; and improved current densities along with flexibility.

8.3.2.3 Multiwalled Carbon Nanotube (MWCNT)

It is made from multiple rolls of graphene or graphite sheet in a cylindrical form, and it is an allotrope of sp^2

TABLE 8.1
Type of SWCNTs, their Conductivity and Linked Chirality

Chirality Vectors (n,m)	Form of SWCNTs	Conductivity
(n,0)	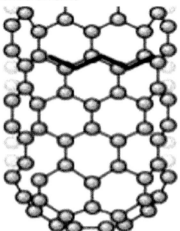 **Zigzag**	Show metallic properties when n is the multiple of 3; otherwise, semiconducting in nature
(n,n)	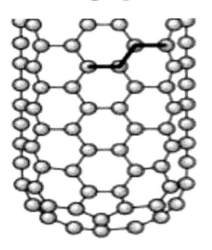 **Armchair**	Show metallic properties
(n,m) When m ≠ 0, and n	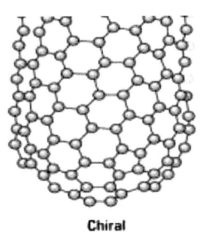 **Chiral**	Show metallic properties when $(2n + m)/3$ is an integer; otherwise, semiconducting in nature

carbon atoms, similar to graphite. The inter-wall distance between each SWCNT unit of MWCNT is ~0.34 nm, and it is not well defined due to its complex nature. One of the major differences between MWCNT and SWCNT is that the former primarily behaves like a metallic semiconductor, although some can behave like a semiconductor with a small bandgap. The outer wall of the MWCNT is mostly conducting in nature. The MWCNT is advantageous over the SWCNT, as they have a low cost of mass production and high thermal and chemical stability. In general, after surface modifications of SWCNT, electrical and mechanical properties are altered because of structural defects in C=C during chemical modification. However, the inherent properties of CNTs can be conserved by the surface functionalization of MWCNTs when the outer wall of MWCNTs is exposed to chemical functionality. A brief comparison is made between SWCNT, DWCNT, and MWCNT based on their properties in Table 8.2.

TABLE 8.2
Comparison of SWCNT, DWCNT, and MWCNT and Their Properties [40]

Carbon Nanotubes	Size	Preparation Methods	Thermal Properties	Mechanical Properties
SWCNT	0.4 to several nm in diameter	Laser ablation, arc discharge, CVD	Stable up to ~1800°C in Ar or 750°C in air Thermal conductivity >3500 Wm^{-1}K^{-1}	Young's modulus ~1.8 TPA, tensile strength ~1.0 TPa
DWCNT	Interwall distance 0.33–0.42 nm	Arc discharge, CVD, Peapod method	Stable up to ~2000°C in Ar or 800°C in air Thermal conductivity >600 Wm^{-1}K^{-1}	Young's modulus ~0.73–1.33 TPA, tensile strength ~13–46 TPa
MWCNT	Interwall distance ~0.34 nm	Laser ablation, Arc discharge, CVD, Peapod method	Thermal stability with diameter and length Thermal conductivity >3000 Wm^{-1}K^{-1}	Young's modulus ~0.27–0.95 TPA, tensile strength ~11–63 TPa

8.3.3 Carbon Nanotube Properties

The CNTs made from graphene or graphite have excellent conductive, strength, chemical inertness, and rigidity compared to their precursors. Some of these electronic characteristics depend on the elasticity and lattice helicity, the arrangement of the rotation, and the linear motion of subatomic units. The densities of MWCNT are 2.6 g/cm^3, and for SWCNT it is in range of 1.3–1.4 g/cm^3, depending on the chirality of CNTs. The densities of SWCNT or MWCNT are smaller than its building block (i.e., graphite) because of the presence of a porous structure at the center of the CNT. The surface area of the CNT is vast due to the nano-sized dimensions such as 10–20 m^2/g. Due to this characteristic, it is commonly used in sensing. Further details on the mechanical, chemical, electronic, and optical properties of CNTs are discussed next.

8.3.3.1 Mechanical Properties

The binding energy between the C-C covalent bond is 607 kJ/mole [41], making CNT the strongest material. Researchers have calculated the elastic modulus of CNT from thermal vibrational amplitudes through transmission electron microscopy (TEM) [42]. It can bear high strain (~40%) and tension devoid of plastic deformation of the band because of the hollow cylindrical closed topology. Under tension, some local bonds are ruptured, and this narrow defect spreads over the entire surface because of the mobility of these defects. This process leads to changes in the helicity of CNTs and ultimately influences its electronic property.

8.3.3.2 Electronic Properties

The CNT is made from graphene or graphite semimetal with a unique Fermi-level surface. Of note, the Fermi level is an energy level that is occupied by the electron orbitals at the temperature of 0 Kelvin, and the level of occupancy determines the conductivity of materials. At the surface of the Fermi level of CNT, it separately occupies an unoccupied electron in reciprocal spaces at zero temperature. Besides this, the electronic conductive properties of CNT depend on their chirality (non-superimposable on its mirror image), degree of twist, or wrapping angle. For example, the armchair tube structure is metallic. On the other hand, a zigzag carbon nanotube generally displays semiconducting properties. The electronic properties of CNTs could be easily modified or altered by doping. The MWCNT becomes metallic when doped with boron and nitrogen, whereas the SWCNT conductivity can be highly improved in the presence of halogen or alkali atoms. They have a high surface area and electronic densities, which make them an ideal candidate for sensing. The conductance (or resistance) of SWCNTs in an ideal condition of the tetra parallel unity transmission channel is ~155 μS (micro mho) or a resistance $R = 1/\text{conductivity} = 6.5$ kΩ, and this idealized value is independent on temperature. The scattering of the electron in the 1D structure of SWCNT allowed in a limited number of vacant space because of it is highly dependent on (or requires) massive momentum transfer [43].

8.3.3.3 Chemical Properties

The CNT tube precursor (i.e., graphite) is chemically inactive; however, CNTs are prone to certain chemical reactions because of the π-orbital mismatch in the arch cylindrical structure. The average charge densities present at the tip of the CNT (pentagon) is three to four times higher than the basal graphite plane (hexagon); hence, the former could act as an electrophilic reaction site. Because of this CNT tips or caps are more reactive compared to the cylindrical parts. The functionalized CNT is broadly used in the field of sensors due to its enhanced electron transfer capability from analyte. The surface modification (wall and cap) or surface defect present over CNTs makes it highly reactive compared to pristine CNT, which is important for sensor design. Purification with strong acid and introduction of carboxylic groups over the surface of CNT, especially over the tips, facilitates the amidation process with amine-modified functional groups.

8.3.3.4 Optical Properties

CNTs with different chirality have different electronic properties, which are correlated with their different optical properties. The CNTs show fluorescence in the NIR (900–1600 nm), as mentioned earlier. In SWCNT, the significant absorptive transition is E22 (V2-C2), and primary fluorescence emission occurs from C1 to V1, as depicted in the Jablonski diagrams (Figure 8.6). The excitation event in the SWCNT semiconductor is an excitonic occurring electron-hole pair with fixed coulombic interaction. This coulombic interaction energy in CNT is in the order of 400 meV with an exciton size of ~2 nm for (6,5)-SWCNTs. The average lifetime of an exciton may vary from 10 to 100 ps, making this type of photoluminescence fall

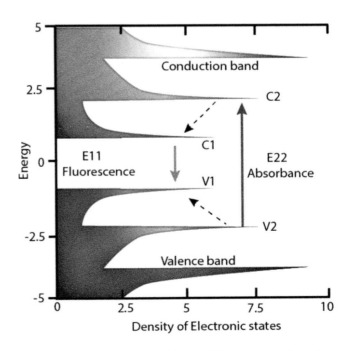

FIGURE 8.6 Electronic transition and fluorescence of semiconducting SWCNTs.

under the fluorescence category by definition [44–46]. The excitons tend to move along the CNT surface. So it can be influenced by any functional modification over the surface of CNT, which makes it an interesting probe for sensing applications.

The SWCNTs exhibit Raman-enhanced resonance signatures. The different vibrational modes of the CNT lattice give rise to various strong bands like radial breathing modes (RBM), G, and G' bands. A very high Raman scattering signal is obtained from a cross-section of SWCNT besides this resonance enhancement in the Raman signal of the optical transition of SWCNT observed [47, 48]. The different mechanism of fluorescence modulation with possible enlightenment on the reason is explained next.

A. Solvatochromism: It is a phenomenon when properties of solvent change in the presence of solute or vice versa. Similarly, when SWCNT is in close proximity to a solvent or analyte, then the dielectric environment present around affects the polarizability of SWCNT. This leads to a change in its absorption and emission spectra. A solute could alter the dielectric environment directly by substituting the solvent molecule or by altering the conformation of the polymer wrapping.
B. Charge transfer: Under this mechanism, analyte orbitals overlap with SWCNT/wrapping orbitals. The transfer of electrons among them leads to a change of excited or ground states, which finally alters the fluorescence spectrum via different rates of exciton (excited electron) quenching.
C. Doping and redox reaction: The defect in the carbon lattice is caused by the presence of analyte through doping (electron or holes) or redox reaction. This causes the routes of the excited electron or excition to decay.

8.3.4 Carbon Nanotube-Based Sensor

Due to the special properties of CNTs, it can be used for sensing applications for a wide array of analytes, such as macromolecules, small molecules, cells, ions, bacteria, etc. CNT-based sensing of different analytes is presented in the following sections.

8.3.4.1 CNT-Based Detection of DNA

The CNT-based DNA sensors could be used to identify changes in DNA structure (single or polynucleotide polymorphisms) and conformation and has significant potential in the diagnosis of disorders such as autoimmune diseases and cancer. The different types of strategies used for DNA detection are as follows.

8.3.4.1.1 Optical-Based Sensor
The ssDNA binds with its complementary strand from a pool of ssDNA sequences with specificity. This property has been exploited to develop DNA sensors, in which ssDNA was wrapped over the SWCNT and then exposed to complementary DNA sequence in the samples, resulting in a hypsochromic shift (shift of a spectral band to a higher frequency or shorter wavelength). This system can detect complementary DNA strands up to nM concentration. The mechanism behind this shift has been attributed to a change in the local dielectric environment of the SWCNT by the enhanced surface coverage of DNA. The interaction was further validated by FRET, in which the SWCNT wrapped ssDNA, and the complementary DNA strand was conjugated with donor and acceptor fluorophore molecules. On binding with the complementary DNA sequence, a reduction in the fluorescence emission of the donor molecule and enhancement of the fluorescence emission of the acceptor molecule was observed [49].

In previously mentioned methods, a solvatochromic shift (the ability of substances to change wavelength by changing solvent polarity) was observed in SWCNT fluorescence in the NIR region upon DNA hybridization. The ssDNA wrapped SWCNT was used for the detection of single nucleotide polymorphism (SNP) in complementary DNA sequences. The ssDNA wrapped SWCNT exposed to complementary and SNP containing DNA sequences as a consequence of this altered wavelength shift was observed on binding with complementary or SNP containing complementary DNA sequence. The final equilibrium shift was influenced by the presence of SNP in the DNA sequence [50].

8.3.4.1.2 Electrochemical-Based Sensor
The CNT-based electrochemical method for DNA detection offers high sensitivity and a fast response as compared to the optical approaches. In a report, a self-assembled SWCNT monolayer was created over a gold electrode. Then the SWCNT was functionalized with the carboxylic group and coupled with amine-modified ssDNA through an amide bond (-CO-NH-) using EDC/NHS chemistry. The hybridization between target complementary DNA and ssDNA/SWCNT/Au probe was monitored by using methylene blue (MB) as a redox indicator, which has a high affinity toward free guanine bases. As a sensing signal, a drop in redox current was observed [51].

8.3.4.2 CNT-Based Detection of Protein Biomarkers

Protein biomarkers present in body fluid such as blood, lymph, cerebrospinal fluid, and over the cell surface for many diseases or disorders are widely known. These biomarkers are extensively exploited for diagnosis purposes following the methods that can be subdivided into the labeled-based and label-free categories. In the label-based method, the query molecule is coupled with a suitable tag (fluorophore, redox dye, radioactive moiety, or enzymes), which provide support in the form of the signal during the detection process. These label-based approaches, though they find various analytical applications, have several drawbacks. Among these drawbacks, the involvement of tedious and time-consuming steps and affecting the native structure of the labeled protein are frequently cited. The label-free method, on the other hand, allows direct detection of the target molecules without the need of coupling the tag, secondary antibody, or washing steps. Hence, these label-free approaches appear simple compared to their complementary labeled-based techniques that encourage their point of care (PoC) and point of need (PoN) applications. Many of the

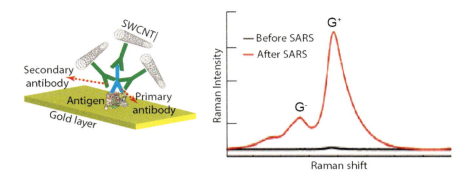

FIGURE 8.7 SWCNTs as Raman tags for protein detection with an example of the Raman spectra with and without the SWCNT Raman label.

label-free methods utilize a variety of nanomaterials such as gold nanoparticles, CDs, and CNTs for the detection of biomarkers and other target analytes of interest. In the following sections some of the prominent works are discussed against each of the detection approaches.

8.3.4.2.1 Optical Method for Biomarker Detection

The intrinsic Raman scattering properties of CNT make it a suitable Raman probe for biomarker detections. CNT provides a high Raman scattering cross-section with characteristic spectral peaks. The SWCNTs, tagged with IgG antibodies, are used as Raman tags for protein detection. A detailed description is depicted in Figure 8.7. The interaction with the target biomarker is analyzed through Raman spectroscopy. The characteristic Raman peaks of SWCNT indicated the binding of the target biomarker protein. This method could detect the target protein up to the femtomolar level, and its sensitivity is attributed to the low background signal from biomolecules [52].

8.3.4.2.2 Electrochemical Method for Biomarker Detection

Insulin is a very important peptide hormone, and it helps in the regulation of blood glucose concentration within a narrow range. The destruction of insulin-secreting beta cells of the pancreas leads to insulin-dependent type 1 diabetes. The ruthenium oxide modified CNT over glassy carbon electrode (RuOx/CNT/GC) offers very sensitive and specific amperometric and voltammetric measurements of insulin. This modified sensor shows an enhanced oxidation rate compared to separated CNT- or RuOx-coated electrodes. A cyclic voltammetry (CV) study revealed that RuOx/CNT/GC was capable of oxidizing insulin at a low potential (0.29 V) compared to CNT/GC (0.48 V) or RuOx/GC (0.67 V) counterparts. The oxidation peak potentials were observed at 0.83 V, 0.6,7 V, and 0.65 V for the RuOx-, CNT, and RuOx/CNT-modified electrodes, respectively. The better response of RuOx/CNT-modified electrodes was attributed to a synergistic enhancement, rather than a combination of the individual coatings or surface area effects. The higher electrocatalytic activity for insulin was indicated from lowering the potential of the oxidation process (starting around 0.35 vs. Ag/AgCl). This system offers a wide linear dynamic range (10–800 nM) with a detection limit of 1 nM [53].

8.3.4.3 CNT-Based Microorganism Detection

Food- and water-borne diseases are a significant concern for developing countries. The CNT-based sensors are an exciting field of research globally for rapid, accurate, and specific identification of pathogens. *Salmonella typhi* is a causative agent for typhoid disease and spreads through contaminated water or food. Guillen and coworkers [54] have developed a CNT-based aptasensor for the detection of *S. typhi*. A self-assembled monolayer (SAM) was created over GC electrode with carboxylic group functionalized SWCNT. The amine-modified *S. typhi* specific-aptamer was covalently linked with carboxylic group functionalized SWCNT through EDC/NHS chemistry over the electrode. The SAM formed between SWCNT and aptamer facilitated л-л stacking between them. The aptamer/SWCNT/GC electrode performed as a sensing probe for the detection of *S. typhi*. The interaction with the target microbes leads to a change in the conformation of the aptamer, separating the ionized phosphodiester group of aptamers from the SWCNT surface. The consequence is the change of surface charge of SWCNT and subsequent alteration in recorded potential. This sensor could detect 0.2 CFU in real time with a sensitivity of 1.87 mV/10 microbial cells. This system is specific, can differentiate between closely related species present in a sample.

8.3.4.4 CNT-Based Small Molecule Detection

8.3.4.4.1 Detection of Toxic Chemicals

There is a growing demand for low-cost, rapid, and reliable sensors for the detection and monitoring of hazardous and toxic chemical agents. The sensors developed following the chemoresisitive principle find practical applications in these environmental and health monitoring sectors. The electrical conductance is considered the sensor response. SWCNTs are broadly used for chemoresponsive sensors because of their important properties such as semiconducting, high thermal and chemical stability, humidity tolerance, and compatibility with solvent-mediated processes. The pristine SWCNT, however, does not provide intrinsic selectivity against a specific analyte molecule. Hence, functionalization or modification of

SWCNT that enables selective responses has been explored. The SWCNT can be functionalized through covalent or noncovalent modification methods. The drawback associated with covalent modification is that it disrupts the π-system, thereby adversely affecting the sensitivity by limiting the range of detection of the target analyte. On the other hand, the noncovalent interaction is weak but less influential over the π- electron system of SWCNT. Ishihara and coworkers [55] developed a chemoresistive sensor by wrapping SWCNT in chemically active metallo supramolecular polymer (MSP) made from anthracene-based ligand (Cu^{+2} or Ni^{+2}). In the presence of target electrophile, analytes Diethychloro Phosphate (DECP) or Thionyl chloride ($SOCl_2$) show concentration-dependent disassembly of MSP from the surface of SWCNT. A consequence of this is a change in conductivity response. The interaction of target with MSP-SWCNT influences the response defined as G^1-G^0/G^0 × 100 (%), where G^0 is initial, and G^1 is final conductance. These findings revealed that wrapping of SWCNTs with chemically degradable polymers is a potent strategy for the development of advanced chemo resistive sensors.

8.3.4.4.2 Adenosine 5′-Triphosphate (ATP) Sensing

ATP is the universal energy currency of the cell in all organisms. The sensing of the ATP level in cells gives useful information about the intercellular signaling cascade, ion channel regulation, metabolic state of cells, and so on. The optical activity of SWCNT in NIR has been reported to be used for the detection of ATP. The SWCNT themselevs do not have specificity for the detection of ATP or ADP. In order to introduce the specific optical response for ATP, SWCNT is wrapped with a phopsholipid-polyethyl glycol (PEG) polymer with luciferase enzyme immoblized over it. In the presence of D-luciferin and ATP substrates, the enzyme luciferase catalyzes the formation of oxyluciferin and AMP. The production of oxyluciferin concurrently quenches the fluoresence of SWCNT, providing an indirect indicator of ATP concentration.The advantage of the SWCNT-based sensor is high photostability, as well as optical activity in NIR region, leading to the minimum autofluoresence signal from the background. Further, the immoblization of luciferase enzyme over a ploymer wrapped SWCNT imparts selectivity to the sensor without significantly affecting the activity of luciferase enzyme. The luciferase/PEG/SWCNT complex is selective for ATP, even in the presence of potentially interfering analogue agents, including GTP, ADP, CTP, and AMP [56]

SWCNT $\longrightarrow$ Fluorescence in NIR region

D Luciferin + ATP $\longrightarrow$ Oxyluciferin + AMP

SWCNT $\xrightarrow{\text{Oxyluciferin}}$ Fluorecence quenching in NIR region

8.3.4.4.3 Reactive Oxygen Species (ROS) Detection

ROS plays an important role in cell defense mechanisms. However, anything higher than the optimum concentration of ROS imparts a negative effect on the cell microenvironment. Aberration of its level is an indication of inflammation, a pathogenic condition of cells, and more. ROS is a collective term referring to oxygen-derived species, including superoxide anion radical ($O_2^{\bullet-}$) and hydrogen peroxide (H_2O_2). Jin and coworkers [57] developed a sensor for the detection of ROS, using SWCNT wrapped with type 1 collagen fiber. Each collagen wrapped SWCNT acted as an individual optical sensing probe for the detection of ROS. The H_2O_2 absorbs over the surface of SWCNT, leading to fluorescence quenching. By using hidden Markov modeling, the authors reported an escalation in the forwarding (adsorption) rate constant as H_2O_2 concentration increased, while the reverse (desorption) rate constant was unaffected by the concentration. The application of a sensing array was designed to measure the rate of H_2O_2 in real time from cultured human carcinoma epidermal A432 cells, and it was observed that epidermal growth factor (EGF) produces 2 nmol H_2O_2 for 50 min in A431 cells, but not in liver 3T3 cells. Through this array, a new mechanism was put forward under which oxygen was first converted to superoxide by NAD (P)H oxidase and then converted to singlet oxygen through a superoxide dismutase. The latter forms H_2O_2 upon binding of EGF to its receptor (EGFR). This platform offers a quantitative way of monitoring cellular production of ROS at a single-cell level in real time.

The concentration of ROS in the internal environment of the cell can be determined through the electrocatalytic method. Rawson and coworkers [58] have found that the ROS level inside the cell is dependent on NADPH oxidase (NOx) and toll-like receptor 4 (TLR4), which are influenced by bacterial endotoxin (lipopolysaccharide-LPS) stimulus. The self-assembled monolayer is formed over ITO (indium tin oxide) electrode with carboxylated SWCNT through carbodiimide chemistry. Further, the SWCNT/ITO sheet can be modified with osmium bipyridine (Osbpy) through carbodiimide chemistry. The Osbpy is biocompatible and can be used as biocatalyst. In the presence of H_2O_2 there was a change in the CV of ITO/SWCNT/Osbpy with an increase in oxidation and reduction peak currents. Moreover, a 30 mV peak shift to the less negative side for the reduction peak and a 60 mV shift of the oxidation peak to a positive direction were observed in the presence of H_2O_2. In this process, the redox behavior of Os (II)/Os (III) was involved.

8.4 GRAPHENE

The study on graphene could be traced back to 1859. Scientists knew about the existence of a one-atom thick, two-dimensional crystal sheet structure of graphene. However, no one had the drive to isolate it from graphite. For the first time, in 2004, two scientists from the University of Manchester, Sir Andre Geim and Sir Kostya Novoselow, isolated a single atomic layer of graphene. The graphene is an allotrope of carbon with six carbons attached in a hexagonal shape in a single layer. The carbon atom has six electrons out of which two are in the inner shell and four are present in the outer shell, which are available for bonding. But in graphene three out of four electrons are involved in covalent bond formation with three other carbon atoms in the two-dimensional hexagonal lattice plane, leaving behind one free electron in the space for conduction. These free electrons are highly mobile in nature and are termed pi electrons (π) and are present as an electron cloud above and below the graphene sheet. These pi orbitals overlap with each other and support the carbon-to-carbon bonds in graphene. Fundamentally, the bonding and antibonding (valence and conduction band) state of these pi orbitals is determined by the electronic properties of graphene. Graphene is theoretically a nonmetal but is generally denoted as a quasi-metal because of its semiconducting metal-like properties, and it does not have space between the conduction and valence band in which electrons act as massless realistic particles. The conduction and valence bands in graphene are conical valleys, attached with each other in high symmetry at points of the Brilluoin zone (an exclusively defined primitive cell in the reciprocal space). The energy near these points follows linear dispersion relation, that is, the energy change is in linear relationship with the magnitude of momentum (vector). As a consequence of this, electrons behave as quasi-relativistic particles described by the originally proposed Dirac equation shown here. It is an electron wave function equation derived by British physicist Paul Dirac in 1928.

$$\left(\beta mc^2 + \sum_{k=1}^{3} a_k p_k c\right) \psi(x,t) = i\hbar \frac{\delta \psi(x,t)}{\delta t}$$

where $\psi = \psi(x, t)$ is the wave function of the electron, of the mass in rest (m) with space-time coordinates x, t. The components of the momentum (p_k) are understood to be the momentum operator in the Schrödinger equation. Also, c is the speed of light and $\hbar$ is the reduced Planck constant. The α and β are Hermitian constants.

The massless Dirac electrons in graphene make it a material with unique characteristics. The electron speed in graphene is 10^6 m/s, about 300 times less than the speed of light. These electrons are not influenced by external electrostatic potentials displaying the Klein paradox and cause the integer quantum hall effect, through which the chance of electrons crossing a potential barrier is always unity. The graphene has high electrical conductivity (~1.0×10^8 S/m), high melting point (4510 K), high thermal conductivity (2000–4000 W m^{-1}K^{-1}), and high current density (~1.6×10^9 A/cm^2), including a high electron mobility (200,000 cm^{2}V^{-1} s^{-1} at electron density ~2×10^{11} cm^{-2}) thus contributing to its application in electrochemical, strain, and electrical sensors [59].

Graphene is incredibly strong – 200 times stronger than steel of the same dimensions – and possesses fascinating optical properties. Graphene-based nanomaterials have great potential for sensor applications. Because of its unique morphology, in which the carbon atom directly interacts with the sensing environment and alters the properties of graphene, along with its other unique physical properties, as discussed elsewhere, makes it possible to use in various types of sensing applications.

8.4.1 Synthesis of Graphene

Graphene is produced mostly from graphite in several ways. Among the different methods, exfoliation and sonication of graphite, electric arc discharge, chemical vapor deposition, reduction of graphene oxide, and reduction of carbon dioxide are well documented. The mechanical exfoliation approach is not very suitable for producing a large quantity of graphene. On the other hand, the chemical vapor deposition method can produce large areas of single-layer graphene in bulk amounts. The reduction of graphene oxide through chemical or thermal treatment is another efficient approach for mass production at a low cost. Moreover, many reports suggest that graphene production through the chemical redox reaction may lead to different kinds of structural defects that are advantageous for electrochemical applications. For the sensing applications, two different forms of graphene, that is, graphene oxide (GO) and reduced graphene (rGO), are widely used. GO is a monolayer of graphene containing different oxygen-containing functional groups such as alcohol, epoxide, and carboxylic acid. When these surface oxygen-containing functional groups are reduced to eliminate oxygen functionalities, rGO is formed. As expected, reduced graphene is less hydrophilic and exhibits less conductivity compared to pristine graphene. The reason is that some oxygenated functional groups are eliminated, leading to the incomplete reduction step and disrupting the sp^2 network. The differences between GO and rGO present various useful properties that can be suitable for sensing applications.

8.4.2 Application of Graphene

8.4.2.1 Graphene-Based Chemical and Electrochemical Sensors

The graphene is a very thin 2D material in which each carbon atom is exposed to the external environment and offers excellent surface-to-volume ratio compared to any other material. This indicates that every carbon atom is a potential target for reactive species and the strength of the interaction depends on the extent of covalent and Van der Waals forces, which finally influence the pristine electrical and chemical properties of graphene. This can be considered a sensing response for the detection of such binding events. The graphene-based gas sensor has been developed in which the interaction of gas molecules with a graphene probe alters the conductivity by adding and removing electrons. In other words, after interaction of graphene with a gas molecule, it is adsorbed over the graphene surface and undergoes weak hybridization and electron or hole coupling, which leads to alteration of the Fermi level and thus changes the graphene conductivity. Such a change in conductivity signal is measured for sensitive detection of gas molecules. This sensitivity response can be extended up to detection of a single molecule (i.e., the smallest alteration in quantum as a resultant conduction can be measured in graphene). The high sensitivity response from graphene is attributed to the extremely low-noise background signal, and even adding or removing a few extra electrons can cause a noticeable change in the carrier concentration or vice versa. The exposure of graphene to various gases such as NH_3 (ammonia) and CO (carbon monoxide) dopes the graphene with electrons, while others such as NO_2 (nitrogen oxide) and H_2O (water) dope it with holes at different degrees for different gases [60]. This leads to a change in the conductivity of graphene, which means that if the system is properly calibrated, it can be used for specific detection of various gases.

In the electrochemical approach, the CV technique of graphene with redox indicators $[Fe(CN)_6]^-$, $[Ru(NH_3)_6]^{3+/4+}$ exhibits well-defined redox peaks, which is an indication of suitable electron transfer behavior of graphene. In CV, both the cathodic and anodic peak current are directly proportional to the square root of the scan rate, indicating that the redox process in graphene is diffusion-controlled [61]. The peak-to-peak potential separation (ΔEp) value in CV with single electron redox dye is close to the ideal value of 59 mV, indicating a rapid and reversible electron transfer process, which makes graphene an ideal electrode for electrochemical studies.

Graphene is also used as a chemical sensor in an electrolyte gated configuration. In an electrolyte, an insulator electrical double layer is formed, and the thickness of this layer is defined as the Debye length. Any interaction taking place beyond the Debye length cannot be sensed by the graphene gate surface of the field-effect transistor (GFET). This thickness depends on the ionic concentration of electrolytes. Upon the addition of electrolyte over graphene, an electrostatic field is formed, depending on the ionic concentration of electrolyte. Any change in ionic concentration of electrolyte leads to a change in electrostatic potential and will induce variation in current density in the graphene, hence ultimately changing the conductivity. This behavior indicates that graphene FETs can detect the pH value by the electrical characteristics using electrolyte gating.

Graphene is emerging as an excellent conductive nanomaterial to support direct electrochemistry (DET) of the redox enzymes without the involvement of any electron transfer mediators (ETMs). However, DET is a challenging task because most of the enzymes' redox center is buried inside the protein matrix. Of note, the DET strategy is highly desirable for developing enzyme-based biosensors. A DET-based glucose biosensor using glucose oxidase (GOD) on the oxidized graphene (GO) was developed [62]. The CV study revealed well-defined redox peaks of FAD on the graphene-GOD bioelectrode, as depicted in Figure 8.8. Moreover, a higher GOD loading efficiency (1.12×10^9 mol/cm^2) was also observed with the graphene electrode than any other electrode. The high enzyme loading has been attributed to the high surface area of graphene nanomaterials, which leads to enhanced sensitivity of the sensor.

8.4.2.2 Graphene-Based Optical Sensor

Graphene acts as an excellent quencher for luminance due to the presence of high surface electron density. In addition to that, GO shows intrinsic fluorescence properties. The NIR, visible, and UV fluorescence are observed from rGO and GO compared to pristine graphene, which is non-luminescent due to the presence of overlapping conductance and valence band or zero bandgap [63, 64]. In GO, fluorescence arises from electron-hole pair recombination in localized electronic states generating from various possible configurations, instead of band-edge transition, just like typical semiconductors. The gold nanoparticles (AuNPs) are an excellent fluorescence quencher, and in conjugation with AuNPs, GO quenches the intrinsic fluorescence of GO. A FRET-based oligonucleotide sensor was developed [65] in which the GO fluorescence quenching signal in the presence of AuNP was exploited to detect the complementary oligonucleotide strands. A specific antigen-antibody interaction of a GO-based immunosensor for parasite detection has also been developed [66]. The specific antibodies against rotavirus were covalently linked with GO through an amidation process. When AuNPs were linked to GO by capture of a target cell, a fall in the fluorescence emission of GO was detected, thus enabling the detection of pathogenic target cells. The multiple steps of chemical modification and a complex labeling process may, however, limit the application of the fluorescence property of GO in optical sensors.

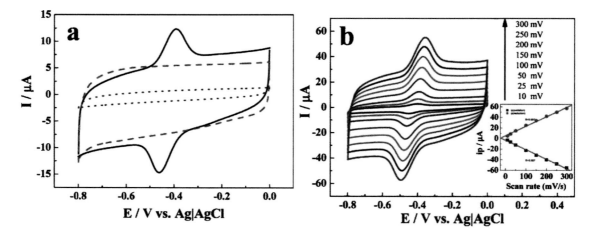

FIGURE 8.8 (A) CV (scan rate 50 mV/s) of graphene (dashed), graphite-GOD (dotted), and graphene-GOD (solid) modified electrodes in N_2-saturated 0.05 M PBS solution (pH 7.4). (B) Cyclic voltammograms at various scan rates on graphene-GOD-modified electrodes (Inset: plot of peak current (ip) vs. scan rate) [62].

8.5 CONCLUSION AND FUTURE PROSPECTS

We witnessed the emergence of diverse carbon-based nanomaterials, their critical properties, and the application potential in various fields. The optical properties of carbon dots and electrical, electrochemical, and optical properties of graphene-based nanomaterials (graphene and different CNTs) are astounding. These unique or superior properties are central for developing various sensors and biosensors with better detection strategies and performance for the detection of diverse target analytes for the industrial, environmental, biomedical, agricultural, and allied fields of interest. With regard to the development of biosensors, a proper and strategic coupling of these properties with the biorecognition elements/biocatalysts is vital to initiate or generate the desired signal for practical use. A prudent approach is vital for facile signal transduction and stabilization of the functional properties of the biorecognition elements following their integration with the sensor surface. We expect an accelerated growth of the biosensor market with carbon-based nanomaterials as a major component of these advance analytical devices in the near future.

REFERENCES

1. Xiaoyou Xu, Robert R, Yunlong Gu, Harry JH, Latha G, Raker G. Electrophoretic analysis and purification of fluorescent single-walled carbon nanotube fragments. J Am Chem Soc. 2004;126(40):12736–37.
2. Tuerhong M, XU Y, YIN XB. Review on carbon dots and their applications. Chinese J Anal Chem. 2017;45(1):139–50.
3. Bao L, Zhang ZL, Tian ZQ, Zhang L, Liu C, Lin Y, et al. Electrochemical tuning of luminescent carbon nanodots: from preparation to luminescence mechanism. Adv Mater. 2011;23(48):5801–6.
4. Du Y, Guo S. Chemically doped fluorescent carbon and graphene quantum dots for bioimaging, sensor, catalytic and photoelectronic applications. Nanoscale. 2016;8(5):2532–43.
5. Li H, He X, Kang Z, Huang H, Liu Y, Liu J, et al. Water-soluble fluorescent carbon quantum dots and photocatalyst design. Angew Chemie Int Ed. 2010;49(26):4430–4.
6. Jin SH, Kim DH, Jun GH, Hong SH, Jeon S. Tuning the photoluminescence of graphene quantum dots through the charge transfer effect of functional groups. ACS Nano. 2013;7(2):1239–45.
7. Lin Z, Xue W, Chen H, Lin JM. Classical oxidant induced chemiluminescence of fluorescent carbon dots. Chem Commun. 2012;48:1051.
8. Xu Y, Wu M, Feng XZ, Yin XB, He XW, Zhang YK. Reduced carbon dots versus oxidized carbon dots: photo- and electrochemiluminescence investigations for selected applications. Chem – A Eur J. 2013;19(20):6282–8.
9. Jin H, Gui R, Sun J, Wang, Y. Facilely self-assembled magnetic nanoparticles/aptamer/carbon dots nanocomposites for highly sensitive up-conversion fluorescence turn-on detection of tetrodotoxin. Talanta. 2018;176:277–83.
10. Wen X, Yu P, Toh YR, Ma X, Tang J. On the upconversion fluorescence in carbon nanodots and graphene quantum dots. Chem Commun. 2014;50(36):4703–6.
11. Chakma B, Jain P, Singh NK, Goswami P. Development of an indicator displacement based detection of malaria targeting HRP-II as biomarker for application in point-of-care settings. Anal Chem. 2016;88(20):10316–21.
12. Yuan C, Liu B, Liu F, Han M-Y, Zhang Z. Fluorescence "turn on" detection of mercuric ion based on bis(dithiocarbamato) copper(II) complex functionalized carbon nanodots. Anal Chem. 2014;86(2):1123–30.
13. Jiang, Y., Wang, Z, Dai Z. Preparation of silicon–carbon-based dots@ dopamine and its application in intracellular Ag+ detection and cell imaging. ACS Appl Mater Inter. 2015;8(6):3644–50.
14. Xue M, Zhang L, Zou M, Lan C, Zhan Z, Zhao S. Nitrogen and sulfur co-doped carbon dots: A facile and green fluorescence probe for free chlorine. Sensors Actuators B Chem. 2015;219:50–6.
15. Mei D, Ke W, Lin L, Huang Z. Carbon dots-involved chemiluminescence: Recent advances and developments. Luminescence, 2019;34(1):4–22.
16. Nandi S, Ritenberg M, Jelinek R. Bacterial detection with amphiphilic carbon dots. Analyst. 2015;140(12):4232–7.
17. Huan XW, Bao YW, Wang HY, Chen Z, Wu FG. Bacteria-derived fluorescent carbon dots for microbial live/dead differentiation. Nanoscale. 2017;9(6):2150–61.
18. Wang R, Xu Y, Zhang T, Jiang Y. Rapid and sensitive detection of Salmonella typhimurium using aptamer-conjugated carbon dots as fluorescence probe. Anal Methods. 2015;7:1701–06.
19. Barati A, Shamsipur M, Abdollahi H. Hemoglobin detection using carbon dots as a fluorescence probe. Biosens Bioelectron. 2015;71:470–5.
20. Qian Z, Chai L, Tang C, Huang Y, Chen J, Feng H. Carbon quantum dots-based recyclable real-time fluorescence assay for alkaline phosphatase with adenosine triphosphate as substrate. Anal Chem. 2015;87(5):2966–73.
21. Kumari S, Solanki A, Mandal S, Subramanyam D, Das P. Creation of Linear Carbon Dot Array with Improved Optical Properties through Controlled Covalent Conjugation with DNA. Bioconjug Chem. 2018;29(5):1500–4.
22. Wang Y, Wang S, Ge S, Wang S, Yan M, Zang D, et al. Facile and sensitive paper-based chemiluminescence DNA biosensor using carbon dots dotted nanoporous gold signal amplification label. Anal Methods. 2013;5(5):1328.
23. Liu C, Zhang P, Tian F, Li W, Li F, Liu W. One-step synthesis of surface passivated carbon nanodots by microwave assisted pyrolysis for enhanced multicolor photoluminescence and bioimaging. J Mater Chem. 2011;21(35):13163.
24. Feng L, Zhao A, Ren J, Qu X. Lighting up left-handed Z-DNA: photoluminescent carbon dots induce DNA B to Z transition and perform DNA logic operations. Nucleic Acids Res. 2013;41(16):7987–96.
25. Qin X, Asiri AM, Alamry KA, Al-Youbi AO, Sun X. Carbon nitride dots can serve as an effective stabilizing agent for reduced graphene oxide and help in subsequent assembly with glucose oxidase into hybrids for glucose detection application. Electrochim Acta. 2013;95:260–7.
26. Song Y, Shi W, Chen W, Li X, Ma H. Fluorescent carbon nanodots conjugated with folic acid for distinguishing folate-receptor-positive cancer cells from normal cells. J Mater Chem. 2012;22(25):12568.
27. Singh NK, Chakma B, Jain P, Goswami P. Protein-induced fluorescence enhancement based detection of plasmodium falciparum glutamate dehydrogenase using carbon dot coupled specific aptamer. ACS Comb Sci. 2018;20(6):350–7.
28. Campos BB, Contreras-Cáceres R, Bandosz TJ, Jiménez-Jiménez J, Rodríguez-Castellón E, Esteves da Silva JCG, et al. Carbon dots as fluorescent sensor for detection of explosive nitrocompounds. Carbon NY. 2016;106:171–8.
29. Iijima, S. Helical microtubules of graphitic carbon. Nature. 1991;354:56–8.

30. Grobert, N. Carbon nanotubes – becoming clean. Materials Today. 2007;10(1–2):28–35.
31. Saito R, Takeya T, Kimura T, Dresselhaus G, Dresselhaus M.S. Raman intensity of single-wall carbon nanotubes. Phys. Rev. B. 1998;7:4145–8.
32. Guo T, Nikolaev P, Thess A, Colbert DT, Smalley R. Catalytic growth of single-walled nanotubes by laser vaporization. Chemical Physics Letters. 1995;243(1–2):49–54.
33. Eatemadi A, Daraee H, Karimkhanloo H, Kouhi M, Zarghami N, Akbarzadeh A, et al. Carbon nanotubes: properties, synthesis, purification, and medical applications. Nanoscale Res Lett. 2014;9(1):393.
34. José-Yacamán M, Yoshida MM, Rendón L. Catalytic growth of carbon microtubules with fullerene structure. Applied Physics Letters. 1993;62(6):657–9.
35. Kumar M, Ando Y. Chemical vapor deposition of carbon nanotubes: A review on growth mechanism and mass production. J Nanosci Nanotechnol. 2010;10(6):3739–58.
36. Chaudhary KT, Rizvi ZH, Bhatti K, Ali I, Yupapin PP. Multiwalled carbon nanotube synthesis using arc discharge with hydrocarbon as feedstock. J Nanomaterials. 2013, Article ID 105145.
37. Novoselov KS, Geim AK, Morozov, SV, Jiang D, Zhang Y, Dubonos SV, Grigorieva IV, Firsov AA. Electric field effect in atomically thin carbon films. Science. 2004;306:666–9.
38. MacKenzie, K., See, C., Dunens, O., Harris A. Do single-walled carbon nanotubes occur naturally?. Nature Nanotech. 2008;3:310.
39. Tison Y, Giusca CE, Stolojan V, Hayashi Y, Silva SRP. The inner shell influence on the electronic structure of double-walled carbon nanotubes. Adv Mater. 2008;20(1):189–94.
40. Shen C, Brozena AH, & Wang, Y. Double-walled carbon nanotubes: Challenges and opportunities. Nanoscale. 2010;3(2):503–18.
41. Luo, Y. R. (2007). Comprehensive handbook of chemical bond energies. CRC Press, London.
42. Treacy MJ, Ebbesen TW, Gibson JM. Exceptionally high Young's modulus observed for individual carbon nanotubes. Nature. 1996;381(6584):678.
43. Collins PG, Avouris P, The electronic properties of carbon nanotubes, Contemporary Concepts of Condensed Matter Science, Chapter 3, 3 (2008) 49–81.
44. Wang F, Dukovic G, Brus LE, Heinz TF. The optical resonances in carbon nanotubes arise from excitons. Science. 2005;308(5723):838–41.
45. Lüer L, Hoseinkhani S, Polli D, Crochet J, Hertel T, Lanzani G. Size and mobility of excitons in (6, 5) carbon nanotubes. Nat Phys. 2009;5(1):54–8.
46. Kruss S, Hilmer A, Zhang J, Reuel N, Mu B, Strano M. Carbon nanotubes as optical biomedical sensors. Advanced Drug Delivery Rev. 2013;65(15):1933–50.
47. Reich S., Thomsen C. Raman spectroscopy of graphite. Philos Trans A Math Phys Eng Sci. 2004;362(1824):2271–88.
48. Zhang Y, Chan HF, Leong KW. Advanced materials and processing for drug delivery: The past and the future. Adv Drug Deliv Rev. 2013;65(1):104–20.
49. Jeng ES, Moll AE, Roy, A. C., Gastala, J. B., Strano, M. S. Detection of DNA hybridization using the near-infrared bandgap fluorescence of single-walled carbon nanotubes. Nano Letters. 2006;6(3):371–75.
50. Lin YW, Ho HT, Huang CC, Chang HT. Fluorescence detection of single nucleotide polymorphisms using a universal molecular beacon. Nucleic Acids Res. 2008;36(19):e123.
51. Wang SG, Wang R, Sellin PJ, Zhang Q. DNA biosensors based on self-assembled carbon nanotubes. Biochem Biophys Res Commun. 2004;325(4):1433–7.
52. Boucetta H, Nunes A, Sainz R, Herrero MA, Tian B, Prato M, et al. Asbestos-like pathogenicity of long carbon nanotubes alleviated by chemical functionalization. Angew Chemie Int Ed. 2013;52(8):2274–8.
53. Wang J, Tangkuaram T, Loyprasert S, Vazquez-Alvarez T, Veerasai W, Kanatharana P, et al. Electrocatalytic detection of insulin at RuOx/carbon nanotube-modified carbon electrodes. Anal Chim Acta. 2007;581(1):1–6.
54. Guillén GA, Blondeau P, Rius FX, Riu J. Carbon nanotube-based aptasensors for the rapid and ultrasensitive detection of bacteria. Methods. 2013;63(3):233–8.
55. Ishihara S, Azzarelli JM, Krikorian M, Swager TM. Ultratrace detection of toxic chemicals: triggered disassembly of supramolecular nanotube wrappers. J Am Chem Soc. 2016;138(26):8221–7.
56. Kim JH, Ahn JH, Barone PW, Jin H, Zhang J, Heller DA, et al. A luciferase/single-walled carbon nanotube conjugate for near-infrared fluorescent detection of cellular ATP. Angew Chemie Int Ed. 2010;49(8):1456–9.
57. Jin H, Heller DA, Kalbacova M, Kim JH, Zhang J, Boghossian AA, Strano MS. Detection of single-molecule H_2O_2 signalling from epidermal growth factor receptor using fluorescent single-walled carbon nanotubes. Nature Nanotechno. 2010;5(4):302.
58. Rawson FJ, Cole MT, Hicks JM, Aylott JW, Milne WI, Collins CM, et al. Electrochemical communication with the inside of cells using micro-patterned vertical carbon nanofibre electrodes. Sci Rep] 2016;6(1):37672.
59. Ohta, T, Bostwick, A, Seyller, T., Horn, K., Rotenberg, E. Controlling the electronic structure of bilayer graphene. Science. 2006;313(5789):951–4.
60. Schedin F, Geim AK, Morozov SV, Hill EW, Blake P, Katsnelson MI, et al. Detection of individual gas molecules adsorbed on graphene. Nat Mater. 2007;6(9):652–5.
61. Lin WJ, Liao CS, Jhang JH, Tsai YC. Graphene modified basal and edge plane pyrolytic graphite electrodes for electrocatalytic oxidation of hydrogen peroxide and β-nicotinamide adenine dinucleotide. Electrochem Commun. 2009;11(11):2153–6.
62. Shan C, Yang H, Song J, Han D, Ivaska A, Niu L. Direct electrochemistry of glucose oxidase and biosensing for glucose based on graphene. Anal Chem. 2009;81(6):2378–82.
63. Sun X, Liu Z, Welsher K, Robinson JT, Goodwin A, Zaric S, et al. Nano-graphene oxide for cellular imaging and drug delivery. Nano Res. 2008;1(3):203–12.
64. Eda G, Lin YY, Mattevi C, Yamaguchi H, Chen HA, Chen IS, et al. Blue photoluminescence from chemically derived graphene oxide. Adv Mater. 2010;22(4):505–9.
65. Liu F, Choi JY, Seo TS. Graphene oxide arrays for detecting specific DNA hybridization by fluorescence resonance energy transfer. Biosens Bioelectron. 2010;25(10):2361–65.
66. Jung JH, Cheon DS, Liu F, Lee KB, Seo TS. A graphene oxide based immuno-biosensor for pathogen detection. Angew Chemie Int Ed. 2010;49(33):5708–11.

9 Photoelectrochemical and Photosynthetic Material-Based Biosensors

Mohd Golam Abdul Quadir and Pranab Goswami
Indian Institute of Technology Guwahati, Assam, India

Mrinal Kumar Sarma
The Energy and Resources Institute, New Delhi, India

CONTENTS

9.1 Introduction 183
9.2 Photoelectrochemistry: A Chronology of Landmark Discovery 183
9.3 Basics of Photoelectrochemistry 184
9.4 Principle of Photoelectrochemical Biosensing 186
9.5 Building Blocks of PEC Biosensors 186
    9.5.1 Biological Components 186
    9.5.2 Photoelectrochemical Transducers 189
9.6 Signal Amplification Strategies 194
9.7 Visible Light- and Solar-Powered PEC Sensor: Recent Reports 194
9.8 Photosynthesis: Molecular Machinery and its Components 196
9.9 Reaction Centers as Components of a Photo-Electrode 196
    9.9.1 Z Scheme Pathway Mimics and the Electrosynthesis System 204
9.10 Conclusion 207
References 207

9.1 INTRODUCTION

A biosensing device must fulfill the fundamental performance criteria of high sensitivity, selectivity, and quick response time, among others, for its practical utility [1]. These performance factors are governed mainly by the biorecognition elements in conjunction with the transducers being used to construct the biosensors. Among the various transducers, the electrochemical and optical transducers are at the forefront in biosensor research due to their high sensitivity, as acclaimed in a vast array of literature. Integrating electrochemical transducers and light-responsive recognition elements into biosensors to function through the principle of photoelectrochemistry could be an effective and inexpensive strategy to develop self-powered sensing devices. Such sensors offer the flexibility of using the natural light sources to activate the recognition elements and transduce the signal through electrochemical platforms imparting high sensitivity and signal-to-noise ratio to the developed device [2]. The strategy can also alleviate the cost of using expensive image processing instruments, as the detection would be based on the current or voltage signal obtained by using the conventional amperometric principle.

Photoelectrochemistry is the phenomenon of incident photons to current conversion as a result of electron-hole pair formation in a photocatalytic material [3]. The electron flux between an electrode, photocatalytic material, and analyte is registered as an amperometric signal. The electron-hole pair can also provide energy for overcoming the activation energy barrier of chemical reactions at the electrode-medium interface. Semiconductor-based materials are most favored in this field due to their excellent photocatalytic property. Discovery of novel photoactive composites, their bandgap engineering, and developing novel strategies for sensing applications is an actively proliferating area of research [4]. Isolation and integration of photosensitive biological proteins like photosynthetic reaction centers into energy harvesting devices is another promising area of research [5]. The promise of such an approach is the abundance of photosynthetic materials in nature, which, if successfully stabilized and optimized into biosensing platforms, could develop into a cost-effective approach for solar-to-energy and solar-to-chemical conversion. These approaches, however, come with their own set of challenges. This chapter describes the principles behind the aforementioned distinct, but fundamentally similar, technological approaches.

9.2 PHOTOELECTROCHEMISTRY: A CHRONOLOGY OF LANDMARK DISCOVERY

French scientist Edmond Becquerel first reported the pioneering discovery of light to current conversion in 1839. His experiments were based on two dissimilar metal electrodes coated with silver halogenide in acidic, neutral, and alkaline electrolytes. Exposure to sunlight caused the flow of electrons observed in a galvanometer. This discovery initiated the field of photoelectrochemistry, and many scientific investigations of the photocurrent at metal-electrolyte interfaces were reported in subsequent years [6]. In 1955, investigation of photoelectrochemistry at the p and n doped germanium-electrolyte interface by Garrett and Brattain revealed anodic and cathodic currents as a function of incident light intensity [7]. This study won the Nobel Prize for 1955 and marked a transition from metal- to semiconductor-based electrodes in photoelectrochemistry research. Many materials with semiconducting properties were investigated in subsequent years, which included the III-V group semiconductors, chalcogenides, and metal oxides. Many of these materials had the problem of etching (self-destruction) during photoelectrochemical reactions. In 1969, A. Fujishima, from the University of Tokyo, discovered titanium dioxide as a semiconducting material that did not undergo etching. In 1972 that group reported on water splitting on TiO_2 electrodes exposed to light [8]. The work attracted worldwide attention and is considered a seminal report in the field of photocatalysis. In a subsequent report, the group formed thin films of TiO_2 and evaluated them for photoelectrochemical water splitting in an attempt to reduce the cost of technology [9].

9.3 BASICS OF PHOTOELECTROCHEMISTRY

Semiconducting materials are the most suitable class of materials for photoelectrochemical and photovoltaic applications. The basis of this property lies in the lattice structure of these solids. A schematic diagram of energy bands corresponding to each orbital is used to represent the energy level of atomic orbitals in an atom. According to the molecular orbital theory, when "n" number of atoms combine to form a molecule, each atomic orbital recombines and splits to form "n" number of corresponding molecular orbitals. When infinite atoms (~Avogadro number) covalently bond to form a monocrystalline solid, each atomic orbital splits to the Avogadro number of molecular orbitals, which are close enough to form a continuum of energy levels, giving the appearance of bands with definite edges. The energy bands corresponding to the highest-occupied molecular orbital (HOMO) and lowest-unoccupied molecular orbital (LUMO) are called valence band and conduction band, respectively (Figure 9.1a). The difference between the upper edge of the valence band and the lower edge of the conduction band is called bandgap energy (E_{bg}) and is the basis of classification of solids into metals, insulators, and semiconductors. For a metallic solid, orbitals with an energy level between that of HOMO and LUMO do exist, bridging the gap, thus leading to free movement of charge carriers (electrons) to unoccupied orbitals at room temperature. This free movement of mobile electrons in the lattice network of a metallic solid is the cause of electrical conductivity. However, semiconductors and insulators are characterized by a void of energy levels between the HOMO and LUMO called the energy bandgap (E_{bg}), forbidden to electrons. For a semiconductor and an insulator, this difference is less than 3 eV and greater than 3 eV, respectively (Figure 9.1b). A supply of energy equivalent to or greater than this bandgap difference through incident photons or thermal energy leads to electronic transition to the conduction band, and a corresponding amperometric signal can be detected. The energy of electronic transition across the bandgap can also be used to drive a redox reaction at the electrode interface. The high bandgap of insulators makes them unsuitable for practical applications of photoelectrochemistry, as extremely high energy photons would be required to facilitate electronic transitions [10].

Unlike metals, where mobile electrons in the lattice are the exclusive cause of conductivity, positively charged mobile holes also contribute to the electrical conductivity of semiconductors. In a true physical sense, a hole is just a space in a lattice devoid of electrons. As an adjacent electron annihilates the hole, a corresponding hole is generated in the space

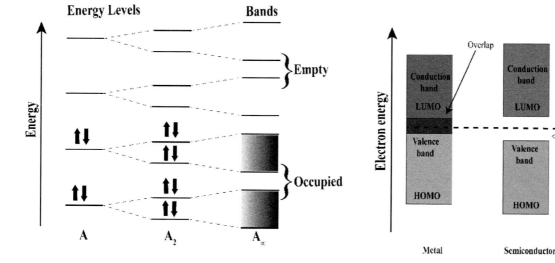

FIGURE 9.1 (a) Representation of orbital energy levels. Reprinted (adapted) with permission from [10]. Copyright (1983) American Chemical Society. (b) Energy bandgap consequence on conductivity

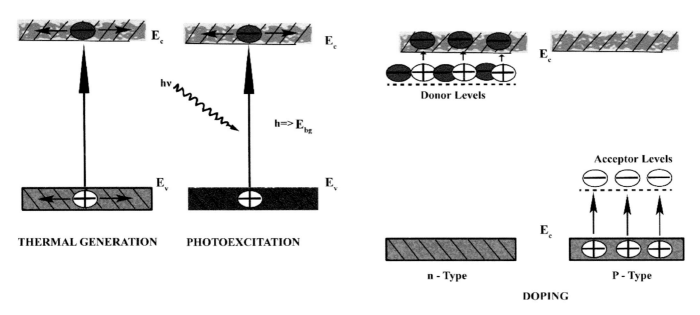

FIGURE 9.2 Energy level of electrons in a semiconductor can be influenced by thermal energy, light exposure, and doping. Reprinted (adapted) with permission from [10]. Copyright (1983) American Chemical Society.

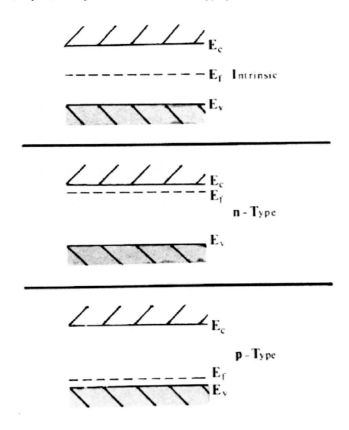

FIGURE 9.3 Representation of Fermi energy level and consequence of doping on its position. Reprinted (adapted) with permission from [10]. Copyright (1983) American Chemical Society.

previously occupied by the electron. In effect, the shifting positively charged space in the medium acts as a mobile charge carrier, which is the essence of its conductivity. For a semiconducting metal, the bandgap energy difference can be tuned by doping. Doping is the process of introducing impurities in the lattice structure of a pure semiconductor that can accept or donate electrons. For example, the introduction of Group III or Group IV element atoms into a semiconductor lattice can cause either holes or electrons to become predominant charge carriers. Doped semiconductors with a predominance of electrons and holes are called n-type and p-type semiconductors, respectively. The dopant entity, being a part of lattice network, does not act as a mobile charge carrier, but rather has a consequence on the energy level of mobile charge carriers (Figure 9.2).

To develop a better understanding of the consequence of semiconductor doping, we introduce a discussion on the Fermi energy level. This concept is based upon the mathematical Fermi function, which gives us the probability of an electron occupying an energy level. In a solid, at a given temperature, a collection of high-energy electrons can be represented by a Fermi energy level. The probability of the presence of an electron at this level is ½. Beyond this, the probability drastically falls to zero. In a solid, supply of energy to electrons in the form of thermal energy or electrochemical potential causes the Fermi energy level to rise in terms of representation. In an undoped semiconductor, this hypothetical Fermi energy level is placed halfway between the upper edge of the valence band and the lower edge of the conduction band. Doping an intrinsic semiconductor with an electron-dense or electron-deficient element can cause the Fermi energy level to drift toward the conduction band or valence band, respectively. Beyond a certain concentration of dopant, the Fermi level merges with the band edges and behaves like a metal. Thus, the conductivity of semiconductors can be tuned by varying the concentration of dopants in the lattice structure. This process is used for tuning the bandgap of an intrinsic semiconductor material and making them useful for specific applications [10] (Figure 9.3).

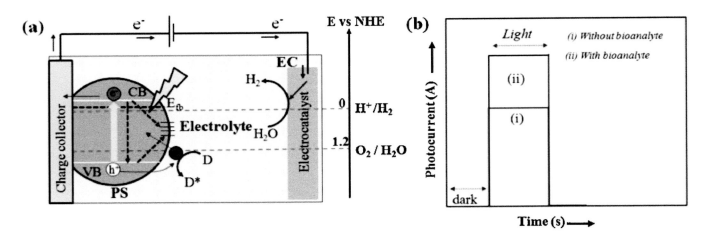

FIGURE 9.4 Principle of photoelectrochemical biosensors. (a) Schematics of electron-hole pair separation and electron. (b) Change in amperometric signal due to analyte detection. Reprinted from [4]. Copyright (2015) with permission from Elsevier.

9.4 PRINCIPLE OF PHOTOELECTROCHEMICAL BIOSENSING

A basic photoelectrochemical cell (PEC) has three fundamental components, which include (i) a light-responsive semiconducting material assembled on an electrode, which is the semiconductor-based photoelectrode; (ii) a counter electrode to receive the photogenerated electrons emanating from the conduction band of the photoelectrode; and (iii) a compatible electrolyte for the analyte of interest. Upon exposure to photons of energy equivalent to the bandgap of semiconducting material, the electrons in the valence band of the semiconductor are excited to the conduction band. This phenomenon is called electron-hole pair (exciton) generation (Figure 9.4a). Under normal conditions, in a semiconducting material, this exciton recombines and the energy is dissipated as thermal energy. However, in a PEC the oppositely charged mobile charge carriers are resolved across the photoanode and cathode. The electron flux is reflected as an amperometric signal, and holes generated in the valence band possess the kinetic energy to drive an oxidative reaction of a chemical species in the electrolyte. This oxidative chemical reaction enhances the amperometric signal and is the basis for quantitative estimation of specific analytes in complex samples (Figure 9.4b).

The landmark discovery of photocatalytic water splitting by Fujishima and Honda in 1972 used TiO_2 as the photocatalytic material that generated bandgap energy high enough to split water molecules [8]. The stoichiometry of the water splitting reaction is $2H_2O + 4h^+$ [holes] $\rightarrow 4H^+ + O_2$ [E_0 ox = -1.2529 V vs. NHE]. The electrochemical potential required to drive this reaction is 1.23 volts, which is equivalent to 237 kJ/mol of thermodynamic Gibbs free energy. For an oxidative reaction of an analyte to proceed at a semiconductor electrode, the oxidation potential of that analyte must be less than that of the photogenerated hole. This technical requirement is also the basis for the sensitivity of a photoelectrochemical biosensor. Another factor that determines the suitability of a semiconducting material for biosensor application is its visible light responsiveness, that is, employing photomaterials with the optimum bandgap energy difference that makes them responsive to visible light. This can enable miniaturization and simplifies the design of the sensing platform, as researchers can do away with a calibrated light source. Thus, the choice of a photoresponsive material with an optimum bandgap energy difference and knowledge of bandgap engineering is crucial for developing a sensitive photoelectrochemical biosensor [4]. Most of the PEC sensors are based on the working principle that affects the light-dependent amperometric signal by enhancing or attenuating it. We will discuss these strategies and give illustrative examples of research papers in the sections that follow.

9.5 BUILDING BLOCKS OF PEC BIOSENSORS

9.5.1 Biological Components

Based on the type of biorecognition element of a photoelectrochemical biosensor, it can be classified into one of the following categories: (i) enzymatic photoelectrochemical sensors, (ii) nonenzymatic photoelectrochemical sensors, (iii) DNA affinity-based sensors, and (iv) immuno-affinity-based sensors. We are devoting a section to this aspect of PEC biosensors, as it influences the functionality of the sensors in terms of the ASSURED criteria [1].

An enzymatic PEC sensor is based on close association of the photoelectrochemical transducer and enzyme. The enzyme catalyzes formation of an electroactive species from the analyte. This electroactive species modulates the amperometric signal of the PEC transducer by reacting on the semiconductor surface. This enzyme-transducer association imparts selectivity of an enzyme to the PEC sensor. However, these systems are plagued with the problem of enzyme inactivation, leading to a short shelf life. These sensors have a fast response time with minimal sample concentration for analysis. The electrical communication between the active site of the enzyme and the hole of the valence band or photoexcited electrons in the conduction band is crucial for functionality of the enzymatic PEC biosensors. This aspect has been the basis of the classification of enzymatic biosensors into first-, second-, and third-generation biosensors. A similar classification can also be introduced for enzymatic PEC biosensors based on the mode

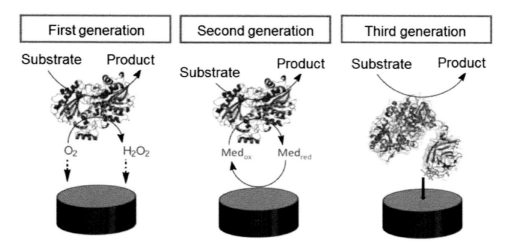

FIGURE 9.5 Classification of enzymatic biosensors [11].

of electron transfer between the enzyme and semiconductor (Figure 9.5) [11].

Work carried out by Tanne and Tang's group illustrate the working principle of a first-generation enzymatic biosensor where semiconductor electron carriers interact with the products of a glucose oxidase (GOD)-mediated reaction like oxygen or hydrogen peroxide. A CdSe/ZnS-based quantum dot photoelectrode assembly was fabricated [12]. The enzyme GOD was immobilized by glutaraldehyde-mediated covalent crosslinking and multilayer assembly of GOD and positively charged poly-[allylamine hydrochloride] (PAH). The quantum dot-electrode constructs responded to light by giving cathodic currents as a result of oxygen reduction by cathodically poised electrodes. This photocathodic signal was attenuated in response to argon purging of the electrolyte, indicating the ability of the quantum dots to reduce oxygen. Upon close association of GOD with quantum dots, the enzyme competed for oxygen as a substrate for glucose oxidation. This caused an attenuation in the cathodic photocurrent signal. Thus the photocurrent attenuation signal was used for analyte sensing.

Tang and his group reported another PEC biosensor that was based on a light-responsive anodic photocurrent signal that was proportional to the concentration of glucose in the solution [13]. The fabrication involved TiO_2 nanowire synthesis on FTO electrodes, followed by crosslinking of GOD to the nanowires. In response to incident light, photocatalytic water splitting on the electrode interface caused by holes led to oxygen evolution. In the presence of glucose, the dissolved oxygen produced acted by oxidizing the reduced enzyme cofactor $FADH_2$. The peroxide produced as a by-product of this reaction enhanced the amperometric signal by reacting at TiO_2 valence band holes. Since the reaction involved oxygen as a mediator, this work is also based on the first-generation enzymatic biosensor principle.

A second-generation enzymatic biosensor involves the use of a diffusible redox mediator to shuttle electrons between the enzyme active site embedded in the enzyme structure and the electrode. To illustrate this principle in the context of PEC biosensors, we discuss the research of Zheng and his group [14]. They prepared a film of TiO_2 on an ITO substrate by spin coating followed by sintering. This film was used as a substrate for layer-by-layer deposition of [poly [ethyleneimine]/[cobalt [o-phenanthroline] $_3]^{2+/3+}$/CdSe@CdS quantum dots]. The number of layers was made to vary from 2 to 10. A final layer of [poly [ethyleneimine]/[cobalt [o-phenanthroline] $_3]^{2+/3+}$/GOD] was deposited on the electrodes. Control electrodes were also fabricated with the absence of quantum dots, [cobalt [o-phenanthroline] $_3]^{2+/3+}$ and GOD, respectively. The performance of the electrodes was compared through photocurrent-voltage curves and photochronoamperometry. Overall results indicated the role of [cobalt [o-phenanthroline] $_3]^{2+/3+}$ as a redox mediator between CdSe@CdS quantum dots and reduced $FADH_2$ cofactor of GOD. The system worked on the principle of a second-generation enzymatic PEC sensor.

Third-generation enzymatic biosensors are based on direct electron exchange between the enzyme active site and electrode interface. In the context of photoelectrochemistry, the electrode interface is replaced with valence band holes or conduction band electrons. To illustrate this category we choose to describe here the research by Chen and his coworkers [15]. A nanotubular TiO_2 forest was grown on titanium substrates by direct anodization. These nanotubes were found to generate photocurrent upon visible light irradiation. Horseradish peroxidase (HRP) enzymes were adsorbed directly on these nanostructures, and the enzyme activity of adsorbed HRP was verified by UV-visible analysis of heme cofactor and colorimetric assay. Upon illumination, it was observed that nanotubular TiO_2 produced cathodic photocurrents as a result of electron ejection from the conduction bands. The HRP immobilization on TiO_2 nanotubes increased the cathodic photocurrent by a negligible magnitude of 0.1 nA. Upon addition of H_2O_2 in the solution, cathodic photocurrent intensity was increased in a linear fashion with an increase in H_2O_2 concentration. This increase in current was attributed to direct electron transfer between the conduction band electrons and the HRP-Fe III/Fe II redox couple followed by reduction of H_2O_2. This system operated on the principle of the third-generation PEC biosensor (Figure 9.6).

Nonenzymatic PEC biosensors, as opposed to enzymatic PEC biosensors, lack any biological component. They are based on direct oxidation of analyte driven by kinetic energy of holes generated in PEC transducers in response to incident

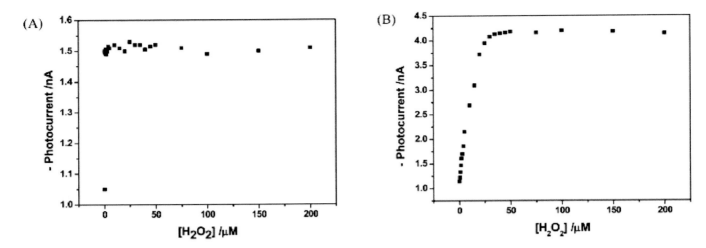

FIGURE 9.6 Photocurrent response of (a) control photoelectrode without enzyme and (b) with enzyme. Reprinted with permission from [15]. Copyright (2010) American Chemical Society.

light. These systems are stable and robust against environmental conditions. However, they lack sensitivity and are based on the destructive oxidation reactions of biomolecules. Thus, scientific research to improve the sensitivity, stability, and selectivity of these systems is a promising field. To illustrate these systems, we discuss two papers here. Zhang and his group developed a platform based on tungsten oxide (WO_3) adorned on the core shell of highly conductive TiC-C nanofibers with a detection limit of 1.12×10^{-8} M and high glucose sensitivity, which is a considerable improvement in limit of detection over earlier developments [16]. WO_3 is highly resistant to photocorrosion and falls in the category of an indirect bandgap semiconductor (2.4–2.8 eV). Also, due to the high photocatalytic activity, it is a potential compound to be used as PEC biosensors. WO_3@TiC–C electrodes achieved a glucose sensitivity with a peak current of 183.4 $\mu A\ mM^{-1}\ cm^{-2}$ in the range of 0.25–10 μM at a light irradiation intensity of 5 mW cm^{-2}. In another representation, Devadoss and coworkers reported glucose detection with simultaneous generation of biohydrogen at the counter electrode, using photo-catalytically active electrodes fabricated from cuprous oxide and titanium dioxide, that is, Cu_2O-TiO_2 [17]. TiO_2 acts as a passivation layer that prevents corrosion of Cu_2O extending its operational time. The resultant reduction in bandgap of the electrode due to this modification is from 2.2 eV to 1.84 eV. This bandgap tailoring made the system capable of harvesting visible light photons. The electrodes achieved a glucose sensitivity of 4.6 mA cm^{-2} when exposed to a light intensity of 100 mWcm^{-2}.

Another class of PEC sensors are affinity sensors that are based on the principle of a selective binding interaction between a biorecognition element and an analyte. A layer of biorecognition entity is formed over a PEC transducer, and its interaction with its corresponding analyte causes a modulation in the photoamperometric response. Two major classes of molecules used in these sensors are DNA and antibodies. A DNA biopolymer, with its endless permutation and combinations of constituent nucleotides, acts as a storehouse of biological information necessary for the propagation of life. DNA strands are selective in their interaction and binding affinities with biological/non-biological entities. Various DNA-based interactions like DNA hybridization, DNA- small molecule interaction, and DNA aptamer-target interaction strategies have been utilized for developing PEC sensors. Diverse DNA-based recognition strategies with electrons originating from PEC indicator molecules, including PEC labels or DNA itself, have been reported. To illustrate these examples, we discuss the work of Zang and coworkers who developed a "signal-on" strategy for selective photoelectrochemical detection of Pb^{2+} [18]. The aptasensor they fabricated was a stepwise modification on the indium tin oxide (ITO) base electrode. Tetraethylene pentamine (TEPA)-functionalized reduced graphene oxide (RGO) was layered on the ITO and introduced many amino functional groups, which are conducive to electrostatic modification of the anchored thioglycolic acid-stabilized CdS quantum dots (QDs). The aptamers were then layered on top of CdS QDs. A competitive interaction of Pb^{2+} with aptamers, signal amplification by RGO, and resonance energy transfer between CdS QD and DNA-labeled gold nanoparticles (DNA-AuNPs) ensued in the platform. In absence of Pb^{2+}, DNA-labeled AuNPs hybridized with aptamers and diminished the current intensity. Upon addition of Pb^{2+} (lead), the system recovered the photocurrent by blocking the hybridization of the aptamer with DNA-labeled AuNPs and changing the conformation of the aptamers to a G-quadruplex structure (Figure 9.7). Another work linked biomolecules to inorganic components by using bridging enediol ligands (dopamine in this case) to facilitate hole transfer across the link, thereby establishing an efficient interaction between metal oxide nanoparticles and biomolecules [19]. They used TiO_2 nanoparticles (~50 Å) and surface modified them to bind DNA oligonucleotides that retained the ability to hybridize with the complementary DNA. This photoelectrochemical system enabled extended charge separation that would pave the way for biosensing platforms for the detection of DNA hybridization. In the presence of donor and acceptor molecules, an amperometric response was generated.

Another class of biomolecules known for their specificity are antibodies raised in response to an antigen in mammalian

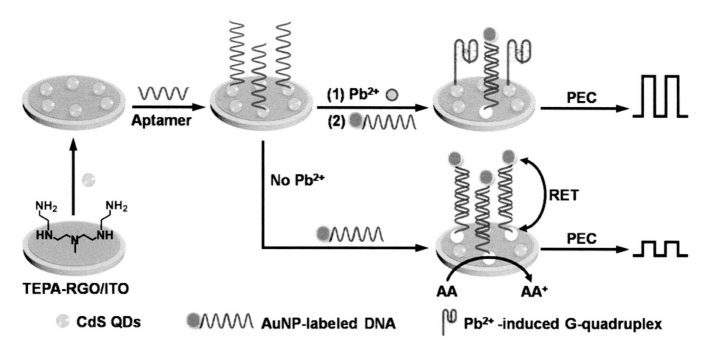

FIGURE 9.7 Stepwise fabrication of a photoelectrochemical aptasensor. The presence of Pb^{2+} ion leads to an increase in the amperometric signal. Reprinted with permission from [18]. Copyright (2014) American Chemical Society.

bodies. Their specificity has been utilized in many biological assays and analytical techniques. Conventionally, immunosensor-based systems require labeling with a biomarker. The PEC transducer-antibody format obviates this requirement of a label and is based on modulation of the photoelectrochemical signal in response to analyte detection. Kang and coworkers fabricated a label-free PEC immunosensor to detect pentachlorophenol (PCP) in contaminated water [20]. It was constructed with ultrasensitive $CdSe_xTe_{1-x}$ nanocrystals (NC) photoelectrodeposited on TiO_2 nanotubes (NT) grown on Ti foil. The PCP antibody was dropped and dried on to a $CdSe_xTe_{1-x}$–NC/TiO_2–NT electrode. The efficient electron injection and hole recovery system of the photoelectrochemical sensor generated a photocurrent density of 1.11 mA cm^{-2}. The immune-interaction between the PCP and antibody decreases the photocurrent density, and a limit of detection (LOD) of 1 pM was recorded. Furthermore, Wang et al devised a sandwich PEC immunosensor for measuring carcinoembryonic antigen (CEA) concentrations using an AuNPs/polydiallyldimethylammonium (PDDA)/graphene-tin dioxide QD solution with capture antibody 2 [AuNP/PDDA/G-SnO_2/Ab_2] electrode and electrodeposited AuNPs on the ITO electrode with a capture antibody 1[AuNp/ITO/Ab_1] [21]. The AuNp/ITO/Ab_1 modified immunosensor was incubated in various concentrations of CEA first and then in an AuNP/PDDA/G-SnO_2/Ab_2 solution. PEC measurements showed an increase in photocurrent density with an increase in CEA concentrations indicating interparticle electron transfer between SnO_2 QDs and AuNPs.

9.5.2 Photoelectrochemical Transducers

An ideal photoelectrochemical transducer should have good photocatalytic property/electron-hole pair formation frequency with minimum charge recombination frequency. Charge injection from a semiconducting material to the adjacent electrode should be efficient. Long-term chemical and physical stability of the photoelectrochemical material is essential for the robustness and stability of the sensor. Moreover, an optimum bandgap energy difference for visible light responsiveness of the sensor is also crucial for practical applications. Many materials with semiconducting properties have been reported for photoelectrochemical biosensing applications. This section describes some of the commonly used PEC materials and illustrates their application as a biosensor transducer.

Semiconducting metal oxides are one of the most widely reported PEC transducers. TiO_2, the metal oxide used for water splitting by Fujishima and Honda, has a wide range of applicability, including photocatalysis, solar cells, battery, sensing, environmental treatment, biomedical applications, etc. The exceptional physical and chemical properties of TiO_2, like corrosion resistance, photochemical stability, biocompatibility, low density, wide bandgap, and strength, make it suitable for these applications. Well-established energetics of the TiO_2 bandgap makes bandgap engineering easy. Scalable electrode fabrication and composite synthesis with TiO_2 are added benefits of TiO_2. The wide bandgap of TiO_2 makes it responsive to UV radiation that is detrimental to biomolecules. Thus, the TiO_2 surface is sensitized with organic molecules or dyes that improve the visible light responsiveness of the system. Optimum relative positioning of the valence and conduction band of the dye and metal oxide enable movement of electrons across conduction bands. Other semiconducting metal oxides like ZnO and SnO_2 are also being employed for photoelectrochemical applications due to their favorable properties like biocompatibility, stability, photocatalytic property, electron mobility, and cost-effectiveness. Here we discuss a few metal oxide-based PEC sensor reports. Tu et al. developed a photoelectrochemical biosensing platform for glutathione (GSH) using water-soluble (meso-tetrakis[4-sulfonatophenyl]porphyrin) iron[III] monochloride modified TiO_2 nanoparticles functionalized on an ITO electrode [FeTPPS-TiO_2/ITO] [22]. The modified electrode displayed an improvement in current density indicating strong electron coupling between the excited state of FeTPPS and the conduction band of semiconductor TiO_2 nanoparticles. In another work, Tu et al developed a ZnO-NP semiconductor-based sensor. They synthesized an ITO-modified electrode through ZnO-NP decorated with carboxylic groups of 4,4′,4″,4‴-[21H,23H-porphine-5,10,15,20-tetrayl]tetrakis[benzoic acid] (TCPP) to detect cysteine photoelectrochemically. The higher level of the energy state of TCPP allowed electrons to transfer to the conduction band of ZnO NPs that intensified the photocurrent density [23].

A second category of widely used PEC transducers is semiconductor nanoparticles/quantum dots (QDs). These nanoparticles are known for their excellent photophysical and tunable fluorescent properties. While the physics of the tunable fluorescence property is discussed extensively in the literature, we emphasize its photoelectrochemical activity here. The fluorescent properties of a QD are a result of radiative decay of the electron-hole pair. The photoelectrochemistry in QD is a result of electron trapping at the surface resulting in an increase of the exciton lifetime. During this time period, an electron can be ejected out either to an anodically polarized electrode, resulting in anodic current, or to a soluble redox molecule with a suitable redox potential, resulting in cathodic current. The hole is neutralized either by a soluble electron donor in a solution, in the former case, or by the electron from the electrode in the latter. Simple surface chemistry of quantum dots makes them suitable for decoration of photoelectrodes through functionalization. A variety of enzymatic, nonenzymatic, and affinity photoelectrochemical sensors have been reported, which have QDs as their components. We are illustrating a few scientific reports demonstrating the use of QD in photoelectrode-based sensing. Wang fabricated near infrared QDs from CdTe decorated on a fluorine-doped tin oxide surface as a working electrode to photoelectrochemically detect glucose [24]. The photoelectrode was cathodically biased at −0.2 V, and this caused reduction of oxygen by conduction band electrons. This oxygen sensitivity was primary for glucose sensing. GOD was taken as the model enzyme to validate the photoelectrochemical mechanism. The glucose oxidation reaction consuming oxygen caused a reduction in the photocurrent density. The reduced amperometric signal was proportional to the glucose concentration and was used for sensing. Vered and his coworkers used nucleic acid scaffolds as a template for the structural ordering of relay units and photoactive CdS semiconductor nanoparticles [25]. The relay unit was constructed by tethering a thiolated DNA template on a gold electrode as a monolayer with sections of the template DNA complementary to N-methyl-N′-[2-carboxyethyl]-4,4′-bipyridinium [V^{2+}]-functionalized and CdS nanoparticle (NP) modified nucleic acids. Triethanolamine (TEOA) was used as an electron donor in different configurations. Similar constructs were made with CdS nanoparticles replaced with a rubidium-based photosensitizer. The photocurrent density with CdS NPs was significantly higher when compared to [Ru[bpy]$_3$]$^{2+}$ due to effective charge separation and interparticle charge transfer in the relay of CdS NP to V^{2+} then to the gold electrode. Upon changing the order of photoresponsive elements with respect to gold electrodes, a corresponding change in the magnitude of photocurrents was observed. Moreover, upon cleaving the aptamer with Dde I endonuclease, a significant change in the photocurrent magnitude was observed, as illustrated in Figure 9.8.

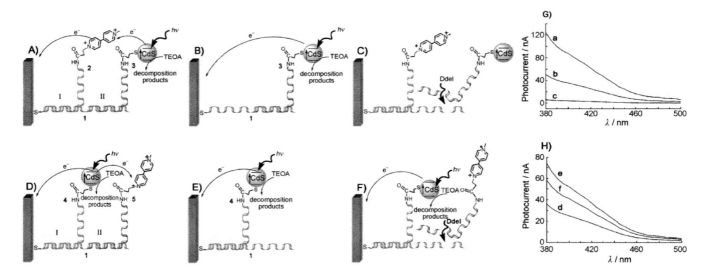

FIGURE 9.8 DNA scaffold-QD construct-based sensor [25]. A and B. Effect of including V^{2+} between the electrode and CdS quantum dots. C. Cleavage of DNA aptamer by endonuclease. G. Effect on photocurrent magnitude. D and E. Effect of changing the order of quantum dots and photosensitizers with respect to gold electrodes. As observed in H. Cleavage of aptamer by endonuclease as observed in F, corresponding change in photocurrent density as observed in H. Reprinted with permission from [25]. Copyright (2008) John Wiley and Sons.

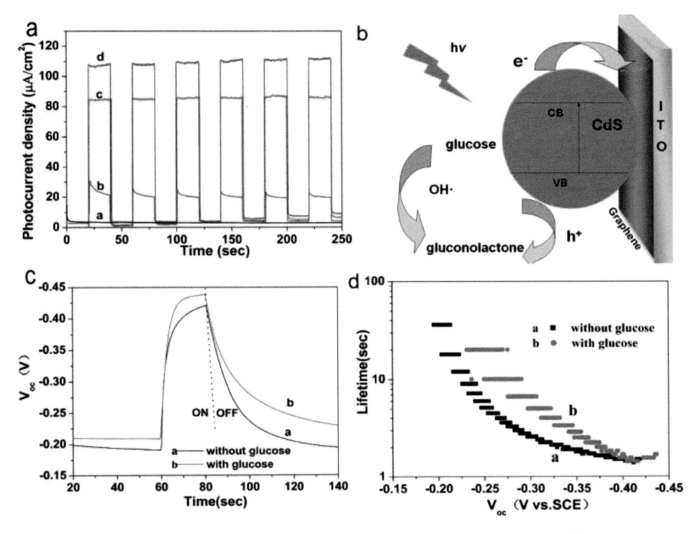

FIGURE 9.9 A. Photocurrent magnitude obtained from a photosensor increased due to the presence of glucose (difference in current density of c and d). B. Schematic of the working principle of a QD-decorated photoelectrode. C. OCP response with and without glucose. D. Exciton lifetime increases with glucose as an analyte. Reprinted from [26]. Copyright (2014) with permission from Elsevier.

Xuyan produced graphene-CdS QDs to non-enzymatically detect glucose using photoelectrochemical conversion [26]. The graphene-CdS QDs were layered on the ITO electrode and were highly light sensitive, and during illumination formed photoexcited holes. The rapid electron flux from the conduction band of CdS to graphene and then to the ITO electrode avoided electron-hole recombination (Figure 9.9). The LOD of glucose achieved during the process was 7 $\mu mol\ dm^{-3}$ at a correlation coefficient of 0.9991, which showed promising application as a biosensor in glucose detection.

Due to the inherent drawbacks of semiconductor- and metal oxide-based photocatalysts such as the strong oxidative property, requirement of a high energy light source, photocorrosion, and poor conductivity, researchers have turned to investigate carbon-based photocatalytic materials. The beautiful, soccer ball-shaped molecules of Buckminster fullerene [C60] are known for their wide electromagnetic absorption spectrum spanning the UV and visible light wavelength range and electron accepting capacity. The electron accepting capacity of a fullerene's ground state is further enhanced upon excitation to singlet and triplet states. The fullerene molecules, upon excitation, stabilize to the excited triplet states from singlet states, which being intersystem crossing, have a longer lifetime. Upon light exposure of electron donor-fullerene constructs, the fullerene's triplet states accept electrons from donor molecules, giving the effect of electron-hole pair formations. Many dyad and triad constructs have been reported with the intent of mimicking photosynthetic reaction center components and construction of photoconversion systems. This aspect of C60 and related molecules have been exploited for the construction of photoelectrochemical sensors. Zhan and coworkers fabricated an assembly of amphiphilic fullerene derivatives assembled inside phospholipid vesicles. These vesicles were assembled on ITO electrodes as bilayers (Figure 9.10a) [27]. Upon exposure to blue light, in the presence of ascorbic acid as an electron donor, anodic photocurrent signals were observed. In saturated oxygen conditions, a cathodic photocurrent was observed (Figure 9.10b).

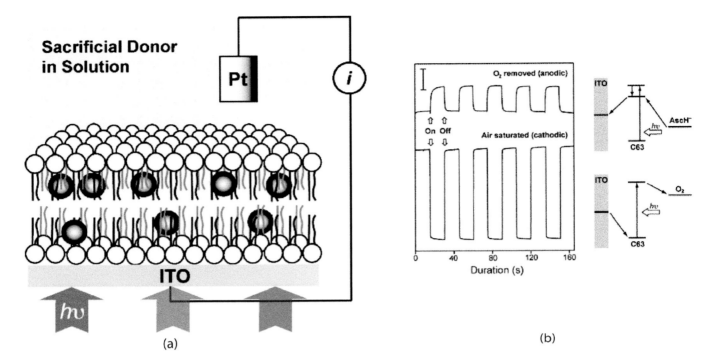

FIGURE 9.10 (a) Lipid bilayers with C60 assembled on ITO electrodes. (b) Photoresponsive anodic currents in the absence of oxygen; air saturation caused cathodic currents. Reprinted with permission from [27]. Copyright (2008) American Chemical Society.

Upon inclusion of fluorescein dye into the system, electronic coupling between fullerenes and fluorescein was detected by fluorescence quenching of the dye. This electron exchange between fullerenes and fluorescein was detected by a higher magnitude of cathodic photocurrent observed in the system.

Hu and his coworkers [28] reported a second significant work we are discussing here. The researchers developed a semiconductor-less photoelectrochemical affinity sensor driven by the photochemical response of fullerene units decorated on a carboxylated carbon nanotube (CNT). Congo red molecules were used for enhancing the solubility of CNT and C60. The CNT–CR–C60 was further added with an antibody specific to CEA. The sensor worked on the principle of a sandwich immunoassay, where a bridge was formed between the CNT–CR–C60-antibody and antibody-p-aminobenzoic acid-ITO via interaction with the analyte CEA. This caused a flow of photocurrent produced from fullerene moieties in the bioprobe. The system responded to illumination by a low-cost and portable green laser light.

A second category of carbon-based photocatalytic material explored for photoelectrochemical sensing application is carbon dots made out of graphene, termed graphene quantum dots (GQDs). Graphene, a two-dimensional honeycombed network of carbon atoms, is known for its electronic properties and has wide applications. Shredding down this two-dimensional graphene network to the nanometer scale (zero-dimension nanoparticle) opens up a bandgap in the electronic structure of the nanoparticle. Thus, it is capable of exhibiting tunable photoluminescence, up converted photoluminescence, and light-responsive electron transfer properties. Due to differences in synthesis procedures widely categorized into top-down and bottom-up approaches, there is a lot of heterogeneity in the properties of synthesized GQD. Often the nanoparticles exhibit a combination of these properties. Unlike carbon dots, which have a combination of crystalline regions and amorphous regions in the core, GQDs have a crystalline core. Manipulating the chemical composition of the lattice through doping, surface functionalization and variation of size cause corresponding modulation in the optical properties. Readers are suggested to go through [29] and [30] for detailed understanding of the mechanism. To give an example of its performance in a photoelectrochemical biosensor, we illustrate the work of [31]. The authors devised an antibody affinity biosensor with antibodies specific to the peptide toxin microcystin. The transducer used in the sensor was one-dimensional nanowires of silicon. In order to cope with the formation of an insulating oxide layer, the nanowires were spin-coated with GQDs. GQD/silicon nanowires exhibited a higher magnitude of photoresponse and reduced charge transfer resistance in response to incident light across the electrode. Upon inclusion of microcyctin-containing analyte into the sensor, formation of a microcystin-antibody complex caused steric hindrance to the flow of analyte to the electrode. This was detected by reduction in the amperometric signal, which was linear for concentration ranges of 0.01 µg L^{-1} to 10 µg L^{-1}.

In many cases direct use of photoelectrochemically active organic molecules or metal complexes as photocatalytic transducers have been reported. Ikeda used 5-, 10-, 15-, 20-tetra (4 pyridyl) porphyrin (TPyP) molecules as photoelectrochemical transducers [32]. The TPyP molecules were functionalized to ITO electrodes via crosslinking through a silane-based moiety. When these fabricated electrodes were exposed to monochromatic light of 420 nm at 0.3 V, photocurrents were observed. Upon introduction of adenosine phosphates into the system, reduction in magnitude of the photocurrent was observed due to steric hindrance to the flow of electrons by the phosphate ions of the adenine nucleotides. The order of responsiveness was AMP > ADP = ATP. The system was reusable after washing with water and ethanol for multiple cycles.

Okamoto reported a hole-mediated photocurrent generation from DNA duplex assemblies on a gold electrode [33]. One strand of the duplex was anchored to the gold electrode through a thiolated moiety. The complementary strand was labeled with anthraquinone. Photostimulation of anthraquinone caused a cathodic photocurrent generation across the electrode interface that was mediated by a flux of positively charged holes. The role of dissolved oxygen in the electrolyte as an electron acceptor from reduced anthraquinone molecules in the DNA duplex was established through a difference in amperometric response in the presence and absence of oxygen. Guanine nucleotides in the DNA played the role of hole mediation by hopping to the electrode. This was established through comparing the photocurrent response of DNA duplexes with various sequences (Figure 9.11).

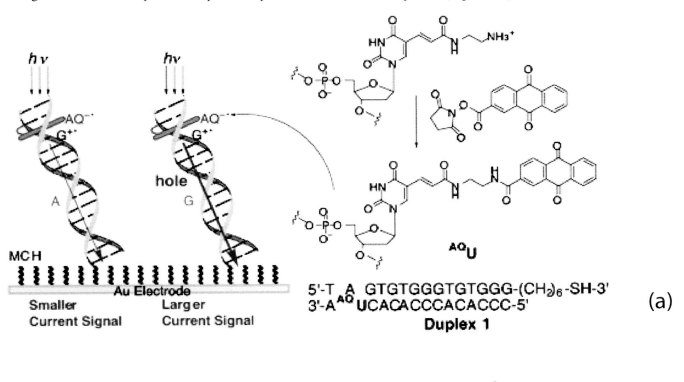

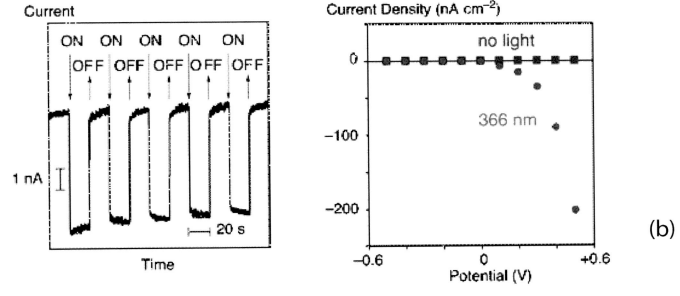

FIGURE 9.11 (a) Affinity-based PEC sensor driven by guanine-mediated hole transfer. (b) Hole-mediated photoresponse; photovoltammetry response of electrode. Reprinted with permission from [33]. Copyright (2004) American Chemical Society.

Researchers [34, 35] exploited this hole-mediated photocurrent generation from the anthraquinone-tethered DNA duplex to develop a photoelectrochemical sensing module for detecting methylated cytosine and a triplex-forming sequence in a DNA duplex. Haddour [36] demonstrated that ruthenium metal complexes, bearing biotin and pyrrole moieties, electropolymerized on electrodes can act as affinity-based photoelectrochemical transducers. Cobalt complexes were used in the system as quenchers for photoactivated ruthenium complexes. Sequential binding of avidin, biotinylated cholera toxin, and antibodies caused steric hindrance to the flow of electrons reflected by a sequential decrease in the magnitude of photoresponse. [37] demonstrated that a ruthenium metal complex tethered to two N, N′-bis (3-propyl-imidazole)-1, 4, 5, 8-naphthalene diimides (PIND) moieties can strongly intercalate with a DNA duplex. The PIND-ruthenium-PIND moiety exhibited a photoelectrochemical response proportional to the increasing intensity of light. The sensitivity of the detection system was much higher than that of the voltammetry-based approach.

9.6 SIGNAL AMPLIFICATION STRATEGIES

Physical separation of a light source from an amperometric sensing platform imparts higher sensitivity and a better signal-to-noise ratio. However, in a quest for higher sensitivity, researchers have introduced a number of amplification strategies in photoelectrochemical sensing platforms. This section will discuss some of these strategies briefly.

Surface plasmon resonance (SPR) is a physical phenomenon that occurs when incident plane polarized light hits a metal surface, which initiates collective oscillation of free surface electrons from the metal surface. Hybrid noble metal oxide nanoparticles of photocatalysts possessing plasmonic properties incorporate new functionalities to improve the physical and chemical properties combined into a single nanostructure. These hybrid nanostructures exceed the overall functionality when compared to the characteristics of the individual components with a wide range of application potential. The possible reasons for improved photocatalytic activity exhibited by plasmonic nanoparticles can be enumerated as follows: (i) formation of a Schottky barrier that prevents recombination of electron-hole pairs, (ii) enhanced activity of the visible light spectrum due to SPR, (iii) high conductivity, (iv) PEC biosensing efficiency is improved by higher exciton-plasmonic interactions. To illustrate the concept of SPR in the context of PEC sensors, we discuss here work reported by Samantara and coworkers. They fabricated a PEC device with raspberry-like ZnO nanostructures hybridized with AuNPs as plasmonic nanocomposites to explore the photoelectrocatalytic detection of NADH at neutral pH [38]. The enhanced electron transfer due to the nanocomposite structure, caused the electrode to achieve a sensitivity of 5.025 nA mM^{-1} and an LOD of 300 nM. Furthermore, Shen and coworkers used Au-SnO$_2$ hybrid nanospheres layered on ITO in a PEC to detect cysteine as a model analyte with a detection limit of 0.1 µM [39]. The AuNPs embedded inside the SnO$_2$ nanostructures aroused the SPR effect and increased the conductivity, thereby increasing the photocurrent.

Resonance energy transfer is the phenomenon of energy exchange between two fluorophores placed at an optimum distance in relation to each other. The emission spectrum of the first fluorophore overlaps with the excitation spectrum of the second fluorophore causing quenching of the first fluorophore. To demonstrate this phenomenon in the context of photoelectrochemical signal amplification, we discuss the work reported by Zhao [40]. This work demonstrated the phenomenon of interparticle energy transfer between CdS QDs and AuNPs. The energy exchange was possible due to a significant overlap of QD emission spectra and AuNPs absorption spectra. This interparticle exchange of energy reduced the magnitude of the amperometric signal. The sensing platform was fabricated by layering an ITO electrode with QD. The QD layer was subsequently crosslinked to aptamers. Complementary aptamers linked to AuNPs were used to sense complementarity in DNA strands, which was detected by a linear decrease in the photocurrent in response to increasing concentration of AuNPs-DNA concentration.

To achieve ultrasensitive detection platforms, high photon to current conversion efficiencies (PCE) of the semiconducting nanomaterials are essential. A significant fraction of excitons formed upon light exposure is lost to electron-hole pair recombination. To deal with this, semiconducting materials with closely spaced energy levels are co-sensitized to enhance electron-hole pair separation and charge collection at the electrode interface. This maximizes the incident photon to current conversion efficiency (IPCE). To illustrate this example, we discuss the work reported by Fan and coworkers [41]. A titanium dioxide layer was formed over a layer of ITO electrode through heat sintering. This layer was further adsorbed with Mn-doped CdS QDs. DNA hairpin loops were covalently crosslinked to this cosensitized semiconducting film. Two differently sized CdTe QDs were further covalently crosslinked to other end of the hairpin loop (Figure 9.12). The energy level of each photoresponsive entity in this assembly was close enough to reinforce the net electron flux flowing into the electrode, giving an amperometric signal readout. Upon introduction of a complementary target DNA into the system, the hairpin structure was denatured, leading to detachment of the CdTe QDs agglomerate from the assembly. The resulting reduction in the magnitude of the photocurrent response was considered a detection signal.

9.7 VISIBLE LIGHT- AND SOLAR-POWERED PEC SENSOR: RECENT REPORTS

Visible light- and solar energy-powered photoelectrochemical sensors are categorized into self-powered sensing platforms for biological and chemical entities, which have attracted considerable interest due to their battery-less design and nonrequirement of an additional physical irradiating source for photocatalysis. Sunlight is harvested as the source of energy for photocurrent generation. Workers have designed and fabricated new PEC sensors that could use solar energy as a stable light source and overcome shortcomings like fluctuating intensity of solar illumination over time and

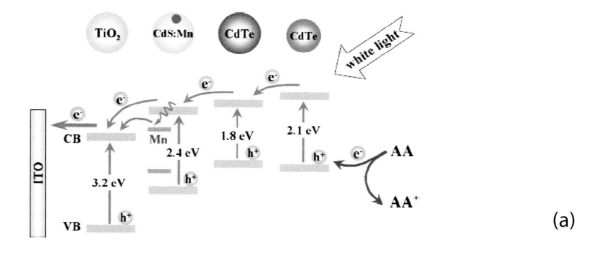

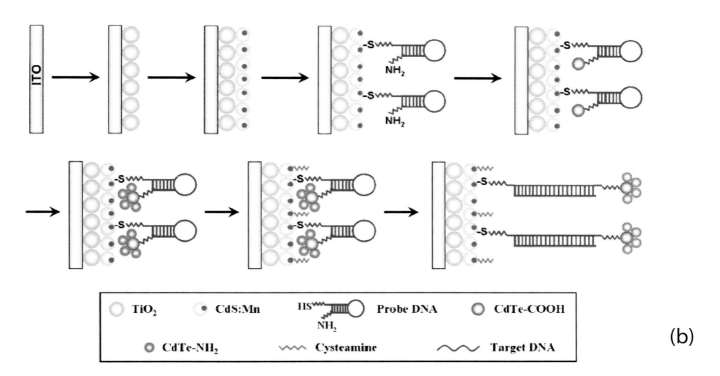

FIGURE 9.12 (a) Cosensitization of photoresponsive elements with matched energy levels. (b) Scheme of photoelectrode synthesis. Reprinted with permission from [41]. Copyright (2014) American Chemical Society.

weather. Soares described the development of a self-powered PEC for tannic acid (TA) detection using a photoanode with TA-sensitized TiO_2 paste layered on FTO (fluorine-doped tin oxide) and a photocathode of Cu_2O/ZnO/FTO to water oxidation. Cu_2O and ZnO were electrodeposited in tandem on FTO. The photoanode and photocathode were placed in separate adjoining chambers to generate a photocurrent upon irradiation, driving the TA sensing process. The principle involved in this design was the oxidation of TA in photoanode and the reduction of water at the photocathode, so the higher the concentration of TA, the higher the photocurrent and the higher the sensitivity. The platform detected a level of TA as low as 0.1 μmol L^{-1} with a photocurrent density of approximately 1.7 μWcm^{-2} [42].

Hao in 2018, using the same principle of using sunlight as power source for application of PEC, fabricated an aptamer-based PEC to detect aflatoxin B1 (AFB1) using the principle of potentiometric resolve ratiometric phenomenon to counter the limitations posed by the intensity of sunlight over time and weather. The photoanode was composed of silver and TiO_2 blended with 3D nitrogen-doped graphene as a hydrogel layered on ITO [43]. Another reported work used TiO_2

nanowire/NiO nanoflakes and the Si nanowire/Pt nanoparticle composites layered on FTO as photoanodes and photocathodes for detection of glucose in a full solar-powered energy conversion–storage–utilization PEC system [44]. Yan in 2016 developed a photofuel cell PEC driven by sunlight for sensing glucose as a model analyte. The photoanode consists of Ni $(OH)_2$/CdS (QD)/TiO_2/FTO and hemin graphene/FTO as photocathode, placed in separate chambers. Upon exposure to solar illumination Ni $(OH)_2$ catalyzed the glucose oxidation reaction, and the electron generated was reduced in the photocathode by H_2O_2 introduced as an electron scavenger, forming water as the end product. The output performance achieved detected a glucose concentration as low as 5.3 μM [45]. Kai and coworkers developed a self-powered PEC platform to detect p-Nitrophenol (p-NP) using 100 mW portable violet laser light. The PEC consisted of PbS QD-modified glassy carbon electrode (GCE) as a cathode and a graphene layered on GCE as the anode. The p-NP binding molecular imprinted polymer was immobilized on the PbS/GCE cathode as a recognition element, and ascorbic acid (AA) was used as an electron donor in the anode chamber. Upon exposure, p-NP in the cathode was reduced at the photocathode and AA was oxidized at the anode, generating a photocurrent with the p-NP minimum detection limit of 0.031 μM [46].

9.8 PHOTOSYNTHESIS: MOLECULAR MACHINERY AND ITS COMPONENTS

Photosynthetic organisms have the capacity to harvest solar energy and convert it into organic molecules like carbohydrate and starch. These molecules are further utilized by other organisms, thus sustaining life on the biosphere. The light-harvesting metabolic machinery of the photosynthetic organisms has been optimized through the process of evolution. The complexity of this molecular machinery increases with more complex life forms as we move across taxonomical heirarchy from monera to plantae. However, the structural motif of the enzymes responsible for actual light to current conversion is quite similar and conserved. These motifs contain redox cofactors and pigments positioned within a protein scaffold to optimize photoinduced electron transfer and subsequent unidirectional electron flow originating from a special pigment pair, passing through a series of redox cofactors, which give the effect of electron-hole pair separation as in a semiconducting material. The idea is to isolate these enzyme motifs in their functional form and integrate them into electrodes and devices as a substitute for photocatalytic materials. Maintaining the functionality of the isolated enzymes during operation and reducing the cost of isolation are the two main bottlenecks to be surmounted for this technology to have any practical application. In recent years significant research has been reported, most of which makes use of enzymes isolated from thermophilic cyanobacterium due to its high thermal stability. The use of complete photosynthetic organisms like algae and cyanobacteria have also been reported and is being actively investigated. The term coined to describe these electrochemical systems in recent literature is "biophotovoltaics" [47]. Before we proceed to describe the working of photosynthetic enzyme–electrode systems, it will be prudent to discuss the functioning of photosynthetic enzymes used in the biophotovoltaic systems in general.

The thylakoid membrane network present in the chloroplasts of algae and plant cells is the site of light-harvesting membrane protein complexes. These membranes are arranged in stacks called "grana," which are interconnected through membrane extensions termed stroma lamellae. For single-celled prokaryotic cyanobacteria, these membranes are arranged as rings of concentric membranes adjacent to the cytoplasmic membrane. These membranes house transmembrane photosynthetic reaction center complexes called photosystem 720 and photosystem 680, which work in tandem with each other. This collaborative action of photosystem 720 and photosystem 680 is also called the "Z scheme pathway." The electron flux of the Z scheme photosynthetic pathway originates by photolysis of water at a manganese-containing oxygen evolution complex close to a special chlorophyll pair of photosystem 680. Contrary to this, purple nonsulfur bacterium reaction centers (photosystem 870) are housed in the bacterial plasma membrane, and electron flux follows a closed loop without oxygen evolution.

The general design of all photosynthetic reaction center complexes can be described as a set of two functional units: light-harvesting complex (LHC) and core reaction center (RC). The protein scaffold of a light-harvesting complex holds a collection of photosynthetic pigments like chlorophyll, carotenoid, phycocyanin, phycoerythrin, etc., in a relative configuration that optimizes the Forster resonance energy transfer (FRET) for harvesting the incident photons. Without this scaffold, these pigments would dissipate the energy of incident photons by fluorescence emission or heat dissipation. The rate of exciton transfer between the pigments of the LHC is reported to be on the order of femtoseconds. The final exciton transfer step is from the LHC to the special chlorophyll dimer of reaction center complexes whose timescale is on the order of picoseconds. This difference in timescale prevents back-transfer of excitation energy as the subsequent photoinduced electron transfer steps are fast enough to compete and ensure efficient trapping of light energy. The ejected electron is transferred through a cascade of redox cofactors within the reaction center protein scaffold giving rise to an intraprotein electron-hole pair formation. The placement of redox cofactors within the reaction center complex is such that it enhances the efficiency of electron transfer across the cascade and makes it a kinetically and thermodynamically efficient process, minimizing the decay of excited chlorophyll dimer to the ground state [48, 49]. Photosystem 870, photosystem 720, and photosystem 680 are the three most widely reported protein complexes used for electrode interfacing with applications in solar to energy conversion and solar to chemical conversion (Figure 9.13).

9.9 REACTION CENTERS AS COMPONENTS OF A PHOTO-ELECTRODE

The purple bacterial reaction center (P870) or type 2 reaction center is the most well-investigated photoenzyme, and its study has significantly contributed to an understanding

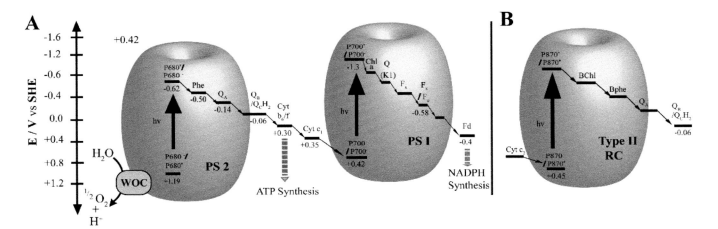

FIGURE 9.13 A schematic representation of the Z scheme pathway and bacterial reaction center.

of the molecular mechanism of photosynthesis. This transmembrane protein complex in a purple bacterium works in association with other complexes, including cytochrome bc1 complex, ubiquinone, cytochrome c2, and ATP synthase, to harvest solar energy [48]. This reaction center is associated with LHCs named core LH1 and peripheral LH2. These LHCs transfer light energy to the core RC special bacteriochlorophyll pair. The functional role of this RC is to pump electrons across its core structure beginning from a special pair of bacteriochlorophyll dimer to bacteriochlorophyll, bacteriopheophytin, quinone A, and finally to quinone B in response to incident light. The energy required for this intra-protein electron-hole pair formation is equivalent to that of the energy of 870-nm wavelength photons. The electrons from the reduced quinone B site are shuttled to the cytochrome bc1 complex via the ubiquinone cycle in the lipid membrane. The soluble cytochrome c2 shuttles electrons back from the cytochrome bc1 complex to the oxidized bacteriochlorophyll pair in the P870. This final step completes the loop and primes the RC for a second round of light-induced electron transfer. The cytochrome bc1 complex in this sequence of electron exchange concomitantly pumps protons across the membrane and contributes to the buildup of the transmembrane proton gradient, which is the source of energy for ATP production by ATP synthase. The timescale of electron-hole pair formation [$P^+Q_B^-$] and stabilization across the bacteriochlorophyll pair and quinone b is approximately 100 µs. This timescale allows the diffusible components like ubiquinones and cytochrome c2 to dock on the RC sites and exchange electrons with it [49].

Many significant works pertaining to P870 as a bioelectronic component in sensors and solar energy harvesting devices have been reported. Herein we are discussing a few of these significant reports and pros and cons associated with P870-electrode interaction and electron transfer mechanism. P870-based electrodes are generally fabricated with their bacteriochlorophyll special pair [P side] oriented toward the electrode. Upon exposure of these electrodes to incident light, electron-hole pair formation within the P870 can only be detected as an amperometric signal if redox mediators mimicking the role of an electron carrier *in vivo* are present in the electrolyte. The amperometric pattern of P870 interacting with electrodes has a characteristic alternate current pattern due to electron transfer kinetics limitation. To illustrate these aspects, we discuss research carried out by Tan and Hollander's group. Tan and coworkers fabricated a bio PEC in a two-electrode configuration using FTO-coated glass electrodes [50]. The cathode was sputtered with platinum. The space between the electrodes was filled with a 10 µL volume of N,N,N',N'-tetramethyl-p-phenylenediamine (TMPD) solution loaded with purified P870 reaction centers. It was assumed that the RC protein is adsorbed on the FTO-coated glass electrode based on the higher hydrophilicity of the FTO electrode, which was determined by contact angle measurement. Upon illumination, a spike in voltage and cathodic current was observed, which stabilized at 7 mV and 0.15 µA cm^{-2}, respectively. External quantum efficiency spectra at monochromatic wavelengths overlapped with UV-visible absorbance of the RCs showing a peak at 875 nm. This proved the origin of photocurrent response to be light dependent electron-hole pair separation in P870. Upon termination of incident light, an interesting phenomenon of reverse anodic current was observed, which stabilized over a period of 20 seconds. This reverse current was attributed to a difference in the rate of Q_B oxidation by TMPD and P^+ reduction by the cathodically biased electrode. This difference in the rate of oxidation and reduction led to an accumulation of charges inside the reaction center cofactors, which acted as an electron sink. Incident light termination led to re-equilibration of charges between these cofactors and oxidized TMPD/TMPD$^+$ couple via an external circuit, which was detected as an anodic current. This pathway was verified by two separate strategies. First, introduction of stigmatellin (an inhibitor of quinone B) led to a decrease in the anodic current amplitude. Second, the inclusion of ubiquinone in molar excess led to an increase in the cathodic current amplitude and complete elimination of the anodic current amplitude. Ubiquinone is the mediator responsible for oxidation of the P870 quinone B site *in vivo*. In effect P870 exhibited alternating current response in a light-dependent fashion due to capacitive properties (Figure 9.14).

Adhesion of reaction centers to an electrode must be unidirectional for efficient electron transfer. To solve this problem,

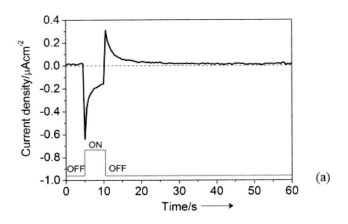
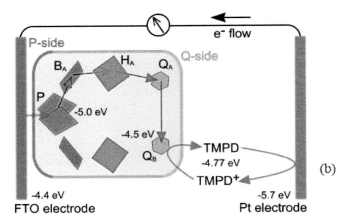

FIGURE 9.14 (a) Alternating current from RC upon illumination. (b) Schematic representation of bioelectrode setup. Reprinted with permission from [50]. Copyright (2012) John Wiley and Sons.

surface functionalization-based linking of proteins to the electrode surface has been reported by many groups. However, surface functionalization reduces the electron transfer kinetics between the reaction center and electrode by increasing the distance for electron tunneling. Work done by Den Hollander and coworkers showed that reaction centers can be directly functionalized on an electrode without the need of a linker [51]. This study investigated P870 in a three-electrode configuration on a 1-mm gold electrode. The reaction center proteins were dropcasted on the electrodes directly and also over surface assembled monolayers covering gold electrodes. The current obtained with the surface assembled monolayer functionalization was less due to a negative effect on electron transfer kinetics. The directly adsorbed reaction centers were sequentially exposed to ubiquinone and cytochrome followed by a second exposure to ubiquinone. Amperometric readings were taken at each step with cathodically biased electrodes. A significant increase in current was observed upon a second round of exposure to ubiquinone. It was concluded that cytochrome provided a conduit between electrodes and reaction centers, causing an increase in current amplitude. Interestingly, the current obtained from unadhered reaction centers was higher than that of adhered reaction centers. This is possible only if the rate of redox exchange between ubiquinone and cytochrome was less in solution. Overall, this study revealed that direct electron transfer is possible between P870 and a gold electrode without the need of functionalization or linking between the P870 and electrodes.

In order to orient the reaction centers uniformly over an electrode, Kamran and his team in 2014, used a Langmuir-Blodgett film deposition of reaction center-light harvesting complexes (RC-LH1) over gold sputtered glass coverslips [52]. Amphiphilic molecules tend to align at the air-water interface with their hydrophilic side facing water and hydrophobic side away from it in a unidirectional fashion. The reaction center proteins tend to show a preferred orientation at an air-water interface. A unidirectional film of these proteins can be collected on a substrate in either orientation by dipping in or dipping out the solid substrate. AFM topography study of these films indicated a densely packed monolayer of reaction center proteins. Photoelectrochemical investigations were carried out in a three-electrode setup using saturated calomel electrode as the reference electrode. An 880-nm LED light was used as the illumination source. The action spectra of the films indicated current contribution by electron-hole pair formation by excitation of a bacteriochlorophyll special pair. Current contribution by carotenoid pigments was poor. Upon inclusion of ubiquinone into the solution, an increase in current was observed. Further addition of cytochrome with ubiquinone enhanced the current magnitude with emergence of characteristic anodic reverse current upon light termination. This feature was similar to that reported by [50]. This pattern of reverse current, indicative of charge storage, was also observed when the concentration of ubiquinone was varied. Both cathodic and anodic currents were saturated beyond 1.6 mM of ubiquinone. In a separate set of experiments, variables like mediator concentrations and potential were fixed and the photocurrent obtained from forward dipped, reverse dipped, and directly adsorbed reaction centers was compared. Forward dipped film produced the maximum photocurrent, which was due to the P side oriented toward the gold electrode. The difference in the orientation of the RC-LH1 complex was also reflected in the I-V curves obtained for forward and reverse dipped films. The oxidative current was higher for forward dipped films, which was other way round for the reverse dipped film. This asymmetry of I-V curves was attributed to a difference in orientation of reaction center protein and its diode-like property. The overall current obtained was 45 μA cm^{-2} with an internal quantum efficiency of 32%. Unlike other redox proteins in which the direction of current is subject to applied bias, current flow through P870 and other reaction centers is unidirectional.

In a subsequent work Kamran and coworkers investigated the route of this unidirectional electron flow across the P870 reaction center in Langmuir-Blodgett films with the aid of conductive atomic force microscopy, genetically engineered P870, and photochronoamperometry [53]. The P870 reaction center has a functional asymmetry in its seemingly symmetric structure. The two polypeptides holding the cofactors in position have 33% identity. This leads to a similar structure but difference in electronic coupling, reorganization energies, and reduction potentials of cofactors. This makes two branches of cofactors, termed "A-wire" and "B-wire" unequal in their capacity to transfer electrons. Films of wild-type reaction centers, broken A wire and B wire reaction centers were used in the experiment. Photocurrents were observed in wild type and broken B wire but were not observed in broken A wire reaction centers. A conductive AFM study of P870 films also revealed electron conduction along wild type and broken B wire samples. The surprising outcome of this experiment was that even under an applied external bias, the A wire was the preferred route for electron conduction.

A photosystem 2 [P680] reaction center is an efficient light-harvesting molecular machinery capable of oxidizing water through the energy of charge separation state between redox cofactors whose transient redox potential difference is 1200 mV. This magnitude is the highest known in the living world and serves as a benchmark of performance in the development of economical water oxidizing catalysts. P680 has a remarkable 100% quantum yield and solar conversion efficiency of 34% due to the utilization of the visible light electromagnetic spectrum region. The downside of such a high oxidizing potential is low stability of the P680 complex, which makes it unsuitable for the long term in *in vitro* applications [54]. P680 has a pair of polypeptide dimers at its core, which acts as a scaffold for holding the redox cofactors at a specific distance and orientations optimized for unidirectional electron transfer. Upon incident light exposure, the energy harvested by the chromophore pigments of LHC are funneled to the P680 special chlorophyll pair which undergoes photo-induced electron transfer to pheophytin. The timescale for this electron transfer step is ~3 picoseconds. Electron transfer follows this from pheophytin to quinone A in ~300 picoseconds. The next step within the reaction center is hole transfer from P680$^+$ to nearby tyrosine Z in a timescale of ~100 nanoseconds. The [TyrZ$^+$ Q$_A^-$] electron-hole pair stabilizes for a timescale of a few milliseconds. During this period of stabilization TyrZ$^+$ extracts electrons from CaO_5Mn_4 containing an oxygen-evolving complex, causing water splitting at the site with evolution of oxygen. Simultaneously electrons are transferred from Q$_A$ to plastoquinone via Q$_B$. The electron flux generated at the Q side is further directed to cytochrome b$_6$f and photosystem 1, causing formation of reducing equivalents like ATP and NADPH, which are used for anabolic pathways by the organism. This pathway of electron transfer is the photosynthetic light reaction famously known as the Z scheme pathway [48].

A P680-based bioelectrode can be used for a solar to chemical conversion unit, water oxidizing photoanode, or mechanistic study of a water oxidizing mechanism. Sequential development and improvement of bioelectrodes for efficient enzyme-electrode wiring and enhanced enzyme loading for better performance and mechanistic studies have been carried out by researchers over the years. Redox polymer-based wiring and mesoporous/microporous electrode structures for high enzyme loading have been the two best-developed strategies to date. In this section, we will discuss the research carried out by these groups in a chronological order that has ultimately led to development and advancement in this field. Zhang and coworkers carried out protein film voltammetry of photosystem 2 core complexes embedded in a lipid film of dimyristoylphosphatidylcholine dispersion and layer-by-layer polyion films of [poly [dimethyldiallylammonium] chloride/poly[4-styrenesulfonate]/ [poly [dimethyldiallylammonium] chloride] on pyrolytic graphite electrodes [55]. The PS2 core complex preparations used in the study were prepared by a protocol, which ensured better oxygen evolving capacity. Availability of crystallography structure with a resolution of 1.9 Å for PS2 preparations following the specific protocol gave a basis for correlating electrochemical and spectroscopic data. Rigorous cyclic voltammogram investigations of PS2 in films revealed three distinct peaks at −0.47 V, −0.12 V, and 0.18 V corresponding to chlorophyll, quinone, and manganese clusters, respectively. The change in voltammetric pattern of redox cofactors in response to changing pH revealed the involvement of protons in the redox reaction mechanisms of individual cofactors.

Badura and coworkers performed photoelectrochemistry of PS2 immobilized in a vinylimidazole-allylamine backbone copolymer with tethered osmium redox moieties on gold electrodes [56]. Use of a redox polymer helped in efficient wiring of the redox enzyme active site and the electrode, irrespective of PS2 enzyme orientation. The photocurrent response achieved upon illumination with 675-nm monochromatic light was about 45 μA cm^{-2}. Cyclic voltammetry investigations of the PS2 immobilized bioelectrode upon illumination with monochromatic 675 nm with increasing light intensities indicated an increase in catalytic anodic current establishing the photoresponsive nature of the anodic current. The long-term stability of the redox polymer-based PS2 bioelectrode upon illumination was improved to an electrode half-life of 18 minutes. This was attributed to the enhanced electron transfer process between PS2 and redox polymer, which might have reduced the rate of reactive oxygen species formation that causes a reduction in PS2 lifetime by oxidative damage. Overall, this immobilization method increased the lifetime of PS2 with efficient enzyme- electrode wiring, irrespective of PS2 orientation at the electrode interface.

Kato and coworkers prepared a water oxidizing photoelectrochemical bioelectrode based on the oxygen evolving photosystem 2 complex from thermophilic cyanobacterium [57]. The photo-enzyme solution was drop-casted on optically transparent, electrically conductive mesoporous ITO electrodes. The mesoporous structure of the ITO electrode offered high enzyme loading and entrapment at the enzyme-electrode interface. Photoelectrochemical investigation was carried out under a bias of +0.5 V vs. NHE upon red light irradiation. A photocurrent response was observed for blank electrodes, Mn complex depleted PS2-based electrodes, oxygen evolving PS2-based electrodes in the absence and presence of quinone-based mediators like 1,4-naphthoquinone-2-sulfonate (NQS) and 2,6-dichloro-1,4-benzoquinone (DCBQ). The blank electrodes and Mn complex-depleted PS2 based photoelectrodes did not show any photoresponse, establishing the origin of electron flux at the Mn complex from the water oxidation reaction. A photocurrent was observed in PS2-based electrodes, both in the presence and absence of mediators. The magnitude of the photocurrent obtained from the mediator-less PS2 electrode was 2.2 µA, from DCBQ it was 22 µA, whereas from that of NQS it was 12 µA. This was due to a difference in the midpoint potentials of the mediator used, which was −60 mV for DCBQ and +21 mV for NQS. This gave a higher driving force for electron flow from the quinone B site to the ITO electrode. In the absence of mediators, only a fraction of PS2 oriented with their electron donor side facing the electrode was capable of transferring electrons. In the presence of a quinone B site inhibitor, a residual current of 0.5 µA was observed. This established the route of electron transfer pathway in the system. It was inferred that the quinone B site is the dominant electron transfer route, as opposed to the quinone A site, though a fraction of electrons pass through the quinone A site. In a subsequent report, PS2 was oriented on negatively functionalized [carboxyl acid] nanostructured ITO electrodes through the electrostatic effect of the inherent dipole moment of PS2 [58]. An EDC [1-ethyl-3-[3-dimethylaminopropyl] carbodiimide] − NHS [N-Hydroxysuccinimide] coupling agent was used to crosslink carboxylic acid and protein amino groups. This strategy greatly enhanced the photocurrent obtained upon illumination in comparison to the previous report, proving the importance of the unidirectional orientation of the reaction center complex for better performance.

Lai and coworkers, in 2014, compared the performance of PS2-based mesoporous ITO electrodes and PS2 inspired synthetic photoelectrodes in terms of water oxidizing activity [59]. The synthetic photoelectrode was made out of WO_3 nanosheets coated with TiO_2 and NiOx. The respective components performed the function of light absorption, charge separation, and water oxidation catalysis. It was observed that highly evolved PS2-based electrodes had a higher turnover frequency and operated at low energy photons as opposed to synthetic photoelectrodes, which operated at high energy ultraviolet and blue end photons. The charge recombination rate in PS2 electrodes was less than that of synthetic photoelectrodes. This was evident from the higher anodic bias required for reaching the saturation current in the case of synthetic photoelectrodes. However, synthetic photoelectrodes were robust with better photostability and current density.

Sokol and coworkers combined the advantage of high enzyme loading and high surface area provided by the mesoporous ITO structure with efficient redox polymer-based enzyme-electrode wiring [60]. A porous scaffold made out of polystyrene beads, coated and sintered with ITO suspension, was developed to be used as a bioelectrode. The pores in the constructed scaffold allowed loading of high-molecular-weight protein complexes and polymers. Polymers and PS2 suspensions were adsorbed through drop-casting on the electrodes. The polymers used for wiring in the study were osmium-modified and phenothiazine-modified redox polymers. The two polymers had redox potential at values of 0.440 V and 0.041 V, respectively (Figure 9.15). The difference in the redox potential of Q_B site (i.e., −0.06 V) and redox polymers rendered different magnitudes of driving force for current flow, which was reflected in photo-chronoamperometry experiments. Action spectra of the PS2 loaded ITO-polymer electrodes revealed a current generation pattern that coincided with the UV-visible absorption spectra of PS2. Stepped potential chronoamperometry under red light illumination revealed different onset potentials for phenothiazine- and osmium-based polymers, which were 0.1 V and 0.3 V, respectively. This result correlated with different redox potentials of respective redox moieties in the polymers used. However, the magnitude of photocurrent obtained from the osmium-based polymer electrode was higher than phenothiazine-based polymer electrodes and without polymer electrodes, which was 230, 45, and 15 µA cm^{-2}, respectively. This was due to a higher driving force for current flow in the osmium-based

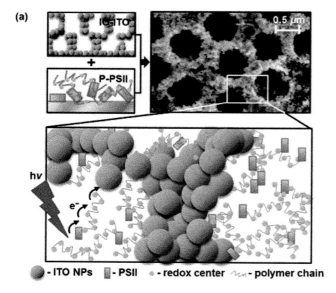
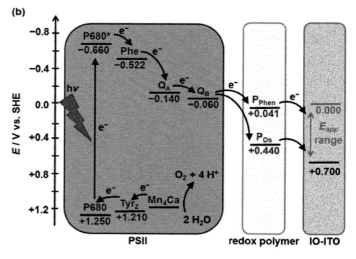

FIGURE 9.15 (a) Mesoporous ITO structure with redox polymer wiring. (b) Schematic representation of electron. Reproduced from Ref. [60] with permission from the Royal Society of Chemistry.

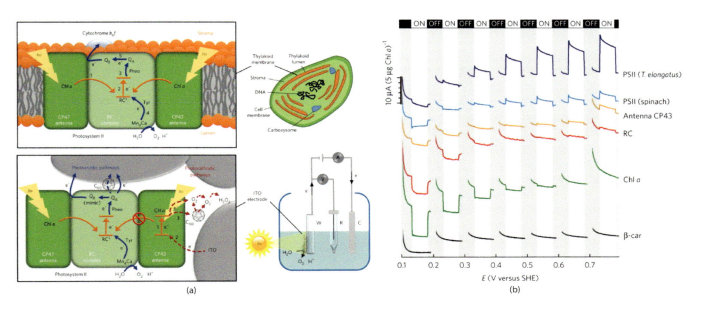

FIGURE 9.16 (a) Buckminster fullerene based interaction of PS2 and ITO electrode. (b) Photo-chronoamperometric response of PS2 components. Reprinted by permission from [61]. Copyright (2016) Springer Nature Customer Service Centre GmbH.

polymer-electrode system. Upon inclusion of dichloro-benzoquinone (DCBQ) mediators in electrodes, the current for the ITO-osmium polymer-PS2 electrode increased to 375 μA cm^{-2}, for the ITO-phenothiazine polymer-PS2 electrode it increased to 236 μA cm^{-2}, and for the ITO-PS2 electrode it increased to 265 μA cm^{-2}. A 1.5-fold increase in the current magnitude of the osmium polymer-based system as opposed to 6- fold and 18-fold increases in bare and phenothiazine polymer-based electrodes further established thatthe wiring was more efficient in the case of an osmium-based polymer electrode system.

Zhang and coworkers in 2016 used the mesoporous ITO-based electrode structure for mechanistic study of electron transfer pathways at the PS2 electrode interface [61]. A fullerene matrix was used in this study to enhance the rate of electron transfer at the electrode-enzyme interface (Figure 9.16). A significant finding of this investigation was the discovery of a competing charge transfer pathway at the electrode interface from high-energy state chlorophyll molecules in the reaction center complex and antenna structure to oxygen, resulting in the formation of O_2^-. This reductive reaction pathway was detected as a cathodic photoresponse, which was enhanced under aerobic conditions. Anaerobicity decreased this cathodic photoresponse and enhanced the anodic photoresponse. A consequence of this finding was the unmasking of the true onset potential of quinone B and quinone A cofactors in the reaction center which is masked due to a competing charge transfer pathway. Though the authors reported a shift of 100 mV in the onset potential in response to the use of quinone B inhibitor, the presence of residual oxygen and the error contributed by it in an accurate determination of quinone redox mediators were also taken into account.

Kornienko and coworkers in 2018 fabricated a mesoporous ITO electrode scaffold on the disc of a rotating ring disk electrode (RRDE) to study oxygen photoreactivity during PS2 mediated water splitting [62]. The products of reaction on the disc of the RRDE are electrochemically detected at the ring. Thus, the water splitting reaction of PS2 was detected on the disc of RRDE and the reactive oxygen species (ROS) formed was detected on the ring of RRDE. The study deciphered two distinct pathways of ROS formation. First was direct reduction of oxygen by chlorophyll anion formed by exposure to a negatively biased electrode. Second was energy transfer to $3O_2$ from the chlorophyll triplet formed as a result of radical pair recombination within the reaction center core.

The photosystem 1 (P700) reaction center is a photoresponsive biomolecular electron pump capable of inducing charge separation of 1 V. Under *in vivo* conditions, it drives low energy electrons through photoexcitation to electrosynthetic pathways for the production of NADPH used in anabolic pathways. The protein exists as a trimer unit in prokaryotic cyanobacteria and as a monomeric unit in eukaryotic plants. The chain of cofactors across which electron-hole pair formation takes place originates from the P700 reaction center special chlorophyll pair. Upon receiving exciton energy from the LHC, P700 is excited and forms a radical pair. The excited P700 chlorophyll is the strongest-known reductant in the biological world, with a magnitude of −1.2 V vs. SHE. The sequence of cofactors that participates in electron exchange thereafter are chlorophyll A_0, phylloquinone A_1 (vitamin K), FeS_x, and FeS_A/FeS_B. These electronic transitions are on the timescale of picoseconds to nanoseconds. The final charge separation and stabilization across $P700:FeS_B$ has a lifetime of 60 milliseconds. This relatively longer lifetime allows the diffusible mediators like ferredoxin and cytochrome c6/plastocyanine to participate in docking and electron transfer reactions *in vivo*. Unlike PS2, where charge separation causes a catalytic activity of water splitting, PS1 charge separation acts as an electron pump that can be harvested in the presence of appropriate diffusible redox mediators. In the absence of redox mediators, an intraprotein electron-hole pair tends to recombine. The relative robustness, stability, and high electron transfer capability of PS1 have been the main motivations for wiring these complexes to the electrode interface for energy harvesting and electrosynthesis applications. In this section we will discuss an instance of the development and improvement of PS1 electrode wiring by some groups [49].

The formation of platinum nanocrystals on the donor site of the PS1 complex as a result of a reduction of $[PtCl_6]_2^-$ salt by reaction $[PtCl_6]_2^- + 4e^- + h\nu \rightarrow Pt\downarrow + 6Cl^-$ is a widely reported phenomenon. The electrons for this redox reaction come from the photoinduced electron pumping reaction of PS1. This PS1-Pt nanocrystal complex has been used for H_2 electrosynthesis and controlling the orientation of PS1 on an electrode. Yehezkeli and coworkers developed three different kinds of PS1-based photoelectrodes [63]. First, a bis-aniline crosslinked monolayer of PS1 on a gold electrode. Second, a monolayer of platinized PS1, thioaniline functionalized platinum nanoparticle. Third, a platinized PS1/Pt nanoparticle monolayer with natural mediator ferredoxin as an electron relay between PS1 and electrode. The bis-aniline linkage exhibited an electron transfer property under an applied bias of 0.01 V. The photocurrent obtained with platinized PS1 crosslinked through bis-aniline linkage was higher than that of the nonplatinized PS1 monolayer. The reason for this was that the Pt nanocrystal crosslinking helped in orienting the PS1 proteins in a unidirectional fashion with their donor sides facing the electrode (Figure 9.17).

Efrati and coworkers crosslinked PS1 to a gold electrode in a random orientation through the electrical relay unit pyrroloquinoline quinone (PQQ), which exists in oxidized and reduced states depending on if anodic or cathodic bias is applied on it through the electrode [64]. Upon applying an oxidative or reductive bias, in the presence of appropriate mediators, the electrode-PQQ-PS1 unit exhibited an anodic or cathodic photocurrent in response to on-off cycle. This established the role of PQQ as a reversible electrical relay unit and also the role of PS1 in generating a photocurrent in the presence of mediators like ascorbic acid, FADH, and oxygen.

Lundgren and coworkers prepared surface assembled monolayers on gold electrodes functionalized with hydrophilic, hydrophobic, and positively charged functional groups [65]. PS1 was deposited on electrodes through electrophoretic deposition at −2 V, which caused them to orient unidirectionally

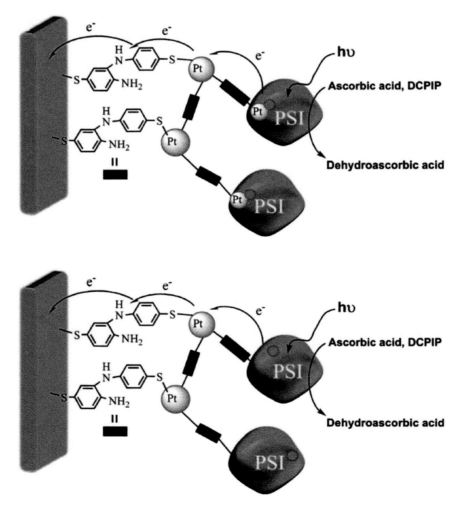

FIGURE 9.17 Bis aniline crosslinking of PS1 to gold electrode via Pt autometallization. Reprinted with permission from [63]. Copyright (2010) American Chemical Society.

due to inherent dipole moment. It was confirmed through AFM that higher electrodeposition occurred on a polar and negatively functionalized monolayer, that is, aminonhexanethiol (AHT), mercaptohexanol (MHO), and mercaptohexanoic acid (MHA), which was attributed to positive charges on the PS1 stromal and luminal faces. The mediator used for shuttling electrons between the electrode and PS1 reaction center was osmium bipyridine complex, and the mediator used for receiving electrons from the PS1 donor side was methyl viologen. Higher photocurrents were recorded on MHA and MHO even though the number of PS1 on AHT was comparable to that of the other two. This was attributed to the positive charge of the osmium complex upon oxidation that caused an electrostatic interaction with the surface assembled monolayer.

Baker et al, in 2014, compared the performance of monomeric and trimeric PS1 complexes on photoelectrodes, as the two forms have different charge transfer kinetics. A conductive redox polymer based wiring method developed by Rogner (discussed below) was used for this study [66]. It was concluded that even though the trimeric form had energy transfer cooperation within its oligomeric units, the process had no bearing on photocurrents generated by the monomeric and trimeric forms. The advantage of using the monomeric form was that it allowed higher loading with dense packing. On the other hand, the trimeric form has the advantage of maintaining unidirectional orientation through fabrication. The only aspect in which the monomeric form differed significantly from trimeric forms was the photocurrent response to incident light intensity. The trimeric form saturated at higher light intensity due to more chlorophyll a per reaction center. Baker et al. performed photoelectrochemical investigation of PS1 entrapped in a nafion ionomer [67]. Nafion polymer has a tendency to retain uncharged and negatively charged species and allows positively charged species to pass through. The redox mediators used to exchange electrons with the PS1 reaction center in the study were a negatively charged osmium complex and positively charged methyl viologen (MV). Based on the properties of nafion, it was assumed that osmium complexes were immobilized in a nafion polymer and MV can freely diffuse across the nafion ionomer. Voltammetric and amperometric experiments indicated the role of the osmium complex in quenching PS1 holes at a cathodic bias. A tradeoff in the amount of nafion used to form a monolayer for optimum performance had to be made, as low concentration caused better diffusion of mediators and high concentration prevented leaching out of PS1. The optimum protein concentration also had to be chosen, as beyond a specific concentration, the photoresponse was attenuated. Overall, this study underlined the importance of choosing an appropriate combination of mediators, protein, and nafion concentrations for bioelectrode fabrication.

Badura and coworkers used osmium redox polymers to wire a PS1 complex with gold electrodes [68]. The redox polymer acted both as electron mediator and immobilization matrix, causing an increase in protein loading and electron transfer kinetics. The approach was similar to the immobilization method used for PS2 by the same author, discussed in a previous section. The high cathodic photocurrent magnitude of 29 $\mu A\ cm^{-2}$ in comparison to previous studies was obtained by using this approach of enzyme wiring. Voltammetric studies indicated the role played by the osmium complex tethered to the redox polymer in carrying out reductive reactions at the reaction center acceptor site. A significant finding of this study was the role played by oxygen as an electron acceptor at the donor site (F_B site) of PS1. Photochronoamperometric investigation in the presence, absence, and different combinations of MV, O_2 indicated the role played by O_2 reduction in the generation of reductive current. It was concluded that assignment of the photocurrent to the electron transfer process between PS1 and the electrode couldn't be made confidently without the removal of oxygen.

The rate of electron transfer across the PS1 reaction center is 47 e^-/sec/PS1, and is a consequence of the diffusion of electron transfer mediators to the donor and acceptor side under *in vivo* conditions. This electron transfer rate was increased three fold to 335 e^-/sec/PS1 by Kothe and coworkers in their experiment with redox polymer (Figure 9.18) [69]. The polymer of choice was poly[vinyl]imidazole Os-[bipy]$_2$Cl with formal redox potential 25 mV less than that of the acceptor site of PS1 fulfilling the requirement of low overpotential. A second aspect of this polymer was its pH tunable configuration alternating between swollen and collapsed states under acidic and alkaline conditions, respectively. Electron transfer kinetics between redox active osmium complex mediators tethered to a polymer backbone is better in a collapsed condition. However, the optimum pH for PS1 photocurrent performance was at acidic pH. The photocurrent magnitude of 25 $\mu A\ cm^{-2}$ obtained in previous work was in a swollen configuration. In order to cope with this situation, the authors crosslinked the imidazole units on the polymer backbone while in a collapsed state at alkaline pH. Thereafter, they changed the pH for optimum PS1 activity. These crosslinked collapsed configurations was amenable to electrochemical investigations, and the bioelectrodes were analyzed through CV and chronoamperometry. It was found that photocurrents were limited by oxygen mass transport, as methyl viologen, upon reduction by the PS1 donor site, is regenerated by oxygen reduction leading to ROS formation. Under oxygen supersaturation conditions, an electron transfer rate of 335 e^-/sec/PS1 was obtained. The study also mentioned the detrimental effect of ROS on PS1 stability and underlined the need of finding alternative terminal electron acceptors to prevent PS1 damage and better utilize electron flux.

Zhao and coworkers achieved this objective by replacing methyl viologen mediator with ubiquinone molecules for accepting electrons from the PS1 donor site [70]. Ubiquinone is a lipid-soluble biological redox mediator active in many metabolic pathways, including photosynthesis. The cathodic photocurrent obtained by using ubiquinone varied linearly with its increasing concentration. Due to the insoluble nature of ubiquinone, a mass transfer effect had a bearing on the electron transfer step between the donor site and ubiquinone, which was the limiting step. Thus, the highest concentration of ubiquinone giving the maximum current was chosen for the study. The optimum pH value for the PS1 response reported

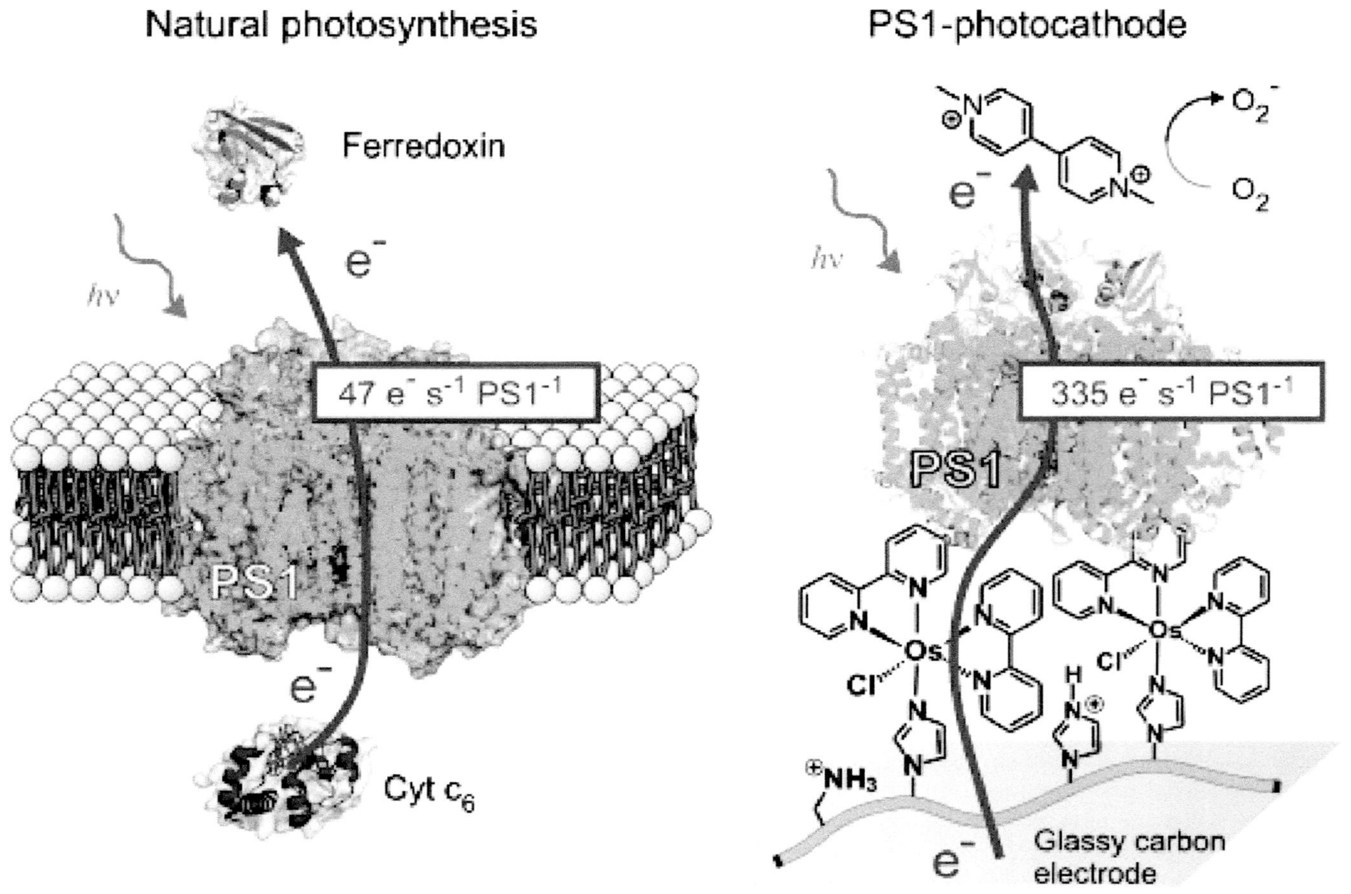

FIGURE 9.18 Schematic representation of PS1 electron transfer rate across the PS1-electrode interface. Reprinted by permission from [69]. Copyright (2014) Springer Nature Customer Service Centre GmbH.

was 4. Since the electron transfer pattern of ubiquinone is pH dependent, voltammetric investigation of ubiquinone at different pH was carried out to choose an appropriate potential for cathodic biasing that ensured low overpotential, sufficient driving force, and prevention of oxidation of reduced electron acceptor on electrode surface (i.e., 200 mV and 300 mV). The photocurrent magnitude under cathodic bias was drastically affected by the presence or absence of oxygen when methyl viologen was used as a cathodic mediator. The photocurrent profiles of ubiquinone mediators remained unaffected by the presence of oxygen. These observations led to the conclusion that ubiquinone can be used as a terminal electron acceptor without the production of detrimental ROS. Long-term amperometric study indicated an increased stability of PS1 when ubiquinone was used as a terminal electron acceptor in the absence of oxygen.

9.9.1 Z Scheme Pathway Mimics and the Electrosynthesis System

The Z scheme pathway is nature's design for directing the low energy electron flux generated from water oxidation at PS2 to the high-energy electron pump PS1 complex. The PS1 complex reduces ferredoxin, which further leads to reductive regeneration of NADPH. The energetic electrons emanating from the PS1 complex can also end up at the hydrogenase enzyme via mediators like ferredoxin, NADP, or NADPH in *in vivo* conditions. This route of hydrogen production in phototrophs is, however, inefficient and difficult to manage predictably for large-scale hydrogen production. Many research groups have successfully been able to mimic this Z scheme by wiring photosynthetic enzymes with electrosynthetic enzymes like hydrogenase, etc., under *in vitro* conditions. Development of these systems and their further improvement is an active field of current research. We devote this section to discussing a few of these significant works.

Yehezkeli and coworkers assembled a Z scheme mimic on a base ITO electrode functionalized with carboxylic moieties [71]. Positively charged redox polymer polybenzyl viologen (PBV^{2+}) was deposited on this by electrostatic interaction. Sequentially deposited PS1 and PS2 between layers of PBV^{2+} yielded a photocurrent higher than that of PS1 or PS2 individually. Inclusion of a second redox polymer poly lysine benzoquinone (PBQ) in a layered assembly in a sequence of PBV^{2+}/PSI/PBQ/PSII yielded a much higher anodic photocurrent. Changing the sequence of deposition to PSII/PBV^{2+}/PSI produced a cathodic photocurrent indicating a vectoral electron transfer pathway between the two reaction center complexes assembled on the electrode. In a subsequent work, Efrati and coworkers introduced cytochrome c hemoprotein between PS1 and PS2 [72]. The PS1 reaction centers were first platinized with nanocrystals and were oriented on electrodes by functionalization and crosslinking. A positively charged pyridine/pyridinium-based polymer interacting with cytochrome c through the pyridine moiety was used to establish electrostatic interaction with the PS1 layer. This step formed a layer of cytochrome c on top of the PS1 layer. Thereafter, the PS2 layer was applied and fixed with glutaraldehyde crosslinking. Photoamperometric investigation revealed a crucial role played by the cytochrome layer in the electron conduit between the water oxidation step in PS2 and electrode via cytochrome c, PS1, and platinum nanocrystals.

Kothe and coworkers resolved the Z scheme configuration across an anode and cathode [73]. The goal of their study was to derive energy for power generation or chemical synthesis from a difference in potential between the reducing side of the PS2 reaction center and oxidizing side of PS1. Wiring in the respective electrodes was performed by using osmium-based redox polymers. The open circuit potential of the system was dictated by the difference in potential of redox polymers in the anode and cathode, which was 505 mV and 395 mV, respectively. The anodic reaction was PS2 mediated water splitting that generated oxygen, and the cathodic reaction was methyl viologen reduction by PS1. The reduced methyl viologen reacted with oxygen by reducing it to ROS that further react with protons to form water. The performance parameters obtained for the system were a short-circuit current of 2.0 μAcm^{-2}, open circuit voltage of 110 mV, power density of 23 $nWcm^{-2}$, and fill factor of 0.128. This system was further improved by [74]. The formal redox potential of polymers was tuned to be close to the redox potential of cofactors at the PS2 donor site and PS1 acceptor site to optimize electron transfer without voltage loss. This caused a significant improvement in the performance parameter. The short-circuit current, open circuit voltage, power density, and fill factor of new system were 9.0 μAcm^{-2}, 372 mV, 1.91 μWcm^{-2}, and 0.564, respectively.

Wang achieved overall water splitting with hydrogen and oxygen evolution in a stoichiometric ratio of 2:1 by developing bioartificial hybrid systems of photosystem 2 and hydrogen evolving artificial photocatalysts [75]. The artificial photocatalysts used in the study were Ru_2S_3/CdS and $Ru/SrTiO_3$:Rh. The electron mediators used to transport electrons from PS2 to photocatalyst were ferricyanide. The choice of this mediator was dictated by the electrochemical potential of ferricyanide, which is 0.358 V, and lies between the electrochemical potential of quinone B and the valence band edge of photocatalysts (i.e., 0 V and 1.5 V/2.2 V). This ensured homogeneous electron transfer between PS2 and the photocatalyst entity (Figure 9.19). The rate of gas evolution in the system showed an increase with increasing concentration of ferricyanide. A significant aspect of this study was that

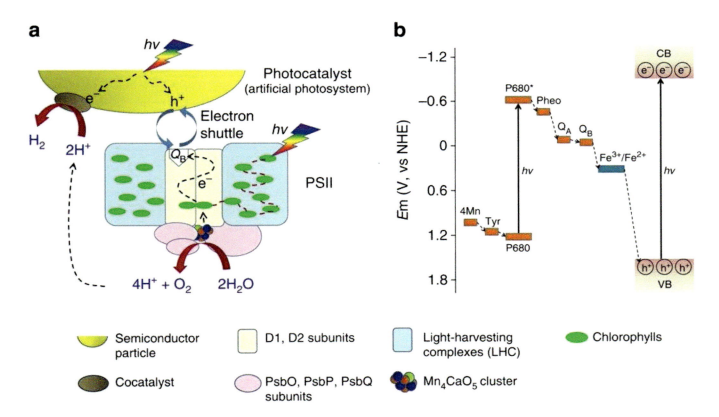

FIGURE 9.19 (a) Schematic diagram representing the electron transfer pathway leading to hydrogen generation. (b) Corresponding energy bandgap diagram. Reprinted by permission from [75]. Copyright (2014) Springer Nature Customer Service Centre GmbH.

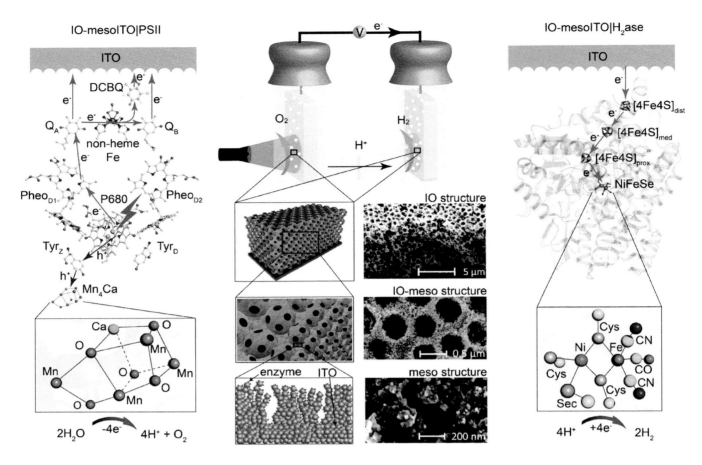

FIGURE 9.20 Device for biocatalytic overall water splitting catalyzed by photosystem 2 and hydrogenase on a hierarchically structured mesoporous ITO electrode. Source: [76] with permission from ACS.

a practical application was demonstrated by measuring gas evolution outdoors in daylight. The system evolved H_2 and O_2 gas in a stoichiometric ratio of 2:1 and showed a stability of 3 hours.

Mersch wired the PS2 reaction center on anode to hydrogenase on a cathode in a Z scheme format for achieving overall water splitting by directing low energy electrons generated in PS2 water oxidation to a cathode for hydrogen generation [76]. PS2 required mediated electron transfer by DCBQ, whereas hydrogenase directly interacted with the electrode to produce a high current (Figure 9.20). The applied bias of 0.6 V to achieve this objective was below the theoretical bias of 1.2 V required for water splitting. The electrode used in the study was the mesoporous ITO structure described earlier. This electrode gave a provision for high enzyme loading, enzyme stability, and activity with quantification of gas evolution due to better faradaic efficiency. The system allowed separation of two incompatible reaction pathways (i.e., oxygenic water splitting and anaerobic hydrogen evolution) across two compartments. The ratio of hydrogen and oxygen evolved was in a stoichiometry of 2:1. The downside of this setup was that it required a monochromatic red light with external bias to operate. In order to be of practical relevance, it is desirable to develop a system capable of operation in daylight without external bias.

Sokol replaced ITO with a semiconducting material TiO_2 to create a mesoporous structure as the electrode [77]. This electrode was then adsorbed by a dye "diketopyrrolopyrrole (dpp)" absorbing light at the green wavelength. The purpose of introducing this dye was to get a substitute for the PS1 reaction center that complements red and blue light absorption by PS2. A tandem photoelectrode was created by introducing an osmium redox polymer-PS2 reaction center into the mesoporous structure. The light source used in this study was UV-filtered simulated solar light. Stepped chronoamperometry measurement under periodic illumination revealed an onset potential at −0.5 V, which was close enough to the onset potential of the TiO_2 conduction band. Control experiments with selective elimination of each component in the tandem construct of TiO_2/dpp/Os polymer/PS2 revealed the indispensability of these sequential fabrication steps in eliciting a significant photoresponse. This photoelectrode was further coupled to a hydrogenase-based mesoporous ITO electrode reported by Mersch. The theoretical estimate of OCP required to initiate the water splitting reaction in this coupled configuration upon illumination was found to be 0.15 V. Significant photocurrents and water splitting were observed beyond 0.1 V. This system evolved hydrogen and oxygen in a ratio of 2:1 (Figure 9.21).

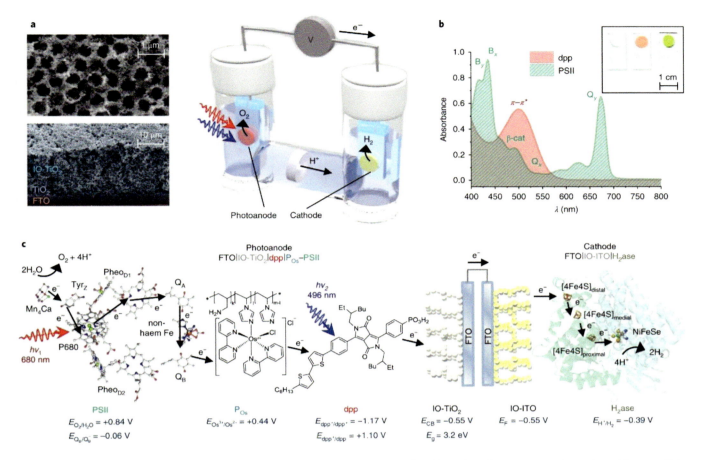

FIGURE 9.21 Device for bias-free overall biocatalytic water splitting with semiconductor-based mesoporous electrodes with green light-responsive dpp dye complementing PS2. Reprinted by permission from [77]. Copyright (2018) Springer Nature Customer Service Centre GmbH.

9.10 CONCLUSION

Significant research on the development of photoelectrochemical biosensors has taken place over the last decade. These research works report many innovative configurations of photosensors, employing various semiconductor-based materials, conductive polymers, carbon-based materials, and photoactive dyes. In many of these cases, the sensitivity of the devices could be enhanced by employing strategies like co-sensitization, resonance energy transfer, and plasmonic signal intensity enhancement. There has been a parallel growth of research in the field of photoelectrochemical biosensors, which mostly focuses on the development of cost-effective, self-sustaining, visible-light-driven sensing devices. Significant progress has been made in developing photosynthetic protein-based bioelectrodes. At this early stage, developments have been seen in fabricating different photoelectrochemical devices with the isolated complex photosystems or their active components to reproduce the natural electron transfer process on the electrode surface. Electrochemical systems mimicking the Z scheme of photosynthetic pathways, capable of splitting water upon direct visible light exposure, have been successfully developed. Such fundamental understanding is crucial for the future development of photosynthetic reaction center-based bioelectrodes and their practical photoelectrochemical biosensing applications for various targets of interest. The concept, however, faces its own set of challenges and drawbacks, such as the requirement of a calibrated visible light source and the discontinuous nature of the solar energy supply. In spite of that, we see significant progress in the field that may help to realize the goal of a renewable energy-based, cost-effective sensor. The future endeavors in this facet of research would be to focus on enhancing the operational stability of these systems for real-world applications.

REFERENCES

1. Peeling RW, Holmes KK, Mabey D, Ronald A. Rapid tests for sexually transmitted infections (STIs): The way forward. Sex Transm Infect. 2006;82(5):1–7.
2. Zhang X, Guo Y, Liu M, Zhang S. Photoelectrochemically active species and photoelectrochemical biosensors. RSC Adv. 2013;3(9):2846–57.
3. Bard AJ. Photoelectrochemistry. Science 1980;207(4427):139–44.
4. Devadoss A, Sudhagar P, Terashima C, Nakata K, Fujishima A. Photoelectrochemical biosensors: New insights into promising photoelectrodes and signal amplification strategies. J Photochem Photobiol C Photochem Rev. 2015;24:43–63.

5. Badura A, Kothe T, Schuhmann W, Rögner M. Wiring photosynthetic enzymes to electrodes. Energy Environ Sci. 2011;4(9):3263–74.
6. B Viswanathan, Scibioh MA. Photoelectrochemistry: Principles and Practices. Narosa Publishing House, New Delhi, India; 2014, pp. 2.7–2.9.
7. Brattain WH, Garrett CGB. Experiments on the interface between germanium and an electrolyte. Bell Syst Tech J. 1955;34(1):129–76.
8. Akira Fujishima KH. Electrochemical photolysis of water at a semiconductor electrode. Nature. 1972;238:37–8.
9. Akira Fujishima KK. Hydrogen production under sunlight with an electrochemical photocell. J Electrochem Soc. 1975;122(11):1487–9.
10. Finklea HO. Photoelectrochemistry: Introductory concepts. J Chem Educ. 2009;60(4):325.
11. Fernández H, Arévalo FJ, Granero AM, Robledo SN, Nieto CHD, Riberi WI, et al. Electrochemical biosensors for the determination of toxic substances related to food safety developed in South America: mycotoxins and herbicides. Chemosensors. 2017;5(3):23.
12. Tanne J, Schäfer D, Khalid W, Parak WJ, Lisdat F. Light-controlled bioelectrochemical sensor based on CdSe/ZnS quantum dots. Anal Chem. 2011;83(20):7778–85.
13. Tang J, Wang Y, Li J, Da P, Geng J, Zheng G. Sensitive enzymatic glucose detection by TiO2 nanowire photoelectrochemical biosensors. J Mater Chem A. 2014;2(17):6153–7.
14. Zheng M, Cui Y, Li X, Liu S, Tang Z. Photoelectrochemical sensing of glucose based on quantum dot and enzyme nanocomposites. J Electroanal Chem. 2011;656(1–2):167–73.
15. Chen D, Zhang H, Li X, Li J. Biofunctional titania nanotubes for visible-light-activated photoelectrochemical biosensing. Anal Chem. 2010;82(6):2253–61.
16. Zhang X, Huo K, Peng X, Xu R, Li P, Chen R, et al. WO_3 nanoparticles decorated core-shell TiC-C nanofiber arrays for high sensitive and non-enzymatic photoelectrochemical biosensing. Chem Commun. 2013;49(63):7091–3.
17. Devadoss A, Sudhagar P, Ravidhas C, Hishinuma R, Terashima C, Nakata K, et al. Simultaneous glucose sensing and biohydrogen evolution from direct photoelectrocatalytic glucose oxidation on robust Cu_2O-TiO_2 electrodes. Phys Chem Chem Phys. 2014;16(39):21237–42.
18. Zang Y, Lei J, Hao Q, Ju H. Signal-On photoelectrochemical sensing strategy based on target-dependent aptamer conformational conversion for selective detection of lead (II) ion. ACS Appl Mater Interfaces. 2014;6(18):15991–7.
19. Rajh T, Saponjic Z, Liu J, Dimitrijevic NM, Scherer NF, Vega-Arroyo M, et al. Charge transfer across the nanocrystalline-DNA interface: Probing DNA recognition. Nano Lett. 2004;4(6):1017–23.
20. Kang Q, Yang L, Chen Y, Luo S, Wen L, Cai Q, et al. Photoelectrochemical detection of pentachlorophenol with a multiple hybrid $CdSe_xTe_{1-x}/TiO_2$ nanotube structure-based label-free immunosensor. Anal Chem. 2010;82(23):9749–54.
21. Wang Y, Li M, Zhu Y, Ge S, Yu J, Yan M, et al. A visible light photoelectrochemical sensor for tumor marker detection using tin dioxide quantum dot-graphene as labels. Analyst. 2013;138(23):7112–8.
22. Tu W, Dong Y, Lei J, Ju H. Low-potential photoelectrochemical biosensing using porphyrin-functionalized TiO_2 nanoparticles. Anal Chem. October. 2010;82(20):8711–6.
23. Tu W, Lei J, Wang P, Ju H. Photoelectrochemistry of free-base-porphyrin-functionalized zinc oxide nanoparticles and their applications in biosensing. Chem – A Eur J. 2011;17(34):9440–7.
24. Wang W, Bao L, Lei J, Tu W, Ju H. Visible light induced photoelectrochemical biosensing based on oxygen-sensitive quantum dots. Anal Chim Acta. 2012;744:33–8.
25. Tel-Vered R, Yehezkeli O, Yildiz HB, Wilner OI, Willner I. Photoelectrochemistry with ordered CdS nanoparticle/relay or photosensitizer/relay dyads on DNA scaffolds. Angew Chemie - Int Ed. 2008;47(43):8272–6.
26. Zhang X, Xu F, Zhao B, Ji X, Yao Y, Wu D, et al. Synthesis of CdS quantum dots decorated graphene nanosheets and non-enzymatic photoelectrochemical detection of glucose. Electrochim Acta. 2014;133:615–22.
27. Zhan W, Jiang K. A modular photocurrent generation system based on phospholipid-assembled fullerenes. Langmuir. 2008;24(23):13258–61.
28. Hu C, Zheng J, Su X, Wang J, Wu W, Hu S. Ultrasensitive all-carbon photoelectrochemical bioprobes for zeptomole immunosensing of tumor markers by an inexpensive visible laser light. Anal Chem. 2013;85(21):10612–9.
29. Wang R, Lu KQ, Tang ZR, Xu YJ. Recent progress in carbon quantum dots: synthesis, properties and applications in photocatalysis. J Mater Chem A. 2017;5(8):3717–34.
30. Hai X, Feng J, Chen X, Wang J. Tuning the optical properties of graphene quantum dots for biosensing and bioimaging. J Mater Chem B. 2018;6(20):3219–34.
31. Tian J, Zhao H, Quan X, Zhang Y, Yu H, Chen S. Fabrication of graphene quantum dots/silicon nanowires nanohybrids for photoelectrochemical detection of microcystin-LR. Sensors Actuators, B Chem. 2014;196:532–8.
32. Ikeda A, Nakasu M, Ogasawara S, Nakanishi H, Nakamura M, Kikuchi JI. Photoelectrochemical sensor with porphyrin-deposited electrodes for determination of nucleotides in water. Org Lett. 2009;11(5):1163–6.
33. Okamoto A, Kamei T, Tanaka K, Saito I. Photostimulated hole transport through a DNA duplex immobilized on a gold electrode. J Am Chem Soc. 2004;126(45):14732–3.
34. Yamada H, Tanabe K, Nishimoto SI. Photocurrent response after enzymatic treatment of DNA duplexes immobilized on gold electrodes: Electrochemical discrimination of 5-methylcytosine modification in DNA. Org Biomol Chem. 2008;6(2):272–7.
35. Haruna KI, Iida H, Tanabe K, Nishimoto SI. Photoelectrochemical evaluation of pH effect on hole transport through triplex-forming DNA immobilized on a gold electrode. Org Biomol Chem. 2008;6(9):1613–7.
36. Haddour N, Chauvin J, Gondran C, Cosnier S. Photoelectrochemical immunosensor for label-free detection and quantification of anti-cholera toxin antibody. J Am Chem Soc. 2006;128(30):9693–8.
37. Gao Z, Tansil NC. An ultrasensitive photoelectrochemical nucleic acid biosensor. Nucleic Acids Res. 2005;33(13):1–8.
38. Samantara AK, Chandra Sahu S, Bag B, Jena B, Jena BK. Photoelectrocatalytic oxidation of NADH by visible light driven plasmonic nanocomposites. J Mater Chem A. 2014;2(32):12677–80.
39. Shen Q, Jiang J, Liu S, Han L, Fan X, Fan M, et al. Facile synthesis of Au-SnO2 hybrid nanospheres with enhanced photoelectrochemical biosensing performance. Nanoscale. 2014;6(12):6315–21.
40. Zhao WW, Wang J, Xu JJ, Chen HY. Energy transfer between CdS quantum dots and Au nanoparticles in photoelectrochemical detection. Chem Commun. 2011;47(39):10990–2.
41. Fan GC, Han L, Zhang JR, Zhu JJ. Enhanced photoelectrochemical strategy for ultrasensitive DNA detection based on

two different sizes of CDTE quantum dots cosensitized TiO2/CdS:Mn hybrid structure. Anal Chem. 2014;86(21):10877–84.
42. Soares da Silva FG, Cerqueira dos Santos GK, Yotsumoto Neto S, de Cássia Silva Luz R, Damos FS. Self-powered sensor for tannic acid exploiting visible LED light as excitation source. Electrochim Acta. 2018;274:67–73.
43. Hao N, Hua R, Zhang K, Lu J, Wang K. A Sunlight Powered Portable Photoelectrochemical Biosensor Based on a Potentiometric Resolve Ratiometric Principle. Anal Chem. 2018;90(22):13207–11.
44. Wang Y, Tang J, Peng Z, Wang Y, Jia D, Kong B, et al. Fully solar-powered photoelectrochemical conversion for simultaneous energy storage and chemical sensing. Nano Lett. 2014;14(6):3668–73.
45. Yan K, Yang Y, Okoth OK, Cheng L, Zhang J. Visible-Light Induced Self-Powered Sensing Platform Based on a Photofuel Cell. Anal Chem. 2016;88(12):6140–4.
46. Yan K, Yang Y, Zhu Y, Zhang J. Highly selective self-powered sensing platform for p-nitrophenol detection constructed with a photocathode-based photocatalytic fuel cell. Anal Chem. 2017;89(17):8599–603.
47. McCormick AJ, Bombelli P, Bradley RW, Thorne R, Wenzel T, Howe CJ. Biophotovoltaics: Oxygenic photosynthetic organisms in the world of bioelectrochemical systems. Energy Environ Sci. 2015;8(4):1092–109.
48. Cox MM, Nelson DL. Lehninger Principles of Biochemistry. New York: W.H. Freeman 2017.
49. Jeuken LJC. Biophotoelectrochemistry: From Bioelectrochemistry to Biophotovoltaics. 2016;158.
50. Tan SC, Crouch LI, Jones MR, Welland M. Generation of alternating current in response to discontinuous illumination by photoelectrochemical cells based on photosynthetic proteins. Angew Chemie - Int Ed. 2012;51(27):6667–71.
51. Den Hollander MJ, Magis JG, Fuchsenberger P, Aartsma TJ, Jones MR, Frese RN. Enhanced photocurrent generation by photosynthetic bacterial reaction centers through molecular relays, light-harvesting complexes, and direct protein-gold interactions. Langmuir. 2011;27(16):10282–94.
52. Kamran M, Delgado JD, Friebe V, Aartsma TJ, Frese RN. Photosynthetic protein complexes as bio-photovoltaic building blocks retaining a high internal quantum efficiency. Biomacromolecules. 2014;15(8):2833–8.
53. Kamran M, Friebe VM, Delgado JD, Aartsma TJ, Frese RN, Jones MR. Demonstration of asymmetric electron conduction in pseudosymmetrical photosynthetic reaction centre proteins in an electrical circuit. Nat Commun. 2015;6:1–9.
54. Milano F, Punzi A, Ragni R, Trotta M, Farinola GM. Photonics and optoelectronics with bacteria: making materials from photosynthetic microorganisms. Adv Funct Mater. 2018;1805521:1–17.
55. Zhang Y, Magdaong N, Frank HA, Rusling JF. Protein film voltammetry and co-factor electron transfer dynamics in spinach photosystem II core complex. Photosynth Res. 2014;120(1–2):153–67.
56. Badura A, Guschin D, Esper B, Kothe T, Neugebauer S, Schuhmann W, et al. Photo-induced electron transfer between photosystem 2 via cross-linked redox hydrogels. Electroanalysis. 2008;20(10):1043–7.
57. Kato M, Cardona T, Rutherford AW, Reisner E. Integrated in a mesoporous indium–tin oxide electrode. J Am Chem Soc. 2012; 134(20), 8332–5.
58. Kato M, Cardona T, Rutherford AW, Reisner E. Covalent immobilization of oriented photosystem II on a nanostructured electrode for solar water oxidation. J Am Chem Soc. 2013;135(29):10610–3.
59. Lai YH, Kato M, Mersch D, Reisner E. Comparison of photoelectrochemical water oxidation activity of a synthetic photocatalyst system with photosystem II. Faraday Discuss. 2014;176:199–211.
60. Sokol KP, Mersch D, Hartmann V, Zhang JZ, Nowaczyk MM, Rögner M, et al. Rational wiring of photosystem II to hierarchical indium tin oxide electrodes using redox polymers. Energy Environ Sci. 2016;9(12):3698–709.
61. Zhang JZ, Sokol KP, Paul N, Romero E, Van Grondelle R, Reisner E. Competing charge transfer pathways at the photosystem II-electrode interface. Nat Chem Biol. 2016;12(12):1046–52.
62. Kornienko N, Zhang JZ, Sokol KP, Lamaison S, Fantuzzi A, Van Grondelle R, et al. Oxygenic photoreactivity in photosystem ii studied by rotating ring disk electrochemistry. J Am Chem Soc. 2018;140(51):17923–31.
63. Yehezkeli O, Wilner OI, Tel-Vered R, Roizman-Sade D, Nechushtai R, Willner I. Generation of photocurrents by bis-aniline-cross-linked Pt nanoparticle/photosystem I composites on electrodes. J Phys Chem B. 2010;114(45):14383–8.
64. Efrati A, Yehezkeli O, Tel-Vered R, Michaeli D, Nechushtai R, Willner I. Electrochemical switching of photoelectrochemical processes at CdS QDs and photosystem I-modified electrodes. ACS Nano. 2012;6(10):9258–66.
65. Lundgren CA, Manocchi AK, Baker DR, Pendley SS, Nguyen K, Hurley MM, et al. Photocurrent generation from surface assembled photosystem i on alkanethiol modified electrodes. Langmuir. 2013;29:2412–9.
66. Baker DR, Manocchi AK, Lamicq ML, Li M, Nguyen K, Sumner JJ, et al. Comparative photoactivity and stability of isolated cyanobacterial monomeric and trimeric photosystem i. J Phys Chem B. 2014;118(10):2703–11.
67. Baker DR, Simmerman RF, Sumner JJ, Bruce BD, Lundgren CA. Photoelectrochemistry of photosystem I bound in Nafion. Langmuir. 2014;30(45):13650–5.
68. Badura A, Guschin D, Kothe T, Kopczak MJ, Schuhmann W, Rögner M. Photocurrent generation by photosystem 1 integrated in crosslinked redox hydrogels. Energy Environ Sci. 2011;4(7):2435–40.
69. Kothe T, Pöller S, Zhao F, Fortgang P, Rögner M, Schuhmann W, et al. Engineered electron-transfer chain in photosystem 1 based photocathodes outperforms electron-transfer rates in natural photosynthesis. Chem - A Eur J. 2014;20(35):11029–34.
70. Zhao F, Ruff A, Rögner M, Schuhmann W, Conzuelo F. Extended operational lifetime of a photosystem-based bioelectrode. J Am Chem Soc. 2019;141(13):5102–6.
71. Yehezkeli O, Tel-Vered R, Michaeli D, Nechushtai R, Willner I. Photosystem i (PSI)/Photosystem II (PSII)-based photo-bioelectrochemical cells revealing directional generation of photocurrents. Small. 2013;9(17):2970–8.
72. Efrati A, Tel-Vered R, Michaeli D, Nechushtai R, Willner I. Cytochrome c-coupled photosystem I and photosystem II (PSI/PSII) photo-bioelectrochemical cells. Energy Environ Sci. 2013;6(10):2950–6.
73. Tim K, Nicolas P, Adrian B, Marc MN, Dmitrii AG, Matthias R, et al. Combination of A photosystem 1-based photocathode and a photosystem 2-based photoanode to a Z-scheme mimic for biophotovoltaic applications. Angew Chemie Int Ed. 2013;52(52):14233–6.
74. Hartmann V, Kothe T, Pöller S, El-Mohsnawy E, Nowaczyk MM, Plumeré N, et al. Redox hydrogels with adjusted redox potential for improved efficiency in Z-scheme inspired biophotovoltaic cells. Phys Chem Chem Phys. 2014;16(24):11936–41.

75. Wang W, Chen J, Li C, Tian W. Achieving solar overall water splitting with hybrid photosystems of photosystem II and artificial photocatalysts. Nat Commun. 2014;5:1–8.
76. Mersch D, Lee CY, Zhang JZ, Brinkert K, Fontecilla-Camps JC, Rutherford AW, *et al*. Wiring of photosystem II to hydrogenase for photoelectrochemical water splitting. J Am Chem Soc. 2015;137(26):8541–9.
77. Sokol KP, Robinson WE, Warnan J, Kornienko N, Nowaczyk MM, Ruff A, *et al*. Bias-free photoelectrochemical water splitting with photosystem II on a dye-sensitized photoanode wired to hydrogenase. Nat Energy. 2018;3(11):944–51.

10 Biofuel Cells as an Emerging Biosensing Device

Sharbani Kaushik
The Ohio State University, Columbus, USA

Caraline Ann Jacob and Pranab Goswami
Indian Institute of Technology Guwahati, Assam, India

CONTENTS

10.1 Introduction 211
   10.1.1 Types of Biofuel Cells..........212
   10.1.2 Working Principle of BFC-Based Biosensors..........213
   10.1.3 Mechanisms of Electron Transfer on Bioelectrodes..........213
   10.1.4 Electron Transfer Theory..........216
   10.1.5 Fundamental Equations Governing the Performance of BFCs..........216
10.2 Materials and Platforms for Bioelectrode Fabrication..........218
   10.2.1 Advanced Materials for Developing Electrodes..........218
   10.2.2 Miniaturized BFCs..........219
   10.2.3 Paper-Based BFC Systems..........220
10.3 EFC Biosensors..........221
   10.3.1 EFC Biosensors for Monitoring Glucose..........221
   10.3.2 EFC Biosensors for Alcohol and Other Analytes..........223
   10.3.3 Wearable EFC Biosensors..........223
   10.3.4 Other Applications..........224
10.4 MFC Biosensors..........224
   10.4.1 Toxic Detection in Water..........225
   10.4.2 Formaldehyde Detection..........225
   10.4.3 Other Applications..........227
10.5 Conclusions..........228
References..........229

10.1 INTRODUCTION

Biofuel cells (BFCs) are a subclass of fuel cells. The first fuel cell that was based on hydrogen fuel came into focus in 1859 through the work of William Grove, who is acclaimed as the father of fuel cells. Later on, the term "fuel cell" was coined by Langer and Mond in 1889 [1]. The earliest reference of bioelectrochemistry can be traced back to the 18th century, to the work of Luigi Galvani on the leg of a frog that made it twitch [2]. However, it was not until 1910 when the experiments conducted by the botanist Michael Cresse Potter established the fact that microorganisms could transfer electrons to the electrode and generate a voltage. Following this invention, many important milestones were achieved in the history of BFC research. Among which, the effort of NASA in the 1960s to convert organic waste into electricity on spacecraft [3], the use of isolated electron transfer proteins on the electrode surface in 1960, the development of complete enzymatic BFCs using three different enzymatic bioanodes in 1964 by Kimble [4], and the application of synthetic electron transfer mediators to generate current in BFC by Bennetto et al. in 1983 [5] are prominent as reviewed in these publications.

BFCs employ biological catalysts such as enzymes, organelles, microorganisms, or even tissues for direct conversion of chemical energy of fuel substrates into electrical energy. This energy conversion occurs at moderate conditions, around room temperature and neutral pH values. Depending on the type of catalyst, BFCs can accommodate a wide range of renewable fuel substrates such as carbohydrates, alcohol, and organic wastewater. The fuels generally used in BFCs are safer and easier to handle than traditional fuels such as methanol and hydrogen gas. Overall, the BFCs are considered a green technology, as these do not typically involve any hazardous substances and conditions in their fabrications and operations [6, 7].

Undoubtedly, the initial spirit of BFC research was mainly incited by the demand for clean and renewable energy [8]. Wastewater treatment and powering small-scale electronic devices emerged as potential areas of application for BFCs. However, the rapid advancement in material sciences, communication technology, and molecular sciences over the last two decades has intensely accelerated the progress in BFC research. These developments have prompted BFC research to develop small-scale devices bearing potential applications in the field of biomedical science, including health care and allied biosensing applications. These self-powered biosensing devices may have a substantial overlap of technical and procedural requirements with the conventional sensors. There are many relevant reviews and experimental literature on BFCs [9, 10]. This chapter highlights the types of BFCs, their working principles, and performance factors, the materials and platforms explored for their fabrications, and biosensing applications.

10.1.1 Types of Biofuel Cells

The basic structure of a BFC shares similarity to a conventional fuel cell. The two electrodes are placed in a suitable electrolyte and are usually partitioned by a proton selective membrane (PEM), also known as the cation exchange membrane. In BFC, biocatalysts are used in the electrodes to produce a bioanode and biocathode. The cells are fabricated using either a bioanode and/or a biocathode. The pure BFCs, however, are made up of a bioanode and a biocathode. The two electrodes are connected via an external circuit (Figure 10.1). The biocatalyst employed at the anode oxidizes the fuel. The biocatalyst in the cathode catalyzes the reduction reaction and is frequently used to catalyze the conversion of oxygen to water. However, in many BFC studies, chemical cathodes are reported, where oxygen reducing materials (e.g., platinum), and chemicals (e.g., ferricyanide, permanganate) are used. The biocatalyst at the anode is responsible for stripping off electrons from the substrate and transferring them to the anode. The generated protons during this oxidation process are moved across the proton exchange membrane. These protons then pair up with oxygen and electron at the cathode, forming water [11]. If the circuit is completed by connecting the anode and cathode across a load, the continuous ferrying of electrons through the closed-loop generates electricity.

Based on the type of biological catalyst used at the electrode (mostly in the anode), BFCs can be primarily divided into enzymatic biofuel cells (EFCs) and microbial biofuel cells (MFCs). As the names suggest, EFCs employ isolated redox enzymes to catalyze the energy transformation reactions, whereas living whole cells/microbes are used as the biocatalyst in MFCs. The gamut of enzymes present inside the whole cell is responsible for catalyzing the reactions in MFCs. The source of "fuel" for BFCs is typically renewable in nature and can be anything – from simple sugar substrates to complex biomass, organic waste, etc. Both these types of BFCs are explored for biosensing applications.

Enzymes exhibit high catalytic rates and substrate selectivity. If the catalyst concentration at a given area in the EFC electrode could be increased, the mass transfer limitation in the electrode reaction would be significantly decreased, and this opens up an avenue to design a miniaturized BFC. Since the enzymes are substrate specific, a membrane-less EFC has also been pursued to simplify the design and technical complexity and to reduce the cost of the devices [12]. However, the stability of the isolated enzymes is usually low due to the denaturation of the enzyme proteins, which grossly affects the lifetime of EFCs. Immobilization of purified enzymes at the electrode surface has evolved as a method of choice to overcome the challenge of limited stability and improve mass transfer kinetics.

Microbes depend on the energy liberated from substrate oxidation (respiration process) to sustain their metabolic

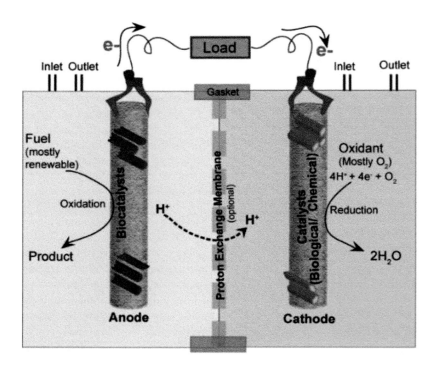

FIGURE 10.1 A general configuration of a biofuel cell.

activities. In aerobic respiration, oxygen acts as an electron acceptor at the terminal of the oxidative phosphorylated chain. In the absence of oxygen, the electrons are exported to ionic species or the conductive solid. There can be three plausible reasons for electron donation by microbes to the ecological niches: (i) Electrical communication between the microbes can be a part of quorum sensing (a mechanism of intraspecies bacterial communication), (ii) a route for protecting themselves from reactive oxygen species by routing the excess reducing power outside the cells, and (iii) reducing metal substrates present in the environment [13]. When acetate is used as a substrate in the MFC, the anodic half-reaction can be represented as $CH_3COOH + 2H_2O \rightarrow 2CO_2 + 8H^+ + 8e^-$. The protons so generated are transferred from the bacterial colony across the PEM membrane through the bulk liquid to maintain the electro-neutrality.

MFCs rely on the capability of microbes to electronically communicate with the electrode. This is often a difficult task, as the bacterial membrane is not conducive to charge transfer. Nanostructured electrodes and synthetic compounds can aid them in ferrying the electrons to the conductive electrode. For the bacteria to give up electrons to the electrode, they need to be in close proximity with a conductive surface. The bacteria tend to assemble around the electrode in clusters known as "biofilms." Biofilms are defined as highly structured and organized cellular aggregates often found to be surface-adherent and encased in an extracellular polysaccharide matrix. This biofilm should be of moderate thickness. Too thin a biofilm may be inadequate to produce sufficient charge species for electron extraction. At the same time, too thick of one may limit mass transfer and even charge transfer due to the inactive cell layer in the cell-electrode interfaces. The biofilm lives on the surface of the conductive solid of the MFC. It hence is termed an "anode respiring bacteria" (ARB) [14]. The ARB is electroactive in nature. Some microbes can readily give up electrons to the electrode. *Geobacter*, *Pseudomonas*, *Shewanella*, and *Escherichia coli* species are some commonly studied exoelectrogenic bacteria [15]. These are essentially gram-negative bacteria. BFCs with pure gram-positive strains are comparatively less studied due to their weak electrogenic nature. The presence of thick non-conducting cell walls is likely to hamper channelization of the metabolic electron from the gram-positive strains to the electrodes. Mixed consortium and heterotrophic cultures are also used as catalysts in the MFCs.

When photosynthetic bacteria are employed as the biocatalyst, the MFC may be termed either a photosynthetic microbial fuel cell (PMFC) [16, 17] (using an organic reducing equivalent) or biophotovoltaics (BPVs) (in the absence of any external carbon substrate other than CO_2) depending upon the conditions, as mentioned [13]. These BPVs, also occasionally called biosolar cells, have recently received wide interest due to various advantages, among which the low operating cost, as no organic fuel substrate is needed for their operation, and scope for miniaturization are prominent factors.

There are some reports on organelle [18, 19] and nucleic acid [20] based BFC. Organelle-based BFCs mostly use mitochondria at the bioanode [18]. Mitochondria contain all the enzymes (the pathways of the citric acid cycle and electron transport chain) necessary for complete oxidation of simple sugars. Mitochondria are extremely sensitive to poisonous toxins, drugs, etc. Hence, such types of BFCs can be used to screen for pharmaceutical drugs, cyanide, etc. Similarly, thylakoids contain the photosystems responsible for photocurrent generation in biosolar cells. The sensitivity of the thylakoids to herbicides may be exploited for sensing various toxicants such as atrazine, diuron, etc., with a limit of detection far below the EPA mandates [21]. However, these types of sensors have yet to gain much popularity due to their lack of selectivity as compared to that of EFCs [22].

10.1.2 Working Principle of BFC-Based Biosensors

Among the various transducers being used, electrochemical sensors are at the forefront for commercial applications due to their high sensitivity, reliability, simplicity, and inexpensive nature. Being electrochemical, BFCs have an additional attractive feature over the conventional electrochemical biosensors and that can be attributed to its stand-alone operational mode. Unlike other electrochemical biosensors, the BFCs are power-generating devices; hence, the presence of the target analyte in a sample could be probed directly from their electrical signal output, thus obviating the need for any external power source [23]. The output signal (current or voltage) may be higher (turn-on sensors) or lower (turn-off sensors) than the control signal depending upon the effect of the target analyte on the electrode biocatalyst of the BFCs. As an example, if the target analyte is an inhibitor of the biocatalyst, the BFC functions as a turn-off sensor; on the other hand, if the target acts as an activator of the biocatalyst, the BFC functions as turn-on sensor. Figure 10.2 displays a simple design of an EFC biosensor [24]. The coupling of the transducer with a redox enzyme (bioreceptor) at anode forms the "molecular transducer." Another enzyme at the cathode catalyzes the reduction of oxygen to water. The current change is proportional to the concentration of the target analyte.

Technically, BFC is a power-generating device. However, while designing BFCs for biosensing applications, the focus is on generating a sensitive and selective electrical signal output rather than the generating power. To obtain a detailed overview of the operation, system design parameters and scale-up of BFCs entirely for energy harvesting, readers may consult some relevant reviews [24–30].

10.1.3 Mechanisms of Electron Transfer on Bioelectrodes

The exchange of electrons between the biocatalyst and electrode is essential to generate an electrical signal in BFCs. Broadly two strategies are involved in exchange of electrons between the biocatalysts and electrodes in BFCs: mediated electron transfer (MET) and direct electron transfer (DET). In the MET approach, a suitable chemical compound is used to shuttle the electrons between the biocatalysts and electrode. The mediator

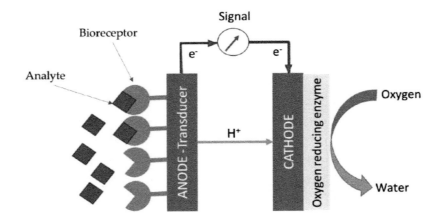

FIGURE 10.2 Schematic of an EFC-based biosensor and its detection mechanism [24]. The analyte is oxidized by the immobilized bioreceptor producing an electrical response signal.

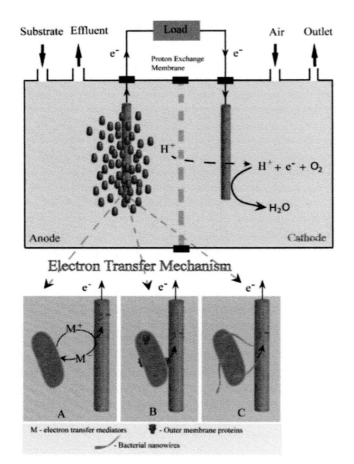

FIGURE 10.3 Electron transfer mechanisms from microorganisms to electrode. It occurs by (**A**) mediated by electron shuttles, (**B**) directly by the cell using outer membrane proteins, or (**C**) by means of "nanowires" [8]. Reproduced from [8], with permission. © Elsevier.

is preferably reduced during the anodic reaction and further re-oxidized at the working anode, which is maintained at high electric potential. The mediator plays the role of a "middleman" shuttling the electrons from the biocatalyst to the electrode and vice versa. Mediators can be synthetic (e.g., methylene blue, neutral blue, ferricyanide, quinones), natural (e.g., vitamin K_3), and secreted by some bacterial species (in MFC) in redox polymer or polymer electrolytes [24]. If the biocatalysts can communicate with the electrode directly, the electron transfer process is termed DET [31, 32]. In case of MFCs, an additional mechanism has come into light, known as nanowire-based electron transfer [33] (Figure 10.3), which has been reported only for a few electrogenic bacterial species, such as *Geobacter metallireducens*. These nanowires are cellular appendages [15] involved in the long-distance extracellular electron transfer (EET) process. The conductivity of such nanowires exhibits semiconducting properties. The metal-reducing bacteria *Shewanella* sp. displays nanowires with resistivity 1 Ω cm [34] and might conjugate cyt-c in the cell wall membrane extensions with electrons hopping along the length. The conductive pili appendages reported in *Geobacter* sp. demonstrate metallic properties similar to carbon nanotubes, which may be credited to the overlapping pi-pi orbitals of aromatic amino acids [35].

In MFCs, the electrons generated from the oxidation of organic substrates are transferred by the bacteria to the anode using a combination of either the DET or MET mechanisms. The DET methods are well exploited for the electrochemically active bacteria *Geobacter* and *Shewanella* sp. due to the presence of redox proteins, c-type cytochromes (cyt-c), on the outer membrane of these cells. An electro-active biofilm (EABF) on the anode usually supports the electron transfer process [14]. The microbes exist as multilayers surrounding the electrode, encased in exopolymeric substances (EPS). Therefore, only the first layer of cells in immediate contact with the electrode can maintain the DET. To overcome the electron transfer barrier across the biofilm matrix, the cells also secrete active redox molecules such as quinones and flavins that facilitate electron transfer (ET). Development of a stable EABF is an important task for generating consistent signals in the MFC biosensor. The composition and structure of EABF depend on the type of bacterial strains. Generally, a pure strain is preferred for determination of single analyte and mixed species for monitoring complex parameters, including toxicity measurements. At the same time, the designed parameters for the biofilms are framed based on the mode of operation such as one-time use/disposable or real-time continuous monitoring. The conductivity and microbial loading of EABFs govern the electron transfer rates and hence the response to the parameters. Although the biomolecules and biopolymers are regarded as insulators, the EPS of an EABF displays semiconductive properties with conductivity between 10^{-9} and 10^3 S cm^{-1} [14].

The photosynthetic cyanobacteria are widely used for PMFC or BPV as compared to algae due to the simple physiology of the former species. An increasing research trend on cyanobacteria-based PMFC has been seen due to various potential benefits that could be accrued by using these photosynthetic bacteria as fuel cell catalysts. One of the unique metabolic features of cyanobacteria that is likely to sustain electron flux on the fuel cell electrode is the interlinked photosynthetic and respiratory electron transport pathways present in their thylakoid membranes [36, 37]. Power generation in PMFC following both MET (externally supplied as well as endogenous) and DET mechanisms are reported. More information on whole cell photoelectrodes for electrosynthesis and the process is provided in Chapter 9.

Oxidoreductases are the key enzymes used in the development of EFCs [38]. These redox enzymes can be divided into two categories: oxygenase and dehydrogenase. Oxygenase uses molecular oxygen as the chemical partner receiving the electrons from the substrate during the catalysis, whereas dehydrogenase oxidizes a substrate by reducing an electron acceptor, usually NAD$^+$/NADP$^+$ or a flavin coenzyme such as FAD or FMN. There are only few redox enzymes such as laccase, peroxidase, cytochrome c, ferredoxin, and alcohol hydrogenase that can communicate directly with the electrode following the DET principles. This is because the catalytic sites of most of the redox enzymes are buried inside the protein's matrix concealed from the environment. This inhibits the transfer of electrons from the redox center of the enzyme to the electrodes. In such cases, redox mediators with suitable properties may be used to shuttle the electrons between the redox centers of the enzymes and the electrode. The redox center is generally composed of different types of molecular entities such as hemes, quinones, copper centers, iron-sulfur clusters, flavins, or NAD(P)H where the exchange of electrons takes place. Eddowes and Hill, as well as Yeh and Kuwana, both independently, demonstrated the capability of redox protein to support DET for the first time [38].

A plausible thumb rule is that if the distance between the enzyme/cell and electrode is greater than 15 Å, the aid of mediators is required for tunneling electrons [24]. The relevant principle of electron transfer is discussed in the next section. The planktonic microbes in the anodic chamber may also undergo ET with the aid of mediators. Commonly used exogenous mediators are ferrocene and its some derivatives, phenazines, flavins, quinones, methyl viologen, neutral red, thionine, etc. Strategies for using such artificial mediators in free form as well as in physically/chemically entrapped systems have been reported.

In MET-based BFC biosensors, the stability of the mediator is one of the key issues to generate a reproducible response signal. The mediator should be stable in both the reduced and the oxidized forms and devoid of any side reactions between the mediator redox states with the enzyme or the environment. To be effective in its role, the mediator must compete with the natural substrate (e.g., molecular oxygen in case of oxidases) of the enzyme to efficiently channelize the flow of electrons to and from the electrode. A mediator should also exhibit a reversible electrochemistry, a large rate constant (k_{et}) for the interfacial electron transfer at the electrode surface. Further, the penetration of the mediator close to the enzyme-active center inside the protein matrix is controlled by the hydrophobic/hydrophilic properties of the mediator and the enzyme, the size and shape of the mediator, and the electrostatic charge interactions between the mediator and the enzyme. Additionally, the mediators used for MET-based MFCs should possess nontoxic, non-biodegradable, and membrane-permeable qualities. The incorporation of mediators limits the potential losses in BFCs, capitalizing the voltage and power outputs. This has resulted in the successful implementation of BFCs as sensitive biosensors.

Conductive nanomaterials are found to be highly effective in bridging the electron transfer gaps, promoting DET between the biocatalysts (redox enzymes and bacterial cells) and the electrodes to develop an efficient bioelectrode. Different highly conductive nanomaterials, particularly graphene and its various forms (such as single and multiwalled carbon nanotubes and graphene nanoplatelets), gold nanoparticles, etc., have received intensive interest to develop bioelectrodes for BFCs and other electrochemical sensors. Various design formats have been evolved to couple these nanomaterials during fabrication of the bioelectrodes. In addition to initiate/improve electron transfer kinetics, nanomaterial-based fabrication greatly increases the electroactive surface area over the electrode to improve bioelectrocatalysis [39]. Some illustration on the subject is given in a later section of this chapter. Based on the design strategies, the bioelectrodes are also termed second-generation and third-generation for the corresponding MET- and DET-based principles. A less widely used strategy to establish a bridge between the redox center of the enzyme and electrode is the co-factor reconstituted approach, particularly in terms of FAD-based redox enzymes [39].

The characteristics of the biocatalyst may be referred to adopt a suitable electron transfer strategy (MET/DET) for developing efficient bioelectrodes. For instance, the redox enzyme whose redox center is located in the periphery of the protein matrix may serve well if the DET principle is adopted to fabricate the bioelectrode. However, the auxiliary factors such as proper orientation/alignment of the enzyme molecules over the electrode surface, loading of the active enzyme molecules, and biocompatible environment for these labile biocatalysts also need to be considered for developing a functionally efficient and stable bioelectrode. The phenotype and genotype of the bacteria that confer the expression of redox proteins on the cell wall surface, extracellular secretion, or conductive nanowires may provide some clues to design rationally MFC bioelectrodes. DET is considered an ideal choice for developing an operationally stable bioelectrode, as MET-based systems usually suffer from the leaching of mediators, which are mostly harmful to the environment, to the electrolyte that may conflict with the electrode stability and environmental safety. Although producing high power is not an objective in BFCs meant for biosensing, electrode stability is a significant concern to achieve reproducible signal outputs. An effective strategy to prolong catalytic activity and improve reproducibility of BFC biosensors is to anchor the microbes/enzymes to a suitable support, a process termed "immobilization," which has already been described in Chapter 1. The mediators co-immobilized along with the enzyme/cell or oriented immobilized (in case of enzymes) may profoundly improve the electron transfer rates [39].

10.1.4 Electron Transfer Theory

The channelization of electrons from the enzyme-active center to the electrode can be defined by the Marcus–Hush–Chidsey equations [38]. Marcus and Hush developed model for homogeneous (Equation 10.1) and heterogeneous (Equation 10.2) bioelectrochemistry models, whereas Chidsey (Equation 10.3) explained the transfer rates on electrode distances. According to this formalism, the electron transfer depends on the enzyme structure and orientation (location of the redox active center) and exponentially on the electron transfer distance. The semi-classical Marcus theory affirms that the ET rate (k_{ET}) is governed by the Gibbs free energy ($\Delta G°$), reorganization energy (λ), and electronic coupling (H_{AD}) between the electron donor (D) and acceptor (A) at the transition state. Equation (10.1) describes a non-adiabatic electron transfer, which occurs for most protein processes [40], according to Marcus theory:

$$k_{ET} = \frac{4\pi^2 H_{DA}^2}{h\sqrt{4\pi\lambda k_b T}} \exp\left[\frac{-(\Delta G° + \lambda)^2}{4\lambda k_b T}\right] \quad (10.1)$$

where k_{ET} is the electron transfer rate, h is the Planck's constant, k_b is the Boltzmann constant, T is the temperature, $\Delta G°$ is the Gibbs free energy, λ is the reorganization energy, and H_{DA} is the electronic coupling between the electron donor (D) and acceptor (A) at the transition state.

The electron transfer in heterogeneous systems is described by the following equation:

$$k_{red/oxi} = \frac{k_{max}}{\sqrt{4\pi\lambda/RT}} \int_{-\infty}^{\infty} \frac{\exp\left[-\left[\left(\lambda \pm F(E-E°)\right)/RT - x\right]^2 RT/4\lambda\right]}{\exp(x)+1} dx \quad (10.2)$$

where E is the applied potential, $E°$ is the standard potential, and k_{max} is the asymptotic value of the rate constant at high overpotential, which is given by Equation (10.3)

$$k_{max} = \frac{4\pi^2 V_0^2}{N_A hRT} \exp(-\beta r) \quad (10.3)$$

where V_0 is the degree of electronic coupling between the donor and the acceptor, β is the decay coefficient, and r is the distance between redox centers.

For an efficient electron transfer to occur, the enzyme, along with its redox center, should be co-aligned to the electrode surface; the redox center should be embedded towards the protein surface; and most importantly, the electron transfer distance should be minimum, as revealed from the analysis of the earlier theory. The issue of alignment may not be a concern in the case of whole cells.

10.1.5 Fundamental Equations Governing the Performance of BFCs

The fundamentals of the fuel cell can be extended to the BFC to evaluate its performance [41]. Most enzymes use co-factors that have lower redox potentials than the reactions they catalyze. So typically, the highest open circuit potential (OCP) for a BFC is equal to the difference in redox potentials between the positive cathode and negative anode.

When the overall reaction of a BFC is thermodynamically favorable, electricity is generated [42]. The concept may be defined by the following equation:

$$\Delta G_r = -E_{emf} \times nF \quad (10.4)$$

where ΔG_r (J) is the Gibbs free energy, nF is the charge transferred in the reaction with n representing the number of electrons per reaction mol, and F is Faraday's constant (9.64853×10^4 C/mol). The electromotive force E (emf) generated in the fuel cell is due to the difference between the cathodic (E_{cat}) and anodic (E_{an}) potential, as shown later [41], and can be calculated from the Gibbs free energy change for the anodic and cathodic reactions [42].

$$E_{emf} = E_{cat} - E_{an} \quad (10.5)$$

Ideally, the cell voltage should be independent of the current drawn. However, in practice, this reversible cell voltage (E_{cell}) is not realized even under infinite load (zero current) conditions due to internal losses and fuel crossover when the cell is operated. The cell voltage at zero current is termed OCP. As current is drawn from the fuel cell (at varying loads), the E_{cell} deviates from OCP as a result of various losses, which are known as overpotential, as depicted by the following equation [41].

$$E_{cell} = E_{emf} - \left(\sum \eta_{act} + \sum \eta_{conc} + IR_\Omega\right) \quad (10.6)$$

where η_{act} is the activation overpotential, η_{conc} is concentration overpotential, and I and R_Ω represent current and resistance (load), respectively. The current discharge pattern with respect to the external resistance can be illustrated by the polarization curve, which is plotted by considering the change in current density versus voltage [43]. The influence of external and internal resistances can be understood by such polarization graphs. There are three distinct regions at different current ranges (Figure 10.4):

a. At low currents, activation (charge transfer) overpotential (η_{act}) dominates. This overpotential arises from the energy barrier to charge transfer or from

Biofuel Cells as an Emerging Biosensing Device

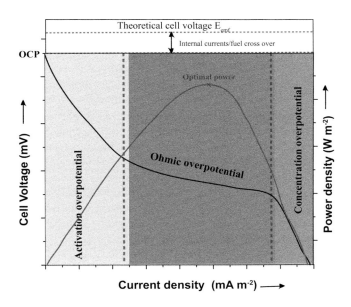

FIGURE 10.4 Polarization and power curves of a typical BFC depicting the possible overpotential losses incurred during the withdrawal of current (at varying load intensities).

the mediator or bacteria/enzyme to the electrodes. The charge transfer overpotentials depend on the nature of the electrode materials, catalysts, reactant activities, electrochemical mediators, biofilm, electrode microstructure, microbial species and their metabolism, and operational conditions. The activation overpotential can be reduced by various means such as facilitating electron transfer, electroactive biofilm formation on the anode, increasing the reaction sites by increasing the electrode surface area, improving the biocatalyst acclimatization, and beneficial gene expression (in case of MFCs) [44].

b. The ohmic overpotential (IR_Ω) is observed at the intermediate current regions. Ohmic losses are due to the resistance to charge transport both ionic and electronic resistances through the current collectors, electrolytes, biofilm, membrane, and electrodes, as well as the interfaces between these components [43]. These overpotentials (separate for the two electrodes) can be approximated if expressions for the reaction rates are known (e.g., a Tafel's or Butler–Volmer's relation).

c. Concentration overpotential (η_{con}) manifests itself due to mass transport limitations at the interface between the electrode surface region and bulk electrolyte prevalent at high current densities. It can be avoided by proper reactor design, mixing/aeration, and eliminating diffusion gradients. This overpotential is dominant in chemical fuel cells where current density is usually high.

d. In MFCs, an additional potential loss occurs that is known as bacterial metabolic loss. Bacteria transport substrate electrons at low potential (e.g., NADH for acetate −0.296 V) through the electron transport chain to the final electron acceptor at a higher potential (e.g., oxygen +0.82 V) to generate metabolic energy. The anode is the final electron acceptor for the bacterial catalysts in an MFC. Thus, the higher the anode potential, the larger the energy gain (and hence growth) for the bacteria, but the lower the maximum attainable MFC voltage (since $E_{emf} = E_{cat} - E_{an}$). However, if the anode potential is lowered to increase the MFC voltage, the emf of the electron transport chain in bacteria declines (and hence ATP production), leading to the lower cell growth [40].

The power generated by a BFC is quantified in terms of power output, $P = V_{cell}$ (V) × I (A). The power density can be calculated by normalizing the power with respect to the electrode cross-sectional area, electrode volume, or working volume of the anodic chamber. The resistance at which both the current and voltage are optimum for the highest power output is called the cell design point [43].

The overpotentials, as discussed earlier, can be analyzed separately for the two electrodes if the expressions for the reaction rates are known with the help of Butler–Volmer's and Tafel's relations. The charge transfer overpotentials are controlled by the rate of heterogeneous electron transfer, and the kinetics of this process are described by the Butler–Volmer equation when the reactants are abundant and the current is small enough that the ohmic and concentration overpotentials are negligible:

$$I = Ai_o \left\{ e^{\left(\frac{\alpha nF\eta_{act,c}}{RT}\right)} - e^{\left(\frac{(1-\alpha)nF\eta_{act,a}}{RT}\right)} \right\} \quad (10.7)$$

where I is the current, A is the electrode active surface area, i_o is the exchange current density, α is the charge transfer barrier (symmetry coefficient), n is the number of electrons involved in the electrode reaction, and η_{act} is the charge transfer overpotential. The Butler–Volmer equation can be simplified in the high overpotential region (>118/n mV), yielding the Tafel equation [44]:

$$\eta_{act} = b \log_{10}\left(\frac{i}{i_o}\right) \quad (10.8)$$

where i is the current density and b is Tafel slope, which is an important experimental parameter commonly used to probe the mechanism of an electrode reaction. Plots of overpotential against $\log_{10} i$ are known as Tafel plots (Figure 10.5); i_o and b are obtained by extrapolation of the linear region of the curve to $\eta_{act} = 0$.

It is necessary to optimize the internal working conditions of a BFC. All these factors contribute to the generation of a stable baseline signal and sensitivity of the response signal. Identification of the overpotential losses using a polarization graph can assist in recognizing the cause of electrode instability at an early stage. Relevant remediation measures can henceforth be taken to optimize the sensor signal.

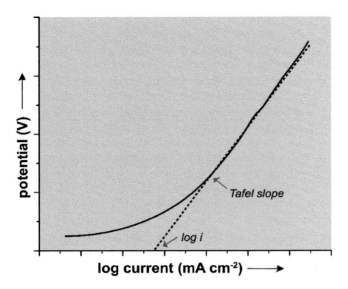

FIGURE 10.5 Tafel plot for calculation of current density.

10.2 MATERIALS AND PLATFORMS FOR BIOELECTRODE FABRICATION

The selection of materials for the construction of a bioelectrode is important for the performance of a BFC and its sensor applications. The electrode materials should be biocompatible and highly conductive to facilitate rapid electron transfer from the biocatalyst to the electrode or vice versa. Additionally, the final cost of the BFC sensor device should be considered from the commercial perspective.

The main abiotic components of a BFC are the anode, the cathode, and the PEM membrane (optional). Many of the studies to date have been related to electrode materials, as these directly influence the BFC power output [1]. Various electrode materials have been used to enhance the performance of BFCs ranging from single metal electrodes to nanomaterials. Based on the choice of biocatalyst and intended target sensing, the components used in the construction of a BFC can be of different sizes, materials, and shapes. This section delineates a few of the advanced materials and support systems that are evolving through the works of the last two decades in the development of BFC and their biosensing applications.

10.2.1 Advanced Materials for Developing Electrodes

A plethora of reports is available on the use of conductive nanomaterials for developing bioelectrodes due to many beneficial effects identified from their use. These materials can be used to build the entire electrode structure, starting from small units, or they can be deposited/created on the surface of the electrode [45]. Many of these nanomaterials assure excellent long-term operational stability for the host molecules, diminish the high overpotential losses commonly associated with BFC systems, improve the electrode surface area, and enhance the electrical conductivity in comparison to conventional electrode materials. Bioelectrodes modified with nanomaterials can be used in various self-sustaining BFC biosensors and medical devices.

Two of the most widely used nanomaterials are carbon nanotubes (CNTs) and graphene. Iijima first discovered CNTs in 1991, and since then both single-walled carbon nanotubes (SWCNTs) and multiwalled carbon nanotubes (MWCNTs) have been used in a myriad of applications [46]. SWCNTs are cylindrical in shape and are formed by rolling a single graphene sheet into a tube. The carbon atoms are present in a honeycomb fashion. MWCNTs have several concentric cylinders of graphene sheets with approximately 3.4°A spacing between the layers [45]. Some of the merits of CNTs are that they have high aspect ratios, increased surface areas, outstanding mechanical strength and robustness, good chemical stability, and remarkable electrical and electronic properties [47]. Their ability to carry current is almost 1000 times greater than normal copper wires [48]. A unique feature of the structure of CNTs is their local anisotropy, which is due to the difference in walls and ends of the tubes. The sidewalls are composed of an inert layer of sp^2-hybridized carbon atoms, which is analogous to the basal plane of pyrolytic graphite. The open ends of nanotubes have carbon atoms bonded to oxygen, similar to the edge planes of pyrolytic graphite.

CNTs pave the way for a variety of biohybrid devices that rely on proper coupling of these nano-sized materials to the biological components. CNTs can be modified to fulfill the requirements of biocatalysts inside BFCs. Various biomolecules like proteins or nucleic acids have been conjugated to CNTs for various applications over the years [47, 48]. Biomolecules can be added to the sidewall of the CNTs via noncovalent or covalent bonding [45]. The covalent bonding involves functionalization with various biochemicals that involve different linkages. Acid treatment is carried out that result in various oxide and carboxylic acid groups to the ends of the tube. Further functionalization with additional groups can also be carried out at these oxide groups [49]. The promising electrochemical properties of SWNT-modified electrodes result from oxygenated carbon species (carboxyl moieties), which are produced on the ends of the nanotubes during acid treatment [50, 51]. CNTs can also be functionalized with various biomolecules without covalent coupling. The sidewalls of CNTs bear graphene, which is chemically inert and hydrophobic. Noncovalent coupling is generally based on hydrophobic interactions between a CNT and a biomolecule. For example, some proteins are adsorbed spontaneously on the sidewalls of acid oxidized CNTs [52].

The impressive properties of CNTs, such as their small size and high conductivity, have allowed the development of various types of electrochemical biosensors [47]. The electron transfer capability and reasonable biocompatibility make the CNTs a viable linkage between the biological component and the electrode [53]. It was shown that a CNT-centered bioelectrode in an EFC enabled direct DET, resulting in an enhanced performance [54]. In MFC, a bacteria-CNT interpenetrated electrode structure was able to harbor bacterial cells, yielding a highly conductive network for charge transfer. The structure of the CNT enabled high values of current in spite of no biofilm formation [55]. A proper charge transfer is integral to maintaining the integrity of the molecular transducer in the BFC.

Even though CNTs are derived from graphene, the latter itself is a very significant nano-sized material. Graphene is the building element of many carbon allotropes, including graphite, CNTs, buckyballs, etc. Graphene consists of a single layer of six-atom rings in a honeycomb network and can be considered a true planar aromatic macromolecule [56]. It was first created in 2004 from highly oriented pyrolytic graphite by mechanical exfoliation [57]. Some of the other methods of graphene preparation include thermal decomposition of SiC wafers under ultrahigh vacuum [58] and chemical, thermal, or electrochemical reduction of graphite oxide [59]. Graphene has high thermal conductivity (5000 W/mK), excellent electrical conductivity (1738 Siemens/m), high surface-to-volume ratio (2630 m^2/g), and biocompatibility. Graphene and functionalized graphene have emerged as successfully implemented materials in electrochemical biosensing [45]. Some of the advantages of graphene over CNTs are low cost, high surface area, ease of processing, and safety. The sensor devices with graphene materials inherit better sensitivity and signal-to-noise ratio [60].

Chemical doping can be used to modify materials intrinsically to tailor the electronic properties. Nitrogen doping can be used to vary the electronic properties of graphene because it is of comparable atomic size [61]. Nitrogen-doped (N-doped) graphene (N-graphene) can be prepared by exposing graphene to nitrogen plasma [62]. By using two or more layers of graphene, its electronic properties can be enhanced. Additionally, some of its properties can be improved by functionalization. Over the last few years, graphene finds enormous applications in the field of MFC technology. Initially, the expensive and rare electrode materials, such as gold, silver, palladium, platinum, etc., were gradually replaced by more cost-effective metals like copper, iron, nickel, aluminum, etc., in BFCs. Unfortunately, these metals undergo corrosion when used for a long time inside MFCs. Graphene has been demonstrated to assist bacterial cells to express signaling molecules and at the same time act as a mediator to improve the electron transfer efficiency [63]. The electron transfer proteins (cytochromes) and nanowires can interact with the graphene nanostructure for ensuring DET. The porous structure of graphene also helps to reduce diffusion and mass transfer resistance in the fuel cell system [64]. The modification/combination of graphene with blending materials, biopolymers, or conducting polymers can alleviate the impairment caused by it, presumably due to the reduction of friction between the nano-sharp edges of the graphene layers and bacterial cell wall membrane [63].

Biopolymers such as chitosan have been extensively used for enzyme immobilization on electrodes, whereas chitosan and silk thin films have been demonstrated to promote biofilm growth on electrode surfaces in MFCs, which otherwise are nonconducive for bacterial growth [16, 65]. Conducting polymers possess a few interesting properties of metals, such as the ability to conduct charge; optical properties; good control of the electrical stimulus; a high conductivity-to-weight ratio; and can be made biocompatible, biodegradable, and porous.

Their chemical, electrical, and physical properties can be tailored to the application needs by incorporating antibodies, enzymes, and other biological moieties [66, 67]. Their properties can be changed by chemical modifications. Two commonly used conducting polymers in MFCs are polypyrrole (PPy) and polyaniline (PANI), due to their high electrical conductivity, ease of processing, and nontoxicity to microorganisms [68]. These properties allow the electrode materials to develop bacterial colonization that eventually leads to more power generation. The copolymerization of conducting monomers combines their individual superior properties [66, 69]. Live bacterial cells coated with PPy maintained the viability of the cells, without degradation, for a long period. Additionally, the bioelectricity generation superseded that of the unmodified cells [70].

Another class of advanced materials worth mentioning for their applications in BFCs are the magnetic nanoparticles (MNPs) such as ilmenite, hematite, and magnetite. MNPs have desirable characteristics such as high coercivity, a low level of toxicity, super paramagnetism, and a low curie temperature [68]. These features make them ideal candidates for various applications such as sensors, batteries, and catalysts. When MNPs are used for BFC applications, it is desirable to fabricate them in smaller sizes and with an ideal particle size distribution that increases the surface area for better catalytic activity. Clustered cells on MNPs can be easily separated under the influence of an external magnetic field [71, 72]. Magnetite NPs have also been used to manipulate bacterial interactions and, to increase the yield of current generation. Magnetite NPs may facilitate the oxidation of substrates whose anaerobic degradation requires syntrophic cooperation among microorganisms. The electrical conductive property of NPs helps in facilitating the extracellular electron transfer and hence the performance (power/sensitivity) of the sensors [73].

10.2.2 Miniaturized BFCs

Translating the BFCs into miniaturized versions can improve the performance of the sensor as well as render its mass production at low cost. If the BFC (especially MFCs) sensor is deployed as an installation device for monitoring of analytes in wastewater, seafloor beds, etc., scaling down may not be a precondition. However, if these are intended for use as portable, compact, and easy-to-handle sensors, then scaling down the BFCs into small/chip-based platform becomes important. The added advantage of designing small-scale BFC sensors is lower internal resistance, which makes them energy dense. This is a critical issue, as the internal resistance affects the sensitivity and response time by contributing to the ohmic overpotential losses of the device. The internal resistance of a BFC has two components: non-ohmic and ohmic resistance. Resistance to the charge transfer and diffusion constitutes the former component. The non-ohmic resistance can be reduced by selecting electrodes that support good catalytic activity and its high surface area. On the other hand, the ohmic resistance affects the output signal. Enlarging the size of the PEM, placing the electrodes in close proximity, and using electrolytes

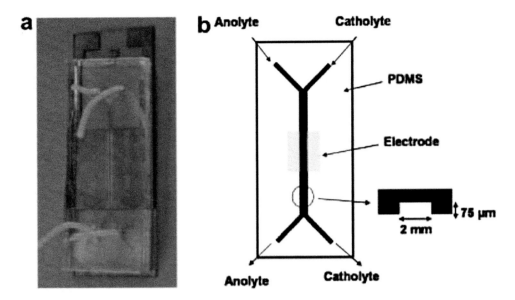

FIGURE 10.6 Glucose/O_2 microfluidic biofuel cell, **(a)** PDMS-glass device, **(b)** scheme of the device consisted in a Y-shaped microfluidic channel with two inlets and two outlets [76]. Reproduced from [76], with permission. © Elsevier.

with high conductivity can help to reduce ohmic resistance. Ohmic resistance can be defined as [74]:

$$\Delta V_\Omega = \frac{d \times I}{K \times A} \qquad (10.9)$$

where ΔV_Ω is the voltage difference between the anode and cathode, d is the electrode separation distance, I is the current, K is the solution conductivity, and A is the cross-sectional area (cm²) through which ionic conduction occurs. The ohmic resistance can be simplified as $R_\Omega = l/(A \times K)$, where l is electrode length (cm). Thus, the ohmic resistance can be tackled by reducing the d/A ratio, which can be addressed by microsystems. Small-scale BFC configurations enable the flexibility of studying the interaction mechanisms within the biological community and between the biomolecules and abiotic conductive surfaces, which is fundamental to the development of a successful sensor [74].

With the help of microfabrication technology, it is possible to design constructs that have a geometrical size regime in the micrometer scale. These are termed microsystems. These are often designed with microfluidic channels that possess a high surface-to-volume ratio, an excellent characteristic to promote surface-based reactions inside a BFC. Additionally, the Reynolds numbers are relatively low at this level, facilitating laminar fluid flow. Thus, convectional mixing is minimized, allowing diffusion to take place. This enables the functioning of the BFC without a membrane. The reduction of physical separation between the anode and cathode helps to implement a compact design.

Microfluidic fuel cells are usually fabricated by two different methods: silicon-based micromachining and soft lithography. In the first method, a silicon substrate is decorated by the lithographic technique and wet or dry etching is carried out on the desired structures. In the final step, additional silicon wafers are bonded onto the substrate for airtight sealing. This is a very cost-effective method. The second method is used for creating prototypes rapidly. The structure of the patterned cavities/channels is prepared in polydimethylsiloxane (PDMS) (Figure 10.6). It is then sealed using glass or silicon [75].

The first microfluidic BFC was reported in 2004. It was an EFC with an immobilized enzyme in a microchannel [77]. The first few studies with respect to microfluidic MFCs did not give very high output values, but helped to realize the miniaturization strategy for MFCs [78]. Further modifications gave better results and enabled long-term operation [79]. In the case of MFCs, even though the structure becomes miniscule the PEM is still widely used due to the sensitive nature of the microbial catalyst. However, in case of EFC the use of PEM can be avoided by adapting a suitable design strategy.

The merging of BFC biosensors into lab-on-chip (LOC) technology using microfluidic systems can offer an alternative to conventional laboratory bioassays. Microfluidic fabrication helps to improve sensitivity of the biosensor devices. Moreover, a single microfluidic biosensor chip can perform many steps of analysis, comprising sampling, sample separation, mixing, etc. With these attractive features of microsystems, along with the inherent specificity of BFC sensor systems, powerful analytical tools can be created [80].

10.2.3 Paper-Based BFC Systems

The advantages of paper for biosensing applications including the properties of paper, its modification and fabrication techniques, and immobilization of the bioreagent on paper have been illustrated in Chapter 3, 12 and in several articles [81–83]. The porous structure and absorbing nature of paper give it the fluid wicking ability, translating it to an autonomous microsystem. Paper, being electrochemically inert, does not interfere with the charge transfer to and from the biomolecules. The biorecognition element (enzyme or cells) can be easily adsorbed/immobilized on paper surface. The conductive electrodes and enzyme biocatalysts can be printed on paper following standard screen-printing technology. It supports the integrity of the biocatalysts used in BFC sensors.

When used as an electrode support, paper can serve significant functions such as regulation of the microfluidic properties of the device, separation of the target analyte from the biological sample matrix, and providing a 3D geometry for the electrode [84]. Most of the BFC paper-based devices employ Whatman#1 filter paper [85]. It also exerts only a meager diffusion resistance. Other kinds of papers commonly used are the Japanese paper, Whatman #4 filter paper, and so forth. Whatman #4 filter paper possesses a high porosity and flow rate, but the diffusion characteristics are comparable [86]. Japanese paper is preferred due to its good absorptive properties [87]. Cellulose and some other derivatives can also substitute for paper. Cellulose itself is a main component of paper. In this respect, cotton fiber, which is a derivative of cellulose, has also been used in analytical devices. Cellulose has been found to support direct diffusion in electrochemical paper-based analytical devices (ePADs), eliminating the need for external equipment [88].

The design of electrodes on paper is much simpler than on a conventional material. The easiest method to obtain a functioning electrode on paper is to draw it with a graphite pencil (graphite in pencils are conductive in nature!) or doped pencils containing the desirable conductive ink, nanoparticles, and binder solution or mediator paste, etc. However, in order to have reproducible results, inkjet-printing and screen-printing techniques are adopted to fabricate electrodes on a paper matrix. These techniques are cost-effective, and hence mass production of the sensor device is possible for commercialization [84]. The wicking properties of paper ensure a continuous supply of fuel to the respective biocatalysts on the electrodes. It can assume the role of the matrix for immobilization of biological components and/or the PEM [89]. The biological materials can be dried and reactivated on paper as and when required. In this respect, paper acts as a stabilizing agent. When paper-based devices are stored even at room temperature, the activity of the biorecognition elements/proteins remains intact for a long time [90].

10.3 EFC BIOSENSORS

The ability of some redox enzymes to specifically act on fuel substrates through oxidation/reduction reactions makes them attractive biocatalysts [91]. The enzyme-electrode interaction may forms a "molecular transducer" that translates the redox activity directly to a readable electric signal without using any ET mediator. The current density developed in EFCs can also cater to the need of power in mini- and micro-scale device electronics [30]. With the advancement in nanomaterials and wiring of enzymes to the electrodes, not only the efficiency of channelizing the electrons directly to the electrode has improved but also the lifetime of the enzymes could be prolonged. This has shifted the research interest of EFCs from renewable energy sources to applications as sensor devices. A wide range of enzymes as BFC catalysts is known and among these the enzymes involve in the oxidation of sugars, alcohols, and hydrogen, the transformation of peroxide, and the reduction of oxygen are prominent [30]. Some of these enzymes are used as cathodic catalysts such as laccase, and bilirubin oxidase (BOx) involved in the reduction of molecular oxygen and contributing potential to the cells. The laccase-based EFC biosensors are also used to monitor sodium azide, which inhibits the laccase activity [92].

10.3.1 EFC BIOSENSORS FOR MONITORING GLUCOSE

Since the pioneering work of Katz et al. [93] self-powered biosensors have received substantial attention. They confirmed the integrity of the bioelectrodes to draw electrons from the glucose in body fluids circulating inside small living animals through *in vivo* studies. Maintaining the blood glucose level within a stringent range is central to the health of diabetic patients in order to avoid medical complications. There has been an advancement of glucose meters for glucose monitoring, as in the case of other medical infrastructures for non-invasive monitoring in developed countries, where diabetes prevalence is also high [94]. Inexpensive glucose monitoring devices that are self-powered can be an easy, accessible solution in the developing countries for routine self-monitoring of glucose.

The simple sugar glucose, a potent biomarker for diabetes, generates 24 electrons upon its complete oxidation during aerobic respiration [91]. The EFC glucose sensors commonly use glucose oxidase (GOx) in the anode. GO_x is commonly derived from the fungus *Aspergillus niger* due to its stability and availability. The redox activity of GO_x can be attributed to the dual flavin adenine dinucleotide (FAD)-active centers in the glycoprotein [95].

EFCs can be created on micro-paper-based analytical devices (μPADs) for sensing glucose and other biomarkers due to several advantages of paper, as discussed elsewhere. Additionally, paper offers the ability to develop origami designs that provide self-assembly options. Moreover, several microfluidic patterns on a single μPADs device allow multiplexing capability. Such EFC μPADs for glucose monitoring offer higher selectivity and sensitivity as compared to conventional amperometric biosensors integrated with optical readouts [94].

An example of the fabrication steps for an origami-designed paper (Whatman #1 filter paper) based EFC is shown in Figure 10.7 [94]. Wax printing on the paper surface can be done using a wax printer that can define the hydrophilic and hydrophobic boundaries for the fluid flow. The graphite ink was screen-printed to form the anodic and cathodic electrodes. GO_x was immobilized using a biopolymer, chitosan, at the anode. The μPAD (size 6.25 cm^2) had anode of surface area 1 cm^2. The pores in the filter paper proved to be a suitable reservoir for the samples with a working volume of 20 μL. The oxygen was reduced at the Ni-bound activated carbon catalyst. The intimate contact of the transducer and catalyst reduces the

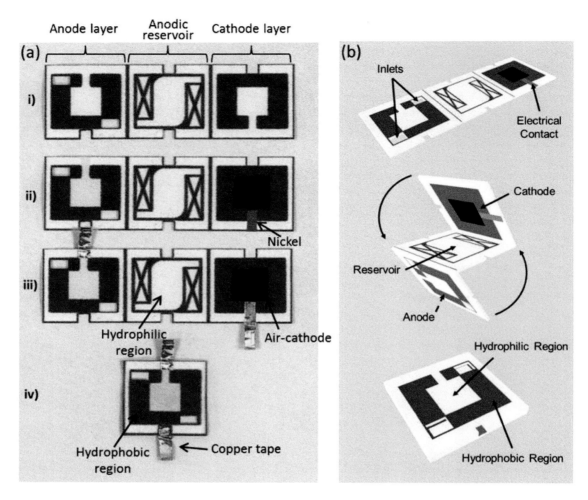

FIGURE 10.7 (a) Intermediate steps for the fabrication process of a single-cell EFC: step (i): wax boundaries are formed; step (ii): electrodes are printed; step (iii) copper tape is pressed on; and step (iv) cell is folded, glued, and flipped to expose the reservoir through the inlet holes. (b) Schematic diagrams of the origami EFC during assembly. The 3D EFC structure can be easily assembled from 2D sheets by simply folding along predefined creases and gluing [94]. Reproduced from [94], with permission. © Elsevier.

start-up time (the time required to generate a stable baseline voltage/current in the BFC after assembling all the components) and improves the electron transfer kinetics. The current response by EFC μPAD exhibited linearity ($R^2 = 0.996$) to 1–5 mM of glucose concentration (Figure 10.8) and limit of detection (LOD) of 0.02 μA mM^{-1}. The reported EFC biosensor was found to be inexpensive for sensing glucose [94].

Enzymes such as laccase and bilirubin oxidase (BO$_x$), which contain copper metal centers, have the capability to reduce oxygen. The BO$_x$ can catalyze at near-neutral pH and ambient

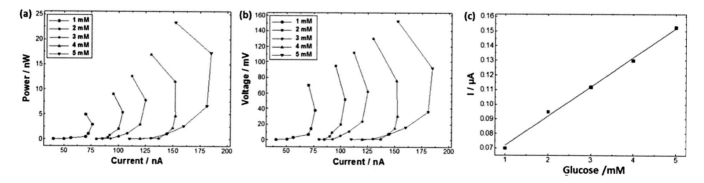

FIGURE 10.8 Power output (a) and polarization curves (b) showing the levels of power generation with varying concentrations of glucose. (c) Calibration curve of glucose biosensor's output current vs. the concentration of glucose in mM with a 1 MΩ resistor connecting the device's electrical leads [94]. Reproduced from [94], with permission. © Elsevier.

temperature. Hence, these redox enzymes are widely used as cathodic catalysts in EFC. The bioelectrochemical reactions at the half cells in the GOx-based anode and BOx-based cathode can be presented, as in the following equations [96].

$$C_6H_{12}O_6 + H_2O \rightarrow C_6H_{12}O_7 + 2H^+ + 2e^- \quad (10.10)$$

$$O_2 + 4H^+ + 4e^- \rightarrow 2H_2O \quad (10.11)$$

To achieve the DET on enzyme electrodes, entrapment of the GO_x and BO_x in carbon nanodots (CNDs) films over glassy carbon electrodes have been reported [95].

There is a report on microfabricated EFC for sensing glucose, where a microneedle device, which is generally used for pain-free drug delivery in the subdermal layers of skin, has been used [80]. The operational principle lies in the fact that a glucose-sensitive and -selective anode in conjunction with platinum (Pt) blended carbon cathode generates power density signals proportional to the transdermal glucose concentrations. The sensor followed MET, where tetrathiafulvalene (TTF) acted as the mediator for the GO_x. Carbon paste was inserted into the hollow microneedle and operated as the current collector. This microneedle EFC was unresponsive to the presence of acetaminophen, bovine serum albumin, uric acid, ascorbic acid, and lactic acid. With the glucose detection range of 0–25 mM, the microneedle-based EFC served as a self-powered glucose biosensor [93]. The sensor contains 70% of its activity even after 1 week of storage at room temperature.

10.3.2 EFC Biosensors for Alcohol and Other Analytes

Selective, sensitive and rapid detection and/or quantification of alcohols is of utmost importance not only in forensic sectors but also in pharmaceutical, brewing, food, and alcohol fuel-based industries. Hence, self-powered EFCs for alcohol and other such analytes have surfaced as a paralleled effort towards cost-effective, miniaturized, and portable versions of alcohol biosensors [97].

Because of its specificity towards alcohol, the enzyme alcohol oxidase (AO_x) is often used as a biocatalyst in electrodes for the detection of alcohol. Most of the amperometric alcohol biosensors rely on the detection of O_2 consumption or H_2O_2 formation based on Equation (10.12) [98].

$$RCH_2OH + O_2 \xrightarrow{AO_x} RCHO + H_2O_2 \quad (10.12)$$

However, in EFC oxygen is not used to mediate the electrons from the biocatalytic reactions in anode, instead MET or DET strategies are used as discussed in Section 10.1.3. As an example, a DET principle based EFC alcohol sensor was developed using AO_x anode and laccase cathode [99]. The biosensor generated a linear current response against the methanol concentration. The DET was enabled by wiring the enzymes AO_x and laccase to Toray carbon paper using synthetic conductive polymers, polyethyleneimine, and osmium tetroxide on poly(4-vinylpyridine), respectively.

In the case of people with heart disease, checking cholesterol levels is of considerable significance. A sol-gel matrix with entrapped cholesterol oxidase in both the anode and cathode was constructed to develop a membrane-free, one-compartment, self-powered EFC [100]. At the anode, electrocatalytic cholesterol oxidation involved electron transfer with the help of a mediator, phenothiazine, whereas the H_2O_2 formed during enzymatic reaction was electrocatalytically reduced with the help of Prussian blue at the cathode. The self-powered biosensor offered enhanced sensitivity (26.0 mA M^{-1} cm^{-2}), compared to either of the two individual electrodes, and a dynamic range up to 4.1 mM cholesterol. The performance of the cholesterol sensor was not interfered by the compounds such as ascorbic acid, uric acid, glucose, and lactic acid, commonly present in blood. The performance of the EFC in the measuring of free plasma cholesterol concentrations was in parallel with that obtained with a standard kit.

10.3.3 Wearable EFC Biosensors

With the onset of the microdevices, the bioelectronic appliances have captured the interest of market niches for providing wearable, routine self-monitoring, and user-friendly biosensors. Body-bound biosensor devices are intended to make routine monitoring much simpler [93]. However, most of the body-bound biomedical sensors suffer from a lack of requisite power input. Whereas, additional power supply from an integrated device increases the size of the biosensor and/or hamper the aesthetic appeal [101]. Moreover, the associated limited charge-discharge cycles and low energy capacity of such devices defeat the intended purpose.

Over the last few years, EFCs have attracted intensive interest in developing portable and wearable biosensors due to their self-powering attributes [96, 102, 103]. However, the development of such small-scale complementary MFCs has encountered some serious challenges that hampered their application for biomedical applications. Among these challenges, the requirement of stringent growth conditions for the microbes and the threat of cytotoxicity for body-integrated applications are prominent. Wearable EFCs can take up glucose, lactate, and pyruvate from body fluids such as blood, sweat, and tears that will act as "fuel" to the anode transducer and generate a concentration-dependent signal response [104]. Several such wearable EFCs have been developed recently; one such example is to monitor lactate in sweat [105].

Lactate can be a decent biomarker to gauge the muscular exertion and fitness level of an individual. The average concentration of lactic acid in human sweat is around 14 mM. Moving forward in this direction, temporary transfer-type tattoo EFCs were developed to determine the fitness level by measuring the lactate concentration in the sweat of the subject. The "UC"-fashioned EFC tattoo consisted of a U-shaped anode and C-shaped cathode. Lactate oxidase (LOx) was immobilized into the CNT along with a mediator, tetrathiafulvalene. The carbon cathode had Pt to reduce oxygen. To prevent efflux and to make the EFC skin-friendly, the anode and cathode were layered with chitosan and Nafion, respectively. The power density increased from 25 to 34 and further 44 µW cm^{-2} at lactate concentrations of 8, 14, and 20 mM, respectively. It could also withstand mechanical strain during physical movements [106].

BFCs need to be flexible, lightweight, and soft so that they do not hamper the body movements. Another wearable EFC lactate biosensor was developed using LOx and naphthoquinone as a mediator on a flexible substrate [107]. Arrays of carbon nanotubes of anodic and cathodic circular pellets were fabricated with the help of screen-printing technology. The electron was reduced at homogeneously distributed silver oxide over the CNT cathode, leading to the formation of silver from silver oxide. The biosensor was coated with hydrogel to prevent skin irritation. Negligible power decay was observed under repeated stretch and strain. When mounted on the human arm coupled with a DC-DC converter, the blinking of the connected LED indicated the production of lactate in human sweat within 2 min of physical activity (when the individual started to sweat) [107].

10.3.4 Other Applications

Essentially a BFC can serve as a biosensor for whatever fuel allows it to operate [6]. EFC biosensors have also emerged for the detection of inactive electrochemical compounds and nonfuel substrates. They are based on the principle of inhibition of the enzymes by the target analytes (such as EDTA, mercury [Hg^{2+}], cyanide [CN^{-1}]) [108–110] or product inhibition (e.g., alcohol dehydrogenase inhibition by acetaldehyde during ethanol oxidation) [111]. For example, in the EFC microchip, the addition of the target compound cyanide binds to the copper cluster (metal redox center) of the laccase enzyme and perturbs the intramolecular electron transfer. This, in turn, led to an immediate decline in the EFC current production. The turn-off sensor exhibited a direct current decay-cyanide concentration relationship [109]. The types of inhibition can be noncompetitive, competitive, uncompetitive, or mixed. Table 10.1 lists recent EFC biosensors used for the detection of various "nonfuel" compounds and the implemented electron transfer mechanism.

The high selectivity of enzymes and a new class of recognition elements with intelligent material properties called aptamers have led to the emergence of a novel BFC: aptasensor [113]. A detailed description of aptamers and their properties can be found in Chapter 4. Integrating the concept of logic gates, biocomputing can result in a very specific selection of target analytes [114]. In a study, the anode included a GO_x and thrombin binding aptamer, whereas the cathode comprised a BO_x and lysozyme binding aptamer. In the presence of either one of the analytes (thrombin or lysozyme), or none of them, the open circuit voltage dropped below the threshold voltage [113]. However, the presence of both the specific analytes (thrombin and lysozyme) led to drop in open circuit voltage below the threshold voltage. The design enables the modulation of the power output from the BFC with the aid of aptamers programmed with the NAND logic gate, in response to the presence of both the targeted analytes. Other affinity binding methods such as antibody-antigen can also be applied at the electrodes. If the biomarker binds to the biorecognition element, either the substrate-enzyme or the enzyme-electrode contact is blocked, and it subsequently affects the BFC power release].

10.4 MFC BIOSENSORS

The high sensitivity of microorganisms to a variety of biological and environmental factors makes them appealing biorecognition elements for biosensors. The two important functionalities – self-power generation of the MFC and sensing by the regenerating microbial catalyst – can be merged to generate auto-powered MFC biosensors. They can prove to be extremely convenient in remote places for detection of the analyte of interest, as they can function autonomously [115]. Microbes are potential candidates to pose as a biocatalyst for sensing because of various attributes: (i) they show broad-spectrum selectivity to toxic compounds, (ii) the "fuel" they metabolize can be anything from simple sugars to complex organic waste and chemical compounds, (iii) they are inexpensive and easy to cultivate and no expensive purification steps are necessary, (iv) they are not susceptible to denaturation under adverse conditions unlike enzymes, (v) they may display good selectivity to analytes, and (vi) they can be genetically modified to tune the sensing property [116]. Microbes degrade organic substrates and reserve the free energy for their catabolic activities. The excess of the energy

TABLE 10.1
A List of EFC Biosensors for the Detection of Various Compounds

Sl No.	Anode Biocatalyst	Cathode Biocatalyst	Electron Transfer Method	Target Compound	LOD	Reference
1	Glucose dehydrogenase	Laccase	MET	Cyanide	1.0×10^{-7} M	[108]
2	Alcohol Dehydrogenase	Bilirubin oxidase	MET	Mercury	1 nM	[109]
3	Glucose oxidase	Platinum	DET	EDTA	-	[110]
4	Alcohol dehydrogenase	Bilirubin oxidase	MET	Acetaldehyde	1 µM	[111]
5	Glucose dehydrogenase	Laccase	MET	Arsenite, arsenate	13 µM 132 µM	[112]
6	Glucose oxidase + thrombin binding aptamer	Bilirubin oxidase + lysozyme binding aptamer	MET	Thrombin and lysozyme	8 nM and 20 nM	[113]

can be channeled to an extracellular conductive solid. Some of the energy produced is retained by the cells to sustain their vitality. This metabolic loss cannot be averted. Hence, a balance needs to be maintained between the MFC energy drawn and the residual energy for sustaining its metabolic activity. The current produced by the MFC apparently indicates the metabolic status of the specific microbes present at the anode. Hence, the electron generation mechanism underpinning the fabrication, applications, and operating procedures of an analytical MFC is of paramount significance [117].

Global acceleration in terms of industrial growth is causing environmental pollution. This needs immediate attention for which real-time monitoring of toxic compounds and their remediation or removal is of utmost importance for a civilized nation. Cost-effective portable sensor systems with rapid detection and quantification of the toxic compounds on-site are in huge demand. There have been several reports of the biosensing potential of MFCs [115, 116, 118, 119]. As compared to the rest of microbial sensors, MFCs are easy to construct, user-friendly, and inexpensive. However, the response time of MFC biosensors is quite lengthy. The microorganisms take time to colonize the anode and display cell potential (start-up time). Once it colonizes, the response time decreases in continuous monitoring systems [119]. Since maximizing power generation is not the prime objective of MFC sensors, an effective way to deal with the long response time and avoid time-dependent changes is the development of small-scale analytical MFCs. Progress of microfabrication techniques, nanosciences, and microfluidics have additionally driven the impetus for improvement in microscale MFC sensors.

Since the demonstration of the first experimental micro-MFC in 2006, upgrading such devices for rapid output signal generation has been a thrust area of research [116]. Microscale devices have since been fabricated using advanced techniques such as etching, metal deposition, photolithography, etc. [74]. Micro-MFC arrays are emerging as bioassay platforms. If the microdevices employ a microfluidic concept, the anodic and cathodic chambers do not need physical separation, since the anolyte and catholyte (the solutions present in anode and cathode chambers, respectively) will stream in a laminar flow. Moreover, small-scale MFCs will consume less reagents and materials. In micro-MFC, close physical contact of the microbes with the anode decreases the start-up time, offers stable baseline signal, and improves electron transfer kinetics. With much of the variabilities disappearing, the reproducibility is expected in small-scale MFCs.

10.4.1 Toxic Detection in Water

The two microbial metabolic pathways involved in energy conversion are respiratory (oxidative pathway) and fermentation. However, it is the oxidative pathway through which the electrons circulate in the respiratory chain and finally releases the electrons through membrane-bound electron acceptors. Some of the Gibbs free energy generated in the process is utilized by the bacteria to sustain their respiration [120]. Any disorder to their metabolic pathways instigated by a change in the immediate environment (abrupt organic load, the existence of a bioactive compound, etc.) manifests itself as an alteration in the electrical signal output. This is the principle behind using MFC as a tool for toxicity monitoring. Due to the simplicity of the inbuilt device transducer and signal output, MFCs can be reasonably employed to monitor complex parameters that are not easy by conventional high-end instruments or modern bioassays [119].

Access to clean drinking water is a necessity to ensure quality health and sustainability of a nation. The biochemical oxygen demand (BOD) is an index for biodegradable organic compounds in water and wastewater, and is widely used for the evaluation of water and wastewater quality. However, the conventional method for determining the BOD is time-consuming (usually 5 days of incubation) and requires experience and skill to achieve reproducible results. Yet for real-time control in the water and wastewater industry, the ability to determine the BOD online is desirable, which has resulted in various extensive experiments. The first MFC as a BOD sensor was demonstrated by Karube et al. in 1977 using *Clostridium butyricum* as the immobilized microbial catalyst [121]. Since then various model strains have been employed for BOD sensing with or without mediators in the MFC designs. MFCs using a single organism have an intrinsic disadvantage due to the limited range of fuel utilization. As such, electrochemically active microbial communities with different nutritional characteristics can be successfully enriched and implemented for developing electrochemical MFC biosensors.

The following example is of an MFC-based BOD sensor engaging electrochemically enriched microbial consortium from activated sludge (Figure 10.9) [122]. The polyacrylic plastic MFC construct has two chambers: an anode and cathode of 20 mL each, separated by Nafion PEM. Graphite felt electrodes operated as the current collectors. Air-saturated tap water acted as the catholyte. Artificial wastewater (AW) supplemented with glucose and glutamic acid served as the anolyte.

The observed current was computed as the ratio of the potential drop and the resistance (10 Ω). With an AW feeding rate less than 0.35 mL/min, the current generation and the wastewater strength followed a linear correlation for BOD concentration up to 100 mg L^{-1}, complying with the DET mechanism (Figure 10.10). Beyond 100 mg L^{-1}, model fitting can be an option to forecast the BOD concentration if dilution is not desired. A 3-hour starvation in the feeding prompted a recovery time of 30 min. The recovery time will depend on the strength of the AW.

10.4.2 Formaldehyde Detection

MFC as a toxicity sensor overcomes the problems of high maintenance, and complicated design. It offers stable operation over a long time, and online monitoring as compared to the existing commercial chemical sensors. Presence of a toxic compound will disrupt the normal functioning of the cellular metabolic pathways, which will be reflected as an anomalous

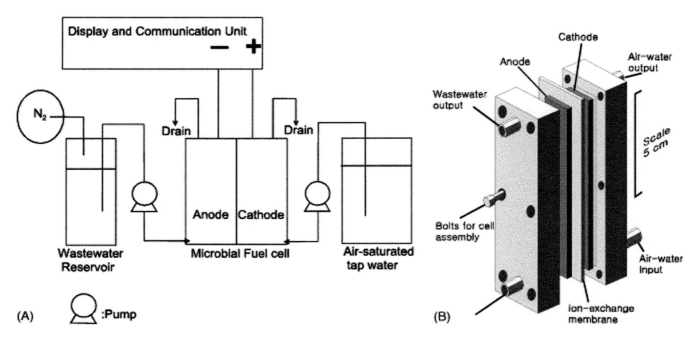

FIGURE 10.9 Schematic diagram of the biosensor system (**A**) and microbial fuel cell (**B**). Reproduced with permission from © Elsevier [122].

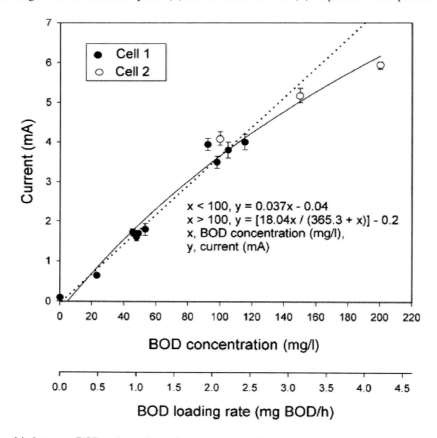

FIGURE 10.10 Relationship between BOD value and steady-state current. Reproduced with permission from © Elsevier [122].

current output. The following microfluidic MFC senses the presence of a common microbial disinfectant formaldehyde (Figure 10.11A) [123]. The micro-sized construct was adopted in this MFC sensor to sustain a stable output voltage and overcome the limitations of mass transport, internal resistance, and ohmic overpotential, which are often associated with a larger-scale MFC. The microdevice assembly had both the anodic and cathodic compartments of Perspex pieces microfabricated onto silicon wafer plates to extend a working reservoir of 144 µL. A PEM was sandwiched between

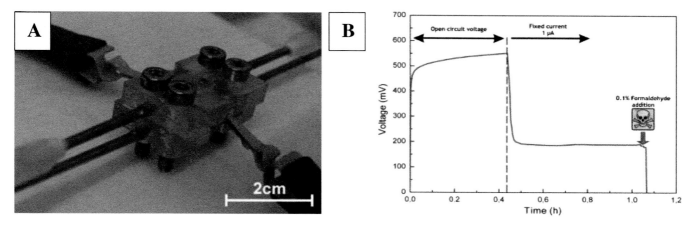

FIGURE 10.11 (A) Final testing assembly of the developed toxicity biosensor based on MFC. (B) Voltage evolution of the MFC biosensor: toxicity measurement response obtained with the microfabricated device when 0.1% formaldehyde was added to the anodic compartment of the cell [123]. Reproduced from [123], with permission © Elsevier.

the silicon plates. The exoelectrogenic bacteria *Geobacter sulfurreducens* was used as the biosensing catalyst in anode in the absence of mediators. Voltage output was recorded at 1 μA of fixed current to monitor the changes administered by formaldehyde injection. Ten mM of acetate containing the culture formed the anolyte and oxygenated ferricyanide solution through the cathode reduced oxygen in water. The electroactive biofilm (colonization of *Geobacter sulfurreducens*) on the anode was used as the electron sink (from acetate oxidation) in the absence of other electron acceptors, which resulted in the observed current generation. On injection of formaldehyde, the voltage dropped, and the voltage signal did not recover. It could be deduced that formaldehyde caused irreversible inactivation of the electroactive biofilm (Figure 11.11B). Another notable advancement for formaldehyde detection was carried out by Chouler et al. [23]. A planar compact MFC contrived on paper was demonstrated using screen-printed technology. Carbon-based conductive ink on paper formed the current collectors. A similar decay in current was observed in response to formaldehyde with a response time of 165 min. Such easy-to-operate, biodegradable MFC devices can offer affordable water testing solutions for on-site deployment in developed countries.

10.4.3 Other Applications

Photosynthetic bacteria such as cyanobacteria are interesting candidates as a microbial catalyst in an MFC. The electrogenic behavior of cyanobacteria has been well established, but comparatively less studied as compared to commonly used exo-electrogenic microbial species such as *Geobacter*, *Shewanella*, *Pseudomonas*, and *E. coli* [8]. The absence of proper dose-dependent response has often clouded the credibility of MFC biosensors [119]. Microscale PMFC devices for biosensors can be fruitful in minimizing the start-up time, the microbe-electrode distance, and their capability to function during both day and night by harnessing the electrons from the interconnected respiratory and photosynthetic electron transport chains [124]. Recently, researchers demonstrated a PMFC for selective alcohol detection [17]. The principle of detection relied upon an alcohol-triggered membrane depolarization that manifested as a burst in the potential (Figure 10.12A). This was a "turn-on" sensor that had heightened the voltage signal in the presence of alcohol. The increase in voltage burst of the PMFC was proportional to the alcohol concentration. The concept was then translated to a paper-based chip of 20 cm^2 size (Figure 10.12B). The chip displayed linear response ($R^2 = 0.99$) to alcohol concentration in the range 0.005–10%. The remarkable short response time (~10 s) for ethanol detection is a breakthrough in the MFC biosensor field. The device fabrication with nano-biocomposite material also supported minimizing the internal resistances [16]. Such strategies of target recognition in contrast to the routine respiration/metabolism inhibition can result in shortening the response time. The use of photosynthetic bacteria has additional advantages, as the bacteria can survive in only light and atmospheric carbon dioxide as the input energy.

Apart from a broad range of toxicity detection, the microbes can be genetically altered to attain sensitivity to a particular analyte. The groundwork of using such a modified organism to construct an MFC sensor for arabinose detection was laid by Golitsch et al. in 2013 [125]. Table 10.2 provides a list of recent key developments of MFCs as biosensors for the detection of various compounds ranging from heavy metals, to toxic gases, to antibiotics and the electron transfer mechanisms. The EABF on the anode can be sensitive to various antibiotics (levofloxacin, neomycin, etc.) [126]. The oxidation of substrates is hampered by the antibiotics, which affect the electron transport from the substrate on biofilm. Based on this detection mechanism, MFCs can be built to monitor the presence of antibiotics [127]. Similarly, toxic gases such as carbon monoxide (CO) can also be detected, which chelate enzymes and interfere with the electron transfer enzyme proteins and the metabolic process of the cells [128].

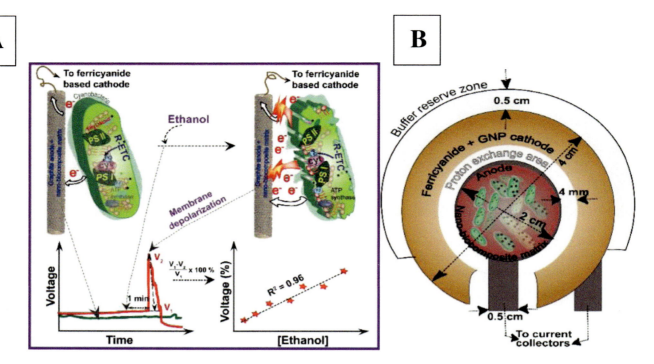

FIGURE 10.12 (A) Scheme showing the working principle of the PMFC as an ethanol sensor. (B) Schematic of the paper-based PMFC system [17]. Reproduced from [17] with permission ACS publications.

TABLE 10.2
A list of the Latest MFCs Implemented for Biosensing a Variety of Compounds

Biorecognition Element	Cathode Catalyst	Electron Transfer Method	Target Compound	Linear Detection Range	Response Time	Reference
Paulschulzia pseudovolvox	Platinum	DET	Copper sulphate, zinc sulphate and glyphosate	8–1000 µM 2.5–1000 µM 4–370 µM	2 h	[13]
Pseudomonas monteilii LZU3	Potassium ferricyanide	DET	p-nitrophenol	15.5–44 4.5 mg L^{-1}	27 ± 4.8 min	[129]
Mixed culture	FePO$_4$ nanoparticles	DET	Levofloxacin	0.1–100 µg L^{-1}	2–8 min	[126]
Enterobacter cloacae	Platinum	MET	Arsenic	0–0.5 mM	>1–24 h	[130]
Mixed culture	Platinum	DET	Neomycin	20–100 mg L^{-1}	>11 min	[127]
Mixed culture	Platinum	DET	Carbon monoxide	10–70 %	1 h	[128]

10.5 CONCLUSIONS

BFCs as self-powered sensors have made significant strides in the global scientific literature. We are witnessing a significant transformation of BFC design from the lab scale to the microscale level as a result of the intensive research carried out over the last two decades. The transformation has been inspired primarily by the self-powered working principle of the BFC devices that supports well the essential requirement of a simple and portable characteristic for a biosensor. This technological innovation has been possible with the advent of advanced knowledge on material sciences (including nanomaterials and polymers), ever-growing understanding of molecular and cell biology, and, finally yet importantly, the massive demand for biosensors in the world markets. In this technological venture, EFC has attained greater progress than its MFC counterpart has, even though the MFC fares well in certain parameters such as long-term stability of the microbial catalyst and low cost. The significant challenges usually encountered in developing MFC-based biosensors are their low selectivity (mostly due to their broad metabolic powerhouse, bacterial contamination, and mutations), long response time, and requirement of defined growth conditions of the microbial catalysts. Additionally, the sensitivity of MFCs requires further improvement to reach the detection limit at the level prescribed by the environmental regulatory bodies. Some of the challenges of MFC are expected to be alleviated by adopting disposable chip-based sensors and by using photosynthetic microorganisms such as cyanobacteria as the electrode catalysts. In the single-use disposable format, the requirement of stringent growth and cell division of the bacterial cells could be avoided by their simple immobilization on solid/semi-solid electrode surfaces. The inclusion of cyanobacteria as catalyst may be likely to remove the substrate complexity in the MFC for long-term operations and to improve cellular activity on the solid electrode surfaces. Contrary to the MFCs, EFCs have different sets of challenges pertaining to the operations. Isolated enzymes are prone to fast denaturation and inhibition by the presence of other compounds. Prolonging the lifetime of enzymes to more than 2–3 months is still a pressing challenge. Efforts are on to mitigate these challenges by using biocompatible support materials for their immobilization on a solid electrode surface and developing sturdy enzymes through genetic engineering techniques. On the device-engineering front, a higher focus has been laid on using DET over the ETM-based principle due to certain advantages endowed by the former that is well suited for developing an operationally stable and environmentally friendly BFC. The body should be able to tolerate the BFC biosensors. Such skin-friendly, portable biosensors are meant to make life more relaxed, with routine health monitoring accessible at one's fingertips.

REFERENCES

1. Santoro C, Arbizzani C, Erable B, Ieropoulos I. Microbial fuel cells: From fundamentals to applications. A review. J Power Sources 2017;356:225–44.
2. Aston WJ, Turner APF. Biosensors and biofuel cells. Biotechnol Genet Eng Rev 1984;1:89–120.
3. Schröder U. Anodic electron transfer mechanisms in microbial fuel cells and their energy efficiency. Phys Chem Chem Phys 2007;9:2619–29.
4. Cooney MJ, Svoboda V, Lau C, Martin G, Minteer SD. Enzyme catalysed biofuel cells. Energy Environ Sci 2008;1:320–37.
5. Bennetto HP, Stirling JL, Tanaka K, Vega CA. Anodic reactions in microbial fuel cells. Biotechnol Bioeng 1983;25:559–68.
6. Meredith MT, Minteer SD. Biofuel cells: Enhanced enzymatic bioelectrocatalysis. Annu Rev Anal Chem (Palo Alto Calif) 2012;5:157–79.
7. Limoli DH, Jones CJ, Wozniak DJ. Bacterial extracellular polysaccharides in biofilm formation and function. Microbiol Spectr 2015;3:MB-0011-2014.
8. Sarma MK, Kaushik S, Goswami P. Cyanobacteria: A metabolic power house for harvesting solar energy to produce bioelectricity and biofuels. Biomass Bioenergy 2016;90:187–201.
9. Bullen RA, Arnot TC, Lakeman JB, Walsh FC. Biofuel cells and their development. Biosens Bioelectron 2006;21:2015–45.
10. Davis F, Higson SPJ. Biofuel cells–recent advances and applications. Biosens Bioelectron 2007;22:1224–35.
11. Logan BE, Regan JM. Electricity-producing bacterial communities in microbial fuel cells. Trends Microbiol 2006;14:512–8.
12. Sekretaryova AN, Eriksson M, Turner APF. Bioelectrocatalytic systems for health applications. Biotechnol Adv 2016;34:177–97.
13. Labro J, Craig T, Wood SA, Packer MA. Demonstration of the use of a photosynthetic microbial fuel cell as an environmental biosensor. Int J Nanotechnol 2017;14:213.
14. Borolea P, Reguera G, Ringeisen B, Wang Z-W, Feng Y, Kim BH. Electroactive biofilms: Current status and future research needs. Energy Environ Sci 2011;4:4813–34.
15. Pankratova G, Gorton L. Electrochemical communication between living cells and conductive surfaces. Curr Opin Electrochem 2017;5:193–202.
16. Kaushik S, Sarma MK, Goswami P. FRET-guided surging of cyanobacterial photosystems improves and stabilizes current in photosynthetic microbial fuel cell. J Mater Chem A 2017;5:7885–95.
17. Kaushik S, Goswami P. Bacterial membrane depolarization-linked fuel cell potential burst as signal for selective detection of alcohol. ACS Appl Mater Interfaces 2018;10:18630–40.
18. Bhatnagar D, Xu S, Fischer C, Arechederra RL, Minteer SD. Mitochondrial biofuel cells: Expanding fuel diversity to amino acids. Phys Chem Chem Phys 2011;13:86–92.
19. Rasmussen M, Minteer SD. Photobioelectrochemistry: Solar energy conversion and biofuel production with photosynthetic catalysts. J Electrochem Soc 2014;161:H647–55.
20. Zhou M, Kuralay F, Windmiller JR, Wang J. DNAzyme logic-controlled biofuel cells for self-powered biosensors. Chem Commun 2012;48:3815–7.
21. Rasmussen M, Minteer SD. Self-powered herbicide biosensor utilizing thylakoid membranes. Anal Methods 2013;5:1140–4.
22. Grattieri M, Minteer SD. Self-powered biosensors. ACS Sensors 2018;3:44–53.
23. Chouler J, Cruz-Izquierdo Á, Rengaraj S, Scott JL, Di Lorenzo M. A screen-printed paper microbial fuel cell biosensor for detection of toxic compounds in water. Biosens Bioelectron 2018;102:49–56.
24. Gonzalez-Solino C, Di Lorenzo M. Enzymatic fuel cells: Towards self-powered implantable and wearable diagnostics. Biosensors (Basel) 2018;8(1):11.
25. Saravanan P. Microbial fuel cell: A prospective sustainable solution for energy and environmental crisis. Int J Biosens Bioelectron 2018;4:2–5.
26. Schröder U. Anodic electron transfer mechanisms in microbial fuel cells and their energy efficiency. Phys Chem Chem Phys 2007;9:2619–29.
27. Logan BE, Rabaey K. Conversion of wastes into bioelectricity and chemicals by using microbial electrochemical technologies. Science 2012;337:686–90.
28. Rosenbaum M, Schröder U. Photomicrobial solar and fuel cells. Electroanalysis 2010;22:844–55.
29. Schröder U. A basic introduction into microbial fuel cells and microbial electrocatalysis. Chem Texts 2018;4:19.
30. Calabrese Barton S, Gallaway J, Atanassov P. Enzymatic biofuel cells for implantable and microscale devices: Chemical reviews. Chem Rev 2004;104:4867–4886.
31. Bond DR, Lovley DR. Electricity production by Geobacter sulfurreducens attached to electrodes. Appl Environ Microbiol 2003;69:1548–55.
32. Rabaey K, Boon N, Siciliano SD, Verhaege M, Verstraete W. Biofuel cells select for microbial consortia that self-mediate electron transfer. Appl Environ Microbiol 2004;70:5373–82.
33. Reguera G, Nevin KP, Nicoll JS, Covalla SF, Woodard TL, Lovley DR. Biofilm and nanowire production leads to increased current in geobacter sulfurreducens fuel cells. Appl Environ Microbiol 2006;72:7345–8.
34. El-Naggar MY, Wanger G, Leung KM, Yuzvinsky TD, Southam G, Yang J, et al. Electrical transport along bacterial nanowires from *Shewanella oneidensis* MR-1. Proc Natl Acad Sci U S A 2010;107:18127–31.
35. Lovley DR, Malvankar NS. Seeing is believing: Novel imaging techniques help clarify microbial nanowire structure and function. Environ Microbiol 2015;17:2209–15.
36. Vermaas WF. Photosynthesis and respiration in cyanobacteria. Encycl. Life Sci., Chichester, UK: John Wiley & Sons, Ltd; 2001.

37. McCormick AJ, Bombelli P, Bradley RW, Thorne R, Wenzel T, Howe CJ. Biophotovoltaics: Oxygenic photosynthetic organisms in the world of bioelectrochemical systems. Energy Environ Sci 2015;8:1092–109.
38. Pereira AR, Sedenho GC, de Souza JCP, Crespilho FN. Advances in enzyme bioelectrochemistry. An Acad Bras Cienc 2018;90:825–57.
39. Das P, Das M, Chinnadayyala SR, Singha IM, Goswami P. Recent advances on developing 3rd generation enzyme electrode for biosensor applications. Biosens Bioelectron 2016;79:386–97.
40. Luz RAS, Pereira AR, de Souza JCP, Sales Fernanda CPF, Crespilho FN. Enzyme biofuel cells: Thermodynamics, kinetics and challenges in applicability. Chem Electro Chem 2014;1:1751–77.
41. Logan BE, Hamelers B, Rozendal R, Schröder U, Keller J, Freguia S, et al. Microbial fuel cells: Methodology and technology. Environ Sci Technol 2006;40:5181–92.
42. Osman MH, Shah AA, Walsh FC. Recent progress and continuing challenges in bio-fuel cells. Part I: Enzymatic cells. Biosens Bioelectron 2011;26:3087–102.
43. Venkata Mohan S, Velvizhi G, Annie Modestra J, Srikanth S. Microbial fuel cell: Critical factors regulating bio-catalyzed electrochemical process and recent advancements. Renew Sustain Energy Rev 2014;40:779–97.
44. Zhao F, Slade RCT, Varcoe JR. Techniques for the study and development of microbial fuel cells: An electrochemical perspective. Chem Soc Rev 2009;38:1926.
45. Walcarius A, Minteer SD, Wang J, Lin Y, Merkoçi A. Nanomaterials for bio-functionalized electrodes: Recent trends. J Mater Chem B 2013;1:4878–908.
46. Sumlo Iijima. Helical microtubules of graphitic carbon. Nature 1991;354:56–8.
47. Yang W, Ratinac KR, Ringer SR, Thordarson P, Gooding JJ, Braet F. Carbon nanomaterials in biosensors: Should you use nanotubes or graphene. Angew Chemie - Int Ed 2010;49:2114–38.
48. Wang J. Carbon nanotube-based electrochemical biosensors: A review. Electroanalysis 2005;17:7–14.
49. Liu Y, Du Y, Li CM. Direct electrochemistry based biosensors and biofuel cells enabled with nanostructured materials. Electroanalysis 2013;25:815–31.
50. Chou A, Böcking T, Singh NK, Gooding JJ. Demonstration of the importance of oxygenated species at the ends of carbon nanotubes for their favourable electrochemical properties. Chem Commun 2005:842–4.
51. Gooding JJ. Nanostructuring electrodes with carbon nanotubes: A review on electrochemistry and applications for sensing. Electrochim Acta 2005;50:3049–60.
52. Zhao Y-L, Stoddart JF. Noncovalent functionalization of single-walled carbon nanotubes. Acc Chem Res 2009;42:1161–71.
53. Kumar A, Sharma S, Pandey LM, Chandra P. Nanoengineered material based biosensing electrodes for enzymatic biofuel cells applications. Mater Sci Energy Technol 2018;1:38–48.
54. Yan Y, Zheng W, Su L, Mao L. Carbon-nanotube-based glucose/O2 biofuel cells. Adv Mater 2006;18:2639–43.
55. Kou T, Yang Y, Yao B, Li Y. Interpenetrated bacteria-carbon nanotubes film for microbial fuel cells. Small Methods 2018;2:1800152.
56. Novoselov KS, Geim AK, Morozov SV, Jiang D, Zhang Y, Dubonos SV, et al. Electric field effect in atomically thin carbon films. Science 2004;306:666–9.
57. Geim AK, Novoselov KS. The rise of graphene. Nat Mater 2007;6:183–91.
58. Jouault B, Camara N, Jabakhanji B, Caboni A, Consejo C, Godignon P, et al. Quantum hall effect in bottom-gated epitaxial graphene grown on the C-face of SiC. Appl Phys Lett 2012;100:1–4.
59. Geim AK. Graphene: Status and prospects. Science 2009;324:1530–4.
60. Sheehan PE, Snow ES, Perkins FK, Robinson JT, Wei Z. Reduced graphene oxide molecular sensors. Nano Lett 2008;8:3137–40.
61. Lee SU, Belosludov R V., Mizuseki H, Kawazoe Y. Designing nanogadgetry for nanoelectronic devices with nitrogen-doped capped carbon nanotubes. Small 2009;5:1769–75.
62. Shao Y, Zhang S, Engelhard MH, Li G, Shao G, Wang Y, et al. Nitrogen-doped graphene and its electrochemical applications. J Mater Chem 2010;20:7491–6.
63. Zhang Y, Mo G, Li X, Zhang W, Zhang J, Ye J, et al. A graphene modified anode to improve the performance of microbial fuel cells. J Power Sources 2011;196:5402–7.
64. Li Y, Fu ZY, Su BL. Hierarchically structured porous materials for energy conversion and storage. Adv Funct Mater 2012;22:4634–67.
65. Chinnadayyala SR, Kakoti A, Santhosh M, Goswami P. A novel amperometric alcohol biosensor developed in a 3rd generation bioelectrode platform using peroxidase coupled ferrocene activated alcohol oxidase as biorecognition system. Biosens Bioelectron 2014;55:120–6.
66. Balint R, Cassidy NJ, Cartmell SH. Conductive polymers : Towards a smart biomaterial for tissue engineering. Acta Biomater 2014;10:2341–53.
67. Kim BD, Richardson-burns SM, Hendricks JL, Sequera C, Martin DC. Effect of immobilized nerve growth factor on conductive polymers : electrical properties and cellular response. Adv Funct Mater 2007;17:79–86.
68. Sarma M.K, Abdul Quadir M.G, Bhaduri R, Kaushik S. Goswami, P. Composite polymer coated magnetic nanoparticles based anode enhances dye degradation and power production in microbial fuel cells. Biosens Bioelectron 2018;119:94–102.
69. Li C, Zhang L, Ding L, Ren H, Cui H. Effect of conductive polymers coated anode on the performance of microbial fuel cells (MFCs) and its biodiversity analysis. Biosens Bioelectron 2011;26:4169–76.
70. Song R, Wu Y, Lin Z, Xie J, Tan CH, Say J, et al. Living and conducting : coating individual bacterial cells with in situ formed polypyrrole. Angew Chemie – Int Ed 2017;56:10516–20.
71. Safarıkova M, Safarik I. Use of magnetic techniques for the isolation of cells. J Chromatogr B 1999;722:33–53.
72. Huang Y-F, Wang Y-F, Yan X-P. Amine-functionalized magnetic nanoparticles for rapid capture and removal of bacterial pathogens. Environ Sci Technol 2010;44:7908–13.
73. Cruz Viggi C, Casale S, Chouchane H, Askri R, Fazi S, Cherif A, et al. Magnetite nanoparticles enhance the bioelectrochemical treatment of municipal sewage by facilitating the syntrophic oxidation of volatile fatty acids. J Chem Technol Biotechnol 2019;94:3134–46.
74. Elmekawy A, Hegab HM, Dominguez-Benetton X, Pant D. Internal resistance of microfluidic microbial fuel cell: Challenges and potential opportunities. Bioresour Technol 2013;142:672–82.
75. Lee J wook, Kjeang E. A perspective on microfluidic biofuel cells. Biomicrofluidics 2010;4:041301.
76. Zebda A, Renaud L, Cretin M, Pichot F, Innocent C, Ferrigno R, et al. A microfluidic glucose biofuel cell to generate micropower from enzymes at ambient temperature. Electrochem Commun 2009;11:592–5.

77. Moore CM, Minteer SB, Martin RS. Microchip-based ethanol/oxygen biofuel cell. Lab Chip 2005;5:218–25.
78. Chiao M, Lam KB, Lin L. Micromachined microbial and photosynthetic fuel cells. J Micromechanics Microengineering 2006;16:2547–53.
79. Siu CPB, Chiao M. A microfabricated PDMS microbial fuel cell. J Microelectromechanical Syst 2008;17:1329–41.
80. Luka G, Ahmadi A, Najjaran H, Alocilja E, Derosa M, Wolthers K, et al. Microfluidics integrated biosensors: A leading technology towards lab-on-A-chip and sensing applications. Sensors (Switzerland) 2015;15:30011–31.
81. Shafiee H, Asghar W, Inci F, Yuksekkaya M, Jahangir M, Zhang MH, et al. Paper and flexible substrates as materials for biosensing platforms to detect multiple biotargets. Sci Rep 2015;5:1–9.
82. Economou A, Kokkinos C, Prodromidis M. Flexible plastic, paper and textile lab-on-a chip platforms for electrochemical biosensing. Lab Chip 2018;18:1812–30.
83. Mahato K, Srivastava A, Chandra P. Paper based diagnostics for personalized health care: Emerging technologies and commercial aspects. Biosens Bioelectron 2017;96:246–59.
84. Desmet C, Marquette CA, Blum LJ, Doumèche B. Paper electrodes for bioelectrochemistry: Biosensors and biofuel cells. Biosens Bioelectron 2016;76:145–63.
85. Henry CS, Dungchai W, Chailapakul O. Electrochemical detection for paper-based microfluidics. Anal Chem 2009;81:5821–6.
86. Delaney JL, Hogan CF, Tian J, Shen W. Electrogenerated chemiluminescence detection in paper-based microfluidic sensors. Anal Chem 2011;83:1300–6.
87. Shitanda I, Yamaguchi T, Hoshi Y, Itagaki M. Fully screen-printed paper-based electrode chip for glucose detection. Chem Lett 2013;42:1369–70.
88. Malon RSP, Chua KY, Wicaksono DHB, Córcoles EP. Cotton fabric-based electrochemical device for lactate measurement in saliva. Analyst 2014;139:3009–16.
89. Fraiwan A, Mukherjee S, Sundermier S, Choi S. A microfabricated paper-based microbial fuel cell. Proc IEEE Int Conf Micro Electro Mech Syst 2013:809–12.
90. Wu G, Srivastava J, Zaman MH. Stability measurements of antibodies stored on paper. Anal Biochem 2014;449:147–54.
91. Slaughter G, Kulkarni T. Enzymatic glucose biofuel cell and its application. J Biochips Tissue Chips 2015;05:1–10.
92. Trudeau F, Daigle F, Leech D. Reagentless mediated laccase electrode for the detection of enzyme modulators. Anal Chem 1997;69:882–6.
93. Katz E, Bückmann AF, Willner I. Self-powered enzyme-based biosensors. J Am Chem Soc 2001; 123, 43: 10752–10753.
94. Fischer C, Fraiwan A, Choi S. A 3D paper-based enzymatic fuel cell for self-powered, low-cost glucose monitoring. Biosens Bioelectron 2016;79:193–7.
95. Zhao M, Gao Y, Sun J, Gao F. Mediatorless glucose biosensor and direct electron transfer type glucose/air biofuel cell enabled with carbon nanodots. Anal Chem 2015;87:2615–22.
96. Liu Z, Cho B, Ouyang T, Feldman B. Miniature amperometric self-powered continuous glucose sensor with linear response. Anal Chem 2012;84:3403–9.
97. Thungon PD, Kakoti A, Ngashangva L, Goswami P. Advances in developing rapid, reliable and portable detection systems for alcohol. Biosens Bioelectron 2017;97:83–99.
98. Azevedo A, Prazeres D, Cabral J, Fonseca L. Ethanol biosensors based on alcohol oxidase. Biosens Bioelectron 2005;21:235–47.
99. Das M, Barbora L, Das P, Goswami P. Biofuel cell for generating power from methanol substrate using alcohol oxidase bioanode and air-breathed laccase biocathode. Biosens Bioelectron 2014;59:184–91.
100. Sekretaryova AN, Beni V, Eriksson M, Karyakin AA, Turner APF, Vagin MY. Cholesterol self-powered biosensor. Anal Chem 2014;86:9540–7.
101. Bandodkar AJ. Review—wearable biofuel cells: Past, present and future. J Electrochem Soc 2017;164:H3007–14.
102. Cheng H, Yu P, Lu X, Lin Y, Ohsaka T, Mao L. Biofuel cell-based self-powered biogenerators for online continuous monitoring of neurochemicals in rat brain. Analyst 2013;138:179–85.
103. Jeerapan I, Sempionatto JR, Pavinatto A, You J-M, Wang J. Stretchable biofuel cells as wearable textile-based self-powered sensors. J Mater Chem A 2016;4:18342–53.
104. Huang X, Zhang L, Zhang Z, Guo S, Shang H, Li Y, et al. Wearable biofuel cells based on the classification of enzyme for high power outputs and lifetimes. Biosens Bioelectron 2019;124–125:40–52.
105. Garcia SO, Ulyanova Y V, Figueroa-Teran R, Bhatt KH, Singhal S, Atanassov P. Wearable sensor system powered by a biofuel cell for detection of lactate levels in sweat. ECS J Solid State Sci Technol 2016;5:M3075–81.
106. Jia W, Valdés-Ramírez G, Bandodkar AJ, Windmiller JR, Wang J. Epidermal biofuel cells: Energy harvesting from human perspiration. Angew Chemie - Int Ed 2013;52:7233–6.
107. Bandodkar AJ, You JM, Kim NH, Gu Y, Kumar R, Mohan AMV, et al. Soft, stretchable, high power density electronic skin-based biofuel cells for scavenging energy from human sweat. Energy Environ Sci 2017;10:1581–9.
108. Deng L, Chen C, Zhou M, Guo S, Wang E, Dong S. Integrated Self-Powered Microchip Biosensor for Endogenous Biological Cyanide. Anal Chem 2010;82:4283–7.
109. Wen D, Deng L, Guo S, Dong S. Self-powered sensor for trace Hg2+ detection. Anal Chem 2011;83:3968–72.
110. Meredith MT, Minteer SD. Inhibition and activation of glucose oxidase bioanodes for use in a self-powered EDTA sensor. Anal Chem 2011;83:5436–41.
111. Zhang L, Zhou M, Dong S. A self-powered acetaldehyde sensor based on biofuel cell. Anal Chem 2012;84:10345–9.
112. Wang T, Milton RD, Abdellaoui S, Hickey DP, Minteer SD. Laccase inhibition by arsenite/arsenate: Determination of inhibition mechanism and preliminary application to a self-powered biosensor. Anal Chem 2016;88:3243–8.
113. Zhou M, Du Y, Chen C, Li B, Wen D, Dong S, et al. Aptamer-controlled biofuel cells in logic systems and used as self-powered and intelligent logic aptasensors. J Am Chem Soc 2010;132:2172–4.
114. Zhou M, Zheng X, Wang J, Dong S. A self-powered and reusable biocomputing security keypad lock system based on biofuel cells. Chem Eur J 2010;16:7719–24.
115. Ivars-Barceló F, Zuliani A, Fallah M, Mashkour M, Rahimnejad M, Luque R. Novel applications of microbial fuel cells in sensors and biosensors. Appl Sci 2018;8:1184.
116. Schneider G, Kovács T, Rákhely G, Czeller M. Biosensoric potential of microbial fuel cells. Appl Microbiol Biotechnol 2016;100:7001–9.
117. Zhou T, Han H, Liu P, Xiong J, Tian F, Li X. Microbial fuels cell-based biosensor for toxicity detection: A review. Sensors (Switzerland) 2017;17:1–21.
118. Su L, Jia W, Hou C, Lei Y. Microbial biosensors: A review. Biosens Bioelectron 2011;26:1788–99.

119. Abrevaya XC, Sacco NJ, Bonetto MC, Hilding-Ohlsson A, Cortón E. Analytical applications of microbial fuel cells. Part II: Toxicity, microbial activity and quantification, single analyte detection and other uses. Biosens Bioelectron 2015;63:591–601.
120. Mahadevan A, Gunawardena D a, Fernando S. Biochemical and electrochemical perspectives of the anode of a microbial fuel cell. Technol Appl Microb Fuel Cells 2014:13–32.
121. Karube I, Matsunaga T, Mitsuda S, Suzuki S. Microbial electrode BOD sensors. Biotechnol Bioeng 1977;19:1535–47.
122. Chang IS, Jang JK, Gil GC, Kim M, Kim HJ, Cho BW, et al. Continuous determination of biochemical oxygen demand using microbial fuel cell type biosensor. Biosens Bioelectron 2004;19:607–13.
123. Dávila D, Esquivel JP, Sabaté N, Mas J. Silicon-based microfabricated microbial fuel cell toxicity sensor. Biosens Bioelectron 2011;26:2426–30.
124. Choi S. Microscale microbial fuel cells: Advances and challenges. Biosens Bioelectron 2015;69:8–25.
125. Golitsch F, Bücking C, Gescher J. Proof of principle for an engineered microbial biosensor based on Shewanella oneidensis outer membrane protein complexes. Biosens Bioelectron 2013;47:285–91.
126. Zeng L, Li X, Shi Y, Qi Y, Huang D, Tadé M, et al. $FePO_4$ based single chamber air-cathode microbial fuel cell for online monitoring levofloxacin. Biosens Bioelectron 2017;91:367–73.
127. Catal T, Yavaser S, Enisoglu-Atalay V, Bermek H, Ozilhan S. Monitoring of neomycin sulfate antibiotic in microbial fuel cells. Bioresour Technol 2018;268:116–20.
128. Zhou S, Huang S, Li Y, Zhao N, Li H, Angelidaki I, et al. Microbial fuel cell-based biosensor for toxic carbon monoxide monitoring. Talanta 2018;186:368–71.
129. Chen Z, Niu Y, Zhao S, Khan A, Ling Z, Chen Y, et al. A novel biosensor for p-nitrophenol based on an aerobic anode microbial fuel cell. Biosens Bioelectron 2016;85:860–8.
130. Rasmussen M, Minteer SD. Long-term arsenic monitoring with an Enterobacter cloacae microbial fuel cell. Bioelectrochemistry 2015;106:207–212.

11 Bioelectrochemiluminescence as an Analytical Signal of Extreme Sensitivity

Vinay Bachu and Pranab Goswami
Indian Institute of Technology Guwahati, Assam, India

CONTENTS

- 11.1 Introduction ... 234
- 11.2 Principles of Bioelectrochemiluminescence ... 234
  - 11.2.1 General Reaction Mechanisms ... 234
    - 11.2.1.1 Annihilation Reaction ... 234
    - 11.2.1.2 Co-Reactant ECL ... 236
    - 11.2.1.3 Oxidative-Reduction ECL ... 236
    - 11.2.1.4 Reductive-Oxidation ECL ... 236
  - 11.2.2 Effect of External Parameters on ECL ... 237
    - 11.2.2.1 Effect of the Magnetic Field ... 237
    - 11.2.2.2 Effect of Quenching on ECL ... 237
- 11.3 Electrodes in ECL ... 237
  - 11.3.1 Metal Electrodes ... 238
  - 11.3.2 Carbon-Based Electrodes ... 238
  - 11.3.3 Metal Oxide Electrodes ... 238
  - 11.3.4 Disposable Electrodes ... 239
- 11.4 ECL Systems Based on Types of Luminophores ... 239
  - 11.4.1 Inorganic Luminophores-Based ECL Systems ... 239
    - 11.4.1.1 Ruthenium Complexes ... 240
    - 11.4.1.2 Iridium Complexes ... 241
  - 11.4.2 Organic Luminophore-Based ECL Systems ... 241
    - 11.4.2.1 Fluorene ... 241
    - 11.4.2.2 Perylene ... 241
    - 11.4.2.3 Lucigenin ... 242
    - 11.4.2.4 Luminol ... 242
    - 11.4.2.5 Di Imides ... 242
    - 11.4.2.6 Organic Polymers ... 242
    - 11.4.2.7 Mechanism of ECL Emission in Organic Compounds ... 242
  - 11.4.3 Nanomaterial-Based Systems ... 242
    - 11.4.3.1 CdTe Semiconductor Nanomaterials ... 242
    - 11.4.3.2 Quantum Dots (QDs) ... 243
    - 11.4.3.3 Carbon Nanotubes (CNTs) ... 244
    - 11.4.3.4 Silole Electrochemiluminescence ... 244
- 11.5 ECL in biology ... 245
  - 11.5.1 ECL-RET ... 245
  - 11.5.2 ECL Quenching ... 246
  - 11.5.3 ECL-ELISA ... 247
  - 11.5.4 ECL Co-Reactant Catalyst ... 247
  - 11.5.5 ECL by Steric Hindrance ... 247
  - 11.5.6 ECL by Nucleic Acid Modification ... 248
- 11.6 Conclusion ... 248
- References ... 248

11.1 INTRODUCTION

The detection of biological samples has always been an arduous task, and researchers worldwide have adopted a variety of strategies to tackle this rather significant barrier. Radioactive labels and fluorescent tags have been traditionally used to analyze a wide range of biological samples, which are otherwise difficult to analyze by other detection strategies. However, the use of radioactive labels conflicts with the issues related to safety, convenience, and sensitivity. On the other hand, the fluorescent tags confront issues like autofluorescence, bleaching, quenching, and a host of other concerns. These inconveniences prompted researchers to explore a novel signal generating system to outweigh the flaws in current detection methods.

An electrochemical reaction-induced conversion of electrochemical energy into radiative energy by emitting measurable luminescent signal is called electrochemiluminescence (ECL) [1]. The ECL-based system offers a spectrum of advantages over its heralded predecessors ranging from high sensitivity, single protein detection to outwitting the safety, and bleaching issues. The first detailed study on ECL began in the mid-1960s [2]. Since then, a considerable amount of effort has been made, which propelled significant progress on the subject over the last decade, as visible from the material presented in this chapter.

The progress in ECL was rather sluggish in its infancy, and works related to ECL were hard to find. The first compound exploited for ECL was tris(2,2'-bipyridine) ruthenium(II) (Ru(by)$_3^{2+}$) which experienced overwhelming success [3] since its inception as an ECL species. A number of works comprising different Ruthenium complexes followed this. Some of them were translated into commercial diagnosis kits. *In vitro* diagnosis with the ECL strategy turned out to be highly successful. ECL-based kits are available for diagnosis of myocardial infarction for both cardiac troponin T and cardiac troponin I.

The interest in the field of ECL surged only in the last one and half decades. A literature survey of the publications on ECL shows that around 74% of all the publications pertaining to ECL are from the 2009 to 2018 period. This has led to an insatiable interest among researchers of late because of certain compelling properties that ruthenium complexes in particular exhibit, like extreme stability, water solubility, high sensitivity, a wide dynamic range, and a host of different analytes. All these characteristics are not found in unison in other techniques like fluorescence, colourimetry, calorimetry, etc. The ECL reactions can be controlled by "turn on" and "turn off," which is due to the electrochemical dependence for the emission of ECL. This characteristic of ECL allows us to control the reaction, which is not possible in a number of optical techniques.

ECL bioassays were performed for the first time in the 1990s. Biosamples like oxalate ions were detected with the ECL technique [4]. Detection of biological molecules by the mechanism of electrochemiluminescence can be termed bioelectrochemiluminescence.

11.2 PRINCIPLES OF BIOELECTROCHEMILUMINESCENCE

The generation of charged species at the electrode surface, which is followed by electron transfer reactions resulting in the formation of excited states, which emit light, is the basic principle on which the process of ECL is based on. ECL can be metaphorically referred to as the combination of chemiluminescence (CL) with electrochemistry. Bioelectrochemiluminescence refers to harnessing ECL active species by tagging them with biological molecules.

ECL is basically a CL reaction, with the main difference being the chemiluminescence reaction that occurs when both reactants are mixed. In the case of ECL, the luminescence is generated at the electrode surface or in the proximity of the electrode surface by electrogeneration of the charged species.

ECL is a CL reaction in which, like all CL reactions, two important interlinked events occur. One is the emission of luminescence and the other is the chemical reaction. The rate of the chemical reaction determines the intensity of the emission. It also defines the quantum efficiency of the ECL reaction. This can be represented by the following equation:

$$I_{CL}(t) = \phi_{CL}\frac{dP(t)}{dt} = \phi_{EX}\phi_{EM}\frac{dP(t)}{dt} \quad (11.1)$$

where

I_{CL}: ECL intensity (emission per second)
φ_{CL}: Quantum efficiency of the reaction
$dP(t)/dt$: Rate of the CL reaction
φ_{EX}: Excitation quantum efficiency
φ_{EM}: Emission quantum efficiency

In general, CL reactions are of low quantum efficiency, ranging from 1% to 30%, and the ECL reaction, in particular, is of further lower quantum efficiency rates, which have been the bottleneck for ECL-based sensing application. Of late, the advent of the silole and other organic polymers has improved the quantum efficiencies of the ECL reactions. Silole, in particular, has shown quantum efficiencies of around 60%. In order to improve the quantum efficiencies of the ECL reaction, it is very important to understand the principle of ECL to its fullest.

11.2.1 General Reaction Mechanisms

The general reaction mechanisms of ECL can be broadly classified into two types: annihilation reaction and co-reactant ECL.

11.2.1.1 Annihilation Reaction

The annihilation reaction was the first type of reaction mechanism by which a detailed ECL study was conducted [5, 39]. The studies involved the generation of a reduced and an oxidized species at the same electrode by the alternate pulsing of the electrode potential. The resultant charged species

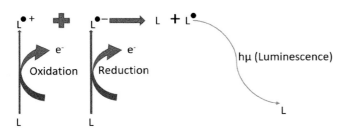

FIGURE 11.1 Schematic illustration of the annihilation reaction.

involved in an electron transfer reaction produced an emission of a particular wavelength.

There is an initial electrochemical generation of two species, which are nothing but the reduced form and oxidized form of the same luminophore, and these are produced within the depletion zone of the electrode formed by the potential sweep. The two generated species interact to generate an electronically excited state, which due to instability at the higher electronic state, emits excess energy of a particular wavelength and relaxes to the ground state. The general mechanism of the annihilation reaction can be illustrated as shown in Figure 11.1.

The free energy (ΔG) for the annihilation energy can be determined with the following equation:

$$\Delta G = nF(E°\text{Donor} - E°\text{Acceptor}) \quad (11.2)$$

where $E°_{\text{Donor}}$ is the potential for the ground state reduction and $E°_{\text{Acceptor}}$ is the potential for the ground state oxidation. The ΔG of the reaction in most cases is way more than the energy required for the generation of an electronically excited state, and hence the emission by the luminophore is observed.

Annihilation reactions were initially studied with compounds with potentials that were easily attainable. The compounds selected were also considered as they were known to electrochemically as well as chemically undergo a one-electron redox reaction in a reversible method. Diphenyl anthracene and rubrene-like polyaromatic hydrocarbons were some of the first complexes to be studied [6, 7].

ECL in the annihilation reaction can be achieved by different approaches based on the type of energy system the luminophore belongs to. They are singlet route (S-route), triplet route (T-route), singlet-triplet route (ST-route), and excimer-exciplex route (E-route).

11.2.1.1.1 Singlet Route

Annihilation ECL materializes in an S-route by the formation of an excited singlet state which is unstable, and hence the excited electron comes back to the ground state by emission of (hν) a specific wavelength of radiation. The singlet systems are the energy-sufficient systems. If the energy available in an annihilation reaction is sufficiently high, then the produced excite state species will occupy the lowest singlet state. Diphenyl anthracene (DPA) is an example of a singlet route ECL (Figure 11.2).

S-route: $L^{•+} + L^{•-} \longrightarrow L + {}^1L^*$ (excited singlet formation)

FIGURE 11.2 Schematic representation of the S-route. L is the luminophore.

11.2.1.1.2 Triplet Route

The annihilation reaction materializes in a T-route by the formation of the triplet-triplet annihilation intermediates (TTA). The energy levels of the luminophores after annihilation are sufficient for an electron transfer reaction but not for achieving the singlet state from which the emission occurs. These kinds of systems are also called energy-deficient systems. The triplets combine together to form the excited singlet state for the luminophore, which eventually results in the ECL emission. TMPD/AN (TMPD N,N,N',N'-tetramethyl-p-phenylenediamine and anthracene) is an example of the triplet route ECL (Figure 11.3).

11.2.1.1.3 ST-Route

Sometimes when the annihilation energy is very close to singlet formation but a fraction short, it forms a T-route. This T-route can contribute to the formation of the excited state singlet species. This, along with other species which have attained the singlet state, facilitates the ST-route. The T-route becomes inefficient when the singlet species significantly outnumbers the high triplet species. But even in this case, the ST-route can exist. The annihilation of the Rubrene anion-cation follows the ST-route.

11.2.1.1.4 E-Route

An annihilation reaction at times can result in the formation of states like excited dimers, which are called excimers, and at times the formation of the excited complexes, which are termed exciplexes [8]. The reactions that result in the formation of the exciplexes and excimer complexes are termed the E-route (Figure 11.4). The excimers and the exciplexes generally tend to exhibit a redshift in comparison with the singlet emissions. The formation of an excimer is also possible by the TTA excimer formation method [9].

T-route: $L^{•+} + L^{•-} \longrightarrow L + {}^3L^*$ (excited triplet formation)

${}^3L^* + {}^3L^* \longrightarrow L + {}^1L^*$ (triplet-triplet annihilation)

FIGURE 11.3 Schematic representation of the T-route.

$L^{•+} + L^{•-} \longrightarrow L + L_2^*$ (excimer formation)

$L^{•+} + L'^{•-} \longrightarrow (LL')^*$ (exciplex formation)

${}^3L^* + {}^3L^* \longrightarrow {}^1L_2^*$ (TTA excimer formation)

FIGURE 11.4 Schematic representation of the E-route formation.

11.2.1.2 Co-Reactant ECL

All known ECL-based commercial products work on the co-reactant ECL mechanism. The major difference between the annihilation and co-reactant ECL is also the advantages of the co-reactant ECL. The most important is the requirement of a single potential step, unlike in the case of annihilation, wherein an extra negative potential step is also required. The potential to be applied is also less in most of the cases, ensuring the commercial viability of the ECL sensors that work on the co-reactant mechanism.

The other advantages of a co-reactant are it can make fluorescent compounds with reversible electrochemical oxidation or reduction to produce an ECL signal. This reaction mechanism is useful when one of the reaction intermediates of annihilation is unstable for the ECL reaction. The co-reactant reaction can also be exploited when the potential window for both the redox states in a particular solvent is very narrow. In cases where the quantum efficiency of the annihilation mechanism of an ECL luminophore is very low, substitution with a co-reactant species improves the quantum efficiency.

In a typical co-reactant mechanism, the luminophore is regenerated after the reaction, whereas the co-reactant is consumed in the reaction which is broken down into products. The co-reactant undergoes electrochemical oxidation or reduction depending on the potential applied and also the compound. An ideal co-reactant should exhibit the following properties: (a) The co-reactant should be easy to oxidize or reduce along with the luminophore at the electrode surface or in its close proximity. (b) The reduced or oxidized form of the co-reactant should be capable of reducing the luminophore. (c) The co-reactant should form a very strong reductant in order to form an unstable luminophore, which in turn produces ECL. (d) Other characteristics include the electrochemical properties and stability of the co-reactant.

In a number of cases, even after the co-reactant ECL pathway develops, a weak ECL signal is produced. The reasons for weak emission can be either the rate of the co-reactant reaction or it can be because of the chemical state of the co-reactant. Co-reactant catalysts are used to overcome the ECL emission intensity issues. The co-reactant catalysts react with the co-reactant and catalyze the ECL reaction by altering the rate of the reaction. Table 11.1 lists a few accepted co-reactant catalysts.

Co-reactant-based ECL can be achieved by oxidative-reduction ECL or by reduction-oxidation ECL. Depending on the redox potential and the working electrode, the first process which takes place is either reduction or oxidation.

TABLE 11.1
Co-Reactant Catalysts for ECL Reactions

Co-Reactant Catalyst	Luminophore	Co-Reactant
Hemin	CdTe QDs	$S_2O_8^{2-}$
Perylene derivatives	PTCA	$S_2O_8^{2-}$
Aniline	PTCA	$S_2O_8^{2-}$
Pt nanomaterials	Rubrene	O_2

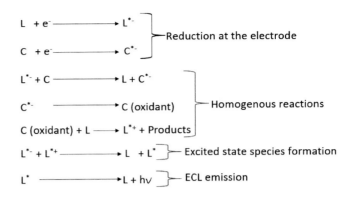

FIGURE 11.5 Schematic representation of oxidative-reduction ECL.

11.2.1.3 Oxidative-Reduction ECL

The first step is the oxidation of both the luminophore and the co-reactant. Once both are oxidized, the homogenous chemical reaction takes place, resulting in the generation of a strong reductant. The reductants interact with the luminophore to form excited-state species and products of the co-reactants. The excited-state species, due to their instability, revert to the ground state with ECL emission. The commonly used oxidative-reduction co-reactants are tri-n-propylamine (TPrA), oxalate ($C_2O_4^{2-}$) and pyruvate/Ce (III). The general reaction mechanism is shown in Figure 11.5.

11.2.1.4 Reductive-Oxidation ECL

The first step is the reduction of the luminophore and the co-reactant. Once both are reduced, the homogenous chemical reaction takes place, resulting in the generation of a strong reductant. The reductants interact with the luminophore to form excited-state species followed by the products of the co-reactants. The excited-state species, due to their instability, revert to the ground state with ECL emission. The commonly used oxidative-reduction co-reactants are hydrazine (N_2H_4), hydrogen peroxide (H_2O_2), and persulfate ($S_2O_8^{2-}$). The general reaction mechanism is shown in Figure 11.6.

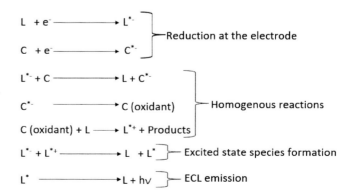

FIGURE 11.6 Schematic representation of reductive-oxidation ECL.

11.2.2 Effect of External Parameters on ECL

11.2.2.1 Effect of the Magnetic Field

In the presence of the applied external magnetic field, an unusual behavior was observed in ECL intensity during its generation. An increase in the luminescence intensity in the case of anthracene and 9,10-diphenylanthracene (DPA) anion radicals by the Wurster's blue cation (the cation radical of N,N,N',N'-tetramethyl-p-phenylenediamine, WB) in 7V, A-dimethylformamide (DMF) was observed [10].

ECL exhibits a dual mechanism for the emission for the chemiluminescent electron transfer processes, which can be divided into two separate categories based on the line of energy sufficiency. In the case of energy-deficient systems, an increasing external applied field strength can result in an increase of up to 27% in luminescence intensity. In systems that are energy deficient, the oxidation of compounds like anthracene, rubrene, DPA, and TPP anion radicals by Wurster's blue cation [11] takes place. The applied external field insulates the triplets in the excited state from quenching to some extent by decreasing the rate of quenching and hence an increase in luminescence intensity in systems with a T-route.

In S-route systems or the energy-sufficient systems, no effect was observed on luminescence when increasing the field strength. This observation also indicates that energy-sufficient systems yield luminescence without required triplet intermediates [12]. Another energy-deficient system, tetracene-N,N'-tetramethyl-p phenylenediamine (TMPD), was found to exhibit an increase of 19% luminescence intensity with increasing strength of the external field [13].

11.2.2.2 Effect of Quenching on ECL

Quenching is the process by which there is a significant reduction in the ECL intensity of the luminophore. Any molecule that aids the quenching process is called a quencher. The presence of a quencher adversely affects the ECL efficiency.

The principle effect of quenching is the reduction in ECL intensity and sometimes complete elimination. Hence, it is important to choose a solvent system and an electrode system that does not aid quenching of the ECL. The phenomenon of quenching can take place in two different routes, one being resonance energy transfer (RET) and the other being electron transfer (Figure 11.7). An overlap of the absorption band of the quencher with the emission band of the luminophore results in Forster transfer, or energy transfer from the excited-state luminophore [14]. This results in the attainment of the ground state of the luminophore without any form of emission. The Forster transfer is only possible when both molecules are in close proximity, and this distance is called the Forster distance. The energy transfer is nonradiative.

Quenching by electron transfer occurs in certain molecules, as it is easier to reduce/oxidize the molecule in the excited state than in the ground state. A reversible electron transfer occurs, and hence at the end of process both molecules retain the same electronic configuration.

11.3 ELECTRODES IN ECL

The electron transfer reactions in ECL are initiated at the electrode surface, which makes it even more important to choose the appropriate electrode. It is pertinent in the context

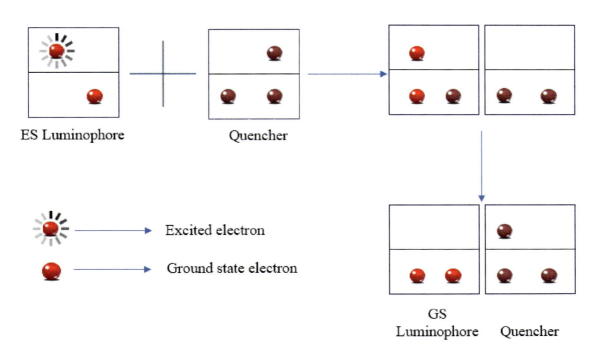

FIGURE 11.7 Illustration of electron transfer quenching.

of ECL to consider the electrode as an important parameter besides the standard ones for better ECL efficiency. There is a change in the interfacial behavior of the electrode during a potential ramp, radically modifying the ECL intensity. Numerous electrodes are deployed in ECL-based biosensors, ranging from the metal electrodes like platinum and gold, to carbon-based electrodes, to paper-based. In this section, the electrodes used in ECL-based sensors will be explored in detail.

The advantages of selecting an appropriate electrode are as follows: (a) The electrode surface characteristics affect the lifetime of the excited singlet and triplet states in the case of certain luminophores, with the standard example being TPrA*. (b) Stability of the ECL systems: In many cases, the redox reactions may result in poisoning of the electrode on repeated emission cycles and thus reducing the ECL intensity.

Electrodes can be broadly categorized into metal electrodes, carbon-based electrodes, metal oxide electrodes, and disposable electrodes.

11.3.1 Metal Electrodes

Platinum and gold electrodes are extensively employed metal electrodes in ECL studies. The principle reasons are the typically fast kinetics, which facilitate the electrochemical reactions. Also, they catalyze the formation of singlet and triplet excited states, which in turn bring about ECL. Most of the annihilation ECL studies are carried out with gold and platinum electrodes, as the majority of the initial annihilation studies were conducted with organic solvent systems.

Gold electrodes have been extensively used in luminol-based studies in both acidic and basic solutions. Gold electrodes are known to produce ten times stronger an ECL signal than platinum electrodes in the same experimental conditions [15]. Gold is a very good electrode in alkaline and acidic solvent systems. But in the aqueous system, the gold is oxidized and forms an oxidizing layer, thus reducing ECL intensity. To overcome this, a number of strategies have been developed, like harnessing a porous gold electrode, surface modifications with a number of polymers, surfactants, and alkaloids. But still, gold electrodes have issues pertaining to ECL efficiency.

Platinum electrodes are one of the most widely exploited electrode systems in electrochemistry. Platinum electrodes perform well in conditions of high temperature and pressure. They work well even in complex solutions. But most of the biological reactions require water as the solvent system, making a high potential of 0.8–1.2 V absolutely necessary for the ECL emission, which makes the use of platinum electrodes less feasible for biological systems. Platinum electrodes form an oxidizing layer under physiological conditions because of the oxidation of platinum. This results in a decrease in ECL efficiency. All these lacunae make platinum electrodes 100 times less effective when compared to glassy carbon electrodes in terms of ECL efficiency.

11.3.2 Carbon-Based Electrodes

Carbon-based materials are extensively used for ECL studies. They are known to be highly efficient in comparison with their metal-based equivalents, with the principle reason being the fast kinetic that these electrodes exhibit due to the oxidation of the amides. Glassy carbon electrodes (GCE) are the most extensively used electrodes. GCE overcomes all the problems that metal electrodes exhibit. They can be used in aqueous solutions, as they work fine with overpotentials. GCE can also be used in organic solvent systems. The TPrA and ruthenium based co-reactant ECL studies are mostly performed with GCE electrodes.

Carbon-based materials have further sensitized the detection limits of ECL systems. The most extensively exploited carbon-based materials include carbon nanotubes (CNTs) and different forms of graphene. CNTs can be very useful in highly complex matrices, as they can deliver results with high reproducibility and reliability.

11.3.3 Metal Oxide Electrodes

One of the biggest disadvantages of the previously mentioned two electrode systems is that the electrode emits their color, which acts as background noise in ECL studies. To overcome this, metal oxide like indium-doped tin oxides (ITO) and fluorine-doped tin oxides (FTO) are used, as they can be used to generate transparent electrodes. Both ITO and FTO are transparent in nature, essentially eliminating the background noise.

ITO has been used extensively in the last decade for ECL-based biosensing [16]. ITO has certain advantages over other electrode materials like good electrical conductivity, wide chemical working window, and high optical transparency. ITO is ternary complexes including indium, tin, and oxygen. The percentage of oxygen determines the physical and chemical as well as electrochemical properties of the ITO. The ITO electrodes are synthesized by sputtering or electron beam vaporization of ITO on glass or quartz surfaces [17].

A number of glucose sensors and nucleic acid sensors are composed of ITO electrodes. A biosensing application is possible only when the ITO molecules are fabricated with biomolecules, and this is possible, as in most cases the ITO electrodes are functionalized with alkyl silanes, which are the molecules on which the linker molecules are coupled with which, by EDC NHS modification, biomolecules can be immobilized [17, 18]. One of the main disadvantages of ITO electrodes is that in the organic electrolytic solution, electrochemical reduction of ITO is observed. To increase the sensitivity of the sensors developed, ITO electrodes co-functionalized with CNT have been developed and are a promising approach to overcome the disadvantages of both the electrodes (Figure 11.8).

Fluorine-doped tin oxide (FTO) has replaced ITO in some of the research studies, as FTO has some advantages over ITO. An FTO electrode is synthesized by coating FTO on the glass surface. It is known to be relatively stable under

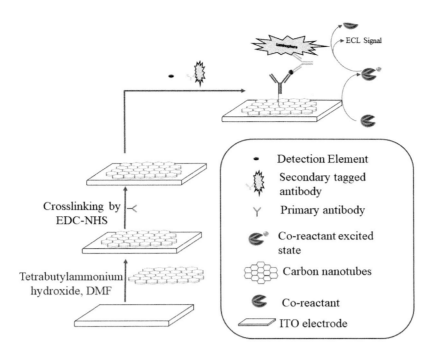

FIGURE 11.8 General scheme of optically transparent electrode (OTE) functionalized with CNTs for ECL detection.

atmospheric conditions. It is chemically inert and can withstand high-temperature conditions [18]. FTO electrodes are more commercially viable than ITO. FTO electrodes can be reduced in acidic electrolytic solutions.

11.3.4 Disposable Electrodes

Environmental awareness has increased in the last two decades, and a lot of interest has been vested in making sensors that are disposable and ecofriendly. Electrodes that are easy to dispose of are a must if electrochemiluminescence-based sensors are to get into the category of ecofriendly sensors. There have been reports of disposable electrodes like paper-based [19] and screen-printed electrodes. The paper-based electrodes work with the microfluidic-based approach, which has been explained in great detail in Chapter 3. The other advantages of paper-based electrodes are low cost and availability, thickness, and light weight. A minimum quantity of sample is enough for sensing, which is another major advantage [20].

Reverse electrodialysis (RED) is used as an external power source. RED patches are generally composed of ion exchange membranes (IEMs). Red patches in combination with bipolar electrodes are harnessed in the construction of ecofriendly and disposable electrodes [21, 22].

11.4 ECL SYSTEMS BASED ON TYPES OF LUMINOPHORES

The selection of luminophores and the co-reactants are equally important in choosing the right electrode for an efficient ECL system. Luminophores are the complexes that emit ECL emission. A good luminophore is the one with a better ECL signal. A broad range of compounds has been studied as an ECL luminophore ranging from organic to inorganic molecules, from polymers to nanoparticles. In the current section, different luminophores that have been used in ECL studies are discussed in detail. The ECL luminophores can be broadly classified into three categories: inorganic luminophores, organic luminophores, and nanoparticle-based luminophore systems.

11.4.1 Inorganic Luminophores-Based ECL Systems

Inorganic luminophores are the fundamental reason for the transformation of the study of ECL into a valuable technology that has been turning out commercially viable products one after the other in the last few years. A number of transition metal complexes are known today for their ECL emission properties.

Paradoxically, the very first metal with which ECL studies were performed turned out to be monumental and unscalable in the brief history of ECL. Ruthenium complexes are the pioneering metal complexes in ECL studies, and a vast majority of our understanding of the ECL phenomenon can be credited to ruthenium metal complexes. There are a number of other metal complexes which have been studied for their ECL properties. Some of them are iridium, chromium, copper, platinum, silicon, cadmium, osmium, and tungsten.

In this chapter only the ruthenium complexes and the iridium complexes are explained in detail, as they are the extensively used metal complexes as luminophores in ECL studies.

11.4.1.1 Ruthenium Complexes

Ruthenium is a transition metal with an atomic number of 44. It is used in a number of chemical reactions as a catalyst, and another application being in the generation of platinum alloys. Once the phenomenon of ECL was introduced, ruthenium was used extensively in ECL. Ruthenium-based ECL sensors are the most well-studied and well-understood ECL-based biosensors. They are known to have good biocompatibility and emit highly stable electrochemiluminescence. Ruthenium-based oxidant and reductant systems are extensively exploited for sensing applications and analytical applications.

Tris(2,2'-bipyridyl) ruthenium (II) [$Ru(bpy)_3^{2+}$] ECL is the most extensively used among the ruthenium complexes and is extremely sensitive at picomolar and femtomolar levels [23]. The $Ru(bpy)_3^{2+}$ is also used in high-performance liquid chromatography (HPLC) as a detection method for the determination of amine compounds and oxalate compounds. The predominant characteristics of $Ru(bpy)_3^{2+}$ such as chemical stability, longer half-life, high luminescence emission, and redox properties are the reasons for its wide range of applications in ECL.

The first $Ru(bpy)_3^{2+}$-based chemiluminescence study was reported in 1966 by Hercules and Lytle. Chemiluminescence of $Ru(bpy)_3^{2+}$ can be attained when the reaction between $Ru(bpy)_3^{3+}$ and $Ru(bpy)_3^{+}$ yields an excited state $Ru(bpy)_3^{2+*}$ – which is an unstable form and forms a stable form (i.e., $Ru(bpy)_3^{2+}$) with the emission of a photon (Figure 11.9). The reduced and oxidized forms can be electrochemically generated [24]. A potential of 1.3 V yields the oxidized form ($Ru(bpy)_3^{3+}$), whereas a potential of −1.1 V yields the reduced form ($Ru(bpy)_3^{+}$). The reaction between this two-redox couple regenerates the stable complex ($Ru(bpy)_3^{2+}$) and emission at a wavelength of 610 nm.

The quantum yield of $Ru(bpy)_3^{2+}$ is very low, but it compensates with a long half-life and very high intensity. The metal-to-ligand triplet charge transfer phenomenon is the reason behind the $Ru(bpy)_3^{2+}$-based ECL. The kind of ECL observed here is the annihilation ECL mechanism. Other strong reducing agents can also successfully produce ECL emission if an excited triplet state ($Ru(bpy)_3^{2+}$) is produced. Some of the commonly used reducing agents are hydrazine, alicyclic amines and aliphatic amines, oxalate, amino acids, and $S_2O_8^{2-}$. $Ru(bpy)_3^{2+}$ can be immobilized on a variety of surfaces like an electrode surface, polymer surface, and fiber-optics and can also be used in solutions, making it a robust ECL system to work with.

$$Ru(bpy)_3^{2+} - e^- \rightarrow Ru(bpy)_3^{3+} \text{ [oxidized]}$$
$$Ru(bpy)_3^{2+} + e^- \rightarrow Ru(bpy)_3^{+} \text{ [reduced]}$$
$$Ru(bpy)_3^{3+} + Ru(bpy)_3^{+} \rightarrow Ru(bpy)_3^{2+*} + Ru(bpy)_3^{2+}$$
$$Ru(bpy)_3^{2+*} \rightarrow Ru(bpy)_3^{2+} + h\nu$$

FIGURE 11.9 Schematic representation of $Ru(bpy)_3^{2+}$ ECL.

$$Ru(bpy)_3^{2+} - e^- \rightarrow Ru(bpy)_3^{3+}$$
$$TPrA - e^- \rightarrow [TPrA^{\bullet}]^+ \rightarrow TPrA^{\bullet} + H^+$$
$$Ru(bpy)_3^{3+} + TPrA^{\bullet} \rightarrow Ru(bpy)_3^{*2+} + \text{Products}$$
$$Ru(bpy)_3^{2+*} \rightarrow Ru(bpy)_3^{2+} + h\nu$$

FIGURE 11.10 Schematic representation of $Ru(bpy)_3^{2+}$ ECL with TPrA.

Other ruthenium systems that are used include $Ru(phen)_3^{2+}$, $Ru(dmbp)_3^{2+}$, $Ru(terpy)_3^{2+}$, $Ru(TPTZ)_3^{2+}$, $Ru(phen)_3^{2+}$, $Ru(dp\text{-}phen)_3^{2+}$, and $(bpy)_2Ru(AZA\text{-}bpy)^{2+}$. Most of the complexes of ruthenium are capable of emitting ECL by the annihilation pathway wherein a co-reactant is not required. Ruthenium complexes are also capable of ECL emission by co-reactant pathways. A wide array of co-reactants are available which are effective co-reactants for the ECL of ruthenium complexes. Tri-n-propylamine (TPrA) is the extensively used co-reactant along with luminol. In the case of TPrA as a co-reactant, there is simultaneous oxidation of both the $Ru(bpy)_3^{2+}$ and TPrA (Figure 11.10). On oxidation, the TPrA cation is unstable and short-lived, which results in the loss of a proton, which in turn results in the formation of a strong reducing agent, which is a reaction intermediate. This reaction intermediate reduces the oxidized $Ru(bpy)_3^{3+}$ into a triplet state of $Ru(bpy)_3^{2+}$, which reverts back to the ground state with the emission of a wavelength of light.

Oxalate can also be used as an effective co-reactant with $Ru(bpy)_3^{2+}$ for the generation of ECL. In ECL studies, oxalate is called an oxidative-reductive co-reactant. Oxalate has the ability to form a strong reductant on electrochemical oxidation. It was also the first co-reactant discovered and is also used in biological systems extensively. In oxalate-based or any co-reactant-based ECL, as explained earlier, a single oxidization step is essential, unlike in the case of the annihilation reaction of ruthenium complexes, where a double potential step is necessary. In oxalate-based ECL, the first step is the simultaneous oxidation of oxalate and $Ru(bpy)_3^{2+}$, resulting in the formation of the oxidized $Ru(bpy)_3^{3+}$ and oxidized oxalate, which upon bond cleavage forms the strong reductant ($CO_2^{\bullet-}$). This reduces the oxidized $Ru(bpy)_3^{2+}$ to form the triplet state of $Ru(bpy)_3^{2+}$, which reverts back to the ground state with the emission of a wavelength of light (Figure 11.11).

Some of the unique features and applications of ruthenium-based ECL include (a) $Ru(bpy)_3^{2+}$ ECL in microfluidic chip-based

$$Ru(bpy)_3^{2+} - e^- \rightarrow Ru(bpy)_3^{3}$$
$$C_2O_4^{2-} - e^- \rightarrow [C_2O_4^{\bullet-}] \rightarrow CO_2^{\bullet-} + CO_2$$
$$Ru(bpy)_3^{3+} + CO_2^{\bullet-} \rightarrow Ru(bpy)_3^{*2+} + CO_2$$
$$Ru(bpy)_3^{2+*} \rightarrow Ru(bpy)_3^{2+} + h\nu$$

FIGURE 11.11 Schematic representation of $Ru(bpy)_3^{2+}$ ECL with oxalate.

analysis, (b) as ECL aptasensors, (c) in HPLC- and FPLC-based applications, and (d) as antibody-based biosensors.

Ruthenium ECL has the potential to convert the ECL biosensing strategies produced in the laboratories into commercially viable products. Ruthenium does have some disadvantages, like the quantum efficiency of the ruthenium ECL. In all ruthenium ECL cases in the early years, a high negative potential had to be applied prior to the ECL emission. This is not suitable for biological systems. The advent of a co-reactant pathway resulted in the development of a number of ECL-based biosensor platforms.

11.4.1.2 Iridium Complexes

The low quantum efficiency of the ruthenium metal complexes has made scientists explore other metals as ECL emitters. Iridium is a silvery white transition metal with an atomic number of 77. It belongs to the platinum group. Iridium (III) complexes exhibit ECL behavior in their neutral state in both aqueous media and in organic solvents [25]. Iridium complexes have been widely used as metal complexes for ECL studies, second only to ruthenium complexes. Only a few iridium compounds have found to be ECL efficient. But these compounds have better quantum efficiencies and they have emission colors that are tunable. At present, the iridium complexes are widely seen as a fitting alternative to the ruthenium complexes [26]. Iridium 2-phenyl pyridine [Ir(ppy)$_3$] is the most widely used iridium complex [27]. These metal complexes are popularly known as the cyclometalated iridium (III) complexes [28, 29]. Iridium 2-phenylquinoline, [Ir(ppy)^2 (dcbpy)]$^+$ (ppy: 2-phenylpyridyl and dcbpy: 4,4'-dicarboxy-2,2-bipyridyl) are the other known complexes of iridium that exhibit ECL.

Iridium complexes exhibit both annihilation and the co-reactant pathway of ECL. Some of the well-known co-reactants are TPrA, 2-cyanofluorene, 4-cyano bi phenyl, 1,4-dicyanobenzene, benzophenone, and 1-acetylnapthalene. TPrA is a very well-understood co-reactant.

The important characteristic of iridium complexes that makes it stand out from other complexes, especially ruthenium, is the ability of these compounds to regulate the emission color. Iridium complexes exhibit a wide wavelength of radiation from green to red. Of late there have been reports of iridium complexes emitting blue radiation. As has been explained, the molecular orbitals of the complexes play a vital role in ECL emission. Therefore, by playing with the energy gap between the highest occupied molecular orbital (HOMO) and the lowest occupied molecular orbital (LUMO), one can tune the emission color of the iridium complexes. One of the widely used strategies is by lowering the HOMO orbital energy levels by the addition of electron-withdrawing groups into the iridium complexes. Some of the ring structures that are employed are the phenyl ring structures. The second strategy is by increasing the energy levels of the LUMO orbitals by introducing the groups, which are electron donating in nature. The popular choice among researchers for electron-donating groups is the heterocyclic ring structures.

Cationic iridium complex with aryl triazole cyclometalated ligands and di amine ligands as an ECL system can be used to tune ECL emissions from 510 to 606 nm. This is achieved by changing the groups in the ligands.

The addition of electron-deficient groups like fluorine and chlorine in the ring structures of the iridium complexes results in stabilizing the HOMO orbitals. The phenyl groups are to be targeted for enhancement of ECL.

The advent of iridium as a luminophore has transformed the ECL sensors, especially for biological applications. In the initial phases of ECL, only ruthenium luminophores were available. Now with the addition of iridium as a luminophore, this has added variety into the ECL systems. In the past few decades, the multiple-color emission property of iridium has turned out to be a shot in the arm for multiplexing. In such multicolor ECL systems, the gradient potential has to be established. More iridium complexes must be explored for ECL to realize its complete potential.

11.4.2 ORGANIC LUMINOPHORE-BASED ECL SYSTEMS

There were a number of detailed studies in the initial years of ECL-based sensing related to organic compounds with ECL emission ability. The initial studies were on polyaromatic compounds such as anthracene and rubrene. The preliminary studies were of annihilation-type ECL studies with organic compounds.

The initial studies were with polyaromatic hydrocarbons, which produce ECL emissions, but the important drawback to compounds was that most organic compounds have very low water solubility [6]. To confront this problem, other compounds were studied. Organic compounds like violanthrone [30], 9-naphthylanthracene [31] 9,10-diphenylanthracene-2-sulfonate (DPAS,) and 1- and 2-thianthrenecarboxylic acid (1- and 2-THCOOH) with TPrA as a co-reactant were used in aqueous solutions [32]. DPAS with thiosulfate is also capable of ECL emissions by co-reactant pathway. The emission maxima were observed at around 420 nm, which shows a blue shift in comparison with ruthenium compounds. Other polyaromatic hydrocarbons that were discovered as ECL emitters were luminol and acridinium esters. Some of the well-understood polycyclic aromatic hydrocarbons are fluorene, perylene, and lucigenin.

11.4.2.1 Fluorene

A polycyclic aromatic hydrocarbon is a class of organic compounds that are fluorescent in nature and capped on polyaromatic hydrocarbons to create an enhanced ECL quantum efficiency [33]. There is an increase in the half-life of the ECL luminophores. The steric hindrance that the fluorene imparts on the polyaromatic hydrocarbons results in an increase in quantum yields and results in the formation of more stable radical ions. The interaction between the chromophore groups is reduced due to the steric hindrance and reduces the electro-decomposition.

11.4.2.2 Perylene

Perylene has been used extensively as ratiometric ECL sensors. Perylene produces ECL by the co-reactant mechanism.

Perylene acts as a donor, and acceptors harnessed here are generally nanocomposites of various compositions like cadmium sulphate/graphene and graphene/gold nanorods.

11.4.2.3 Lucigenin

Lucigenin (bis-*N*-methylacridinium nitrate) is the most extensively used chemiluminescent probe. The conventional application is for the detection of superoxide in animal cell culture. Riboflavin detection was made for human chorionic gonadotropin, and the ECL is achieved by substituting lucigenin with hydrogen peroxide.

Other important organic compounds that are used in ECL-based sensing are described next.

11.4.2.4 Luminol

Luminol is a crystalline solid that is white to yellowish in color. It is soluble in polar organic solvents and is insoluble in water. Luminol is one organic compound that has been used more extensively than any other in biological ECL systems. It has a characteristic emission at 425 nm. Luminol in the presence of silver nanoparticles as an electrode material generates high-intensity ECL emission under neutral pH conditions. The luminol facilitates enzyme-based ECL sensors. Luminol and its derivatives are used extensively in aptsensors and DNA sensors where the luminol groups bind covalently to biomolecules. Luminol with H_2O_2 allows detection of a variety of bioanalytes. A number of approaches to the mechanism have been proposed based on the potential bias applied. Luminol undergoes irreversible oxidation, which means it is consumed in the ECL reaction.

11.4.2.5 Di Imides

A number of non-polyaromatic hydrocarbons have also been studied, like the perylene di imide cations and anions, which emit ECL emission. The ECL emission here is by the TTA route (T-route). The reason for following the TTA route is that the excited-state perylene di imide does not have sufficient energy to attain the singlet energy levels. They have emission wavelengths, which are observed in a longer wavelength region (red). Perylene dicarboxylic imide (PI), perylene tetracarboxylic di imide (PDI), terrylenetetracarboxylic di imide (TDI), and quaterrylenecarboxylic di imide (QDI) are some of the ECL-emitting di imides. Di imides undergo reversible one-electron redox reactions as part of ECL emission.

11.4.2.6 Organic Polymers

Recently organic polymers have also created great interest in the field of ECL. The most widely used polymers are the derivatives of polyphenylene vinylene (PPV). 4-methoxy-(2-ethylhexoxy)-2,5-polyphenylenevinylene (MEH-PPV) is another organic polymer that can produce ECL by both annihilation and co-reactant reaction mechanisms. The co-reactant ECL of MEH-PPV is with co-reactants like thiosulphate and TPrA. TPrA is another widely used organic compound that is ECL active. TPrA acts as the co-reactant in a number of ECL systems with different luminophores, from ruthenium to iridium, and also other organic luminophores and nanoparticle-based luminophores. TPrA is oxidized on applied potential and acts as a strong reductant, which can excite the luminophore into singlet state.

11.4.2.7 Mechanism of ECL Emission in Organic Compounds

Electron transfer mechanisms that occur in certain organic compounds are capable of emitting ECL by inducing changes in the conjugate organic structures by application of a potential bias. The potential changes the redox state of the molecule, resulting in the transfer of an electron from the HOMO to the LUMO and in turn, emission of ECL. This phenomenon is experienced in organic molecules in the solvent state as well as in the condensed state. The potential bias makes extreme modification on the ring/conjugated π orbitals, which changes the energy levels of the HOMO and/or LUMO orbitals initiating ECL emission. This very property has been exploited in the last two decades for configuring devices such as light-emitting diodes (LEDs). An excited conjugate organic molecule should attain an energy range of 1.5–3.5eV to lead to an emission in the visible range. This range can be achieved when the organic molecules have an alternative conjugated π electron system, which can have ionization energies in the range of 6.5–8.5 eV, whose electronic transitions can lead to ECL emissions. All these properties of organic ECL emitters have led to a new class of light-emitting devices that are called light-emitting electrochemical cells (LECs). Different organic emitters have been studied and tried as LECs – some of them are poly-(vinyl-9,10-diphenylanthracene), sulfonated poly(p-phenylene), poly(9,9-dioctylfluorene), poly(3-hexyl-thiophene), and poly[triethyleneoxy-p-(diethoxy)phenylenevinylene].

Organic compounds along with nanoparticles or organic polymers are hot research topics for those who are in constant search of new ECL luminophores. It has also been found that organic compounds in association with transition metal complexes also exhibit tunable ECL emission properties, making organic ECL emitters the single largest source for the ECL studies.

11.4.3 Nanomaterial-Based Systems

Nanomaterials have ruled the scene in the last decade and a half in different frontiers of science and technology. The innumerable and diversified pros of nanomaterials have moved from one field of science to the other. They have also significantly contributed to the progress of ECL as an extremely sensitive sensing platform. The initial uses of nanomaterials in ECL were mainly as electrode materials and subsequently found use as co-reactants, as luminophores, and as catalytic materials, which aid the enhancement of the ECL intensity and its quantum efficiency. Various nanomaterials are being explored for ECL activity, and some of the well-established nanomaterials are discussed in the following sections.

11.4.3.1 CdTe Semiconductor Nanomaterials

Cadmium telluride (CdTe) is one of the most widely used classes of nanostructures of any form for ECL studies [34]. CdTe nanostructures like CdTe-based nanoparticles, quantum dots, and microspheres are studied in detail for ECL studies [35]. This class of quantum dots has a narrow fluorescence range but compensates with high luminescence quantum yield because of its resistance to photodegradation.

11.4.3.2 Quantum Dots (QDs)

Quantum dots are nanoparticles with dimensions of 2–10 nm. QDs as ECL lumiphores have been studied since 2002. QDs have optical, chemical, and electronic features different from the bulk materials or even nanoparticles. The evolution of ECL and QDs can be depicted through Figure 11.12.

The phenomenon of QDs can be appreciated by looking at band theory. The photoexcitation of the valence electrons into the conduction band is normal in semiconducting materials. This excitation results in the formation of a hole, the electron-hole pair, also called an exciton, which is bound by electrostatic attraction forces. The wavefunction of the exciton forms the Bohr radius, which is measured over the crystal lattice. Formation of QD is possible at a size smaller than the Bohr radius. QDs allow fine-tuning their emission in a size-dependent manner.

The QDs for ECL have to be water-compatible, and the basic steps for the complete fabrication procedure and post-fabrication events are assembly on to the electrode surface, functionalization, bioconjugation with both the luminophore and biorecognition molecule, and finally amalgamation of the QDs with bioconjugate molecules into bioassays. QDs juxtaposed with ECL can be categorized into semiconducting QDs, which include germanium crystals, carbon QDs, CdTe QDs, and Si QDs.

The QD-based ECL systems are highly sensitive, and single-molecule event detection is possible. This can be achieved by reducing the background noise by combining monochromators with charge-coupled devices (CCDs), followed by proper functionalization of the QDs [36, 37].

11.4.3.2.1 Semiconductor Quantum Dots

11.4.3.2.1.1 Silicon Quantum Dot (SiQD)
Silicon and its compounds, along with germanium, can be categorized as semiconductor quantum dots. The exceptional optical properties and the ease of synthesis make SiQDs a preferred luminophore. SiQD is nontoxic, and surface modifications can be readily achieved. The photoluminescence emission of SiQD is at 420 nm, but an ECL emission is observed at 640 nm. This significant redshift is because the surface states in SiQD have an easier charge injection electrochemically than by photoinjection [38].

SiQD synthesis uses an etching strategy and microemulsion procedure. Etching allows predetermining the size of the SiQD. One of the major limitations of SiQD is the unstable luminescence and its poor colloidal stability.

11.4.3.2.1.2 Germanium Nanocrystals (GeNC)
Germanium (Ge) is a semiconductor like silicon (Si). GeNCs are of more interest than Si because Ge has a better quantum confinement effect in comparison with Si. Ge has a larger Bohr radius because Ge has 4p valence electrons, whereas Si has 3p valence electrons. Hence, more interest is shed on GeNC than Si nanocrystals. GeNC has photoluminescence in the visible region, thus making it a candidate for ECL reactions. GeNC can be synthesized with tetraethyl germane by decomposing it in a 3:1 mixture of hexane and octanol at 450°C. The reduced forms of GeNC are found to be more stable. GeNC follows the annihilation reaction of ECL. The GeNC, when pulsed between −2.5 V and +1.5 V in steps of 0.5 s, yields ECL [1, 39].

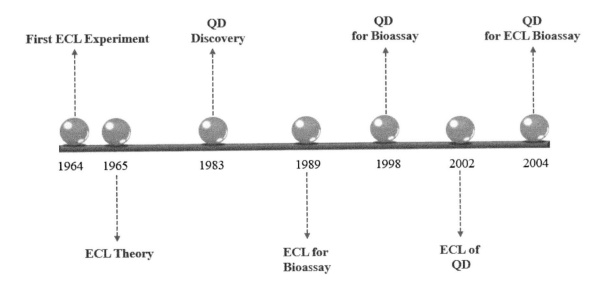

FIGURE 11.12 Timeline of ECL and QD evolution.

A GeNC ECL peak is observed at around 700 nm, where it exhibits photoluminescence at 550 nm and has excitation at 380 nm. A redshift of approximately 200 nm is observed between photoluminescence and ECL. The reason behind the redshift is attributed to a strong role of the nanocrystal's surface states in case of ECL. The ECL spectrum of GeNC is more sensitive than the photoluminescence spectrum. GeNC-based biosensing still has to be explored. But GeNC, due to its inert behavior and high ECL signal, promises to be a very good luminophore material, as it is considered a green material and is economically viable.

11.4.3.2.1.3 Carbon Quantum Dots (CQDs) CQDs are emerging ECL luminophores and still in their infancy in terms of biosensing applications. CQDs are nanocrystals synthesized and composed of carbon with dimensions ranging between 2 nm and 10 nm [40]. CQDs are spherical in nature. They are composed of an amorphous or crystalline core. The core is composed of sp^2 carbon orbitals along with oxidized carbon shells and oxygen-containing groups. CQDs are synthesized by two approaches: top-down and bottom-up.

A variety of carbon precursors ranging from fruit juices, grass, and leaves to carbon nanotubes and graphene oxide are used in the synthesis of CQDs. In the top-down approach, a carbonization method is employed. The major disadvantage of this strategy of synthesis of CQD is that there is no control over the size of the CQDs synthesized. In contrast, the bottom-up approach is far more controlled, which in turn converts to fewer defects. The interaction is controlled by regulating the elemental precursor's ratio. The CQD synthesized by this approach is composed of uniform chemical compositions. CQD has several advantages over its counterparts like chemical inertness, easy labeling, low toxicity, water solubility [41], and ease in synthesizing. CQDs can replace conventional ECL reagents due to the aforementioned compensations. The ECL mechanism by which CQDs emit luminescence is the co-reactant pathway [42]. The most widely used co-reactant is $S_2O_8^-$ but SO_3^{2-} can also be used. The major drawback of CQD is the lack of variety of co-reactants. More studies need to be carried out to overcome the present drawback, and more co-reactants need to be discovered in order to make CQD-based ECL systems more robust.

11.4.3.2.1.4 Metal Quantum Dots This is the most extensively utilized class of QDs. A wide range of metal QDs has been reported. Some of the widely used metals include elements from group VI, which includes Cd, Se, Te, and Cs. Doping metal quantum dots is known to increase the luminescence lifetime and enables researchers to work with a wide spectrum of luminescence wavelength [43].

11.4.3.3 Carbon Nanotubes (CNTs)

CNTs are the allotropic forms of carbon, which find their application in fields like electronics and nanotechnology due to their excellent conducting properties and tensile strength. These nanomaterials are made of a hexagonal carbon lattice, which in turn attains a hollow cylindrical form in its 3D structure. The strength of the CNTs is due to the orbital hybridization forming sp^2 hybrid orbitals. CNTs are discussed in detail in Chapter 8. CNTs, due to their hybrid orbital structures and conductive nature, have drawn the attention of the ECL scientists, and a good number of publications have been reported with CNTs as an electrode material. It has also been noticed that CNTs not only are good conductors but they can also catalyze various electrochemical reactions by the co-reactant pathway [44]. They are also known to facilitate the electron transfer reactions of the luminophores.

CNT-based ECL electrode systems are in their infancy. But there is a wide scope because of the ability of CNTs to immobilize biomolecules. Both single-walled carbon nanotubes (SWCNTs) and multi-alled carbon nanotubes (MWCNTs) are used currently in ECL, especially with aptamers [45]. Most of the CNT-based sensors are used in biosensing. The MWCNTs are capable of successfully accelerating the electron transfer reactions and also amplify the ECL signal [46]. CNTs are generally used as dopants in polymers like Nafion, polystyrene, silicon dioxide, and sol-gel. In one such study, it has been found that wrapping the ruthenium complex [Ru(bpy)$_3^{2+}$] with CNT-doped Nafion for the detection of TPrA yielded in ECL reactions magnitudes three times higher than in similar ECL systems with no CNTs [47]. Furthermore, the ECL enhancement was not affected by the change in analyte concentration. A luminol-H_2O_2-based ECL, when introduced with CNTs, also yielded an increase in ECL intensity. This aspect of CNTs needs to be further studied, as it can open a new pathway in ECL systems. Furthermore, applications of CNTs are a must to explore the complete potential of CNTs in ECL. It has also been observed that CNTs at times acts as a catalyst in some ECL systems.

11.4.3.4 Silole Electrochemiluminescence

Nanoparticles as ECL emitters are a hot research area considering the apparent advantages nanoparticles exhibit over their inorganic and organic counterparts. The majority of the semiconducting nanoparticle-based ECL emitters are QDs. QDs can be categorized as hazardous materials, as most of them contain heavy metals, which are an environmental concern and also have health issues. This has led to shift the focus to a variety of semiconducting materials that are capable of ECL emissions. Semiconducting polymers are still in their infancy, but this is an excellent direction to look forward, considering that these polymers are not known as toxic material and exhibit all the characteristics that a QD would. A polymer nanoparticle or polymer dot (PD) has synthesis procedures, which are versatile and can be easily functionalized [48].

One such polymer that has been exploited as an ECL emitter is silole. Siloles are the silicon congeners of cyclopentadiene-based units. In the case of silole-based compounds, the C atom is substituted by the Si atom. Silole chemistry has considerably expanded since the 1990s due to the advances made in terms of synthesis and the potential applications. Silole polymers can be fine-tuned as both ECL donors and acceptors. The biggest advantage of silole as an ECL luminophore is the development of sensors, which requires a low potential.

Silole-based PDs have unique electronic and photophysical properties, which can be explained as a result of the silole molecular orbitals. The LUMO orbitals drop significantly because of the mixing of the σ* of the silylene scaffold and the π* orbital of the butadiene moiety [49].

Siloles have very quick electron mobility, and they exhibit a very high electron transferring capability. These properties are due to the lower LUMO orbitals in comparison to heterocyclic systems like pyrrole and furan.

A highly stable and efficient ECL signal can be achieved by adding aryl and ethynyl substituents to the silole groups. Even extending the silole rings with groups such as thiophene groups also demonstrations the same properties as aryl and ethynyl substituents do.

The efficiency of the silole luminophores depends on the co-reactant. TPrA is an attuned co-reactant, which results in a very strong anodic ECL emission at +0.78 V. A silole-containing polymer (SCP) TPrA ECL mechanism was utilized to create a biosensor with dopamine as the ECL quencher in 2016 [50].

The silole-based ECL sensors are still in their infancy and exhibit all the qualities to overthrow ruthenium as the most extensively used ECL luminophore. The main reasons are the low potential requirement for ECL generation, ability to couple into biological systems, ecofriendly nature of these polymers, tunability of the polymers (which ultimately results in increased ECL efficiency), and higher quantum yields that can be attained with silole polymers.

11.5 ECL IN BIOLOGY

Research on ECL began in the 1960s and since then has revolutionized not just as a science but also as a detection technology. ECL-based analysis is rapidly developing and has been applied in a number of areas in science, especially biology. Biosensing and bioimaging are the two prominent capacities of ECL in biology. Although still new in terms of use in biology, ECL has been exploited as a detector in flow injection analysis (FIA), HPLC, and capillary electrophoresis. ECL-based detection systems are more reliable than other optical detection strategies. The ECL bioassays came into being in 1989. Since then, thousands of publications have been made, and a few sensors have also made it to the commercial scene.

The most commonly used strategy for ECL-based biosensing is similar to the construction of an electrochemical biosensor. The first thing to consider is the electrode material. Different kinds of electrodes are considered, which were discussed earlier. This is followed by the immobilization matrix on the electrode. The general immobilization matrices are polymers, nanomaterials, a self-assembled monolayer (SAM), and sol-gels [51]. The method of immobilization depends on the electrode and the immobilization matrices. The general immobilization techniques include the entrapment, adsorption, or formation of chemical bonds. Once the immobilization is achieved, the penultimate task is the addition of the bioreceptor, which can be an enzyme or antibody and, of late, aptamers (Figure 11.13).

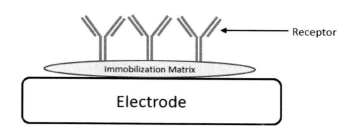

FIGURE 11.13 Components of typical ECL biosensors.

A number of biosensors have been reported on, especially aptasensors for the detection of thrombin. The thrombin-aptamer interaction is very well understood, and hence it is considered a model aptamer-ligand conjugate for most of the ECL-based autosensing studies.

ECL has vast applications in biosensing, and in order to understand and explore this more, we need to categorize ECL sensing. Even though most of the sensing platforms reported till date are a combination of a number of ECL principles, for the sake of understanding, ECL biosensors can be classified into the following types: ECL-RET (resonance energy transfer), ECL quenching, ECL ELISA, ECL co-reactant catalyst, ECL by steric hindrance, and ECL by nucleic acid degradation.

11.5.1 ECL-RET

RET is a physical phenomenon that occurs between two molecules. A high-energy donor molecule emits a characteristic emission at shorter wavelengths, which overlaps with the absorption spectra of an acceptor molecule. In an ECL-RET phenomenon, there is no photon transfer. The dipole interactions of the acceptor and donor are the actual cause for the phenomenon. RET is an efficient phenomenon, and the efficiency of RET depends on the extent of spectral overlap. The higher the overlap between the donor emission spectra and the acceptor absorption spectra, the more efficient the RET. The RET in ECL is very similar to the FRET. The major advantage of ECL-RET is that there is no excitation light required, and this ensures there is no background interference of any form. A number of donor-acceptor ECL-RET pairs (Table 11.2) are well established and have been employed in the construction of sensors.

A luminol-based thrombin aptasensor was reported [52]. The sensor platform works on the RET pair of luminol and CdSe-ZnS quantum dots. The distance between the ECL-RET pair is pivotal for the ECL emission, and this is successfully achieved by the formation of a sandwich of aptamers between the RET pair. This is disrupted when thrombin is introduced into the system and hence the loss of ECL emission occurs. The reduction of ECL emission inversely detects the concentration of thrombin present.

ECL-RET pairs like luminol-QDs and luminol-Ru are used in the design of sensor platforms, which works on ECL emission as a measure of analyte concentration.

TABLE 11.2
List of Acceptor-Donor Pairs for ECL-RET

Donor	Acceptor
CdS QDs	$Ru(bpy)_3^{2+}$
g-C_3N_4 NS	$Ru(bpy)_3^{2+}$
CdS NRs	$Ru(bpy)_3^{2+}$
CdS QDs	Cy5
BSA-AuNC	$Ru(bpy)_3^{2+}$
Luminol	$Ru(bpy)_3^{2+}$
Luminol	CdSe@ZnS QDs
$Ru(bpy)_3^{2+}$	AuNPs/GO
BSA-AuNC	$Ru(bpy)_3^{2+}$
$Ru(bpy)_3^{2+}$	CdTe QDs
CdS QDs	Au NPs

11.5.2 ECL Quenching

The loss of ECL emission can be termed quenching. A huge array of ECL quenchers is available, and they have been extensively explored in terms of biological applications. ECL quenching has been studied by two important methods in particular, the first being the direct loss of ECL emission due to the addition of the quencher. The other method is by eliminating the quencher and hence a resultant increase in ECL emission signal. Both methods work well, and depending on the sensor platform designed, either of these two methods can be harnessed. There have been reports of utilizing both methods in a single sensor. This kind of sensor is called as a multiple signal sensor (MSS). The advantages of MSS are it increases the specificity of the system. It also reduces significantly the false-positive or false-negative results that can be observed in a single signal sensor (SSS). The general principle of each of the ECL-quenching methods is described in detail next, with a few case studies.

ECL quenching with signal loss: An increase in analyte concentration results in a decrease in ECL signal, which can be quantified by measuring the change in ECL intensity vs. the analyte concentration. The central element is the quencher, which is the basis of detection. A number of sensors are reported to work on this principle (Figure 11.14). Ferrocene is an efficient and widely used ECL quencher. Ferrocene uses an electron transfer mechanism to quench the ECL of $Ru(by)_3^{2+}$ complexes. QDs, as mentioned, can be tuned to achieve ECL emissions of the desired wavelength. The QDs are also biocompatible. Using ferrocene or its complexes, one can quench QDs also.

Case 1: Zhao [53] group proposed an aptasensor for thrombin detection. An intense ECL signal is quenched when thrombin and a bioconjugate linked with the thrombin aptamer are introduced. The proposed sensor works on a multistep quenching mechanism. The ECL signal here is achieved by a ruthenium complex (PTCA-PEI-Ru(II)) and its interaction with nicotinamide adenine dinucleotide (NADH). The ECL emission here is by the co-reactant pathway. The quenching of the ECL signal is achieved by consuming the co-reactant NADH from the reaction concoction. Hemin/G quadruplex DNAzymes

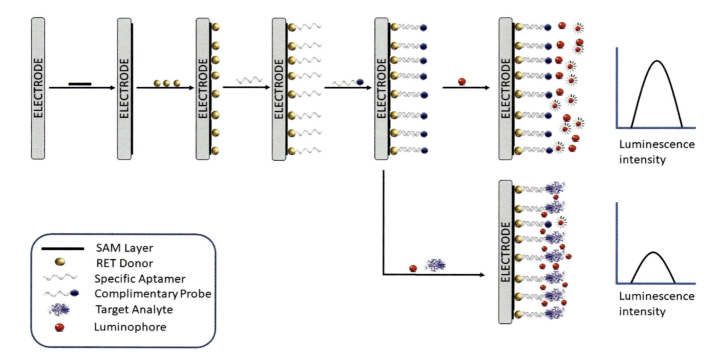

FIGURE 11.14 Schematic representation of the general ECL-RET principled aptasensors.

Bio-ECL as an Analytical Signal

mimics NADH oxidase, consuming the co-reactant NADH. Thrombin binding aptamer II (TBA II) bioconjugate is the quencher.

Case 2: A thrombin aptasensor was reported by Wang et al [54], which is an integration of ECL emission based on ECL-RET, and the detection of thrombin is achieved by quenching the emitted ECL. The ECL platform is based on the RET pair of CdS nanocrystals and gold nanoparticles. The RET distance is achieved by linking the gold nanoparticles to the thrombin aptamer. When thrombin is introduced into this system, it replaces gold nanoparticles, resulting in a logarithmic decrease in ECL intensity and in turn the thrombin detection.

11.5.3 ECL-ELISA

ECL-ELISA (enzyme-linked immunosorbent assay) works on the same principle as regular ELISA, the only difference being the signal is not enzyme-based and is based on the ECL luminophore. This is an assay that is highly efficient and used extensively. ECL-ELISA is used in the recognition of a number of biological molecules.

The principle is based on the selective introduction of the luminophore. The general principle of ECL-ELISA is illustrated in Figure 11.15. A coat of immobilization matrix is applied to the electrode. On the matrix, a recognition element is affixed, which is a biological molecule like an aptamer or antibody. The recognition element specifically binds to the target analyte. The specificity of the recognition element determines the specificity of the sensor. Once the target analyte binds to the recognition element, an ECL probe is introduced into the system. The ECL probe is a secondary antibody or aptamer, which is tagged with the ECL luminophore. Then a co-reactant is added, followed by a potential bias, which results in excited states of both luminophore and co-reactant, resulting in the ECL emission.

11.5.4 ECL Co-Reactant Catalyst

The annihilation reaction is not feasible for all ECL-based sensors. Hence, the co-reactant pathway was discovered, and the advantages of a co-reactant are increased ECL intensity and, in several cases, increased biocompatibility of the ECL sensor. However, co-reactants at times have a slow reaction rate, and sometimes even after the introduction of a co-reactant the ECL intensities do not attain the desired levels. Hence, the need for the introduction of co-reactant catalyzers, which are capable of increasing the rate of the reaction and thereby higher ECL emission.

In ECL sensors, which work on the principle of ECL co-reactant catalysts, there are two important steps to be charted. Initially, the co-reactants have to be generated in the reaction concoction. This is followed by the addition of the co-reactant catalysts. This technique improves the ECL intensity by several-fold.

11.5.5 ECL by Steric Hindrance

Steric hindrance is an intermolecular nonbonding interaction caused by the molecular structure. It is the principle cause for slowing down the chemical reactions in crowded chemical environments. A number of polymers are known to be chemically inert and work as excellent agents exhibiting

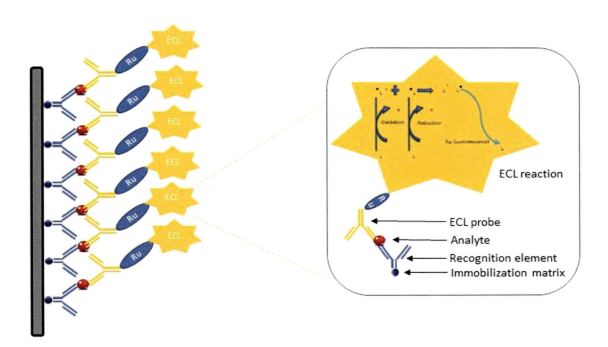

FIGURE 11.15 Schematic representation of ECL-ELISA.

steric hindrance – these polymers or molecular complexes are called crowding agents. Crowding agents do not just slow down the chemical reactions but at times also affect the bond angles and polarity of molecules. Steric hindrance is caused by steric effects, which are nothing but opposing/repulsive forces occurring due to the proximity of the electron clouds of the ions and molecules.

Steric hindrance is measured in terms of the R-rates and the A-values. R-rates are the relative reaction rates that a given chemical reaction undergoes in the presence and absence of the crowding agents. A-value gives the conformational analysis of the molecule. The A-value determines the stable orientation of the molecule, which is vital for attaining high-energy conformation.

ECL biosensors exploit the steric behavior of ECL reactions. The advent of nanostructures has successfully increased the surface area and hence an increase in ECL intensities. The ECL signal arising at the electrode surface is quelled due to steric hindrance [55]. A number of "signal-off" sensors have been proposed where the biorecognition reaction induces steric hindrance.

11.5.6 ECL by Nucleic Acid Modification

ECL emission of these biosensors is constructed based on nucleic acid degradation and ECL emission by amplification of nucleic acid. Endonucleases and exonucleases are used extensively for nucleic acid degradation. The nucleases widely used in the construction of ECL biosensors are EcoRI, T7 exonuclease, Nt.AIwI, and Bam H1. Aptamers are a potential replacement for antibodies, and highly sensitive aptasensors can be constructed with nucleic acid modification. Nucleases are enzymes with very high catalytic efficiency. Hence, ECL signals obtained from this kind of sensor can be sharp.

An aptasensor developed by Lou's group [56] for hepatitis C virus (HCV) uses the BamH1 restriction endonuclease for nucleic acid degradation. The advantage of endonucleases is the site-specific cleavage. The recognition element in the sensor is a dithiocarbamate DNA, which is specific for HCV 1b genotype complementary DNA. When both the probe strand and the analyte strand form a duplex, intense ECL emission is observed. ECL emission is followed by the introduction of BamH1. BamH1 in the presence of dsDNA cleaves at 5'-GGATCC-3', resulting in a sharp decrease in ECL signal. A nonspecific dsDNA, if formed, will yield an ECL signal that will not decrease with the introduction of BamH1.

The development in nucleic acid amplification strategies has opened a new avenue of nucleic acid-based emission sensors. Targets like microRNA, cDNA, and aptamers can be detected by this technique. This technique is still in its infancy, and there is a lot of scope, especially because of the availability of a number of enzyme-free amplification techniques.

11.6 CONCLUSION

ECL as a science has established itself since the 1960s; however, as an analytical platform, we find its applications mostly in the last two decades. In this short duration, ECL has proved to be a noteworthy alternative to the conventional fluorescence and other optical techniques. The major stimuli for its rapid growth in the field of analytical science have been attributed to their extraordinary sensitivity and specificity. To further expand the field, however, some hurdles need to be overcome. Among these, the availability of a limited range of ECL active compounds, including co-reactants, expensive instruments, and large sensing platforms, could be tackled first. The last two issues, namely, cost of the instruments and size of the sensing platforms, are essential factors that need to be adequately addressed for their applications in point-of-care and point-of-need applications. Hence, with its inherent high sensitivity and selectivity, a portable, low-cost format with suitable reaction strategy will undoubtedly accelerate the deployment of ECL-based sensors for on-site applications, including in resource-limited countries.

REFERENCES

1. N. Myung, X. Lu, K. P. Johnston, and A. J. Bard, "Electrogenerated chemiluminescence of Ge nanocrystals," *Nano Lett.*, vol. 4, no. 1, pp. 183–185, 2004.
2. M. M. Richter, "Electrochemiluminescence (ECL)," *Chem. Rev.*, vol. 104, pp. 3003–3036, 2004.
3. D. Dini, "Electrochemiluminescence from organic emitters," *Chem. Mater.*, vol. 17, no. 8, pp. 1933–1945, 2005.
4. N. Egashira, H. Kumasako, and K. Ohga, "Fabrication of a fiber-optic-based electrochemiluminescence sensor and its application to the determination of oxalate," *Anal. Sci.*, vol. 6, no. 6, pp. 903–904, 1990.
5. D. M. Hercules, "Chemiluminescence resulting from electrochemically generated species," *Science*, vol. 145, no. 3634, pp. 808–809, 1964.
6. R. E. Visco and E. A. Chandross, "Electroluminescence in solutions of aromatic hydrocarbons," *J. Am. Chem. Soc.*, vol. 86, no. 23, pp. 5350–5351, 1964.
7. K. S. V. Santhanam and A. Bard, "Chemiluminescence of electrogenerated 9,10-diphenylanthracene anion radical," *J. Am. Chem. Soc.*, vol. 87, no. 1, pp. 139–140, 1965.
8. K. Itaya and S. Toshima, "Formation of intramolecular exciplexes in electrogenerated chemiluminescence," Chemical Physics letters, *Chem. Phys. Lett.*, vol. 5, no. 3, pp. 447–452, 1977.
9. A. J. Bard, "Electrogenerated chemiluminescence. XV. On the formation of excimers and exciplexes in ECL," *Chem. Phys. Lett.*, vol. 24, no. 2, 1974.
10. C. P. Keszthelyi, N. E. Tokel-Takvoryan, H. Tachikawa, and A. Bard, "Electrogenerated chemiluminescence. XIV. Effect of supporting electrolyte concentration and magnetic field effects in the 9,10-dimethylanthracene-tri-p-tolylamine system in tetrahydrofuran." *Chem. Phys. Lett.*, vol. 23, no. 219, 1973.
11. L. R. Faulkner and A. J. Bard, "Electrogenerated chemiluminescence. IV. Magnetic field effects on the electrogenerated chemiluminescence of some anthracenes," *J. Am. Chem. Soc.*, vol. 91, no. 1, pp. 209–210, 1969.
12. H. Tachikawa, A.J. Bard, "Magnetic field effects on ECL in the tetracene-TMPD system: evidence for triplet-triplet annihilation of tetracene," *Chem. Phys. Lett.*, vol. 19, no. 2, pp. 287–289, 1973.

13. L. R. Faulkner, H. Tachikawa, and A. J. Bard, "Electrogenerated chemiluminescence. VII. the influence of an external magnetic field on luminescence intensity," *J. Am. Chem. Soc.*, vol. 94, no. 3, pp. 691–699, 1972.
14. K. N. Shinde, *Basic mechanisms of photoluminescence*, Springer Series in Materials Science 174, Springer-Verlag, Berlin Heidelberg, 2013. DOI: 10.1007/978-3-642-34312-4_2.
15. G. Valenti, A. Fiorani, H. Li, N. Sojic, and F. Paolucci, "Essential role of electrode materials in electrochemiluminescence applications," *Chem. Electro. Chem.*, vol. 6, pp. 1–9, 2016.
16. E. Kanata, M. Arsenakis, and T. Sklaviadis, "Caprine PrP variants harboring Asp-146, His-154 and Gln-211 alleles display reduced convertibility upon interaction with pathogenic murine prion protein in scrapie infected cells," *Prion*, vol. 10, no. 5, pp. 391–408, 2016.
17. D. Yu and K. Kim, "Electrochemically directed modification of ITO electrodes and its feasibility for the immunosensor development," *Bull Korean Chem. Soc.*, vol. 30, no. 4, pp. 955–958, 2009.
18. S. Bouden, A. Dahi, F. Hauquier, and H. Randriamahazaka, "Multifunctional indium tin oxide electrode generated by unusual surface modification," *Nat. Publ. Gr.*, no. October, pp. 1–9, 2016.
19. A. T. Singh, D. Lantigua, A. Meka, S. Taing, M. Pandher, and G. Camci-Unal, "Paper-based sensors: Emerging themes and applications," *Sensors (Basel).*, vol. 18, pp. 2038, 2018.
20. L. Chen, C. Zhang, and D. Xing, "Paper-based bipolar electrode-electrochemiluminescence (BPE-ECL) device with battery energy supply and smartphone read-out: A handheld ECL system for biochemical analysis at the point-of-care level," *Sensors Actuators B. Chem.*, vol. 237, pp. 308–317, 2016.
21. S. Baek, S-R. Kwon, S. Y. Yeon, S-H. Yoon, C. M. Kang, S. H. Han, et al., "Miniaturized reverse electrodialysis-powered biosensor using electrochemiluminescence on bipolar electrode," *Analytical Chem.*, vol. 90, pp. 4749–4755, 2018.
22. E. M. Gross, H. E. Durant, K. N. Hipp, and R. Y. Lai, "Electrochemiluminescence detection in paper-based and other inexpensive microfluidic devices," *Chem. Electro. Chem.*, vol. 4, no. 7, pp. 1594–1603, 2017.
23. J. Hou, N. Gan, F. Hu, L. Zheng, Y. Cao, and T. Li, "One renewable and magnetic electrochemiluminescence immunosenor based on tris (2,2'-bipyridine) ruthenium (II) modified magnetic composite nanoparticles labeled anti-AFP," *Int. J. Electrochem. Sci.*, vol. 6, pp. 2845–2858, 2011.
24. R. S. Glass, and L. R. Faulkner, "Electrogenerated chemiluminescence from the tris(2,2'-bipyridine)ruthenium(II) system. An example of S-route behavior," *J. Phys. Chem.*, vol. 85, no. 9, pp. 1160–1165, 1981.
25. Kalen N. Swanick, Sebastien Ladouceur, Eli Zysman-Colman, and Zhifeng Ding, "Bright electrochemiluminescence of iridium(III) complexe," *Chem. Commun.*, vol. 48, 3179–3181, 2012.
26. S. Yuyang, Kai Xie, Ruimei Leng, Lingyan Kong, Chengbao Liu, Qingqing Zhang, Xiaomei Wang, "Highly efficient electrochemiluminescence labels comprising iridium(III) complexes," *Dalton Trans.*, vol. 46, p. 355, 2017.
27. D. Bruce and M. M. Richter, "Green electrochemiluminescence from ortho-metalated tris (2-phenylpyridine) iridium (III)," *Anal. Chem.*, vol. 74, no. 6, pp. 1340–1342, 2002.
28. M. A. Haghighatbin, S. E. Laird, and C. F. Hogan, "Electrochemiluminescence of cyclometalated iridium (III) complexes," *Curr. Opin. Electrochem.*, vol. 7, pp. 216–223, 2018.
29. J. II Kim, I. Shin, H. Kim, and J. Lee, "Efficient electrogenerated chemiluminescence from cyclometalated iridium (III) complexes," *J. Am. Chem. Soc.*, vol. 127, no. 6, pp. 1614–1615, 2005.
30. T. I. Quickenden and K. Hansongnern, "Electrogenerated Chemiluminescence from Violanthrone," *J Biolumin Chemilumin.*, vol. 10: pp. 103–106, 1995.
31. J. Suk, Z. Wu, L. Wang, and A. J. Bard, "Electrochemistry, electrogenerated chemiluminescence, and excimer formation dynamics of intramolecular π-stacked 9-naphthylanthracene derivatives and organic nanoparticles," *J. Am. Chem. Soc.*, vol. 133, pp. 14675–14685, 2011.
32. A. Fiorani, G. Valenti, M. Iurlo, M. Marcaccio, and F. Paolucci, "Electrogenerated chemiluminescence: A molecular electrochemistry point of view," *Curr. Opin. Electrochem.*, vol. 8, pp. 31–38, 2018.
33. K. M. Omer, S. Ku, K. Wong, and A. J. Bard, "Zuschriften efficient and stable blue electrogenerated chemiluminescence of fluorene-substituted aromatic hydrocarbons," *Angewandte Chemie*, vol. 121, no. 49, pp. 9464–9467, 2009.
34. Y. Bae, N. Myung, and A. J. Bard, "Electrochemistry and electrogenerated chemiluminescence of CdTe nanoparticles," *Nano Lett.*, vol. 4, pp. 1153–1161, 2004.
35. X. Liu, H. Jiang, J. Lei, and H. Ju, "Anodic electrochemiluminescence of CdTe quantum dots and its energy transfer for detection of catechol derivatives," *Anal. Chem.*, vol. 79, no. 21, pp. 8055–8060, 2007.
36. H. Huang, J. Lia and S. Zhu, "Electrochemiluminescence based on quantum dots and their analytical application," *Anal. Methods*, vol. 3, pp. 33–42, 2011.
37. P. Bertoncello and P. Ugo, "Recent advances in electrochemiluminescence with quantum dots and arrays of nanoelectrodes," *Chem. Electro. Chem.*, vol. 4, pp. 1663–1676, 2017.
38. X. Chen, Y. Liu, and Q. Ma, "Recent advances in quantum dot-based electrochemiluminescence sensors," *J. Mater. Chem. C.*, vol. 6, pp. 942–959, 2018.
39. D. Carolan and H. Doyle, "Size controlled synthesis of germanium nanocrystals: effect of Ge precursor and hydride reducing agent," *J. Nanomater.*, pp. 1–9, 2015.
40. Y. Dong, C. Chen, J. Lin, N. Zhou, Y. Chi, and G. Chen, "Electrochemiluminescence emission from carbon quantum dot-sulfite coreactant system," *Carbon N. Y.*, vol. 56, pp. 12–17, 2013.
41. W. Xue, Z. Lin, H. Chen, C. Lu, and J. Lin, "Enhancement of ultraweak chemiluminescence from reaction of hydrogen peroxide and bisulfite by water-soluble carbon nanodots," *J. Phys. Chem. C*, vol. 115, pp. 21707–21714, 2011.
42. W. Zhao, J. Wang, Y. Zhu, J. Xu, and H. Chen, "Quantum dots: electrochemiluminescent and photoelectrochemical bioanalysis," *Anal. Chem.*, vol. 97, pp. 9520–9531, 2015.
43. M. Tinkham, "Metallic quantum dots," *Philos Mag B*, vol. 79, no. 9, pp. 1267–1280, 1999, DOI: 10.1080/13642819908216970.
44. A. Sanginario, D. Demarchi, P. Civera, M. Giorcelli, M. Castellino, and A. Tagliaferro. "Carbon nanotube electrodes for electrochemiluminescence biosensors," *Procedia Eng.*, vol. 5, pp. 808–811, 2010.
45. R. Huang, L. Wang, Q. Gai, D. Wang, and L. Qian, "DNA-mediated assembly of carbon nanotubes for enhancing electrochemiluminescence and its application," *Sens. Actuators B. Chem.*, vol. 256, pp. 953–961, 2018.
46. X. Zhang, H. Ke, Z. Wang, W. Guo, A. Zhang, and C. Huang, "An ultrasensitive multi-walled carbon nanotube–platinum–luminol nanocomposite-based electrochemiluminescence immunosensor," *Analyst.*, vol. 142, pp. 2253–2260, 2017.

47. A. Sanginario, D. Demarchi, M. Giorcelli, M. Castellino, and A. Tagliaferro, "Carbon nanotube electrodes for electrochemiluminescence biosensors carbon nanotube electrodes for electrochemiluminescence," *Procedia Eng.*, vol. 5, pp. 808–811, 2010.
48. Y. Feng, C. Dai, J. Lei, H. Ju, and Y. Cheng, "Silole-containing polymer nanodot: an aqueous low-potential electrochemiluminescence emitter for biosensing," *Anal. Chem.*, vol. 88, pp. 845–850, 2016.
49. C. Booker, X Wang, S. Haroun, J. Zhou, M. Jennings, B. L. Pagenkopf, "Tuning of electrogenerated silole chemiluminescence," *Angewandte Chemie.*, vol. 47, pp. 7731–7735, 2008.
50. L. Feng, Z. Zhang, J. Ren, and X. Qu, "Graphene platform used for electrochemically discriminating DNA triplex," *ACS Appl. Mater. Interfaces*, vol. 6, pp. 3513–3519, 2014.
51. M. Rizwan, M.U. Ahmed, N.F. Mohd-Naim, "Electrochemiluminescence nanobiosensors," *Sensors.*, vol. 18, p. 166, 2018.
52. Y. Zhuo, H. Wang, Y. Lei, P. Zhang, J. Liu, Y. Chai, R. Yuan, "Different modes of switching signals," *RSC Analyst.*, vol. 143, p. 3230, 2018.
53. M. Zhao, N. Liao, Y. Zhuo, Y. Chai, J. Wang, and R. Yuan, "Triple quenching of a novel self-enhanced Ru(II) complex by hemin/G-quadruplex DNAzymes and its potential application to quantitative protein detection, *Anal. Chem.*, vol. 87, 7602–7609, 2015.
54. J. Wang, Y. Shan, W. Zhao, J. Xu, and H. Chen, "Gold nanoparticle enhanced electrochemiluminescence of CdS thin films for ultrasensitive thrombin detection," *Anal. Chem.* vol. 83, pp. 4004–4011, 2011.
55. Y. Wei, X. Liu, C. Mao, H. Niu, J. Song, and B. Jin, "Highly sensitive electrochemical biosensor for streptavidin detection based on CdSe quantum dots," *Biosens. Bioelectron.*, vol. 103, pp. 99–103, 2018.
56. J. Lou, S. Liu, W. Tu, and Z. Dai, "Graphene quantums dots combined with endonuclease cleavage and bidentate chelation for highly sensitive electrochemiluminescent DNA biosensing," *Anal. Chem.* vol. 87, pp. 1145–1151, 2015.

12 Paper Electronics and Paper-Based Biosensors

*Federico Figueredo, María Jesús González-Pabón,
Albert Saavedra, and Eduardo Cortón*
Universidad de Buenos Aires (UBA), Ciudad Universitaria, Argentina

Susan R. Mikkelsen
University of Waterloo, Waterloo, Ontario, Canada

CONTENTS

12.1 Introduction 251
12.2 Paper Electronics 252
12.3 General Overview: Paper 252
    12.3.1 Paper Composition and Modified Paper-Like Products 253
    12.3.2 Paper Properties Related to Biosensor Performance 253
    12.3.3 Fabrication Methods Used to Produce Paper-Based Analytical Devices 254
12.4 Methods to Immobilize the Bioreagent and Improve Biosensor Performance 255
    12.4.1 Physical Immobilization 255
    12.4.2 Chemical Immobilization 255
    12.4.3 Entrapment 256
12.5 New Approaches to Improve Paper Performance 256
    12.5.1 Incorporation of Nanomaterials 256
12.6 Bioreagents Used for Biorecognition 257
    12.6.1 Enzymes 257
    12.6.2 Antibodies 257
    12.6.3 Nucleic Acids 258
    12.6.4 Microorganisms 258
12.7 Detection Methods Used for Transduction 258
    12.7.1 Electrochemical 258
    12.7.2 Optical 258
12.8 Applications 259
    12.8.1 Clinical Chemistry 259
    12.8.2 Environmental Analysis 260
    12.8.3 Food Security 261
12.9 Perspectives 261
References 261

12.1 INTRODUCTION

The development of new, cost-effective, and simple analytical systems, useful for environmental analysis as well as human, animal, and plant diagnostics, is greatly needed, especially in developing countries where budgets and infrastructures are very limited [1–3]. However, most of the traditional quantitation methods (i.e., chromatography, mass spectrometry, nuclear magnetic resonance spectrometry, enzyme-linked immunosorbent assays) not only demand large investments in equipment and infrastructure but also possess high operating costs that include reagents, maintenance, and well-trained professionals for operation, calibration, and servicing. In comparison with traditional laboratory-based instruments, biosensors (with or without microfluidic systems) possess significant differences that make them suitable for applications onsite and in developing countries. In clinical diagnosis or environmental analysis, the determination of the target analytes can be produced in short periods with high accuracy in centralized laboratories or specialized facilities such as clinics, hospitals, or monitoring stations. Biosensors, on the other hand, deliver a limited number of results, but can be used near the analytical problem, the so-called point-of-care (POC) or point-of-need (PON) testing. Although precision and accuracy are generally somewhat poorer with biosensors in comparison with large testing facilities, sample testing or screening can be performed *in situ* by personnel with less expertise, in locations with poor infrastructure, and without delays due to transport or environmental conditions that would challenge the use of complex and expensive instrumentation [4–7].

Various inert support materials have been used for the fabrication of biosensors and have been selected on the basis of their availabilities and the intended applications. During early microfluidic development, silicon and glass were used as support materials for the fabrication of microfluidic devices [8]. Although these materials are compatible with a wide range of operational properties, the processing fees and environmental impacts incurred during the manufacturing of the final devices make these materials unreasonable options for accessible and single-use devices [9, 10].

12.2 PAPER ELECTRONICS

Technological innovations and consumer habits are powerful forces that limit the durability of most goods. This is especially true when electronic devices are considered; new and more powerful communication, control, and computing devices reach the market every year, making earlier innovations obsolete. Recycling technologies lag behind in utility and expense, so many unused electronic devices continue to be directed to landfills. Modern electronics have a variety of complex materials in circuit boards and interfaces, and a convenient recycling technology that allows recovery of all the materials at a reasonable cost is still not available.

This new paradigm has emerged over the past two decades and has led to thoughts of alternative materials. It may be possible soon to replace most short-lived small electronic devices with paper-printed devices by adapting well-known industrial printing technologies that have been used for decades to print newspapers and books, as well as Rotaprint, flexography, and others. In this way, low-cost, mass-scale printing of new electronic devices could make them usable for weeks, with relatively easy disposal and replacement. All necessary electronic components could, in the future, be printed using different modified conducting or insulating inks. Electronics made using paper as the support material are viewed as a potentially cost-effective alternative to current materials. Even at this early stage, devices such as simple sensors and microfluidic systems can be printed; both of these are fundamental components of biosensors.

Paper-based devices are based on a very abundant and inexpensive material. In comparison with plastics, the fabrication process is environmentally friendly. Paper is biodegradable and may also be degraded by composting or burning. It allows the incorporation of flexible electronic circuits. It is also nontoxic and hypoallergenic. The porous structure of paper allows the absorption and movement of liquids through channels in simple microfluidic systems.

To date, relatively simple devices have been constructed. Typical applications include packaging in the food industry that can assess product safety or freshness; novel disposable, single-use diagnostic systems; and personal hygiene products.

The applications and recent advances of paper functionalized with conducting inks (sometimes modified with nanomaterials) have been recently reviewed [11], and a brief summary of these are presented in Figure 12.1. Conductive materials printed on paper are very useful as new electrode materials, with production costs a fraction of the standard for the electrochemical biosensors industry (screen printing onto plastics). Another interesting possibility involves batteries and capacitors that use biological components to provide energy; these could be based on living microorganisms or isolated enzymes, with well-known operating principles for microbial and enzymatic fuel cells.

12.3 GENERAL OVERVIEW: PAPER

There is strong evidence showing that paper was probably invented in China in 105 AD [12]. In modern times, paper is a multipurpose material used for packaging, printing, sanitation, and creative arts, among other applications. One recent application area for paper involves low-cost and disposable analytical devices for POC and PON testing [13]. Paper manufacturing involves cellulose extraction from raw materials such as wood, textiles, or grasses, followed by the compression and drying of cellulose fibers. As well, POC or PON devices can be fabricated using various paper materials, including nitrocellulose (NC) membranes, filter paper, graphite paper,

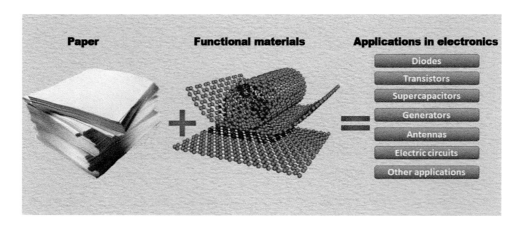

FIGURE 12.1 Building electronic devices on paper supports have proven feasible for small and simple components and circuits. Further research could allow the complete fabrication of POC and PON analytical devices. Reprinted with permission from ref. [11]. Copyright 2016 American Chemical Society.

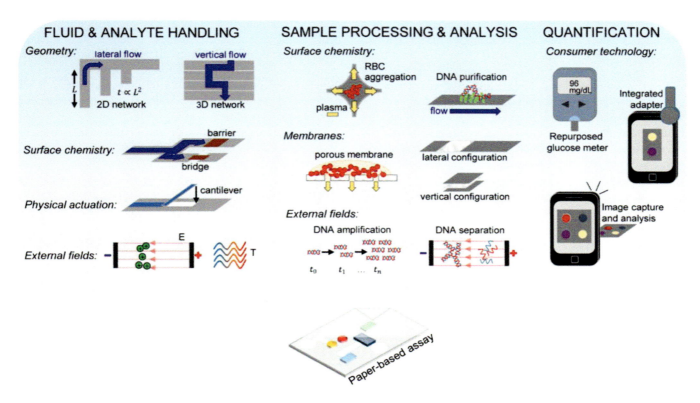

FIGURE 12.2 Overview of possible analytical capabilities employed in paper-based assays. Paper can provide simple and affordable solutions for fluid and analyte handling, sample processing, analysis, and quantification. Reprinted with permission from ref. [19]. Copyright 2017 American Chemical Society.

or chromatography paper [14]. Paper selection for fabricated devices depends upon the procedure and the field of application of the particular platform [15]. Some characteristics that depend on the type of paper make it more suitable for some uses. For example, NC membrane is commonly used for lateral flow test strips [16]; filter papers have a wide range of use in microfluidic paper-based analytical devices (μPADs) [14], and printer papers can be employed as electrode supports [17, 18].

The intrinsic properties of various papers and their derivatized or modified products allow it to perform most of the tasks required for simple and disposable analytical devices, from sample treatment to analysis, as shown in Figure 12.2.

12.3.1 Paper Composition and Modified Paper-Like Products

Cellulose $(C_6H_{10}O_5)_n$ is a material consisting of β-D-glucopyranose units joined by $(1 \rightarrow 4)$-glycosidic bonds [20]. It is the major component of many paper types and the most abundant natural biopolymer on Earth. Its inherent properties (i.e., reactivity, lightness, biodegradability, biocompatibility) make paper an ideal solid for portable and disposable sensor and biosensor devices [21]. Additional features like porosity, surface chemistry, and the optical properties of cellulose fibers are key factors for paper-based biosensor design. Surface chemistry must facilitate bioreceptor immobilization, minimize nonspecific binding, and be compatible with enzyme-catalyzed reporting strategies [22]. Paper porosity (solid matrix with an average fiber diameter of 1–100 μm and an average pore size of 1–10 μm), along with surface chemistry, influence the wet properties of bioactive-paper fabrication. In addition to surface chemistry possibilities, the cellulose structure can be functionalized, thereby changing properties such as hydrophilicity, permeability, and reactivity [23] in order to obtain high adhesion of bioreceptors or to produce well-defined channels for microfluidic devices used for the transportation, separation, and/or immobilization of the various substances used in analytical determinations [24]. Optical properties may also be adjusted to accommodate optical-based sensing strategies, based on transmitted or reflected light [12].

12.3.2 Paper Properties Related to Biosensor Performance

After the group led by George M. Whitesides proposed in 2007 that paper be used as an inexpensive platform for portable bioassays [25], paper-based devices have emerged as a promising technology providing rapid POC and PON sensing/biosensing applications [26, 27]. In many ways, paper-based devices can outperform other microfluidic devices.

Paper allows simple, inexpensive manufacturing and ease of large-scale production [13]. It has mechanical flexibility, hydrophilicity, and desirable dielectric properties [18, 28]. Moreover, the paper microstructure allows fluid transport through capillary action, so that external pumping is not required. Paper is easily modified by printing, coating, and impregnating [22]. Paper has filtering properties to remove particulates from samples [29]; for example, filter paper with appropriate pore sizes can separate erythrocytes from blood [30]. Three-dimensional (3D) paper configurations can simulate cell microenvironments and benefit cell development [13].

Paper also possesses surface properties that can easily be modified by some chemical or physical treatments to add flexibility to designs and applications of the resulting structures [15]. Of additional great importance are paper's characteristics of being lightweight, renewable, and biodegradable – relevant characteristics towards commercialization [31].

12.3.3 Fabrication Methods Used to Produce Paper-Based Analytical Devices

Paper-based biosensors are analytical devices fabricated that use, partially or entirely, paper supports for device components and are used to detect analytes contained in complex matrices and small volumes of biochemical samples (10^{-9} to 10^{-18} L) [31]. A bioreceptor is immobilized on a paper support for analyte recognition, capture, and detection. According to their detection mechanisms, these tests can be categorized into biochemical, immunological, and molecular types. Paper-based biosensors include lateral-flow test strips (LFTSs) and paper microfluidic devices [32]. Many fabrication methods have been described, most of which are presented in Figure 12.3.

Most of the published work and commercially available devices are of the lateral-flow test strip type, consisting of prefabricated strips, containing dry (lyophilized) reagents that are hydrated by the application of a fluid sample; the sample migrates along the strip via capillary action, crossing specific regions (Figure 12.4). Bioreceptors are immobilized on the paper support (usually nitrocellulose) to form test and control lines. Adjacent components overlap slightly in order to coordinate the fluid flow. Growing in applications, but still weakly represented in commercial devices, are the millifluidic or microfluidic paper analytical devices (mPADs or μPADs, respectively); these are analytical devices manufactured with

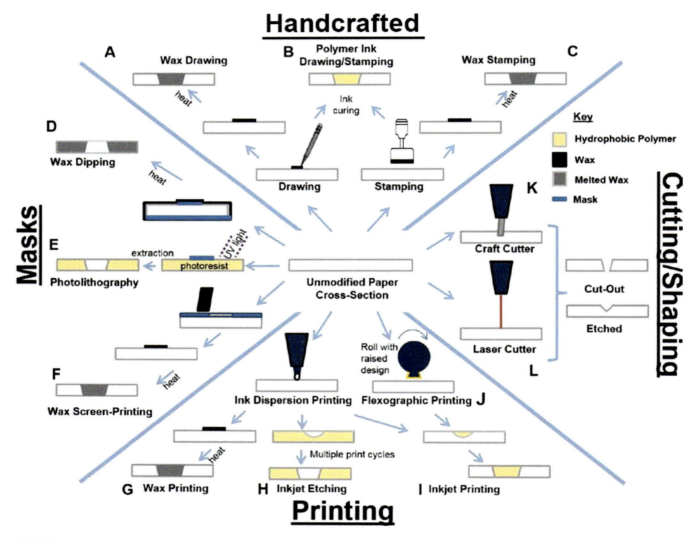

FIGURE 12.3 Four principal methodologies are used to fabricate paper-based analytical devices. Reprinted with permission from ref. [29]. Copyright 2014 American Chemical Society.

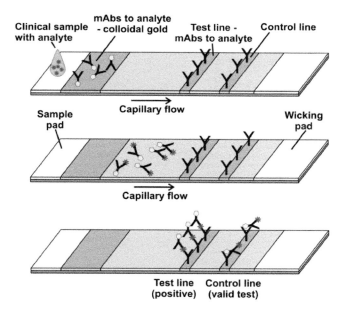

FIGURE 12.4 Schematic diagram of a standard lateral flow test strip. The main components are shown, and the sequence of operation from top (where the sample is introduced) to the bottom, where a positive result (two lines) is shown. Reproduced with permission from ref. [34]. Copyright 2019.

2D or 3D designs composed of hydrophilic macro- or microchannels and/or assay regions separated by hydrophobic boundaries [7]. Many fabrication procedures have been used to create combinations of hydrophilic and hydrophobic areas on paper supports (Figure 12.3), including inkjet printing, silanization using a paper mask, screen printing, wax dipping, paper cutting, flexographic printing, plasma treatment, photolithography, and direct drawing, among others [10, 33].

The design, construction, and demonstration of μPAD devices encompass very active research and technological fields, with a myriad of very interesting and creative models. It is possible to incorporate both a power source and a detection system into a single paper-based device, since some biological systems can be used to produce energy (enzymatic and microbial fuel cells), allowing data to be collected, processed, and presented/transmitted.

Figure 12.5 shows a proof-of-concept device that uses aptamer-based biosensors to detect adenosine. The device was fabricated by wax printing on chromatography paper, with electrodes made of carbon, deposited using a screen-printing technique. The yellow color corresponds to the ferricyanide used for the electrochemical assay. When the sample is included, its flow into the so-called oPAD (origami made) is split into two channels. One is used as a control, and in the other an aptamer (raised against adenosine) is immobilized on microbeads; if the aptamer binds to the target, glucose oxidase is released and catalyzes the oxidation of glucose, which in turn reduces the ferricyanide, generating a reoxidation current as the analytical signal. The current can easily be monitored using a simple digital multimeter and can also be used to charge a capacitor [35].

12.4 METHODS TO IMMOBILIZE THE BIOREAGENT AND IMPROVE BIOSENSOR PERFORMANCE

The literature suggests widespread biosensing strategies for biosensor design and fabrication. Although molecular interactions between the target analyte and the bioreceptor are the main focus of biosensor development, interactions between the bioreceptor and the support material (paper in this chapter) are important too. The physical and chemical properties of paper allow surface modification and the immobilization of bioreceptors and dry chemical reagents [7]. The hydroxyl groups (naturally present) and carboxyl groups (artificially introduced by pulping and bleaching processes), as well as other groups in processed cellulose, make the paper surface hydrophilic with a negative polarity. As a result, paper physically adsorbs cationic molecules. To immobilize bioreceptors with a negative charge, many methods have been developed. In general, immobilization methods are classified as physical adsorption, bioactive ink entrapment, bioaffinity attachment, or covalent chemical bonding immobilization [36, 37]. In the following sections, we describe the most widely used techniques for bioreceptor immobilization on paper supports.

12.4.1 Physical Immobilization

Physical adsorption occurs when biomolecules without external influence adhere to the surfaces of paper fibers by hydrogen bonding, Van der Waals, or electrostatic forces. Cationic biomolecules freely adsorb onto cellulose from aqueous solution, which can be easily deposited via contact printing or printed by noncontact methods (i.e., inkjet printing) onto paper [22]. Nonetheless, most paper surfaces are slightly anionic; thus, bioreceptors that are not highly cationic are only weakly adsorbed. Furthermore, proteins, phages, and DNA are sensitive to pH, ionic strength, and specific ion effects (e.g., high-molecular-weight DNA does not adsorb onto cellulose at pH 6 or pH 8, although it does at pH 4) [38].

Two main strategies have been described to improve physical adsorption. Treatment with a wet-strength resin allows paper to function under wet conditions. These resins are reactive polymers that can react with each other or with the paper surface. One common resin used for longer-term paper products is polyamide-epichlorohydrin (PAE), and this is used for chemically binding the bioreceptor (protein and DNA) to the polymer and binding the polymer to the paper [22, 36].

A second method is called layer-by-layer (LBL) assembly. This methodology enhances electrostatic adsorption of biomolecules (enzymes and bacteriophages) by the consecutive adsorption of polyelectrolyte multilayers with opposite charges. This method requires a multistep coating/printing method to adsorb the various layers, making for more complicated manufacturing [39].

12.4.2 Chemical Immobilization

Covalent bonding is commonly used to produce robust attachment of bioreceptors on paper [37]. The insufficient reactivity

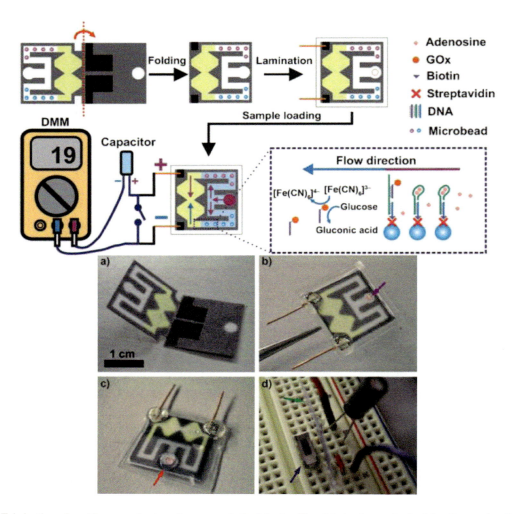

FIGURE 12.5 Fabrication of a self-powered origami paper analytical device. Top: fabrication and principle of operation. Bottom: (a) physical appearance of the device; (b) after lamination, the arrow shows the opening for sample introduction; (c) incorporation of the sample; and (d) protoboard used to measure the produced current and capacitor charge. Reproduced with permission from ref. [35].

of cellulose hydroxyl groups in water at ambient temperature has led to surface activation methods that are used prior to the immobilization reaction. This approach involves multiple chemical steps, which are generally not compatible with manufacturing low-cost devices, and some examples are shown in Figure 12.6. Typically, this process takes the form of derivatization of the surface hydroxyl groups to aldehyde or epoxy groups, the introduction of poly(carboxybetaine) followed by 1-ethyl-3-(3-dimethylaminopropyl) carbodiimide (EDC) alone or in combination with N-hydroxysuccinimide (NHS), or the modification of the hydroxyl groups using divinyl sulfone. An alternative route exists with biochemical coupling techniques, where biomolecules express cellulose-binding modules (CBM), which spontaneously attach to cellulose [40].

12.4.3 Entrapment

Bioactive paper can also be produced by entrapping biomolecules in paper fibers [8]. This technique is noncovalent and corresponds to entrapment of biomolecules in polymer microspheres/beads, made of a semi-permeable polymeric material, protecting the bioactive material during the printing process. The main drawbacks of this technique are that pore size cannot be controlled, and then some loss of enzyme activity with time, biomolecule leaching, and dehydration occurs [40]. This approach has been applied to the immobilization of enzymes and their substrates in different layers, using sol-gel procedures [41, 42] or polymer (poly L-arginine) and silica combinations [43].

12.5 NEW APPROACHES TO IMPROVE PAPER PERFORMANCE

12.5.1 Incorporation of Nanomaterials

Electrochemical and optical detectors are the most commonly used for biosensing applications [7, 14], and paper modifications must consider compatibility with one or both transduction methods. Even though light absorbance is readily measured using optically transparent support materials, enhanced colorimetric signals may be achieved by the incorporation of nanotechnology approaches. Carbon-based,

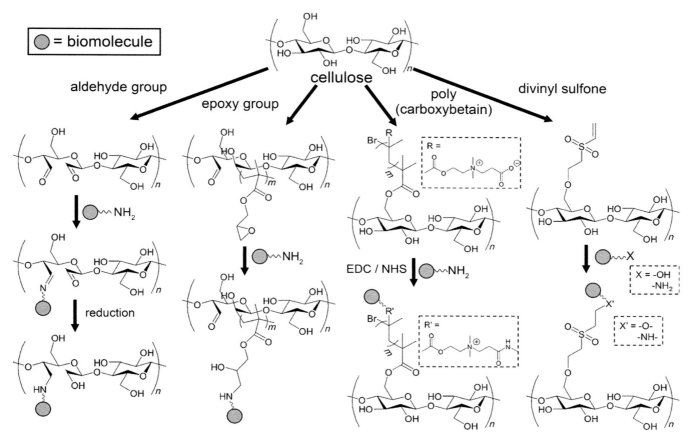

FIGURE 12.6 Strategies applied to the activation of cellulosic hydroxyl groups and the covalent attachment of biomolecules. Reproduced with permission from ref. [37].

metal, and quantum dot nanomaterials have been shown to improve the sensitivities of new devices [32].

12.6 BIOREAGENTS USED FOR BIORECOGNITION

12.6.1 Enzymes

Enzymatic paper-based biosensors have been used to detect target molecules such as glucose, creatinine, and phenylalanine, among others. The most common enzymes used in paper-based biosensors correspond to oxidase, transferase, and hydrolase enzymes [44, 45]. Recent trends regarding electrochemical enzyme paper-based biosensors propose new electrode materials, support modifications, and immobilization methods to guarantee reproducible and effective enzyme-support coupling and detector response (i.e., optical or electrochemical), according to analyte concentration [46]. For example, Cinti et al. have reported a paper-based nanomaterial-modified electrochemical biosensor for ethanol detection in beers. They used a nanocomposite formed by carbon black and Prussian blue nanoparticles as an electrocatalyst [17]. Mohammadifar and co-workers reported a paper-based electrochemical biosensor for the detection of glucose levels in urine by glucose oxidase enzyme and poly(3,4-ethylenedioxythiophene) polystyrene sulfonate (PEDOT:PSS) to functionalize the paper structure; a simple setup allowed the paper color change to be related to glucose concentration, and so the device was proposed for POC diagnostics [47].

12.6.2 Antibodies

Immunosensors have been employed for a broad range of medical, agricultural, and environmental applications [46]. Recent paper-based immunosensor advances also encompass the use of nanomaterials (e.g., carbon nanomaterials) and strategies to improve fabrication techniques. Ji and co-workers reported the fabrication of a paper-based biosensor for early-stage prostate cancer detection based on the prostate-specific antigen (PSA) antibody attached to multiwalled carbon nanotubes deposited on a microporous filter paper [48]. Sun and co-workers developed a sensor platform employing branched zinc oxide nanorod arrays on a reduced graphene oxide-paper working electrode to increase sites for antibody loading and constructed an electrochemical immunosensor in a 3D origami device [49]. Khan et al. proposed a biosensor to monitor psychological stress by cortisol detection in human saliva by an electrode design using a poly(styrene)-block poly(acrylic acid) polymer and graphene nanoplatelet suspension on Whatman filter paper; graphene "platelet" morphology increased the sensitivity and improved other analytical characteristics [50]. The research-based construction of individual biosensors allows proof-of-concept testing to be done rapidly for presentation to the academic community. For example, crayon and pencil have been proposed as practical tools for immunosensor device construction. Wax pens and graphite pens are simple and widely available for the creation of hydrophobic barriers and electrodes by hand [51, 52].

12.6.3 Nucleic Acids

Nucleic acid biosensors (or genosensors) use DNA or RNA fragments as molecular biorecognition elements [44]. Recent advances in paper-based nucleic acid biosensors include the immobilization of aptamers on paper. Aptamers are single-stranded DNA or RNA with high affinity and specificity to a target molecule that is not another nucleic acid. This type of device is called an aptasensor or aptamer-based biosensor [53]. Aptasensors have been applied to the detection of low-molecular-weight pollutants (e.g., metals, toxins, hormones, drugs, and pesticides) in environmental, food, and water samples. Aptasensor applications for the detection of small-molecule contaminants have been well-reviewed recently [54, 55]. Contemporary important advances in paper-based nucleic acid biosensing strategies and applications include devices for adenosine detection [35, 56] and a paper-based fluorescence aptasensor for multiplexed cancer cell detection [57]. Teengam and co-workers used an anthraquinone-labeled pyrrolidinyl peptide nucleic acid probe combined with graphene-polyaniline to modify the surface of an electrode for early detection of human papillomavirus [58].

Significant challenges around paper-based nucleic acid devices are related to the integration of nucleic acid amplification procedures with the paper diagnostic. In fact, early studies of paper-based biosensors incorporating nucleic acid extraction, amplification and detection, or quantification have been reported. Whitesides's group developed a prototype paper microfluidic device that integrates sample preparation and loop-mediated isothermal amplification (LAMP) and detection of *Escherichia coli* DNA from whole cells [59]. Choi et al. described a prototype of an integrated paper-based biosensor platform to sequentially perform nucleic acid extraction, amplification, and detection in about 1 h using *E. coli* as the target analyte [60].

12.6.4 Microorganisms

Biosensors or bioassays based on living cells include bacteria, cyanobacteria, algae, fungi, or bacteriophages as recognition elements, and these are well-known as bioreporters [61–63]. These can mimic the biological effects of polluted samples on diverse populations of living organisms [64].

Most work involving paper-based systems have explored one of the following two possibilities. Microfluidic cell (MFC)-based biosensors have been used to monitor toxic compounds in water samples [65–67]. The output of an MFC-based biosensor consists of amperometric signal changes generated by changes in bacterial metabolism that could occur due to alterations in pH, temperature, or the presence of toxic compounds, for example. The wide applications of these types of biosensing systems have recently been reviewed [68, 69].

Bioreporter cells equipped with intentionally designed genetic circuits have emerged from synthetic biology. These methodologies involve genetic circuits implanted inside living cells (including microorganisms) to enable the targeting of specific compounds (e.g., toxic compounds or compound groups). Thus, exposure to the target compound induces a proportional increase in the expression of a reporter protein [64]. These devices employ genetic engineering techniques, where a reporter gene with a readily detected product (such as *lux*) is associated with an inducible promoter gene or operon (such as *mer*, which is sensitive to the mercuric ion).

12.7 DETECTION METHODS USED FOR TRANSDUCTION

Many aspects must be addressed in paper-based biosensor fabrication in order to detect the target analytes and design an appropriate detection range and sensitivity: sample type and volume, reagent, paper support requirements, device designs (layout, 2D or 3D design), operating range, and detection method. In this section we summarize and discuss recent trends regarding the most common detection methods.

12.7.1 Electrochemical

Electrochemical paper-based biosensors mainly link with cyclic voltammetry (CV), amperometry, coulometry, and potentiometry techniques. For example, bromide, iodide, chloride, nitrite, and even *Staphylococcus aureus* have been detected by CV with electrochemical paper-based biosensors. Chronoamperometry has been used for phosphates and glucose detection, and electrochemical impedance spectroscopy has been used for total and glycosylated hemoglobin quantitation. Electrochemical biosensing performance with these systems could be enhanced using ion-selective electrodes, ion-exchange membranes, convective mass transfer, flow injection, and/or other methods [14].

A wide range of studies has been reported for the use of strategic electrode materials and surface modifications for working electrodes, mainly using micro- or nano-sized materials, to improve electrode conductivity and/or increase the electrode surface area in contrast to planar surfaces [48]. Incorporation of nanoparticles, such as gold and platinum nanoparticles, as well as multiwalled carbon nanotubes, has been the most frequently reported strategy for working electrode modification [70]. Recent advances using gold nanoparticles involve the detection of carcinoma antigen 125 [71], recognition of chronic myelogenous leukemia cell lines (K-562 cells) [32], multiple cancer cell detection [72], and the early detection of lung cancer by recognition of the tumor marker carcinoembryonic antigen [73].

12.7.2 Optical

Optical detection methods used for signal transduction in biosensors typically measure absorbance or fluorescence changes at a fixed wavelength in the visible spectral region in response to analyte presence or concentration. Full ultraviolet-visible spectra may be used in development stages for selection of an appropriate fixed wavelength, but once selected, simple colorimetric detection can be used in a single-point or timed measurement protocol. Colorimetric detectors in paper-based biosensors provide rapid results with simple and inexpensive instrumental components. Analyte quantitation is achieved using image-processing software [45] and scanners, cell phone cameras, digital cameras, or complementary metal-oxide sensors (CMOS) embedded in cameras [14]. Even simple benchtop scanners used routinely in office administrative work and open-source image processing software have been used for colorimetric detectors [74]. Soni et al. devised a smartphone-based optical biosensor for urea determination in saliva samples by the immobilization of the enzyme urease along with a pH responsive dye on a filter-paper-based strip [75].

12.8 APPLICATIONS

Biosensors have been widely and creatively reported for the detection and quantitation of many analytes in a variety of sample matrices, following the introduction in the early 1960s of the first biosensor by Leland Clark and Champ Lyons [76]. They used an oxygen electrode, initially to measure oxygen depletion (consumed by glucose oxidase enzymatic reaction) and later for hydrogen peroxide measurement (also produced by the action of the same enzyme as a by-product of glucose oxidation). Glucose oxidase was simply immobilized/trapped over the electrode and used for glucose quantitation in blood. These first biosensors would now be called enzyme-based amperometric biosensors for glucose. For decades, new biosensing strategies and transducer designs have been tested against this pioneering work, or modern commercially available versions that exploit the same principles for the detection of many different analytes.

In this section, we provide a more focused examination of applications and detection schemes that make use of paper as the common component for single-use, disposable test strips in POC and PON biosensors. Glucose remains a very important analyte, and the main commercial biosensor successes employ disposable test strips, as they are widely used for POC devices, allowing simple, fast, and convenient blood glucose control by diabetic patients, allowing them to control insulin doses carefully without the need for clinical assistance.

Glucose monitoring is also very important in the food and beverage industries, including most fermentation processes. An example of a colorimetric, enzyme-based device has been presented recently by the Coltro and Cortón groups, where the use of nanomaterials allows better color distribution and thus better analytical performance [77]. The principle of operation and some analytical characteristics of this device are shown in Figures 12.7 and 12.8.

12.8.1 CLINICAL CHEMISTRY

Paper-based biosensors have become a powerful tool for the diagnosis and prevention of disease. Enzymatic biosensors, immunosensors, and nucleic acid paper-based biosensors are the most widely used of these devices in the biomedical field [46]. Classical electrochemical biosensor examples correspond to glucose [74] and human chorionic gonadotropin detection [78], while other electrochemical paper-based biosensors include those for acetylcholinesterase [79], cholesterol [80], and metals such as Pb^{2+}, Cd^{2+}, and Zn^{2+} in human serum [81]. Furthermore, paper-based devices for some analytes with biomedical interest have become of commercial interest with easy-to-use POC devices and detection kits (e.g., glucose, pregnancy test, uric acid, lactate, antibiotics) [53].

Recent paper-based biosensing advances encompass multiplexed detection as well as detection of chronic tumor cells [71]. Zhao and co-workers have developed a paper-based electrochemical biosensor array for simultaneous detection of physiologically relevant metabolic biomarkers, including glucose, uric acid, and lactate, taking advantage of the paper properties for an easily accomplished design (AUTOCAD software) of electrodes and channels [82]. Wu et al. have

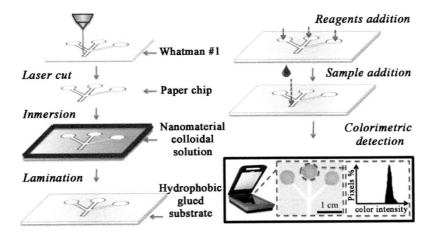

FIGURE 12.7 Schematic representation of the construction procedure of the μPADs showing the treatment step with magnetic nanoparticles (MNPs), multiwall carbon nanotubes (MWCNTs), and glucose oxidase (GOx), as well as the procedure required for colorimetric detection involving scanner and pixel intensity analysis. Reprinted with permission from ref. [77]. Copyright 2015 American Chemical Society.

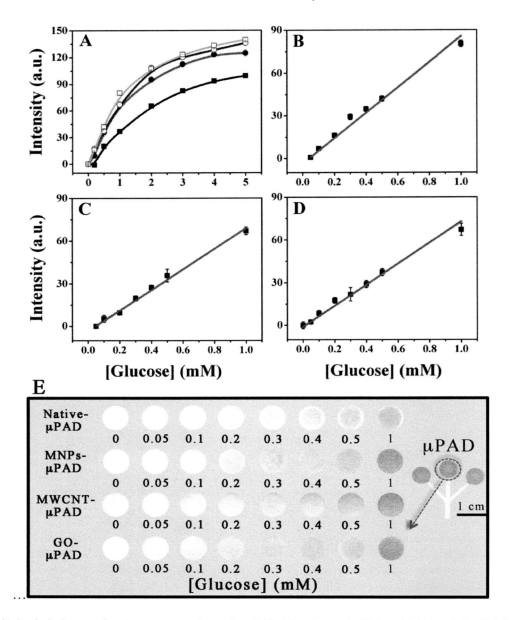

FIGURE 12.8 (A) Analytical curves for glucose assay using native μPAD (■) and treated μPADs with 100 μg/mL of MNPs (□), 10 μg/mL of MWCNT (●), and 100 μg/mL of GO (○). Linear ranges of the analytical curve for glucose using (B) MNPs-μPAD, (C) MWCNT-μPAD, (D) GO-μPAD. (E) Optical images of the detection zones corresponding to glucose assays in concentration ranging from 0 to 1 mM for native μPAD, MNP-μPAD, MWCNT-μPAD, and GO-μPAD. Reprinted with permission from ref. [77]. Copyright 2015 American Chemical Society.

designed a 3D paper-based sensor for multiplexed protein detection that employs kirigami and origami codes [72]. In an application related to tumor cell detection, Ge and co-workers have presented a device to detect the human chronic myelogenous leukemia cell line, K-562, using a concanavalin A protein immobilization matrix to capture cells [83].

12.8.2 Environmental Analysis

Many paper-based biosensors have been developed to monitor environmental pollutants due to their ease of handling and portability for easy *in situ* monitoring. Some analytes of interest that have been studied are heavy-metal ions such as Pb^{2+}, Cd^{2+}, Cl^-, Zn^{2+}, and Bi^{3+}. Hossain and Brennan presented a β-galactosidase-based colorimetric paper sensor, which, with the inclusion of standard chromogenic metal sensing reagents, was able to detect specific metal ions such as Ag^+, Ni^{2+}, and Cr^{4+} in complex samples [41]. An impedimetric paper-based biosensor has been reported for the detection of bacterial contamination in water, fabricated by screen printing carbon electrodes onto the paper supports [84].

Paper-based biosensors that have used microfluidics principles have been used in environmental applications, mainly for monitoring toxic compounds (see Section 12.6.4). Furthermore, a self-supporting paper-based microfluidic system, integrated with signal acquisition subsystem and an alert subsystem, has been designed to function as a real-time "shock" biosensor for online biological treatment process monitoring [85].

12.8.3 Food Security

Food and beverage control systems are necessary to ensure compliance with national regulations that include nutritional and chemical properties, as well microbial load and absence of pathogens according to specified values. In this area, paper-based biosensors have been fabricated mainly to detect pesticides, foodborne pathogens, and water and beverage quality. For example, Cinti et al. fabricated a paper-based, screen-printed biosensor with drop-cast electrodes for the detection of ethanol in beer samples [17]. Paper-based biosensors for the detection of pathogens and infectious diseases in animals and plants, as well as bacterial contamination in food and drinking water samples, have also been reported [3].

12.9 PERSPECTIVES

Paper-based devices have been of great interest for screening and POC and PON tests due to their disposable and inexpensive characteristics [7, 53]. With these benefits, however, come challenges. The detection challenge appears to have been addressed with reusable systems. Off-sensor detection has become common, and test strip results have recently been measured and evaluated/processed using mobile technology such as smart cell phones, portable scanners, digital cameras, and other lightweight, off-sensor devices with significant perceived needs for health care, environmental monitoring, and food quality control, especially in developing countries and remote locations. A second challenge exists in the elimination and/or simplification of sample pretreatment steps that are commonly performed in centralized analytical laboratories. Solvent extractions, precipitations, and centrifugations, for example, must be eliminated for POC or PON tests. Sample matrices may be quite complex, with large solid or semi-solid components, particulate matter, or cells in addition to a (generally) aqueous portion. Analytes may be present, entirely or partially, in any of these fractions. While it may be tempting to design a sensing concept around one specific analyte, perhaps classes of analytes might be considered instead. On-site or on-strip preparations and treatments could be designed for broad classes of analytes that are present only in solids, partitioned between solids and liquids, adsorbed on surfaces with or without being partially present in the aqueous component, or present only in the aqueous component, for example, and this might allow more general design strategies. Following this stage, the incorporation of testing steps into the paper sensor design may follow from very promising recent advances, for example, with blood samples used for nucleic acid amplification-based diagnostics.

Any chemical, biochemical, or biological reagents used in paper-based biosensors must demonstrate long-term stability under conditions appropriate for their fabrication, transport, storage, and usage. Dried or lyophilized single-use test strips with immobilized and/or dehydrated reagents present an excellent foundation for the extensive optimization that is needed for commercial success from both liability and customer satisfaction perspectives. The reproducibility of paper-based biosensor measurements may well be limited, at least currently, by the reproducibility of strip fabrication and storage. Comprehensive testing for reproducibility is very expensive, and careful consideration in the early design stages can provide enormous advantages later in the commercialization process, when full validation and/or regulatory testing is necessary and large-scale fabrication is occurring.

REFERENCES

1. Lv M, Liu Y, Geng J, Kou X, Xin Z, Yang D. Engineering nanomaterials-based biosensors for food safety detection. Biosens Bioelectron 2018; 106: 122–8.
2. Manisha H, Priya Shwetha PD, Prasad KS. Low-cost paper analytical devices for environmental and biomedical sensing applications. In: Bhattacharya S, Agarwal A, Chanda N, Pandey A, Sen A (Eds.) Environmental, Chemical and Medical Sensors. Singapore: Springer 2017; pp. 315–41.
3. Neethirajan S, Ragavan KV, Weng X. Agro-defense: Biosensors for food from healthy crops and animals. Trends Food Sci Technol 2018; 73: 25–44.
4. Almeida M Inês GS, Jayawardane BM, Kolev SD, McKelvie ID. Developments of microfluidic paper-based analytical devices (μPADs) for water analysis: A review. Talanta 2018; 177: 176–190.
5. Raghavulu SV, Goud RK, Sarma PN, Mohan SV. *Saccharomyces cerevisiae* as anodic biocatalyst for power generation in biofuel cell: Influence of redox condition and substrate load. Bioresour Technol 2011; 102(3): 2751–7.
6. Sackmann EK, Fulton AL, Beebe DJ. The present and future role of microfluidics in biomedical research. Nature 2014; 507(7491): 181–189.
7. Sharma N, Barstis T, Giri B. Advances in paper-analytical methods for pharmaceutical analysis. Eur J Pharm Sci 2018; 111: 46–56.
8. Ahmed S, Bui MN, Abbas A. Biosensors and bioelectronics paper-based chemical and biological sensors: Engineering aspects. Biosens Bioelectron 2016; 77: 249–63.
9. Nge PN, Rogers CI, Woolley AT. Advances in microfluidic materials, functions, integration, and applications. Chem Rev 2013; 113(4): 2550–83.
10. Xia Y, Si J, Li Z. Fabrication techniques for microfluidic paper-based analytical devices and their applications for biological testing: A review. Biosens Bioelectron 2016; 77: 774–89.
11. Wang Y, Guo H, Chen J, *et al*. Paper-based inkjet-printed flexible electronic circuits. ACS Appl Mater Interfaces 2016; 8(39): 26112–8.
12. Hu J, Wang SQ, Wang L, *et al*. Advances in paper-based point-of-care diagnostics. Biosens Bioelectron 2014; 54: 585–97.
13. Ng K, Gao B, Yong KW, *et al*. Paper-based cell culture platform and its emerging biomedical applications. Mater Today 2017; 20(1): 32–44.
14. Fu L-M, Wang Y-N. Detection methods and applications of microfluidic paper-based analytical devices. Trends Anal Chem 2018; 107: 196–211.
15. Bhattacharya S, Agarwal AK, Chanda N, *et al*. Introduction to environmental, chemical, and medical sensors. In: Bhattacharya S, Agarwal A, Chanda N, Pandey A, Sen A (Eds.) Environmental, chemical and medical sensors. Singapore: Springer 2017; 3–6.

16. Mak WC, Beni V, Turner APF. Lateral-flow technology: From visual to instrumental. TrAC – Trends Anal Chem 2016; 79: 297–305.
17. Cinti S, Basso M, Moscone D, Arduini F. A paper-based nanomodified electrochemical biosensor for ethanol detection in beers. Anal Chem Acta 2017; 960: 1–8.
18. Li Z, Liu H, He X, Xu F, Li F. Pen-on-paper strategies for point-of-care testing of human health. Trends Anal Chem 2018; 108: 50–64.
19. Gong MM, Sinton D. Turning the page: advancing paper-based microfluidics for broad diagnostic application. Chem Rev 2017; 117(12): 8447–80.
20. O'Sullivan AC. Cellulose: the structure slowly unravels. Cellulose 1997; 4(3): 173–207.
21. Mangiante G, Alcouffe P, Gaborieau M, et al. Biohybrid cellulose fibers: Toward paper materials with wet strength properties. Carbohydr Polym 2018; 193: 353–61.
22. Pelton R. Bioactive paper provides a low-cost platform for diagnostics. TrAC – Trends Anal Chem 2009; 28(8): 925–42.
23. Bracher PJ, Gupta M, Whitesides GM. Patterning precipitates of reactions in paper. J Mater Chem 2010; 20(24): 5117–22.
24. Swerin A, Mira I. Ink-jettable paper-based sensor for charged macromolecules and surfactants. Sensors Actuat B Chem 2014; 195: 389–95.
25. Martinez AW, Phillips ST, Butte MJ, Whitesides GM. Patterned paper as a platform for inexpensive, low-volume, portable bioassays. Angew Chemie Int Ed 2007; 46(8): 1318–20.
26. Figueredo F, González-Pabón MJ, Cortón E. Low cost layer by layer construction of CNT/chitosan flexible paper-based electrodes: a versatile electrochemical platform for point of care and point of need testing. Electroanal 2018; 30(3): 497–508.
27. Tian T, Bi Y, Xu X, Zhu Z, Yang C. Integrated paper-based microfluidic devices for point-of-care testing. Anal Methods 2018; 10(29): 3567–81.
28. Chen YH, Kuo ZK, Cheng CM. Paper – a potential platform in pharmaceutical development. Trends Biotechnol 2015; 33(1): 1–9.
29. Cate DM, Adkins JA, Mettakoonpitak J, Henry CS. Recent developments in paper-based microfluidic devices. Anal Chem 2015; 87(1): 19–41.
30. Martinez AW, Phillips ST, Whitesides GM, Carrilho E. Diagnostics for the developing world: microfluidic paper-based analytical devices. Anal Chem 2010; 82: 3–10.
31. Akyazi T, Basabe-Desmonts L, Benito-López F. Review on micro fluidic paper-based analytical devices towards commercialization. Anal Chim Acta 2018; 1001: 1–17.
32. Ge S, Zhang L, Zhang Y, Lan F, Yan M, Yu J. Nanomaterials-modified cellulose paper as a platform for biosensing applications. Nanoscale 2017; 9(13): 4366–82.
33. Sriram G, Bhat MP, Patil P, et al. Trends in analytical chemistry paper-based micro fluidic analytical devices for colorimetric detection of toxic ions: A review. Trends Anal Chem 2017; 93: 212–27.
34. Dxdiscovery.com [homepage on the Internet]. Dx Discovery. [updated 2019; cited 9 October 2019]. Available from: http://www.dxdiscovery.com/lateral-flow-assay.html
35. Liu H, Xiang Y, Lu Y, Crooks RM. Aptamer-based origami paper analytical device for electrochemical detection of adenosine. Angew Chemie Int 2012; 51(28): 6925–28.
36. Kong F, Hu YF. Biomolecule immobilization techniques for bioactive paper fabrication. Anal Bioanal Chem 2012; 403(1): 7–13.
37. Yamada K, Henares TG, Suzuki K, Citterio D. Paper-based inkjet-printed microfluidic analytical devices. Angew Chemie Int Ed 2015; 54(18): 5294–310.
38. Halder E, Chattoraj DK, Das KP. Adsorption of biopolymers at hydrophilic cellulose-water interface. Biopolymers 2005; 77(5): 286–95.
39. Zhang D, Tanaka H, Pelton R. Polymer assembly exploiting three independent interactions. Langmuir 2007; 23(17): 8806–9.
40. Sicard C, Brennan JD. Bioactive paper: Biomolecule immobilization methods and applications in environmental monitoring. MRS Bulletin 2018; 38(4): 331–4.
41. Hossain SMZ, Brennan JD. β-Galactosidase-based colorimetric paper sensor for determination of heavy metals. Anal Chem 2011; 83(22): 8772–78.
42. Luckham, RE, Brennan JD. Bioactive paper dipstick sensors for acetylcholinesterase inhibitors based on sol – gel/enzyme/gold nanoparticle composites. Analyst 2010; 135(8): 2028–35.
43. Wang J, Bowie D, Zhang X, Filipe C, Pelton R, Brennan JD. Morphology and entrapped enzyme performance in inkjet-printed sol – gel coatings on paper. Chem Mater 2014; 26(5): 1941–7.
44. Morales MA, Halpern JM. Guide to selecting a biorecognition element for biosensors. Bioconjug Chem 2018; 29(10): 3231–39.
45. Yang Y, Noviana E, Nguyen MP, Geiss BJ, Dandy DS, Henry CS. Paper-based microfluidic devices: emerging themes and applications. Anal Chem 2017; 89(1): 71–91.
46. Karimzadeh A, Hasanzadeh M, Shadjou N, De M. Peptide based biosensors. Trac-Trend Anal Chem, 2018; 107: 1–20.
47. Mohammadifar M, Tahernia M, Choi S. An equipment-free, paper-based electrochemical sensor for visual monitoring of glucose levels in urine. SLAS Technol 2019; 24(5): 499–505.
48. Ji S, Lee M, Kim D. Detection of early stage prostate cancer by using a simple carbon nanotube@paper biosensor. Biosens Bioelectron 2018; 102: 345–50.
49. Sun G, Yang H, Zhang Y, et al. Branched zinc oxide nanorods arrays modified paper electrode for electrochemical immunosensing by combining biocatalytic precipitation reaction and competitive immunoassay mode. Biosens Bioelectron 2015; 74: 823–9.
50. Khan MS, Misra SK, Wang Z, et al. Paper-based analytical biosensor chip designed from graphene-nanoplatelet-amphiphilic-diblock-co-polymer composite for cortisol detection in human saliva. Anal Chem 2017; 89(3): 2107–15.
51. Narang J, Malhotra N, Singhal C, et al. Point of care with micro fluidic paper based device integrated with nano zeolite–graphene oxide nanoflakes for electrochemical sensing of ketamine. Biosens Bioelectron 2017; 88: 249–57.
52. Yang H, Kong Q, Wang S, et al. Hand-drawn & written pen-on-paper electrochemiluminescence immunodevice powered by rechargeable battery for low-cost point-of-care testing. Biosens Bioelectron 2014; 61: 21–7.
53. Mahato K, Sravastava A. Paper based diagnostics for personalized health care : Emerging technologies and commercial aspects. Biosens Bioelectron 2017; 96: 246–59.
54. Nguyen V-T, Kwon YS, Gu MB. Aptamer-based environmental biosensors for small molecule contaminants. Curr Opin Biotechnol 2017; 45: 15–23.
55. Zhang W, Xiu Liu Q, Hou Gu Z, Sheng Lin J. Practical application of aptamer-based biosensors in detection of low molecular weight pollutants in water sources. Molecules 2018; 23(2): 344–69.
56. Zhang Y, Gao D, Fan J, et al. Naked-eye quantitative aptamer-based assay on paper device. Biosens Bioelectron 2016; 78: 538–46.

57. Liang L, Su M, Li L, et al. Aptamer-based fluorescent and visual biosensor for multiplexed monitoring of cancer cells in microfluidic paper-based analytical devices. Sensors Actuat B Chem 2016; 229: 347–54.
58. Teengam P, Siangproh W, Tuantranont A, Henry CS, Vilaivan T, Chailapakul O. Electrochemical paper-based peptide nucleic acid biosensor for detecting human papillomavirus. Anal Chim Acta 2017; 952: 32–40.
59. Connelly JT, Rolland JP, Whitesides GM. 'Paper machine' for molecular diagnostics. Anal Chem 2015; 87(15): 7595–601.
60. Choi JR, Hu J, Gong Y, et al. An integrated lateral flow assay for effective DNA amplification and detection at the point of care. Analyst 2016; 141(10): 2930–39.
61. Anany H, Brovko L, El Dougdoug NK, et al. Print to detect: a rapid and ultrasensitive phage-based dipstick assay for foodborne pathogens. Anal Bioanal Chem 2018; 410(4): 1217–30.
62. Farooq U, Yang Q, Ullah MW, Wang S. Bacterial biosensing: Recent advances in phage-based bioassays and biosensors. Biosens Bioelectron 2018; 118: 204–216.
63. Figueredo F, Cortón E, Abrevaya XC. In situ search for extraterrestrial life: a microbial fuel cell–based sensor for the detection of photosynthetic metabolism. Astrobiology 2015; 15(9): 717–27.
64. Roggo C, van der Meer JR. Miniaturized and integrated whole cell living bacterial sensors in field applicable autonomous devices. Curr Opin Biotechnol 2017; 45: 24–33.
65. González-Pabón MJ, Figueredo F, Martínez-Casillas DC, Cortón E. Characterization of a new composite membrane for point of need paper-based micro-scale microbial fuel cell analytical devices. PloS One 2019; 14(9): 1–22.
66. Chouler J, Cruz-Izquierdo Á, Rengaraj S, Scott JL, Di Lorenzo M. A screen-printed paper microbial fuel cell biosensor for detection of toxic compounds in water. Biosens Bioelectron 2018; 102: 49–56.
67. Zhou T, Han H, Liu P, Xiong J, Tian F, Li X. Microbial fuels cell-based biosensor for toxicity detection: A review. Sensors 2017; 17(10): 2230–51.
68. Abrevaya XC., Sacco NJ, Bonetto MC, Hilding-Ohlsson A, Cortón E. Analytical applications of microbial fuel cells. Part I: Biochemical oxygen demand. Biosens Bioelectron 2015; 63: 580–90.
69. Abrevaya XC., Sacco NJ, Bonetto MC, Hilding-Ohlsson A, Cortón E. Analytical applications of microbial fuel cells. Part II: Toxicity, microbial activity and quantification, single analyte detection and other uses. Biosens Bioelectron 2015; 63: 591–601.
70. Lee VBC, Mohd-Naim NF, Tamiya E, Ahmed MU. Trends in paper-based electrochemical biosensors: from design to application. Anal Sci 2018; 34(1): 7–18.
71. Wang S, Ge L, Yan M, et al. 3D microfluidic origami electrochemiluminescence immunodevice for sensitive point-of-care testing of carcinoma antigen 125. Sensors Actuat B Chem 2013; 176: 1–8.
72. Wu L, Ma C, Zheng X, Liu H, Yu J. Paper-based electrochemiluminescence origami device for protein detection using assembled cascade DNA-carbon dots nanotags based on rolling circle amplification. Biosens Bioelectron 2015; 68: 413–20.
73. Wang Y, Xu H, Luo J, et al. A novel label-free microfluidic paper-based immunosensor for highly sensitive electrochemical detection of carcinoembryonic antigen. Biosens Bioelectron 2016; 83: 319–26.
74. Soni A, Jha SK. A paper strip based non-invasive glucose biosensor for salivary analysis. Biosens Bioelectron 2015; 67: 763–8.
75. Soni A, Surana RK, Jha SK. Smartphone based optical biosensor for the detection of urea in saliva. Sensor Actuat B Chem 2018; 269: 346–53.
76. Clark LC, Lyons C. Electrode systems for continuous monitoring in cardiovascular surgery. Ann N Y Acad Sci 1962; 102: 29–45.
77. Figueredo F, Garcia PT, Cortón E, Coltro WKT. Enhanced analytical performance of paper microfluidic devices by using Fe_3O_4 nanoparticles, MWCNT, and graphene oxide. ACS Appl Mater Interfaces 2015; 8(1): 11–15.
78. Cao L, Fang C, Zeng R, Zhao X, Jiang Y, Chen Z. Paper-based microfluidic devices for electrochemical immunofiltration analysis of human chorionic gonadotropin. Biosens Bioelectron 2017; 92: 87–94.
79. Panraksa Y, Siangproh W, Khampieng T, Chailapakul O, Apilux A. Paper-based amperometric sensor for determination of acetylcholinesterase using screen-printed graphene electrode. Talanta 2018; 178: 1017–23.
80. Ruecha N, Rangkupan R, Rodthongkum N, Chailapakul O. Novel paper-based cholesterol biosensor using graphene/polyvinylpyrrolidone/polyaniline nanocomposite. Biosens Bioelectron 2014; 52: 13–9.
81. Ruecha N, Rodthongkum N, Cate DM, Volckens J, Chailapakul O, Henry CS. Sensitive electrochemical sensor using a graphene–polyaniline nanocomposite for simultaneous detection of Zn(II), Cd(II), and Pb(II). Anal Chim Acta 2015; 874: 40–48.
82. Zhao C, Thuo MM, Liu X. A microfluidic paper-based electrochemical biosensor array for multiplexed detection of metabolic biomarkers. Sci Technol Adv Mater 2013; 14(5): 054402.
83. Ge S, Zhang L, Zhang Y, et al. Electrochemical K-562 cells sensor based on origami paper device for point-of-care testing. Talanta 2015; 145: 12–19.
84. Rengaraj S, Cruz-Izquierdo Á, Scott JL, Di Lorenzo M. Impedimetric paper-based biosensor for the detection of bacterial contamination in water. Sensor Actuat B Chem 2018; 265: 50–8.
85. Xu Z, Liu Y, Williams I, et al. Disposable self-support paper-based multi-anode microbial fuel cell (PMMFC) integrated with power management system (PMS) as the real time 'shock' biosensor for wastewater. Biosens Bioelectron 2016; 85: 2329.

13 Strategies to Improve the Performance of Microbial Biosensors
Artificial Intelligence, Genetic Engineering, Nanotechnology, and Synthetic Biology

Natalia J. Sacco, Juan Carlos Suárez-Barón, and Eduardo Cortón
Universidad de Buenos Aires (UBA), Ciudad Universitaria, Argentina

Susan R. Mikkelsen
Department of Chemistry, University of Waterloo, Waterloo, Ontario, Canada

CONTENTS

13.1 Introduction 265
13.2 The Pros and Cons of Using Life as Part of an Analytical System... 266
13.3 Immobilization Methods 267
    13.3.1 Characteristics of a "Good" Method 267
    13.3.2 New Strategies 268
13.4 Common Transducers and New Transduction Strategies 268
    13.4.1 Electrochemical 268
        13.4.1.1 Amperometric 268
        13.4.1.2 Conductometric... 269
        13.4.1.3 Impedimetric: EIS... 270
        13.4.1.4 Potentiometric... 270
        13.4.1.5 Microbial Fuel Cell 270
    13.4.2 Optical 271
        13.4.2.1 Bioluminescence 271
        13.4.2.2 Colorimetric.. 272
        13.4.2.3 Fluorescence 272
    13.4.3 Others... 272
13.5 Microbial Life Used as Bioreceptors: From Nature Toward Engineered Organisms 272
    13.5.1 Single Species to Community-Based Microbial Biosensors 272
    13.5.2 Genetically Engineered Organisms... 272
    13.5.3 Synthetic Biology Applied to Develop New Biosensors.. 274
13.6 Platforms to Integrate, Analyze, and Use Microbial Biosensor Data.. 274
    13.6.1 Artificial Intelligence... 275
        13.6.1.1 Methods and Strategies: The Role of Artificial Intelligence in Biosensor Applications...................... 275
        13.6.1.2 Applications of Machine Learning to Biosensor Performance 275
    13.6.2 Empirical Models 277
        13.6.2.1 Smart Biosensors and Their Relationship with Pattern Recognition 277
        13.6.2.2 Artificial Neural Networks Applied to Smart Biosensors 277
13.7 New Developments Toward Commercial Analytical Applications 279
13.8 Future Trends... 279
References.. 280

13.1 INTRODUCTION

The intentional coupling of viable microorganisms with the surfaces of measurement devices began in the late 1970s. One report proposed two devices for the measurement of biological oxygen demand (BOD): a bacterial consortium cultured from soil, entrapped in a collagen membrane covering the Teflon membrane of an oxygen electrode, and a defined *Clostridium butyricum* strain, entrapped in polyacrylamide, covering the anode of a biofuel cell [1]. With both devices, the steady-state currents measured after 15 and 40 min, respectively, were linearly related to wastewater BOD values measured by a standard five-day method.

Later the same year, a potentiometric microbial biosensor was reported that used an ammonium-selective electrode in contact with *Streptococcus faecium* confined to the electrode's Teflon membrane surface with a dialysis membrane; the biosensor was applied to the quantitation of L-arginine [2]. Limited testing showed some selectivity toward arginine, with glutamine and asparagine showing significant responses. Linear potentiometric calibration was achieved with a 20-min response time over the arginine concentration range of 0.050–1.0 mM. In 1980, Grobler and Rechnitz reported another potentiometric device, a pH electrode, modified with a dental plaque microbial consortium that showed selectivity toward four sugars in a mixture of hexoses and pentoses [3]. In addition, in 1977, Mattiasson et al. [4] reported the concept of a thermistor modified with polyacrylamide-entrapped *Saccharomyces cerevisiae* and demonstrated its use by monitoring localized heat-induced signal changes due to glucose consumption by the yeast cells. According to this report, a temperature change of 0.02° C gave a 100 mV change in output signal, and the immobilized cells remained viable for 7 days during continuous operation. A second report of a thermistor-based microbial biosensor was published in 1991 and suggests their use for monitoring exposure to environmental pollutants [5].

In 1984, the use of *Escherichia coli* and a CO_2-sensing potentiometric electrode was demonstrated in a toxicity-screening device for water pollution [6]. The microbial biosensor was sensitive to a wide range of toxicants, including metal ions, gases, anions, and organic compounds. Results were found to agree with those of a much slower commercially available toxicity assay. An early review focuses on amperometric and potentiometric measurements and indicates the nearly overwhelming use of electrochemical transducers during the early years of microbial biosensor exploration [7].

Following their pioneering first paper on microbial biosensors, members of the Karube collaboration published a report of an oxygen-electrode-based ethanol sensor that exploited the acetic acid bacterium *Gluconobacter suboxydans* [8]. The bacteria were adsorbed to a nitrocellulose filter, which was then sandwiched between the gas-permeable Teflon membrane of an oxygen electrode and an external porous Teflon membrane. Under the reported conditions, the selectivity of the microbial biosensor for ethanol was better than that of a previously reported enzyme electrode constructed with alcohol oxidase. Members of the same group subsequently reported a microbial biosensor for mutagenicity screening [9]. The surface of an oxygen electrode was modified with a lysogenic (phage-hosting) strain of *E. coli* using a dialysis membrane. Exposure to a mutagen caused phage induction and cell lysis, resulting in decreased respiration and higher local oxygen concentration. A non-lysogenic culture of the same strain of *E. coli* yielded a biosensor that did not respond to the two tested mutagens. Antibiotics and bactericides caused reduced respiration at both the lysogenic and non-lysogenic biosensors, allowing facile screening for mutagens.

During the same year, the first use of a photosynthetic cyanobacterium, a defined strain of *Synechococcus*, in an amperometric ferricyanide-mediated microbial biosensor was reported by Turner and his collaborators. This biosensor was applied to herbicide screening in environmental samples [10]. Ferricyanide accepts electrons readily and replaces oxygen as the terminal electron acceptor in aerobic bacteria; it is also readily reoxidized and quantitated by amperometry. Sensitivity was found to be highest for those herbicides that target the photosynthetic electron-transport chain.

The introduction of optical transduction to microbial biosensors occurred in 1994 and exploited a strain of *Pseudomonas fluorescens* genetically modified to contain naphthalene and salicylate sensitivities as well as bioluminescent reporting [11]. Bacteria entrapped in strontium alginate beads were fixed at the surface of a liquid light guide in a flow cell containing a photomultiplier to detect bioluminescence intensity. As expected, the biosensor signal was selective for the presence of naphthalene and salicylate, with increased bioluminescence signals. Step changes of the flow stream to glucose solutions had no effect, while step changes to toluene-saturated water completely, but reversibly, eliminated the signal.

A review by Bousse made the formal distinction between analytical and functional microbial biosensors [12]. Analytical devices are designed to recognize and quantitate individual chemical analytes and provide, ideally, perfect selectivity toward only those analytes. Functional devices, on the other hand, provide information about the overall effect of a sample on a living system. Before and after this review was published, the critical distinction between analytical and functional devices has guided the efforts of many researchers in how they approach goals for their sensing devices. For example, sensor arrays coupled with chemometric and pattern recognition software can aid in selectivity improvements for analytical microbial sensors, while engineered microorganisms with reporter genes can be exploited for enhanced signals and compatibility with selected transducers for functional devices. Three more recent reviews cover selected aspects of new ideas in the microbial biosensor field and indicate the very wide spectrum of applications of these devices to modern analytical challenges [13–15]. The following sections of this chapter address separate but connected aspects of modern microbial biosensor research.

13.2 THE PROS AND CONS OF USING LIFE AS PART OF AN ANALYTICAL SYSTEM

Many unique properties critical to the survival of living systems, such as growth and reproduction, developed during geological times. These and other properties are adapted to the physical and chemical conditions of Earth. Evolutionary forces have selected from the available genetic pool characteristics that allow competition for the limited resources that different ecosystems offer, whereas negative characteristics – and the genes responsible for them – are normally eliminated from populations. It may therefore be assumed that it would be very difficult (or impossible) to develop a microbial biosensor that responds to substances that are not naturally present on Earth, or when their concentrations are too low to be a resource or a toxin. Another negative characteristic can be that hundreds of substances can often affect microbial metabolism in a positive or negative way, so selectivity is normally low. This has been improved in some cases by using genetically engineered microorganisms, where a genetic construct can guarantee a unique response to a predetermined substance.

On a more positive note, microbial biosensors are very easy to make and are inexpensive, given that microorganisms can grow rapidly in simple microbial growth culture media under mild chemical and physical conditions, and simple procedures can be used to separate and immobilize them. After immobilization, they can survive weeks or more and reproduce. A long-lasting biosensor able to cope with environmentally diverse and changing conditions (that could denature an isolated enzyme) can therefore be developed.

Perhaps the most interesting and useful feature of the microbial biosensor is that it can interpret environmental conditions as a whole; this fundamental feature allows development of early warning toxicity biosensors able to monitor water quality in natural environments, water supplies, and water treatment facilities. Like the "canary in a cage" used by coal miners, known or unknown toxic substances can be detected, and their overall effects summarized by the biological agent. Early warning systems detect toxic concentrations that are relatively high because they rely on acute toxicity mechanisms, at least for nonengineered organisms. This feature of integrating the effects of multiple substances – in adequate concentrations – has led to the main applications presented in the literature to date, as well as some commercialized biosensors for biochemical oxygen demand (BOD) and toxicity measurements.

13.3 IMMOBILIZATION METHODS

The analytical properties of microbial biosensors are influenced by the way the recognition elements (microbial cells) are retained near the sensing surfaces of the transducers. The relationship must be intimate and stable in order to facilitate fast response times, reasonable biosensor stability, and eventually the continuous or repeated use of the device. Immobilization plays a key role in achieving these goals, and several methods, usually classified as physical or chemical approaches, have been developed. As the viability of the microorganisms must be guaranteed, physical immobilization methods (including adsorption, entrapment, and encapsulation) are widely used, since these are considered less aggressive than chemical methods and allow higher survival rates when compared with covalent bonding or crosslinking approaches. An appropriate immobilization method must allow the transport of analyte(s) and substances required for viability from the bulk solution to the surface microenvironment and allow the elimination of any generated waste products from the surface to the bulk solution. Moreover, the immobilization and operation processes must be compatible with the transducer and transduction method. For example, irreversible chemical damage may be incurred during immobilization, or fouling of the surface could result from adsorption or locally high concentrations of microbial medium components or metabolic waste products. Although many options exist for both transducers and immobilization methods, careful consideration often narrows the field of possibilities. For example, the entrapment of a highly colored microorganism on the surface of an inexpensive optical absorbance transducer would be misguided, as would an attempt to employ a polyacrylamide gel with an inadequate pore size for viable microbial immobilization.

Commercialization is often the goal of biosensor research and may be aimed at specific applications for highly trained personnel or at wider applications for which user expertise is less critical. In both cases, low productions costs are priorities, and simple, stepwise construction using stable and safe components, including chemicals needed for microbial growth, immobilization, maintenance, and storage, are key factors. Chemical and biohazardous waste disposal costs may be minimized by planning in the initial research stages. Mitigation of environmental risks and the need to conform to industrial regulations concerning routine large-scale use, as well as preparedness for accidental chemical and microbial liberation, can also contribute to production costs. The design of a convenient immobilization step is therefore crucial to the development of a suitable biosensor [16].

Most of the methods used in immobilization processes are simple and well known, and most of them have also been used to construct other types of biosensors, involving purified enzymes and antibodies, for example. Several protocols are well established, and we direct the readers to books and reviews that deal with the more widely used methods [16, 17].

13.3.1 CHARACTERISTICS OF A "GOOD" METHOD

The selection of a microbial immobilization method depends on several factors. The first of these is the goal or intended use, including the analyte(s), desired response/concentration range(s), and the sample matrix. The intended operation mode, be it single-use, repetitive, or continuous, must be determined. Is the sample static/stagnant or is there flow/agitation? Consideration must also be given to maintaining the viability of the immobilized microorganism(s). Nutrients and their waste products must be transported along with analyte(s). Microorganisms show great inter- and intra-species variability in cell size, shape, membrane permeability and metabolic needs, and the optimization of post-immobilization viability is generally done empirically, if at all.

In addition to these scientific considerations, planning must consider several factors related to commercialization. These include scale-up and automation of processes, setup and maintenance of facilities, adherence to regulatory frameworks, and the costs associated with these.

13.3.2 New Strategies

While many examples exist of the use of physical methods, like gel entrapment, or chemical immobilization, such as crosslinking cells with bifunctional reagents (as described in the references given earlier), most microorganism immobilization methods are selected through consideration of the analyte(s) and the transducer. What kind of signal can the analyte or its product generate, and what kind of transducer is needed to measure that signal?

Historically, the first reported biosensor was an enzyme electrode based on a gas-sensing amperometric transducer. New electrode materials are being introduced and incorporated into biosensors both simple, like the enzyme electrodes, and more complex, like microbial biosensors.

New nanomaterials, which provide strong signals due to their large surface areas and allow miniaturization due to their small dimensions, can also enhance electrical conductivity and catalytic properties, allowing the fabrication of better electrodes. Carbon-based nanomaterials, such as nanotubes (CNT) and graphene are promising materials, as recently reviewed [18].

Gold and other metallic nanoparticles (NPs) have also been used to improve electrode characteristics. A relatively new application is the use of magnetic NPs to capture bacteria: using a magnetic force, certain bacteria can be immobilized easily and without modifying the transport of materials from and to the immobilized material. Gold NPs have been shown to be useful to provide better sensor sensitivity by confining electric fields at the surface of the biosensor when conductivity is used as the transduction method and phenol-adapted bacteria are used as bioreceptors [19].

An interesting new development not yet widely used to develop biosensors is the use of single-cell coatings that can offer advantages from various functional materials, including protection from matrix components, electrical conductivity, heat resistance, responsiveness to magnetism, and adaptive behavior. Some of the materials used as protective layers are polyelectrolytes assembled layer by layer, and bioinspired coatings (such as those from some planktons like diatoms and radiolarians) by *in situ* mineralization of a shell made of calcium, silica, or other minerals. Both methods create a protective cover outside each individual cell [20]. Further developments are needed to exploit the protective functions of the coating alongside the necessarily good diffusive properties needed for microbial survival and diffusion of analyte/s and other substances.

13.4 COMMON TRANSDUCERS AND NEW TRANSDUCTION STRATEGIES

The transducer is the element that converts the variations of physical or chemical properties that are produced by the interaction between the recognition element (here, microorganism) and the analyte in a signal that can be amplified, recorded, and used to determine the analyte concentration in the sample. The selection of a good transducer depends on the microorganism but also the intended use and cost. Several reviews present comprehensive lists of published work in this area; selected examples are shown here [21, 22].

13.4.1 Electrochemical

Electrochemical transducers can be very simple, such as a pair of conductive materials inside a solution, or more complex, involving 2D or 3D structures, or diffusive/protective membranes. One of the first electrodes used to develop microbial biosensors is the amperometric Clark oxygen electrode, which typically involves a Pt or Au cathode and an Ag or AgCl anode. When the cathode is polarized (0.7–0.9 V negative of the anode), the O_2 that diffuses through a protective membrane reaches the cathode and becomes reduced. The current is proportional to the dissolved oxygen concentration outside the electrolyte chamber, where the anode and cathode are immersed. The Clark electrode was used to develop respirometric microbial biosensors, where the metabolic activity of the biorecognition layer was easily and continuously followed. Electrochemical transducers are relatively simple and allow several different measurement modes, from simple (potentiometry, amperometry) to more sophisticated (impedance analysis).

13.4.1.1 Amperometric

Amperometry is the most widely used technique in microbial biosensors. The amperometric working electrode (usually an inert metal) operates at fixed potential with respect to a reference electrode and is used to detect current as a function of time, generated by the oxidation or reduction of species at the surface of the electrode [23]. Most amperometric microbial biosensors are based on the measurement of the respiratory activity of bacteria. This allows the measurement of compounds that negatively affect the respiratory rate, such as toxic chemical compounds like BTE (benzene, toluene, ethylbenzene), heavy metals, and pesticides. Moreover, compounds that positively affect the metabolism (acting as an energy source, for example) can also be detected, as shown in early work where *Pseudomonas* strains able to use haloaromatic compounds were associated with an O_2 electrode [24].

Amperometric microbial biosensors have been widely developed for the determination of BOD in aqueous samples. BOD measurements allow easy monitoring of biodegradable substances that can became organic pollutants if present at high concentrations by depleting dissolved O_2 through localized microbial respiration. The standard method to measure BOD involves the incubation of water samples (previously diluted if necessary and saturated with O_2) over a period of 5 days in dark bottles (to avoid photosynthesis) and is reported as BOD_5 [25].

When BOD is measured by means of a microbial biosensor ("short-term BOD," BOD_{st}), results are typically obtained in less than 1 hour and are related to BOD_5 values (discrepancies are attributed to different incubation times). Most BOD biosensors are based on amperometric O_2 electrode transducers modified with microorganisms that can degrade/metabolize organic substances present in samples such as wastewater [9, 25–27]. Most published work reports a single-strain approach for the microbial component; this allows reproducible results but narrows the spectrum of metabolizable substrates, so applications to real samples can be jeopardized. To avoid this limitation, [28] used a synergetic microbial consortium, obtaining the very interesting detection limit of 1 mg/L.

As most of the proposed BOD systems are based on electrochemical (amperometric) or optical (fluorescence quenching) O_2 measurements, they are limited by the low O_2 solubility in water, so the dynamic calibration range is very small (typically 1–8 mg/L). Normally, successive dilutions of the sample are required to ensure that at least one dilution falls within the calibration range. To overcome this limitation, some authors have proposed the use of soluble mediators (such as ferricyanide) that can replace O_2 in the microbial reductive metabolism. The reduced mediator (ferrocyanide) can diffuse outside the microbial cell and be easily measured electrochemically by re-oxidation at the amperometric transducer; the kinetics and mechanisms of this process have been demonstrated [29]. This approach has been reported by researchers who used a single bacterial strain, several strains, or an entire microbial community [30, 31]. Some work presents the attractive possibility of using lyophilized bacteria, which could be a step towards the introduction of BOD_{st} devices in the market [32].

Many of the BOD biosensors presented in the literature can be adapted for use as toxicity sensors by adjusting the sample preparation procedure. BOD samples are usually mixed with a diluent to regulate pH (among other properties) and depleted of carbon sources because they are metabolized. When toxicity measurement is the goal, diluent solutions contain a known carbon source in a relatively high concentration (i.e., 10 g/L), so the metabolism of the microorganisms will consume O_2 (or reduce mediators) at maximum rate and be unaffected by any other carbon source if present in the sample. With this procedure, the diminution of O_2 consumption is related to the presence and concentration of toxic or inhibiting substance(s) in the sample.

Besides BOD and toxicity biosensors, amperometric transducers have been used to design devices able to measure other chemicals. Neurotoxic organophosphate (OP) compounds have found wide applications as pesticides and insecticides in agriculture, so their detection is relevant to environmental and food applications. Lei et al. [33] have developed a biosensor based on genetically engineered *Moraxella* sp. and *P. putida* with surface-expressed organophosphorus hydrolase (OPH) for sensitive, selective, and cost-effective detection of OPs.

Recently a rapid and sensitive amperometric microbial biosensor for determination of vitamin B_{12} in untreated samples was reported. In this biosensor, the recognition element was the bacterium *Tetrasphaera duodecadis*, which oxidizes vitamin B_{12} with O_2 consumption. The results show a linear response from 10^{-7} to 10^{-5} mol/L [34].

13.4.1.2 Conductometric

As the conductivity changes with the concentration of ionic species in media or electrolyte, many reactions can be measured using this method. Conductivity can change in response to results of multiple reactions, resulting in low analyte selectivity related to the properties of the biorecognition element. A simple conductometry device consists of two metal electrodes of approximately equal surface area separated by a fixed distance. A low-amplitude alternating current (AC) voltage is applied, which causes a typically sinusoidal current flow between the electrodes [35]. Biosensors based on the conductometric principle present several advantages because they are inexpensive and easy to miniaturize and microfabricate. Moreover, they do not require a reference electrode. However, conductometric transducers are less sensitive and selective compared to most of the other electrochemical transducers.

There are numerous reports of the determination of toxic compounds using microbial biosensors based on this transducer. For example, Korpan et al. [36] developed a conductometric microbial biosensor for alcohol quantitation in beverages using immobilized yeast cells. The yeast cells were immobilized in an alginate gel attached to planar gold electrodes. The response time of the constructed microbial biosensor was less than 5 min, and linearity was observed in the range of 5–100 mM ethanol concentration. Good correlation between their results and gas chromatography results was observed.

A microbial biosensor based on whole cells of *Rhodococcus ruber* has been applied to the detection of acrylonitrile in solution. The bacteria were immobilized by entrapment in a disc of dimethyl silicone sponge. The biosensor was capable of the detection and quantitation of acrylonitrile, with a linear response to concentrations between 2 mM and 50 mM [37].

A biosensor based on *Chlorella vulgaris* microalgae was assembled to detect heavy metal ions and pesticides in water samples [38]. *C. vulgaris* was immobilized onto bovine serum albumin deposited on Pt interdigitated electrodes (IDEs). The biosensors were sensitive to Cd^{2+} and Zn^{2+} with a limit of detection of 10 ppb for both ions. In the same report, the determination of the presence of organophosphorus pesticides was demonstrated. A similar conductometric biosensor using *C. vulgaris* as the sensing element was fabricated by Guedri and Durrieu [39] to detect Cd^{2+}, and the detection limit was as low as 1 ppb.

By entrapping lyophilized *Brevibacterium ammoniagenes* in a polystyrene sulfonate–polyaniline (PSS–PANI) conducting polymer on a Pt twin wire electrode, a device to detect urea was fabricated [40]. The bacteria catalyze the production of ammonia (from urea), thereby increasing the pH locally. The pH variation resulted in changes to the resistivity of the conducting polymer, which was detected by the twin working electrodes. The sensor response was linear over a range of 0–75 mM urea.

Recently, Kolahchi et al. [19] presented a fast and sensitive miniaturized whole-cell conductometric biosensor for the determination of phenol. The biosensor consists of *Pseudomonas* sp. (a phenol degrading bacterium isolated from the environment) immobilized on the surface of gold IDE, where the conductivity increases proportionally with the phenol concentration. The results show the sensitive detection of phenol in the range of 1–300 mg/L. Furthermore, the bacterial biosensor was successfully applied to the determination of phenol in spiked river water samples.

13.4.1.3 Impedimetric: EIS

Electrochemical impedance spectroscopy (EIS) is a sensitive technique for the investigation of both bulk and interfacial changes to electrical properties related to biorecognition events and can be used for quantitative analysis. EIS-based biosensors have been designed for monitoring biological reactions at electrode surfaces using binding proteins, lectins, receptors, nucleic acids, whole cells, antibodies, or antibody-related substances. Recently the use of this transducer has been comprehensively reviewed [41]. The accuracy of EIS measurement depends not only on the operating procedures but also on the technical precision of the instrumentation and the use of noise-reducing techniques, as well as the working electrode itself. The various requirements to obtain valid impedance spectra make this technique challenging to use at present [42].

EIS transduction has multiple advantages. It is a nondestructive technique because small-amplitude perturbations are used to interrogate the sample, usually very close to the steady state. EIS signals can be measured in the absence or presence of a redox couple, which is referred to as nonfaradic and faradic impedance measurements, respectively. It is a sensitive method, so low analyte concentrations can be measured. Finally, EIS is suitable for real-time monitoring, since it can provide label-free/reagentless detection [43, 44]. Microbial biosensors based on impedimetric detection are based either on changes produced by metabolites released by cells as a result of growth in the presence of the analyte or on other changes that occur when the analyte and the microbe-modified electrode interact.

Hnaien et al. [45] developed a particular and novel type of microbial impedimetric biosensor for the detection of trichlorethylene (TCE), one of the most common organic pollutants found in soil and groundwater. The device is based on the immobilization of the strain *Pseudomonas putida* F1 on gold microelectrodes functionalized with single-walled CNT covalently linked to anti-*Pseudomonas* antibodies. The response of the biosensor was linear with a TCE concentration up to 150 µg/L and a detection limit of 20 µg/L. This microbial biosensor was applied to TCE determination in natural water samples spiked at the 30, 50, and 75 µg/L levels. Recoveries were very good, ranging from 100% to 103%.

13.4.1.4 Potentiometric

pH and other potentiometric ion-selective electrodes (notably CO_2, NO_3^-, and NH_4^+) have been used as transducers for microbial biosensors since their early beginnings in the 1970s (see the introduction of this chapter). The metabolic activity of the organisms consuming the analyte generates a change in the local concentration of the chemical species that the transducer selectively detects. As potentiometric electrodes have a logarithmic response, a wide detection range is often observed. However, these devices require stable references electrodes, and these are challenging to miniaturize and fabricate at low cost. These issues have limited the practical applications of this type of transducer [46].

Examples of relatively new developments have been presented by Kumar et al. [47]. A pH electrode modified by permeabilized *Pseudomonas aeruginosa* was used for detecting antibiotics from the cephalosporin group. The operational and storage stabilities were studied for the detection of cephalosporin C, and good results with high accuracy were obtained. Similarly, Ferrini et al. [48] developed another potentiometric microbial biosensor to detect the presence of β-lactam antibiotic residues in milk using a CO_2 electrode to measure the decrease of metabolic activity and growth of the test microorganism *Bacillus stearothermophilus var. calidolactis*.

A system for BOD determination where the microbial cells are attached to the gas-permeable membrane of a carbon dioxide ion-selective potentiometric electrode has also been reported. The biosensor is claimed to be insensitive to reduced nitrogen or metal compounds present at nontoxic concentrations. Oxygen is not measured, but, instead, estimated by means of CO_2 production [49]. The use of a CO_2 transducer instead of an O_2 electrode has some advantages, such as overcoming the limited calibration range related to the low O_2 solubility; it also allows the direct determination of the carbonaceous biochemical demand (CBOD). CBOD is the O_2 consumption exclusively related to organic matter degradation and is more difficult to measure by traditional O_2-based methods, given that there are other biochemical and chemical reactions (such as the O_2 consumed by reduced forms of nitrogen and inorganic ions like ferrous iron), which are considered sources of error in BOD_5 measurements.

A potentiometric pH electrode with *Thiobacillus thioparus* immobilized in a gelatin matrix was also successfully used to develop a microbial biosensor for the determination of hydrogen sulfide with an extended response range [50].

13.4.1.5 Microbial Fuel Cell

Recently, bioelectrochemical systems that employ electrodes colonized by electrogenic microorganisms and microbial fuel cells (MFCs) have been proposed as new categories of microbial biosensors. The electrogenic microorganisms are capable of growth on electrodes and can transfer electrons to the electrode surfaces; the potentials of these electrodes can be measured with respect to a reference electrode or can become a DC power source in an appropriate configuration. MFCs are typically configured to include a proton-exchange membrane that separates the anodic and cathodic chambers, a bioanode, and a cathode where O_2 or another dissolved chemical component is reduced. Since bacterial metabolism is directly proportional to the electrical parameters of the MFC (voltage, current, or power, depending on the configuration), they can become simple and convenient metabolic transducers. The development of MFC biosensors has recently been reviewed [51, 52], and most published uses involve BOD and toxicity biosensors. Figure 13.1 shows the variety of MFC designs presented in the literature; most of these can be adapted for

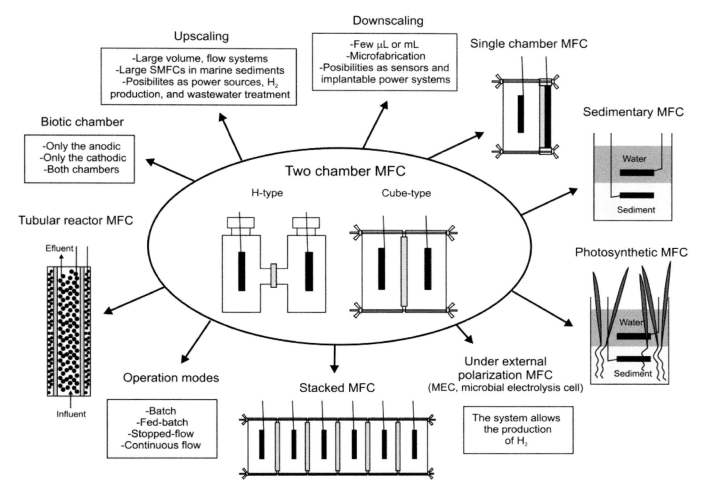

FIGURE 13.1 MFC architecture, construction, and operation modes. From the two basic designs (center), new devices, operation modes, and applications have been proposed. Electrodes are in black and separation membranes in gray. Reproduced with permission from Abrevaya et al. [51].

use as microbial biosensors, explaining – at least in part – the popularity of this type of transducer.

13.4.2 Optical

Optical transducers are based on the measurement of the variations that occur in the properties of light as a result of the physical or chemical interaction between the analyte and the biological component of the biosensor. Bioluminescence, fluorescence, and absorption-based biosensors have been widely investigated due to their compactness, selectivity, flexibility, sensitivity, resistance to electrical noise, and small probe size [22]. In addition, optical sensing techniques are very attractive in high-throughput screening methods, since simultaneous, multiplexed analyte detection can be accomplished with parallel detection, using CCD cameras and pixel analysis, in real time in the absence or presence of labels [53].

13.4.2.1 Bioluminescence

Microbial bioluminescence biosensors are based on the luminescence change that occurs when metabolically active microorganisms are exposed to a target analyte and provide sensitive, nondestructive, and real-time monitoring [21]. The direct relationship between viability and light emission allows the use of naturally bioluminescent bacteria to assess the effects of various chemical and biological compounds, such as toxins that negatively affect metabolism and decrease the intensity of light production. These biosensors, based on naturally bioluminescent organisms, are not normally very selective, and low sensitivity is also expected.

On the other hand, specific and sensitive biosensors can be designed using genetically engineered organisms. The bacterial luminescence *lux* genes have been widely used in reporter systems, making constructs in which the genes are expressed in either the inducible or constitutive mode. When the inducible mode is chosen, *lux* genes are fused to a promoter regulated by the concentration of a compound of interest; for example, in a mercury sensor, the *mer* operon, involved in bacterial resistance to Hg^{2+}, can be used. In this case, the concentration of the mercuric ion can be quantitatively analyzed by detecting the bioluminescence intensity (in the light-on approach). In the constitutive manner, the reporter genes are fused to promoters that are continuously expressed as long as the organism is alive and metabolically active [54], so in this configuration (the light-off approach), the evaluation of the total toxicity of contaminants can be achieved, as when using naturally luminescent microorganisms.

Most of the work in this area has been reviewed previously and presented in a tabular form [21], where analyte, microorganisms, and the promoter::reporter combinations that have been used are listed. For example, a bioluminescent microbial biosensor was developed for the detection of water toxicity, based on genetically engineered *E. coli* bacteria, carrying a *recA::luxCDABE* promoter-reporter fusion; the promoter genes are related to the SOS response in bacteria, involving the production of the protein RecA as part of a global defense mechanism against DNA damage. So, this biosensor will, in principle, be sensitive to any DNA-damaging agent [55].

New developments in this area include the use of CMOS (or smart phone cameras) to interrogate disposable cartridges or pads containing the microorganisms [56]. Another interesting possibility is to use genetically engineered human embryonic kidney cells constitutively expressing a green-emitting luciferase mutant; in this work, a toxicity biosensor mimicking microbial luminescent biosensors has been designed [57].

13.4.2.2 Colorimetric

Colorimetric microbial biosensors are based on the measurement of absorbance change of a particular compound to determine the presence or concentration of the target analyte. The presence of methyl parathion (an organophosphorus pesticide) can be determined if hydrolyzed by *Sphingomonas* sp. bacteria, able to catalyze the reaction to release an absorbing product, *p*-nitrophenol, which can also be detected electrochemically. Based on this principle, Kumar and D'Souza [58] built a microbial biosensor immobilizing *Sphingomonas* sp. by adsorption and crosslinking onto the inner epidermis of onion bulb scales used as a natural support; the colorimetric measurements were done on polystyrene microplates (96 wells). The membranes were proposed for use in biosensor applications; the detection range using the colorimetric microplate reader was 4–80 μM methyl parathion.

13.4.2.3 Fluorescence

Fluorescence spectroscopy it is a sensitive technique that can detect very low concentrations of analyte because emission is measured against a dark background; fluorescence emission intensity at low concentrations is directly proportional to fluorophore concentration [17].

One straightforward way to develop a fluorescent microbial biosensor is to fuse an inducible reporter gene, such as *gfp* coding for the green fluorescent protein (GFP), to the host gene to allow reporter activity to be examined in individual cells. This type of biosensor has been widely applied due to ease of construction using standard molecular biology techniques. Some advantages of using this system is the stability of the genetic construct, sensitivity to low concentrations, and excellent selectivity, given the impossibility of other microorganisms to synthesize GFP. However, the delay (several hours) between exposure to analyte and fluorescence (observed after a determined amount of GFP synthesized) is a significant disadvantage [21, 59]. Most of these works have been reviewed previously and presented in a tabular form [21], where analyte, microorganisms, and promoter::reporter combinations are listed.

13.4.3 Others

Some rarely used transduction strategies, such as thermal transduction, are interesting because they can be used with any microorganism, since heat is released as a by-product of most metabolic activity, including growth and maintenance, following thermodynamic laws. New thermistor-based biosensors are rarely presented, but recent developments, including chip calorimetry (also called miniaturized, integrated circuit, nano calorimetry), high-throughput calorimetry (enthalpy arrays), ultra-sensitive calorimetry, and photocalorimetry [60] could offer new possibilities to develop heat-based microbial biosensors.

13.5 MICROBIAL LIFE USED AS BIORECEPTORS: FROM NATURE TOWARD ENGINEERED ORGANISMS

13.5.1 Single Species to Community-Based Microbial Biosensors

The analytical characteristics of a microbial biosensor are dictated by several factors, including the identity of the selected organism. The simpler systems involve the immobilization of a well-known single species onto the transducer; in this way reproducible systems can be developed, and stability can be expected over time. When two or more species are combined into a single biosensor, a change in the proportion of each kind of microorganism can be expected, given the differential reproduction and death rates. Changes in composition would affect most analytical characteristics, producing irreproducible results. This is especially relevant if the device is designed to be functional for times long enough to allow reproduction.

It has been shown that mixed microbial populations can adapt much better to environmental stress, while consuming a wider range of organic substances, than isolated species [61, 62]. Mixed populations are thus preferred when environmental or food samples are measured, and the analyte is or can be a complex mixture of substances, such as in BOD and toxicity analysis. Some studies have used natural communities found, for example, in wastewater or sludge, to develop BOD biosensors, even though the diverse nature of the unknown bacterial community in sludge has been recognized as a limiting factor in terms of the repeated preparation of the biosensor system [63].

13.5.2 Genetically Engineered Organisms

The use of microorganisms that are genetically modified in a way to link a regulatory gene able to sense the concentration of a given substance or mixture with a reporter gene presents an interesting approach to improve the analytical characteristics of microbial biosensors. The principles of recombinant

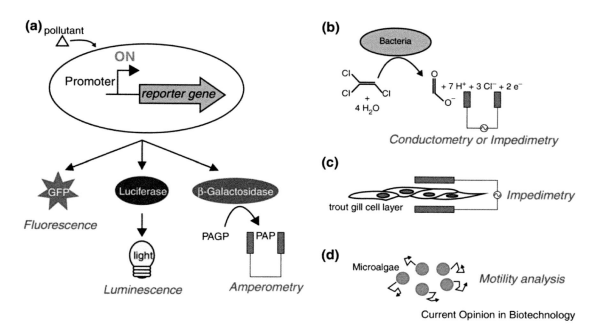

FIGURE 13.2 Some possibilities for coupling biological reactions and transducers. (a) Bioreporter cells are genetically engineered to carry a synthetic DNA construct to permit sensing of a polluting chemical (or condition) and turn on expression of a reporter gene. The output of the reporter protein can be measured by, for example, fluorescence, luminescence, or electrochemistry. (b) Electron release from native enzymes, such as toluene dioxygenase, can be measured by conductivity or impedance and can be used to measure the presence of trichloroethylene. (c) Release of tight cell-cell junctions as a measure of toxicity response can be detected by impedance. (d) Changes in random cell motility can be used for detecting the presence of heavy metals. Reproduced, with permission, from C. Roggo and J.R. van der Meer [64].

bacteria construction and some possible transduction systems are presented in Figure 13.2.

The use of recombinant organisms has several advantages. First, the selectivity and sensitivity can reach much better levels when compared with biosensors based on unmodified microorganisms and close (sometimes better) to those obtained using enzymatic biosensors. Second, as the modified microorganisms are not natural, some technological protection, perhaps patenting, of the developed devices can be achieved. Major disadvantages include regulation of the use and release of these organisms due to potential risks. A second inconvenience is that there are limited natural regulatory genes, and this points to a likely limitation on the number of analytes that could eventually be measurable with such a system. Some of the developed systems have been reviewed recently [65] and are presented in Table 13.1.

TABLE 13.1

A Comparison of Sensitivities for Different Types of Microbial Biosensors Based on Genetically Engineered Organisms. Modified from Gui et al. [65] under the Creative Commons Attribution License, which Permits Unrestricted Use, Distribution, and Reproduction

Host	Reporter Gene	Target Analyte	Detection Sensitivity	References
E. coli	luxCDABE	Arsenic	0.74–69 µg/L	[66]
E. coli	lacZ	Arsenic	<10 µg/L	[67]
D. radiodurans	lacZ	Cadmium	1–10 mM	[68]
	crtI		50–1 mM	
E. coli	Gap	Chromate	100 nM	[69]
E. coli	Gfp	Zinc	16 µM	[70]
		Copper	26 µM	
E. coli	Luc	Benzene, toluene and xylene	40 µM	[71]
E. coli	luxAB	Benzene, toluene and xylene	0.24 µM	[72]
P. putida	luxAB	Phenol	3 µM	[73]
B. sartisoli	luxAB	Naphthalene and phenanthrene	0.17 µM	[72]
E. coli	luxAB	C6–C10 alkanes	10 nM	[74]
E. coli	luxCDABE	Tetracyclines	45 nM	[75]

Although genetically engineered microbial biosensors may be useful tools for the detection of a wide range of analytes, these biosensors face several technical and social/regulatory challenges that have limited their adoption and use. The natural (and limited) promoters used in many biosensors may show off-target reactivity, responding not only to the molecule of interest but also to a group of compounds that interfere with the function of the promoter; in addition, the long response times needed for cell growth and the production of reporter gene products complicates use for real-time monitoring [76]. Moreover, negative concepts regarding the uses and dangers of recombinant organisms for human health and the environment have stimulated resistance to their use.

Some solutions have been proposed for one of the main problems with microbial biosensors: the sluggish response time (from several minutes to hours). The delay between analyte contact and response is related to the time needed for diffusion (mass transport) from the bulk solution to the immobilized cells, and then for penetration through microbial walls and membranes, to reach the metabolic intracellular machinery, where a change will occur and later be detected by the transducer. If chemically, physically, or genetically the external envelope of the microbial cells was altered while maintaining cell viability, faster transport would occur, resulting in shorter response times. Another approach involves cell surface display technology that can redesign cell surfaces with functional proteins and peptides (such as enzymes) to endow cells with unique features [77]. In this way, mass transport limitations, and enzyme purification, can be avoided. The uses of this technique and its applications to microbial biosensors have recently been reviewed in the aforementioned work.

13.5.3 Synthetic Biology Applied to Develop New Biosensors

Synthetic biology is a relatively new area that exploits developments in molecular biology and genetic engineering (among others) to rapidly and inexpensively focus on new products. Early work in this area was reported near the turn of the century, and the discipline has been defined: "Synthetic biology is the engineering of biology: the synthesis of complex, biologically based (or inspired) systems, which display functions that do not exist in nature. This engineering perspective may be applied at all levels of the hierarchy of biological structures, from individual molecules to whole cells, tissues and organisms. In essence, synthetic biology will enable the design of 'biological systems' in a rational and systematic way" [78].

Synthetic biology uses the "standard bioparts" nomenclature (promoters, operators, regulators, pathways, etc.) that are available from public websites such as iGEM Foundation (iGEM's main program is the iGEM Competition, where university students design and build a biological system based on standard bioparts). The final results achieved using this approach are recognized to probably be similar to those obtained by traditional biotechnology methods, but are obtained more rapidly and inexpensively [78].

One important and well-developed area of synthetic biology is the construction of heavy metal biosensors, given that metabolic detoxification of environments containing these toxins is important, and they are genetically regulated. Several regulation mechanisms are well-studied and used as sensing modules, such as transcription factors (activators/repressors) or sensor kinases of two-component systems, in conjunction with regulatory elements such as promoters, operators, and ribosomal-binding sites (RBSs). Relevant work has recently been reviewed [79]. Microorganisms able to detect several metals, including antimony, arsenite, cadmium, cobalt, copper, gold, lead, mercury, nickel, and zinc, have been described, but few of them have been experimentally assembled in a biosensor; most were tested in a bioassay configuration, in which luminescence, fluorescence, or colorimetric changes are delivered by the reporter genes used (such as *lux*, *gfp*, etc.).

13.6 PLATFORMS TO INTEGRATE, ANALYZE, AND USE MICROBIAL BIOSENSOR DATA

Sensors are important elements in the world of measurement, since they provide a convenient way to obtain information from a physical or chemical environment. They are used in a variety of applications, including automation and process control, environmental monitoring, and measurement of biological variables [80]. Alongside sensors, transducers (sometimes used synonymously) share importance in measurement systems. Therefore, a formal definition is important. For Song and Lee [81], a transducer is an element that converts energy from one form to another; in this way a transducer can be a sensor or an actuator, and therefore the sensor must be considered as transducer that can generate a signal (such as electricity) in proportion to a physical, biological, or chemical parameter [82].

The development of sensors has resulted in the incorporation of qualities and characteristics that turn them into intelligent devices with added functionalities beyond an output measurement [83]. These characteristics can be viewed as integrated components, and such as transducers, analog-to-digital converters, signal conditioners, and microprocessors, within the same chip [84]. In turn, these elements give shape to a system that provides options for performing tasks such as acquisition, signal processing, configuration verification, compensation, elimination, communication, and, optionally, multisensing [85].

Also, from a functional perspective, adaptation can be incorporated using the information obtained from the sensors' environment and operation. This can be achieved by providing the sensors with decision-making capabilities that lead to the development of self-diagnosis, self-validation, self-control, self-adaptation, self-identification, self-calibration, and self-compensation [86], for which adaptation plays a transcendental role [87]. Current technology trends, such as machine learning, data mining, and artificial intelligence, can thus be key factors in the facile development of intelligent new sensors designed to perform different functions or tasks.

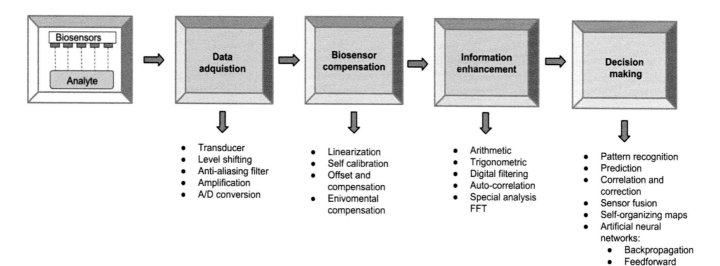

FIGURE 13.3 Block diagram of a smart biosensor.

The integration of biosensor systems with actuators and control modules can produce useful devices for multiple measurement and control applications, allowing lower costs for many of the processes, analyses, and control functions that are currently carried out without the assistance of intelligent systems. This improves efficiency and provides solutions to many current problems in medicine, industrial, and environmental control. Biosensors can be considered intelligent systems if they make use of numerical algorithms and analysis techniques based on signal processing, machine learning, pattern recognition, and/or artificial intelligence [88].

The purpose of this section is to review the reported applications, trends, and challenges in the development of intelligent biosensors, with special emphasis on microbial biosensors. The integration of different areas of knowledge forms a synergy of elements that lead to systems in which artificial intelligence oversees and adds value to performance capabilities.

13.6.1 Artificial Intelligence

13.6.1.1 Methods and Strategies: The Role of Artificial Intelligence in Biosensor Applications

Artificial intelligence (AI) emerged as part of the technological revolution at the end of the 20th and the beginning of the 21st century. It is usually defined as an area of computer science that emphasizes the creation of intelligent machines that work and react, in some way, like humans. Artificial intelligence has been applied to speech recognition, learning, planning, and problem solving.

Biosensors have evolved from contributions from biology, chemistry, engineering, and more recently, computing. Contributions from computing have led to advances in areas as diverse as materials, sensor calibration, data analysis, instrument control, and rapid reporting, providing the biosensors with intelligent characteristics. This section presents a review and discussion of intelligent biosensors as well as trends, potential applications, and future challenges in this area. Specifically, the discussion focuses on the development of microbial biosensors supported by machine learning (ML) and AI techniques. These devices conform to the definition of intelligent systems: they employ numerical algorithms and data analysis techniques based on signal processing, machine learning, pattern recognition, and artificial intelligence [88].

One well-known scenario for technological integration related to intelligent biosensors is shown in Figure 13.3. This concept, originally proposed by Knopf and Bassi [87], presents a basic structure with a set of functional building blocks for intelligent biosensors. Different branches of science, engineering, and information technologies converge to provide a unique perspective on intelligent biosensor design. In addition to this, reported performance specifications of biosensors have improved alongside the incorporation of microprocessors and/or microcontrollers into most of the building blocks shown in Figure 13.3. Although this scenario implies a linear design with unidirectional data flow, feedback and real-time adjustment of component properties are characteristics incorporated into intelligent biosensors and are considered in this section.

13.6.1.2 Applications of Machine Learning to Biosensor Performance

Langley [89] considered ML to be a branch of AI, in which algorithms are used to improve performance through experience. A more traditional approach considers ML to be focused solely on solving a problem of pattern recognition, using regression or classification. For Knopf and Bassi [87], ML refers to the extraction of significant information from sensor data using statistical, analytical, and correction algorithms. These algorithms can be potentially used to perform ML tasks in processes such as separation chemistry, biosensing, genomics, and other areas [90].

The functional and reporting capabilities of an individual biosensor or sensor array can be significantly enhanced by ML, according to Kimet et al. [91], since pattern recognition algorithms can be used to help categorize and analyze a large quantity of data from real-time and stored signals. ML can play a relevant role in the development of new or better functionality of sensors and intelligent biosensors through the incorporation of algorithms that can determine the temporal or spatial occurrence of a specific instance of a desired pattern. For example, algorithms could aid in detecting the presence of increased blood glucose; geographic regions with high flood risk; or types of contaminants detected in lakes, rivers, or areas of natural conservation.

At present, ML applications have taken on new challenges, including the design and development of smart prediction and clustering models applied to nanotechnology and electronic instrumentation design. Complementary work by Zhang et al. [92] has established that ML is an effective empirical approach for regression and/or classification (supervised or unsupervised) of nonlinear systems, specifically in the training of learning models for detection of pollutants.

A series of steps, with feedback, has been presented for the application of ML in the estimation of biochemical parameters in wastewater using microbial biosensors, as shown in Figure 13.4. First, an adjustment is carried out, corresponding to the characteristics, including extraction and selection, to identify the conditional attributes that are related to the prediction of the parameter in question. Second, the formation of a preliminary model occurs, in which the mapping relationship between the conditional factors and the decision attributes is defined. Finally, the trained model is be used for the prediction of the desired parameter [93].

The different algorithms used to perform tasks related to ML can be organized in relation to a taxonomy that is based on the desired output of the system. They can be divided into two main classes: supervised and unsupervised learning. This taxonomy introduced by Kapitanova and Son [94] and followed later for other authors organizes these algorithms according to their applications in biosensors. Figure 13.5 shows the classification of ML algorithms that may be applied to learning and training models for the analysis of logged (stored) data and information. Each algorithm in Figure 13.5 is represented within a node, where machine learning algorithms form the main node. Taxonomy allows the following classification four key areas: supervised, unsupervised, semi-supervised, and reinforcement learning algorithms. Supervised and semi supervised learning algorithms require labeled data models, because the models must possess one specific meaning or tag. On the other hand, unsupervised learning algorithms employ unlabeled data, since their final purpose is information analysis. Among the best-known algorithms are decision trees, Bayesian networks, support vector machines (SVM), artificial neural networks (ANN), and Markov chains.

In supervised learning, tagged training data is used, where both the correct inputs and outputs are given; this focuses mainly on finding a function that relates a set of data inputs to a set of information outputs. In unsupervised learning, the correct output is not provided – input and training data are

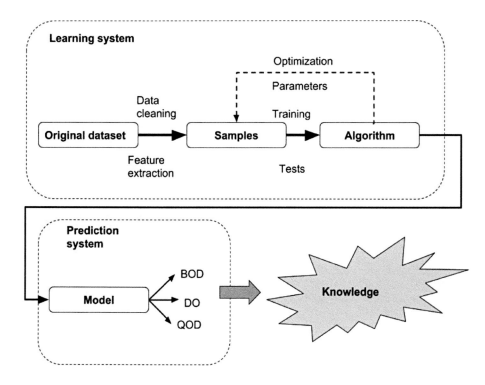

FIGURE 13.4 Framework for the application of ML in the estimation of water and wastewater quality parameters using microbial biosensors.

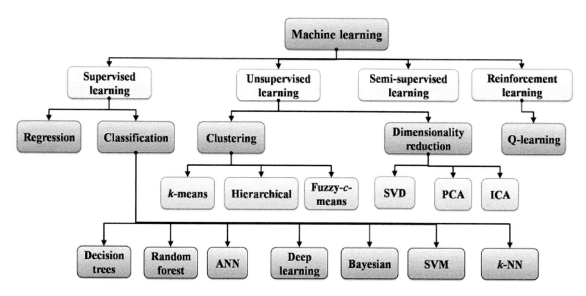

FIGURE 13.5 Potential machine learning algorithms in the development of biosensors. Reproduced with permission from Kumar et al. [95]

not available; instead, the learning algorithm relies on other sources of feedback to determine whether it is learning correctly. These learning algorithms are used to find the underlying structure of data sets from different sources, such as data logs from sensors or biosensors, continuous monitoring and control system data, and even processed images of different types. There is a third class of machine learning technique, called semi-supervised learning, that uses a combination of tagged and untagged data for training, and where applications have focused on applied geo-referenced databases or geographic positioning systems.

13.6.2 Empirical Models

13.6.2.1 Smart Biosensors and Their Relationship with Pattern Recognition

Pattern recognition is part of ML and aims to classify data or models based on the *a priori* information acquired [96] or information of statistical origin extracted directly from the data set object of analysis. The analysis involves or takes into account the pre-processing of signals to obtain data from the transducer used, the extraction of the relevant characteristics to carry out their classification, grouping of data into classes with similar characteristics, and the interpretation of the final assignment of the class.

From the perspective of Botta et al. [97] pattern recognition is the process of classifying the analyte, detecting the chemical composition, or estimating the descriptive parameters of a pattern. Thus, the functional characteristics of a biosensor or a set of biosensors can be improved using pattern recognition algorithms to analyze and categorize the characteristics extracted from the signals obtained [98]. Pattern recognition is a process that goes from the signal generated by the sensor to the identification of the pattern or group to which the signal belongs. It consists of several stages, usually called classification, detection, and estimation of parameters. The purpose of the classification stage is to assign a pattern of the signal coming from the sensor, or in its absence a vector of characteristics to a predefined class [99]. The detection focuses on determining the temporal or spatial occurrence of a specific instance of the desired pattern.

The last step, the estimation of the parameters, involves the determination of specific characteristics that help define a given pattern, that is, the identification of spatial-temporal information, such as speech waveforms, seismic waves, ECG, EEG, images, text, symbols, and other information types [92]. In large data sets, parameter estimation is used to eliminate noise and redundant information in the original data or to measure similarities between patterns [100]. Data compression and data fusion algorithms are often used to reduce the number of data vectors for interpretation or analysis, while pattern discrimination techniques help to establish measures of similarity between two or more patterns [101].

13.6.2.2 Artificial Neural Networks Applied to Smart Biosensors

The traditional approaches used to solve problems related to the measurement and estimation of biochemical parameters, for example, of water quality, usually have limitations due to highly complex and nonlinear systems. Different techniques have been proposed for such types of processes [102–104]. Artificial neural networks (ANNs) are recommended computational tools for this purpose, mainly because they do not require knowledge of physical or chemical laws involved in the processes. In addition, after being developed and trained, an ANN can make predictions easily, and this is useful for biological or chemical studies [105].

A graphical approach to the estimation of biochemical variables from the use of biosensors, among which BOD is highlighted, is shown in Figure 13.6. This approach consists of a

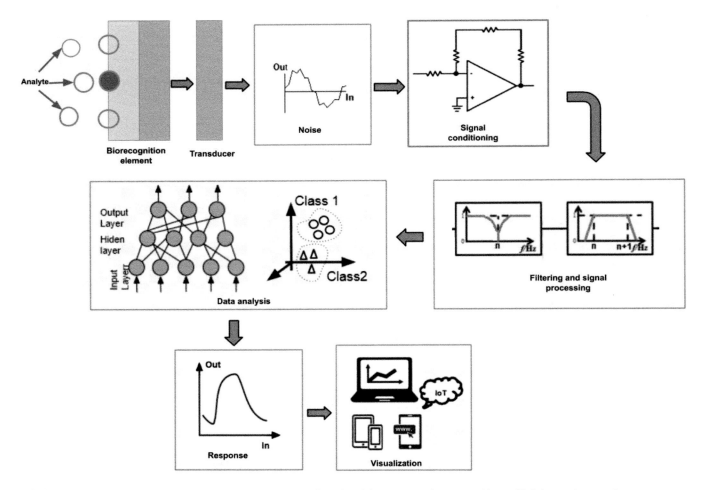

FIGURE 13.6 Biochemical parameter measurement system based on biosensors and supported by artificial neural networks.

flow diagram showing the components of a measurement system that uses ANNs to perform the sensor calibration, the classification of patterns, and multidimensional data visualization.

There are mathematical approaches for the analysis of biosensor readings that employ partial differential equations. The use of ANN and vector support machines (VSMs) to model the responses received by biosensors have been explored [106]. The results show that by using ML techniques to understand biosensor data, the accuracy is improved in comparison with mathematical models.

An important application of microbial biosensors is water quality testing as the global need increases for safe water supplies for human and animal consumption. These devices can detect unique chemicals with great specificity (specific biosensors) or a combination of multiple substances (nonspecific biosensors).

Microbial fuel cell (MFC) biosensors, relatively recently introduced and of the nonspecific type, are well-suited to water quality monitoring. They are constructed with inexpensive and simple materials such as carbon electrodes and microorganisms. Recent reviews describe their applications to BOD and toxicity measurements [51, 52].

King et al. [107] describe the integration of ANNs with an MFC to analyze water samples. This particular work assesses biosensors that are responsible for detecting three organic pollutants that do not degrade easily in water and can accumulate to reach toxic concentrations. Classification is performed by the ANN. This work demonstrates that ANNs can be very helpful in identifying concurrent chemicals in water, relating and differentiating both input and output data.

Feng et al. [108] exploit ANNs in novel smart biosensors. The ANNs are trained with algorithms that allow automating tasks such as predicting contaminating agents and react rapidly to any situation to obtain efficient and expected results. Thus, smart biosensors are evolving to a great extent with the incorporation of ML techniques, providing knowledge and innovation in areas such as bioengineering and information technology. This is described by Feng et al. [109] who emphasize that MFCs can be implemented in a successful manner due to the power that ANNs possess; they especially recognize the utility of back-propagation algorithms, since at the moment of programming the mathematical models, the algorithms can immediately recognize the entered patterns and start the testing process.

Gomolka et al. [110] created a smart system to monitor and provide real-time control of the water quality at Wisłok River, located in the southeastern part of Poland. The system is composed of a two-layer neural network of feed-forward type, using 25 neurons in the hidden layer and trained by the back-propagation algorithm, which is responsible for estimating BOD and dissolved oxygen (DO). In addition to demonstrating greater efficiency than the Kalman-Bucy filter, the system allows predictions of future values to control the aeration of the river using techniques such as optimal and predictive control.

13.7 NEW DEVELOPMENTS TOWARD COMMERCIAL ANALYTICAL APPLICATIONS

Several hundred microbial biosensor systems have been developed and presented to the scientific community, some of them just presented as a novelty, others in more or less developed conditions. But several barriers must be overcome to deliver a commercial device. Initially, the market must be identified and defined, and the proposed device must perform better and/or be less expensive than competing methods. Modern instrumentation can process thousands of samples a day, at very economic prices per sample, and with sensitivities and specificities impossible to achieve with microbial biosensors. Thus, the market niche should be related to sample types, locations, and needs not covered by centralized analytical laboratories. Point-of-need (PON) and point-of-use devices, designed as single-use disposable cartridges, with integrated microfluidics, prefilled and containing preserved bioreporter cells that are activated when hydrated by the sample, could provide a real competitive advantage for handheld devices [64]. User training would be minimal or unnecessary, perhaps better than the home glucose monitoring devices used by diabetics, because all necessary training and knowledge would be contained within the algorithms of the smart device.

BOD and toxicity tests have historically been performed using time-consuming and relatively expensive biological tests, but it is challenging to replace these tests with other nonbiological analytical methods. These were chosen as the first targets for commercial developments, and some have become commercial successes. The ARAS BOD sensor, based on the electrochemical measurement of O_2 consumed by a microbial community, is considered the first commercial success, and is produced by the company Dr. Lange GmbH, Germany [111]. Several BOD microbial biosensors have now been patented and produced, such as the commercial online system available from Biosensores SL Moncofar, Spain, and Isco GmbH, Gross Umstadt, Germany. To detect toxicity, a biosensor named the Green Screen Environmental Monitoring (EM) with a yeast cellular sensing element was designed for the simultaneous detection of genotoxicity and cytotoxicity by Gentronix Ltd., Manchester, UK. In addition, bioluminescent whole-cell toxicity biosensors based on genetically modified bacteria with SOS-*lux* genes have been developed by Remedios Ltd., Aberdeen, UK, and the amperometric mediated toxicity device CellSense, is fabricated by Euroclone Ltd., West Yorkshire, UK [112]. BOD and toxicity sensors based on ferricyanide-mediated amperometric detection, which typically use microorganisms, blends or whole communities from sludge, are named Micredox and commercialized by LincolnVentures Ltd, New Zealand.

An early example of a complete instrument with disposable cartridges named Lumisens 2 was presented in 2007 [113]. This is a small luminometer including thermostatting and fluidic subsystems and is $60 \times 40 \times 35$ cm in size. It was designed for easy operation, and is based on a recombinant (*lux*) *E. coli*, which is immobilized in a disposable card (35×17 mm); the system is claimed to specifically detect organotin compounds (tributyltin or dibutyltin), with a detection limit of around 26 µg/L for tributyltin. Possible interferents were not assayed, and calibration curves were not presented.

13.8 FUTURE TRENDS

Whole-cell biosensors have some important limitations as analytical devices when compared to modern instrumentation, such as gas and liquid chromatography with mass spectrometry detection; concentrations below 0.1 µM are generally undetectable for bacterial biosensors. Efforts to improve the sensitivities of biological systems are difficult to envision, but preconcentration or amplification methods may be possible. The present sensitivities are adequate for many applications and have been improved through the use of genetically engineered microorganisms (GEMs).

The design of portable, single-use, economic kits could be an interesting market; given that PON devices are necessary to monitor environmental pollution and water or wastewater quality. For example, arsenic-contaminated water (of geological origin) is a problem in several countries, including India and Argentina, and several systems based on recombinant bacteria have been proposed to measure this (and other) heavy metals.

The use of microarrays containing several different microbial strains, or including just a single cell per array element, could be a future means to recognize different samples. Using computational methods to analyze the responses to a single sample obtained from hundreds of different microbial strains, a kind of fingerprinting could be possible, and the use of artificial intelligence and machine learning could allow precise identification of single components in a sample. Lab-on-a-chip disposable microfabricated systems employing microbial biosensors could have many applications in drug screening, disease diagnosis, and cellular biology.

The main market for the development of useful microbial biosensor is presently environmental chemistry, with food chemistry in second place. Additional areas of application exist, but have been more challenging for development, given regulatory practices and some technical limitations of microorganisms as biorecognition agents.

Unique and powerful advantages of the use of living organisms for recognition in analytical systems include the interpretation of the chemical and physical characteristics of the sample as a whole. This is ideal for the measurement of the toxic effects of combined multiple chemical substances, which can lead to different effects than the simple summation of the toxicities of the components. Microbial biosensors and bioassays allow the development of analytical systems for complex parameters such as toxicity and BOD (and perhaps others, such as hormone effects), which could be very difficult to estimate using conventional analytical systems. Microbial biosensors can be used as effective early warning systems to monitor water resources, allowing rapid corrective action when contamination or sabotage is detected. Given that a microorganism is sensitive to most substances that are also toxic to animal/human life, provided that a given threshold concentration is present (acute toxicity response), the chemical nature(s) of the toxicant(s) is not a concern. Problematic samples can be preserved for detailed chemical analysis later in a centralized laboratory facility to determine the chemical identities and concentrations of toxic substances detected by the microbial biosensor.

REFERENCES

1. Karube I, Tadashi M, Satoshi M, Shuichi S. Microbial electrode BOD sensors. Biotechnol Bioeng 1977; 19(10): 1535–1547.
2. Rechnitz GA, Kobos RK, Riechel SJ, Gebauer CR. A bioselective membrane electrode prepared with living bacterial cells. Anal Chim Acta 1977; 94(2): 357–365.
3. Grobler SR, Rechnitz GA. Determination of D(+)glucose, D(+)mannose, D(+)galactose or D(-) fructose in a mixture of hexoses and pentoses by use of dental plaque coupled with a glass electrode. Talanta 1980; 27(3): 283–285.
4. Mattiasson B, Larsson PO, Mosbach K. The microbe thermistor. Nature 1977; 268, 519–520.
5. Thavarungkul P, Håkanson H, Mattiasson B. Comparative study of cell-based biosensors using *Pseudomonas cepacia* for monitoring aromatic compounds. Anal Chim Acta 1991; 249(1):17–23.
6. Dorward EJ, Barisas BG. Acute toxicity screening of water pollutants using a bacterial electrode. Environ Sci Technol 1984; 18(12): 967–972.
7. Corcoran CA, Rechnitz GA. Cell-based biosensors. Trends Biotechnol 1985; 3(4): 92–96 and references therein.
8. Kitagawa Y, Ameyama M, Nakashima K, Tamiya E, Karube I. Amperometric alcohol sensor based on an immobilised bacteria cell membrane. Analyst 1987; 112(12): 1747–1749.
9. Karube I, Sode K, Suzuki M, Nakahara T. Microbial sensor for preliminary screening of mutagens utilizing a phage induction test. Anal Chem 1989; 61(21): 2388–2391.
10. Rawson DM, Willmer AJ, Turner APF. Whole-cell biosensors for environmental monitoring. Biosensors 1989; 4(5): 299–311.
11. Heitzer A, Malachowsky K, Thonnard JE, Bienkowski PR, White, DC, Sayler GS. Optical biosensor for environmental on-line monitoring of naphthalene and salicylate bioavailability with an immobilized bioluminescent catabolic reporter bacterium. Appl Environ Microbiol 1994; 60(5): 1487–1494.
12. Bousse, L. Whole cell biosensors. Sens Actuators B Chem 1996; 34(1–3): 270–275.
13. Park M, Tsai SL, Chen W. Microbial biosensors: engineered microorganisms as the sensing machinery. Sensors 2013; 13(5): 5777–5795.
14. Lim JW, Ha D, Lee J, Lee SK and Kim T. Review of micro/nanotechnologies for microbial biosensors. Front Bioeng Biotechnol 2015; 3(61): 1–13.
15. Chang H-J, Voyvodic PL, Zúniga A, Bonnet J. Microbially derived biosensors for diagnosis, monitoring and epidemiology. Microb Biotechnol 2017; 10(5): 1031–1035.
16. D'Souza SF. Immobilization and stabilization of biomaterials for biosensor applications. Appl Biochem Biotechnol 2001; 96(1–3): 225–238.
17. Mikkelsen SR, Cortón E. Bioanalytical Chemistry, 2nd ed. John Wiley and Sons, New Jersey, 2016.
18. Holzinger M, Le Goff A, Cosnier S. Nanomaterials for biosensing applications: a review. Front. Chem 2014; 2(63): 1–10.
19. Kolahchi N, Braiek M, Ebrahimipour G, Ranaei-Siadat SO, Lagarde F, Jaffrezic-Renault N. Direct detection of phenol using a new bacterial strain-based conductometric biosensor. J Environ Chem Eng 2018; 6(1): 478–484.
20. Dai B, Wang L, Wang Y, Yu G, Huang X. Single-cell nanometric coating towards whole-cell-based biodevices and biosensors. Chem Select 2018; 3(25): 7208–7221.
21. Su L, Jia W, Hou C, Lei Y. Microbial biosensors: a review. Biosens Bioelectron 2011; 26(5): 1788–1799.
22. Lei Y, Chen W, Mulchandani A. Microbial biosensors. Anal Chim Acta 2006; 568(1–2): 200–210.
23. Bard, A.J., Faulkner, L.R. Electrochemical Methods: Fundamentals and Applications, 2nd ed., John Wiley & Sons, Inc., Hoboken, NJ, 2001.
24. Riedel K, Naumov AV, Boronin AM, Golovleva LA, Stein HJ, Scheller F. Microbial sensors for determination of aromatics and their chloroderivatives I. Determination of 3-chlorobenzoate using a Pseudomonas-containing biosensor. Appl Microbiol Biotechnol 1991; 35(5): 559–562.
25. Chan C, Lehmann CM, Chan K, Chan P, Chan C, Gruendig B, Kunze G, Renneberg, R. Designing an amperometric thick-film microbial BOD sensor. Biosens Bioelectron 2000; 15(7–8): 343–353.
26. Nakamura H, Suzuki K, Ishikuro H, et al. A new BOD estimation method employing a double-mediator system by ferricyanide and menadione using the eukaryote *Saccharomyces cerevisiae*. Talanta 2007; 72(1): 210–216.
27. Kara S, Keskinler B, Erhan E. A novel microbial BOD biosensor developed by the immobilization of *P. syringae* in microcellular polymers. J. Chem Technol Biotechnol 2009; 84(4): 511–518.
28. Dhall P, Kumar A, Joshi A, et al. Quick and reliable estimation of BOD load of beverage industrial wastewater by developing BOD biosensor. Sens Actuators B Chem 2008; 133(2): 478–483.
29. Ertl P, Unterladstaetter B, Bayer K, Mikkelsen SR. Ferricyanide reduction by *Escherichia coli*: kinetics, mechanism, and application to the optimization of recombinant fermentations. Anal Chem 2000; 72(20): 4949–4956.
30. Pasco N, Baronian K, Jeffries C, Webber J, Hay J. MICREDOX-development of a ferricyanide-mediated rapid biochemical oxygen demand method using an immobilised *Proteus vulgaris* biocomponent. Biosens Bioelectron 2004; 20(3): 524–532.
31. Bonetto MC, Sacco NJ, Hilding-Ohlsson A, Cortón E. Assessing the effect of oxygen and microbial inhibitors to optimize ferricyanide-mediated BOD assay. Talanta 2011; 85(1): 455–462.
32. Bonetto MC, Sacco NJ, Hilding-Ohlsson A, Cortón E. Metabolism of *Klebsiella pneumoniae* freeze-dried cultures for the design of BOD bioassays. Lett Appl Microbiol 2012; 55(5): 370–375.
33. Lei Y, Mulchandani P, Chen W, Mulchandani, A. Direct determination of p-nitrophenyl substituent organophosphorus nerve agents using a recombinant *Pseudomonas putida* JS444-modified Clark oxygen electrode. J Agric Food Chem 2005; 53(3): 524–527.
34. Ovalle M, Arroyo E, Stoytcheva M, Zlatev R, Enriquez L, Olivas A. An amperometric microbial biosensor for the determination of vitamin B_{12}. Anal Methods 2015; 19: 8185–8189.
35. Rogers KR, Mascini M. Biosensors for field analytical monitoring. Field Anal Chem Technol 1998; 2(6): 317–331.

36. Korpan YI, Dzyadevich SV, Zharova VP, El'skaya AV. Conductometric biosensor for ethanol detection based on whole yeast cells. Ukr Biokhim Zh 1994; 66(1): 78–82.
37. Roach PCJ, Ramsden DK, Hughes J, Williams P. Development of a conductimetric biosensor using immobilized *Rhodococcus ruber* whole cells for the detection and quantification of acrylonitrile. Biosens Bioelectron 2003; 19(1): 73–78.
38. Chouteau C, Dzyadevych S, Chovelon JM, Durrieu C. Development of novel conductometric biosensors based on immobilised whole cell *Chlorella vulgaris* microalgae. Biosens Bioelectron 2004; 19(9): 1089–1096.
39. Guedri H, Durrieu C. A self-assembled monolayers based conductometric algal whole cell biosensor for water monitoring. Microchim Acta 2008; 163(3–4): 179–184.
40. Jha SK, Kanungo M, Math A, D'Souza SF. Entrapment of live microbial cells in electro polymerized polyaniline and their use as urea biosensor. Biosens Bioelectron 2009; 24(8): 2637–2642.
41. Bahadır EB, Sezgintürk MK. A review on impedimetric biosensors. Artif Cell Nanomed B 2016; 44(1): 248–262.
42. Yuan X, Song C, Wang H, Zhang J. Electrochemical Impedance Spectroscopy in PEM Fuel Cells. Fundamentals and Applications. Springer-Verlag, London, 2010.
43. Barsoukov E, Macdonald JR., Editors. Impedance Spectroscopy: Theory, In: Experiment, and Applications, 2nd ed., Wiley-Interscience, Hoboken, New Jersey, 2005.
44. Pejcic B, De Marco R. Impedance spectroscopy: over 35 years of electrochemical sensor optimization. Electrochim Acta 2006; 51(28): 6217–6229.
45. Hnaien M, Bourigua S, Bessueille F, Bausells J, Errachid A, Lagarde F, Jaffrezic-Renault N. Impedimetric microbial biosensor based on single wall carbon nanotube modified microelectrodes for trichloroethylene detection. Electrochim Acta 2011, 56(28):10353–10358.
46. Wang J. Analytical Electrochemistry, 3rd ed., John Wiley & Sons, Inc., Hoboken, NJ, 2006.
47. Kumar S, Kundu S, Pakshirajan K, Dasu VV. Cephalosporins determination with a novel microbial biosensor based on permeabilized *Pseudomonas aeruginosa* whole cells. Appl Biochem Biotechnol 2008; 151(2–3): 653–664.
48. Ferrini AM, Mannoni V, Carpico G, Pellegrini GE. Detection and identification of beta-lactam residues in milk using a hybrid biosensor. J Agric Food Chem 2008; 56(3): 784–788.
49. Chiappini SA, Kormes DJ, Bonetto MC, Sacco NJ, Cortón, E. A new microbial biosensor for organic water pollution based on measurement of carbon dioxide production. Sens Actuators B Chem 2010; 148(1): 103–109.
50. Ebrahimi E, Amoabediny FYG, Shariati MR, Janfada B, Saber M. A microbial biosensor for hydrogen sulfide monitoring based on potentiometry. Process Biochem 2014; 49(9): 1393–1401.
51. Abrevaya XC, Sacco NJ, Bonetto MC, Hilding-Ohlsson A, Cortón E. Analytical applications of microbial fuel cells. Part I: biochemical oxygen demand. Biosens Bioelectron 2015; 63: 80–590.
52. Abrevaya XC, Sacco NJ, Bonetto MC, Hilding-Ohlsson A, Cortón E. Analytical applications of microbial fuel cells. Part II: toxicity, microbial activity and quantification, single analyte detection and other uses. Biosens Bioelectron 2015; 63: 591–601.
53. Bhatta D, Stadden E, Hashem E, Sparrow IJG, Emmerson GD, Multi-purpose optical biosensors for real-time detection of bacteria, viruses and toxins. Sens Actuators B Chem 2010; 149(1): 233–238.
54. Rensing C, Maier RM. Issues underlying use of biosensors to measure metal bioavailability. Ecotoxicol Environ Safety 2003; 56(1): 140–147.
55. Ramiz D, Almog R, Ron, A, Belkin S, Diamand YS. Modelling and measurement of a whole-cell bioluminescent biosensor based on a single photon avalanche diode. Biosens Bioelectron 2008; 24(4): 882–887.
56. Eltzov E, Pavluchkov V, Burstin M, Marks RS. Creation of a fiber optic based biosensor for air toxicity monitoring. Sens Actuators B Chem 2011; 155(2): 859–867.
57. Cevenini L, Calabretta MM, Tarantino G, Michelini E, Roda A. Smartphone-interfaced 3D printed toxicity biosensor integrating bioluminescent 'sentinel cells'. Sens Actuators B Chem 2016; 225: 249–257.
58. Kumar J, D'Souza SF. Immobilization of microbial cells on inner epidermis of onion bulb scale for biosensor application. Biosens Bioelectron 2011; 26(11): 4399–4404.
59. Eltzov E, and Marks RS. Whole-cell aquatic biosensors. Anal Bioanal Chem 2011; 400(4): 895–913.
60. Maskow T, Kemp R, Buchholz F, Schubert T, Kiesel B, Harms H. What heat is telling us about microbial conversions in nature and technology: from chip- to megacalorimetry. Microb Biotechnol 2010; 3(3): 269–284.
61. Tan TC, Wu C. BOD sensors using multi-species living or thermally killed cells of a BODSEED microbial culture. Sens Actuators B Chem 1999; 54(3): 252–260.
62. Anam M, Yousaf S, Sharafat I, Zafar Z, Ayaz K, Ali N. Comparing natural and artificially designed bacterial consortia as biosensing elements for rapid non-specific detection of organic pollutant through microbial fuel cell. Int J Electrochem Sci 2017; 12: 2836–2851.
63. Liu J, Olsson G, Mattiasson B. Short-term BOD (BODst) as a parameter for on-line monitoring of biological treatment process; Part II: instrumentation of integrated flow injection analysis (FIA) system for BODst estimation. Biosens Bioelectron 2004; 20(3): 571–578.
64. Roggo C, van der Meer JR. Miniaturized and integrated whole cell living bacterial sensors in field applicable autonomous devices. Curr Opin Biotechnol 2017; 45:24–33.
65. Gui Q, Lawson T, Shan S, Yan L, Liu Y. The application of whole cell-based biosensors for use in environmental analysis and in medical diagnostics. Sensors 2017; 17(7): 1623–1640.
66. Sharma P, Asad S, Ali A. Bioluminescent bioreporter for assessment of arsenic contamination in water samples of India. J Biosci 2013; 38(2): 251–258.
67. De Mora K, Joshi N, Balint BL, Ward FB, Elfick A, French CE. A pH-based biosensor for detection of arsenic in drinking water. Anal Bioanal Chem 2011; 400(4): 1031–1039.
68. Joe MH, Lee KH, Lim SY, et al. Pigment-based whole-cell biosensor system for cadmium detection using genetically engineered *Deinococcus radiodurans*. Bioprocess Biosyst Eng 2012; 35(1–2): 265–272.
69. Branco R., Cristóvão A, Morais PV. Highly sensitive, highly specific whole-cell bioreporters for the detection of chromate in environmental samples. PLoS One 2013; 8, e54005.
70. Ravikumar S, Ganesh I, Yoo IK, Hong SH. Construction of a bacterial biosensor for zinc and copper and its application to the development of multifunctional heavy metal adsorption bacteria. Process Biochem 2012; 47(5): 758–765.
71. Willardson BM, Wilkins JF, Rand, TA, et al. Development and testing of a bacterial biosensor for toluene-based environmental contaminants. Appl Environ Microbiol 1998; 64(3): 1006–1012.
72. Tecon R, Beggah S, Czechowska K, et al. Development of a multistrain bacterial bioreporter platform for the monitoring of hydrocarbon contaminants in marine environments. Environ Sci Technol 2009; 44(3): 1049–1055.

73. Shingler V, Moore T. Sensing of aromatic compounds by the DmpR transcriptional activator of phenol-catabolizing *Pseudomonas* sp. strain CF600. J Bacteriol 1994; 176(6): 1555–1560.
74. Sticher P, Jaspers MC, Stemmler K, *et al*. Development and characterization of a whole-cell bioluminescent sensor for bioavailable middle-chain alkanes in contaminated groundwater samples. Appl Environ Microbiol 1997; 63(10): 4053–4060.
75. Korpela MT, Kurittu JS, Karvinen JT, Karp MT. A recombinant *Escherichia coli* sensor strain for the detection of tetracyclines. Anal Chem 1998; 70(21): 4457–4462.
76. Yagi K. Applications of whole-cell bacterial sensors in biotechnology and environmental science. Appl Microbiol Biotechnol 2007; 73(6): 1251–1258.
77. Han L, Zhao Y, Cui S, Liang B. Redesigning of microbial cell surface and its application to whole-cell biocatalysis and biosensors. Appl Biochem Biotech 2018; 185(2): 396–418.
78. European Commission, 2005. At www.synbiosafe.eu/uploads///pdf/EU-highlevel-syntheticbiology.pdf. Last time acceded January 2018.
79. Kim HJ, Jeong H, Lee SJ. Synthetic biology for microbial heavy metal biosensors. Anal Bioanal Chem 2018; 410(4): 1191–1203.
80. Bedekar VN., Tantawi KH. MEMS sensors and actuators. In: Advanced Mechatronics and MEMS Devices II, Zhang D, Wei B. (Eds.). Springer, New York, 2017.
81. Song EY, Lee K. Understanding IEEE 1451-Networked smart transducer interface standard - What is a smart transducer? IEEE Instru Meas Mag 2008; 11(2): 11–17.
82. Frasco MF, Truta LA, Sales MG, Moreira FT. Imprinting technology in electrochemical biomimetic sensors. Sensors 2017; 17(2): 523.
83. Gutierrez Soto M, Adeli H. Recent advances in control algorithms for smart structures and machines. Expert Syst 2017; 34(2): e12205.
84. Ranky PG. Smart sensors. Sensor Rev 2002; 22: 312–318.
85. Powner ET, Yalcinkaya F. From basic sensors to intelligent sensors: definitions and examples. Sensor Rev 1995; 15(4): 19–22.
86. Boltryk PJ, Harris CJ, White NM. Intelligent sensors – a generic software approach. J Phys Conf Ser 2005; 15(15): 155–160.
87. Knopf GK. Smart biosensor functions-a machine learning perspective. In: Smart Biosensor Technology, 1st ed., Knopf, GK, Bassi, AS (Eds.). CRC Press, Boca Raton, FL, 2006, pp. 173–198.
88. Tang D, Huang D, Yang Z, Ji Q. Developmental trend of microfluidic chip and biosensor technologies and the integration mode with machine learning model and wearable device. Int J Biomed Eng Tech 2017; 23(2–4): 281–302.
89. Langley P. The changing science of machine learning. Mach Learn 2011; 82(3): 275–279.
90. Mendoza-Madrigal AG, Chanona-Perez JJ, Hernández-Sánchez, H., *et al*. Mechanical biosensors in biological and food area: a review. Revista Mexicana de Ingeniería Química 2013; 12: 205–225.
91. Kim J, Campbell AS, de Ávila BEF, Wang J. Wearable biosensors for healthcare monitoring. Nat Biotechnol 2019; 37(4): 389–406.
92. Zhang J, Williams SO, Wang H. Intelligent computing system based on pattern recognition and data mining algorithms. Sustain Comput-Infor 2017; 20: 192–202.
93. Liu Y, Zhao T, Ju W, Shi S. Materials discovery and design using machine learning. J Materiomics 2017; 3: 159–177.
94. Kapitanova K, Son SH. Machine Learning Basics. In: Intelligent Sensor Networks: the Integration of Sensor Networks, Signal Processing and Machine Learning, 1st ed., Fei, H., Qi, H. Eds., CRC Press, Boca Raton, FL, 2012.
95. Kumar DP, Amgoth T, Annavarapu CSR. Machine learning algorithms for wireless sensor networks: a survey. Inform Fusion 2019; 49: 1–25.
96. Qiu J, Wu Q, Ding G, Xu Y, Feng S. A survey of machine learning for big data processing. EURASIP J Adv Sig Pr 2016; 67.
97. Botta A, De Donato W, Persico V, Pescapé A. Integration of cloud computing and internet of things: a survey. Future Gener Comp Sy 2016; 56: 684–700.
98. Long F, Zhu A, Shi H. Recent advances in optical biosensors for environmental monitoring and early warning. Sensors 2013; 13(10): 13928–13948.
99. Pasolli E, Truong DT, Malik F, Waldron L, Segata N. Machine learning meta-analysis of large metagenomic datasets: tools and biological insights. PLoS Comput Biol 2016; 12: e1004977.
100. Zhang L, Tian F, Pei G. A novel sensor selection using pattern recognition in electronic nose. Measurement 2014; 54: 31–39.
101. Spiegel S, Gaebler J, Lommatzsch A, De Luca E, Albayrak, S. Pattern recognition and classification for multivariate time series. Proceedings of the Fifth International Workshop on Knowledge Discovery from Sensor Data 2011; 34–42.
102. Shacham M, Brauner N. Preventing oscillatory behavior in error control for ODEs. Comput Chem Eng 2008; 32(3): 409–419.
103. Precup RE, Preitl S, Petriu EM, *et al*. Generic two-degree-of-freedom linear and fuzzy controllers for integral processes. J Frankl Inst 2009; 346(10): 980–1003.
104. Cole WJ, Powell KM, Edgar TF. Optimization and advanced control of thermal energy storage systems. Rev Chem Eng 2012; 28:81–99.
105. Pirdashti M, Curteanu S, Kamangar MH, Hassim MH, Khatami MA. Artificial neural networks: applications in chemical engineering. Rev Chem Eng 2013; 24(9): 205–239.
106. Gonzalez-Navarro FF, Stilianova-Stoytcheva M, Renteria-Gutierrez L, Belanche-Muñoz LA, Flores-Rios BL, Ibarra-Esquer JE. Glucose oxidase biosensor modeling and predictors optimization by machine learning methods. Sensors 2016; 16(11): 1483–1496.
107. King ST, Sylvander M, Kheperu M, Racz L, Harper WF. Detecting recalcitrant organic chemicals in water with microbial fuel cells and artificial neural networks. Sci Total Environ 2014; 497–498: 527–533.
108. Feng Y, Barr W, Harper Jr. WF. Neural network processing of microbial fuel cell signals for the identification of chemicals present in water. J Environ Manage 2013; 120: 84–92.
109. Feng Y, Kayode O, Harper Jr. WF. Using microbial fuel cell output metrics and nonlinear modeling techniques for smart biosensing. Sci Total Environ 2013; 449: 223–228.
110. Gomolka Z, Twarog B, Zeslawska E, Lewicki A, Kwater T. Using artificial neural networks to solve the problem represented by BOD and DO indicators. Water 2018; 10(4): 1–26.
111. Riedel K, Kloos R, Uthemann R. Minutenschnelle Bestimmung des BSB. LWB. Wasser Boden Luft 1993; 11: 35–38.
112. Farré M, Brix R, Barceló D. Screening water for pollutants using biological techniques under European Union funding during the last 10 years. TrAC Trends Anal Chem 2005; 24(6): 532–545.
113. Horry H, Charrier T, Durand MJ, Vrignaud B, Picart P, Daniel P, Thouand, G. Technological conception of an optical biosensor with a disposable card for use with bioluminescent bacteria. Sens Actuators B Chem 2007; 122(2): 527–534.

14 FET-Based Biosensors (BioFETs)
Principle, Methods of Fabrication, Characteristics, and Applications

J. C. Dutta
Tezpur University, Napaam, Assam, India

CONTENTS

14.1 Introduction 283
14.2 Definition and Nomenclature for FET-Based Biosensors 284
14.3 Working Principle 284
14.4 Classification of FET-Based Biosensors 286
    14.4.1 ENFET 287
    14.4.2 ImmunoFET 288
    14.4.3 GENFET 288
    14.4.4 Cell-Based FET 289
    14.4.5 "Beetle/Chip" FET 289
14.5 Fabrication Technology 289
    14.5.1 Conventional ISFET and BioFET Fabrication Technology 289
    14.5.2 ISFET and BioFET Fabrication Using a Chemical Solution Process 289
14.6 Determination of the Characteristics of Biofets 291
    14.6.1 Sensitivity 291
    14.6.2 Drift and Hysteresis 292
    14.6.3 Selectivity 293
    14.6.4 Linearity 293
    14.6.5 Limit of Detection 293
    14.6.6 Stability 294
    14.6.7 Repeatability 294
    14.6.8 Reproducibility 294
14.7 Instrumentation 294
14.8 Applications 295
14.9 Conclusions 295
References 296

14.1 INTRODUCTION

Biorecognition materials are found to be very sensitive and specific, leading toward the development of biosensors having unique detection properties. In the last few decades, a vast amount of research has been carried out toward the development of biosensors. The first biosensor, developed by Clark in 1962, was an amperometric oxygen electrode biosensor [1]. This innovation gave a kick-start to the development of different types of biosensors found today and still under research. With the era of device miniaturization, the task of biosensor miniaturization threw a challenge for the researchers. With this challenge, a new type of potentiometric biosensor was emerged using the ion-sensitive field effect transistor (ISFET) as the basic structure. ISFET was first introduced by Bergveld in 1970 and considered the first miniaturized silicon-based chemical sensor [2]. Using ISFET as the fundamental block, the idea of ENFET (enzyme FET) was first introduced by Janata and Moss in 1976 [3]. It was the first ISFET-based biosensor reported. Taking this idea a step further, Caras and Janata developed the first ENFET biosensor in 1980 which was used for detection of penicillin [4].

The ENFET biosensor consists of an enzyme layer, that is, a biological layer placed on top of ISFET. Thus, ISFET-based biosensors are also termed biologically modified field-effect transistors (BioFETs). These FET-based biosensors have attracted great attention because of the widely used silicon-based FET as their basic structure. The silicon-based substrates helped in the miniaturization of devices. A lot more advantages came such as low weight, fast response, better fabrication scopes and packaging, high reliability, and low cost. Moreover, it offers a broad field of applications, including medicine, biotechnology, environmental monitoring, defense, and security, among others.

14.2 DEFINITION AND NOMENCLATURE FOR FET-BASED BIOSENSORS

A biosensor can be thought of a system consisting of two blocks connected in series along or integrated with other components such as an amplifier, processor, and display. The first block is a biochemical recognition system, and the second is a physicochemical transducer. The biochemical recognition system consists of bioreceptor, which can be either biological molecular species like enzymes, proteins, antigens, antibodies, etc., or living biological systems like cells, tissues, plants, etc. The main task of the biological recognition system is to convert the information that is received from the biochemical part, which is an analyte, into a physical or chemical or biochemical response signal. A specific molecular interaction occurs which thereby changes one or more physico-chemical parameters. As a result of these changes electrons, ions, gases, heat, light, etc., are produced. These quantities are then quantified (generally converted to an electrical signal) with the use of transducer, whose output is fed to the signal processor to give the desired result.

The biochemical interaction mechanism that occurs between the receptor and analyte may be considered to classify biosensors into two fundamental types: biocatalytic sensors and biocomplexing or bioaffinity sensors. Apart from these, hybrid biosensors have also been invented. Biosensors which work on the principle of electrochemical transducer are the most familiar sensor devices because of their simple measurement principle and on-chip integrated signal processing capability. As recommended by IUPAC [5], the official nomenclature of electrochemical transducers includes potentiometric, amperometric, impedimetric, conductometric, and semiconductor field-effect principles. In this chapter, the biosensors based on potentiometric and semiconductor field-effect principles are discussed.

14.3 WORKING PRINCIPLE

The working principles of BioFETs are directly related to that of ISFETs, as the BioFET is a structure with a biological layer on top of the ISFET surface. Again, the working principle of an ISFET can be best described by referring to a conventional IGFET (insulated-gate field effect transistor) or MOSFET (metal-oxide-semiconductor field-effect transistor), as ISFET is formed by replacing the metal gate of a MOSFET with a reference electrode, which is connected to the gate dielectric via an aqueous solution. An electrolyte solution and a reference electrode, along with a chemically sensitive membrane (gate surface), perform in the same way as the metal gate electrode of MOSFET does. The bare gate dielectric layer or thin film is used as the ion sensing layer.

The site-binding model shown in Figure 14.1 explains the response of the ISFET to pH. Yate and co-workers first explained the site-binding model in 1973 [6]. It explains how protonated and deprotonated sites are formed in the insulating surface according to the concentration of hydrogen ions in the

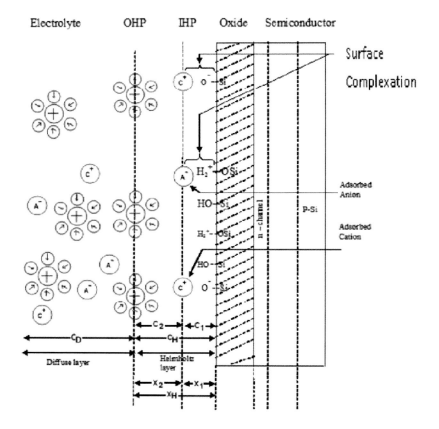

FIGURE 14.1 Site-binding model with electrical double layer (Stern layer and diffuse layer).

electrolyte. When hydroxyl groups bind to the hydrogen ions, they are called binding sites. For silicon dioxide, the binding sites formed are $Si-OH$ (neutral), $Si-O^-$ (negative), and $Si-OH_2^+$ (positive). The ionization reactions are:

$$Si-OH \leftrightarrow Si-O^- + H^+ \quad (14.1)$$

$$Si-OH + H^+ \leftrightarrow Si-OH_2^+ \quad (14.2)$$

where H^+ are the protons in the surface. Thus, the original neutral surface may be a positive site or negative site by accepting or donating protons from or to the electrolyte solution, respectively. Here, the acid and base equilibrium constants are:

$$K_a = \frac{[Si-OH][H^+]_s}{[Si-OH_2^+]} \quad (14.3)$$

$$K_b = \frac{[Si-O^-][H^+]_s}{[Si-OH]} \quad (14.4)$$

According to Equations (14.1) to (14.4), $Si-OH_2^+$, $Si-OH$, and $Si-O^-$ represent positive, neutral, and negative sites, respectively, and $[Si-OH_2^+]$, $[Si-OH]$, and $[Si-O^-]$ are the number of sites per surface area. The subscript "s" in $[H^+]_s$ signifies the concentration of proton near the surface of the insulator. If $[H^+]_b$ represents the proton concentration in the bulk electrolyte, then the relationship between $[H^+]_s$ and $[H^+]_b$ is given by the Boltzmann equation:

$$[H^+]_s = [H^+]_b * e^{\left(\frac{-q\psi_0}{kT}\right)} \quad (14.5)$$

where

ψ_0 = Electrolyte insulator surface potential

In fact, this surface potential is generated by the net surface charge σ_s, given by:

$$\sigma_s = q\left([Si-OH_2^+]-[Si-O^-]\right) \quad (14.6)$$

Knowing the total number of sites per unit area, that is,

$$N_s = [Si-OH]+[Si-O^-]+[Si-OH_2^+] \quad (14.7)$$

the expression for surface charge density σ_s can be determined in terms of K_a, K_b, $[H^+]_s$, and N_s.

Moreover, the active sites might react not only with the hydrogen ions but also with other ions present in the electrolyte. These electrolyte ions form ion pairs with oppositely charged surface sites or groups – a process known as surface complexation. The formation of surface complexes also readjusts the acid–base equilibrium and affects the surface charge by partly compensating the charged sites. Of course, the distribution of ions in the electrolyte solution can be well explained by using Gouy–Chapman–Stern theory. According to this theory, a double layer is formed in the electrolyte solution, namely the Stern inner layer and diffuse layer. The inner layer is made up of two planes: the inner Helmholtz plane (IHP) and outer Helmholtz plane (OHP). The IHP is the locus of the centers of the adsorbed ions, which forms pairs with charged surface sites. The OHP is the locus of the centers of the hydrated ions. The diffuse layer extends from the OHP to the bulk solution which contains nonspecifically absorbed ions. These ions behave as an ionic cloud and balanced by the uncompensated surface sites. The electrical double layer consists of two capacitors: the Helmholtz capacitance (C_H) and diffused capacitance (C_D).

An n-type ISFET structure is shown in Figure 14.2. It consists of an n-doped source and drain region on a p-type substrate with the two regions separated by a channel. The gate insulator, generally SiO_2, is deposited on the top of the channel. In some cases, one more layer is also deposited on the top of the SiO_2 layer for performance enhancement. Si_3N_4, Al_2O_3, and Ta_2O_5 are used as the upper layer. These serve as a sensitive layer for pH-sensitive ISFETs. The gate voltage V_G is applied through the Ag/AgCl reference electrode, which is also responsible for fixing the potential of the electrolyte

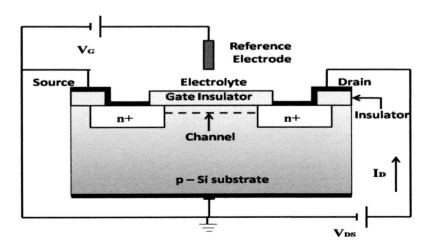

FIGURE 14.2 Structure of an n-channel ISFET.

solution. The reference electrode closes the gate to the source circuit via an electrolyte solution. An n-type inversion layer (i.e., channel) is created in the region between the gate and source at a sufficiently large positive gate voltage.

Since ISFET is analogous to MOSFET, the drain to source current of an ISFET (I_D or I_{DS}) can be deduced from that of a MOSFET and can be expressed as:

$$I_D = \mu C_i \left(\frac{W}{L}\right) V_{DS} \left[(V_{GS} - V_{TH(ISFET)}) - 0.5\, V_{DS}\right] \quad (14.8)$$

where C_i is the oxide capacitance per unit area, W and L are the channel width and length respectively, and μ is electron mobility in the channel.

Here the term $\mu C_i \left(\frac{W}{L}\right)$ is constant. In ISFET, V_{GS} and V_{DS} are kept constant through biasing; therefore, $V_{TH(ISFET)}$ is the only input parameter, that is, unlike MOSFET (in case of MOSFET, V_{GS} is the only input parameter), it is a threshold voltage-controlled device. The threshold voltage of ISFET is given by:

$$V_{TH(ISFET)} = (E_{ref} - \psi_0 + \chi_{sol} - (\Phi_{Si}/q) \\ - (Q_i + Q_{SS} + Q_B)/C_i + 2\Phi_f) \quad (14.9)$$

In this equation, E_{ref} is the potential of the reference electrode; Φ_{Si} is the silicon electron work function; q is the elementary charge; C_i is the capacitance of the gate insulator; Q_i, Q_{SS}, and Q_B are the charges located in the insulator, in the surface and interface states, and in the depletion region, respectively; χ_{sol} is the surface dipole potential of the solution, which is constant; Φ_f is the potential difference between the Fermi level of doped and intrinsic silicon; $2\Phi_f$ is the surface inversion voltage; and ψ_0 is the potential at the electrolyte-insulator interface that depends on the activity of ions in the analyte. According to the site-binding model, the pH dependence electrolyte-insulator interface potential is given by the following equation:

$$\psi_0 = 2.3 \left(\frac{kT}{q}\right)\left[\frac{\beta}{\beta+1}\right](pH_{pzc} - pH) \quad (14.10)$$

where pH_{pzc} is the pH value for which the oxide surface is electrically neutral, called the point of zero charge; k is the Boltzmann constant; T is the absolute temperature; and β determines the final sensitivity of the gate insulator and depends on surface hydroxyl groups and surface reactivity.

Moreover, based on the site-binding model, R.E.G. Van Hal [7] has developed a model. According to this model, the surface potential ψ_0 is expressed as

$$\Delta\psi_0 = -2.3\, \alpha\, \frac{kT}{q} \Delta pH_{bulk} \quad (14.11)$$

with

$$\alpha = \frac{1}{\left(2.3\, kTC_{diff} \Big/ q^2\beta\right) + 1} \quad (14.12)$$

where β is the buffer capacity of the oxide surface (i.e., the ability to take up or deliver protons); C_{diff} is the differential double-layer capacitance, and its value is determined from the ion concentration of the bulk solution; and α is a dimensionless sensitivity parameter, and its value varies between 0 and 1.

The surface buffer capacity plays an important role in determining the sensitivity of an oxide layer. The sensitivity of SiO_2 is much less compared to other oxides such as Si_3N_4, Al_2O_3, and Ta_2O_5. These oxide layers have higher surface buffer capacity β. Among the three, Ta_2O_5 has the highest buffer capacity and so has a pH sensitivity of 58 mV/pH over a pH range from 1 to 12 and independent of C_{diff}. High sensitivity can be achieved with high intrinsic buffer capacity oxides, which are found in oxides with high surface states.

The resulting pH-dependent surface charge of the gate insulator modulates the drain current, I_D, of the ISFET in accordance with the field effect concept. Therefore, the pH value of the solution can be determined quantitatively by measuring the changes in the drain current.

ISFET, being a sensing device, is very sensitive to any type of electrical interaction at or nearby the gate insulator/electrolyte interface. If the pH-dependent surface charge of the gate insulator or the interface potential can be brought into existence by coupling ISFET with a bioreceptor, then such a device is called a BioFET. Here, the pH variations have different causes, depending on the selection of bioreceptors. The layer containing the bioreceptor is chemically or electrostatically bound to the ISFET's surface. Thus, BioFET is fundamentally an ISFET whose surface charge variation in the gate insulator is caused either by analyte/bioreceptor interaction or directly by analyte/insulator binding. The choice of bioreceptors is done on the basis of the application of the device and classified likewise. Thus, in general, Equations (14.8) to (14.12) are also applicable to BioFET. However, it is essential to correlate the ISFET's output signal (i.e., I_D) with the analyte concentration or biological measurand, which can be done by performing experiments (by performing the linearity test).

14.4 CLASSIFICATION OF FET-BASED BIOSENSORS

Based on the biorecognition element used, BioFETs can be categorized into various types [8], as shown in Figure 14.3.

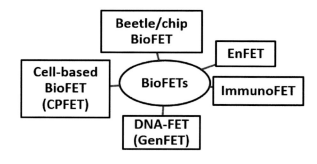

FIGURE 14.3 Classification of BioFETs.

FET-Based Biosensors (BioFETs)

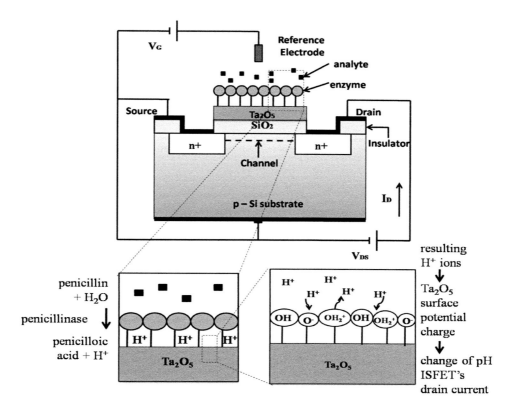

FIGURE 14.4 PENFET (penicillin-sensitive ENFET) structure and functional principle.

14.4.1 ENFET

ENFET is a bioelectronic device which is formed by incorporating an enzyme layer with an ISFET. The enzyme stimulates the biocatalytic processes, thereby altering the pH value at the ISFET's gate surface. This variation in pH occurs because of either consumption or generation of protons in accordance with the site-binding theory. The change in pH alters the potential at the surface of the ISFET and results in variation in current flowing through the channel. Hence, from a signal transduction point of view, an ENFET can be termed as a bioelectronic device. Its main purpose is to convert a biochemical or biological signal into an electrical signal. The structure of the first ENFET, which was a penicillin-sensitive biosensor [4], is shown in Figure 14.4.

In this BioFET, the hydrolysis of penicillin was used as the basic enzymatic reaction catalyzed by penicillinase enzyme. This resulted in the production of penicilloic acid as a product, which thereby became the cause of pH variation of the ISFET. Ta_2O_5 was used as the pH-sensitive membrane on top of the gate insulator. The enzyme penicillinase was immobilized on top of Ta_2O_5. The output directly varied with the concentration of penicillin in the given solution. In a similar way, various other enzymes can be utilized as bioreceptors for ENFET development. A few enzyme-analyte combinations, which are mostly used for the development of biosensors, are shown in Table 14.1.

TABLE 14.1
The Bioreceptor Enzymes and Analytes Used in ENFETs [8]

Sl. No.	Analyte	Enzymes Used as Bioreceptors
	Urea	Urease, penicillin G acylase, penicillin penicillinase
	Glucose	Glucose dehydrogenase, glucose oxidase/MnO_2 powder, glucose oxidase
	Sucrose	Invertase/glucose dehydrogenase, invertase/glucose oxidase/mutarotase
	Ethanol	Aldehyde dehydrogenase, alcohol dehydrogenase
	Maltose	Maltase, glucose dehydrogenase
	Lactose	β-Galactosidase, glucose dehydrogenase
	Ascorbic acid	Peroxidase
	Acetylcholine	Acetylcholinesterase
	Creatinine	Creatinine deiminase
	Formaldehyde	Alcohol oxidase
	Fluorine containing organophosphates	Organophosphorus acid anhydrolase
	Organophosphate compound (paraoxon)	Organophosphate hydrolase

14.4.2 ImmunoFET

The antibody-antigen (Ab-Ag) interaction shows the unique, specific, and highly sensitive recognition ability of biomolecules. An antibody is a complex biomolecule formed from a highly ordered amino acid sequence. The defensive mechanism of an organism is called the immune response, as a result of which antibodies are generated against the antigen (which can be any macromolecule). The recognition capability of antibodies is very high and specific for proteins with molecular weights higher than 5000 Daltons. Such proteins are called immunogenic. A small change in the chemical structure of an antigen can excessively reduce its attraction toward the specific antibody. Because of this feature, immunosensors can serve as an excellent tool for clinical diagnostics. When the antibodies are placed in a layer over the surface of FET, it is called ImmunoFET. Basically, direct label-free ImmunoFETs are used for immunoenzymatic surveys (indirect ImmunoFETs for immunoenzymatic surveys are less used). Schenck in 1978 gave the direct immunosensing concept by an ISFET [9]. Such an ImmunoFET is shown in Figure 14.5, where the gate is immobilized by antigens or antibodies (generally in a membrane). The antigen-antibody interaction changes the surface potential and, hence, the drain current.

14.4.3 GENFET

The development of genosensors and DNA chip technology has picked up pace with the Human Genome Project. Such sensors use genetic blocks (i.e., DNA and RNA) as biological recognition elements. Such devices enable simple, speedy, and cheaper nucleic acid sample analysis in real-time medical applications like the genetic disease diagnosis, infectious agent detection, screening of drugs, etc.

Mostly the DNA detection is done using a DNA hybridization process. Here, the target (i.e., unknown single-stranded DNA [ssDNA]) is detected by another molecule with which it forms a double-stranded helix structure called dsDNA. Many complementary and noncomplementary nucleic acids are present, but the probe molecule identifies the complementary one with high efficiency and specificity. This unique complementary nature, called base pairing, helps in the biorecognition process. A few examples of base pairs are adenine–thymine and cytosine–guanine.

A DNA-FET (also called GENFET) for hybridization detection is shown in Figure 14.6. A GENFET is obtained by immobilizing a membrane consisting of sequences of ssDNA onto a transducer (say ISFET) [10]. After hybridization, dsDNA is formed, which is detected by the ISFET. The miniaturization of these devices has picked up pace along with recent micro-fabrication technology. This can make them very promising tools for DNA diagnosis applications. Less work has been done on DNA sensing using ISFET.

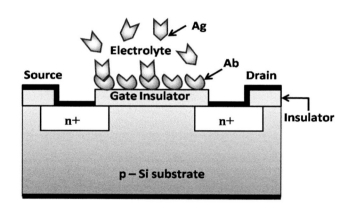

FIGURE 14.5 The schematic structure of an ImmunoFET.

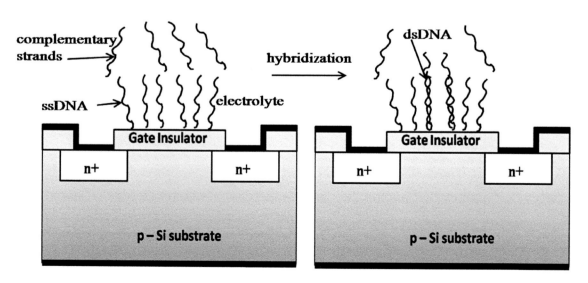

FIGURE 14.6 The structure and DNA hybridization principle of a DNA-FET.

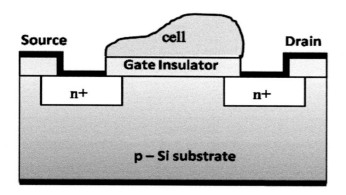

FIGURE 14.7 "Cell-transistor" hybrid.

14.4.4 Cell-Based FET

In cell-based BioFETs, the biological cells are used as the recognition element. If a single cell or a system of cells is directly coupled to the gate insulator of a FET, a "cell-transistor" hybrid is formed, as shown in Figure 14.7. The outputs of such sensors are highly informative and efficient from the clinical diagnostics point of view [11, 12] as these sensors can be used both for cell metabolism detection and extracellular potential measurement. A plethora of studies on the detection of toxic elements and compounds, pollutants, etc., that affect cell metabolism activities can be conducted using the cell-based BioFETs. Referring to extracellular potential measurement, the first attempt to apply an ISFET in neurophysiological measurements was made by P. Bergveld in 1970 [2].

14.4.5 "Beetle/Chip" FET

The idea of using the whole organism or a part of the organism (i.e., the sensory organs) as a biorecognition element was first forwarded by Rechnitz in 1986 [13]. Using his idea, a new type of bioelectronic sensor was developed. For instance, the specialty of insects is their sense of smell. Insects are very selective and sensitive toward sensing different odors. If the insect antenna is coupled with a FET device, a new biosensor is formed called the beetle chip FET sensor. Such a sensor was first realized in 1997 [14]. Here, the voltage generated in the antenna was used for driving a FET device. The voltage in the antenna changes when it detects some odor. This voltage alters the potential of the test solution and thereby varies the drain current of the FET sensor [15]. For this purpose, the FET device must be connected with the insect antenna, forming an interface. This is done in two possible ways, either by using the whole organism (say the whole beetle) and dipping the tip of the insect antenna in the electrolyte, as shown in Figure 14.8(a) with the reference electrode (say Pt wire) placed at an appropriate point on its body (generally between the neck and the head), or only using the isolated antenna placed on both sides of the plate with the electrolyte solution, as shown in Figure 14.8(b).

14.5 FABRICATION TECHNOLOGY

14.5.1 Conventional ISFET and BioFET Fabrication Technology

ISFET is formed by few modifications over MOSFET. If the metal gate of a MOSFET is replaced by an electrolyte solution and a reference electrode, we get ISFET. So, the conventional ISFET fabrication technology follows the p-channel or n-channel MOSFET fabrication steps, excluding gate metallization. For fabrication, silicon films on sapphire wafers (SOS) are taken as a substrate on which the gate SiO_2 film is thermally grown at about 1000°C. But unlike the MOSFETs, the selection of a gate dielectric coating for ISFETs is important, as protonation/deprotonation of this material is influenced by the pH of the electrolyte. To make the ISFET pH sensitive, another insulator layer is deposited on top of the SiO_2 layer [16]. Materials like Si_3N_4, Al_2O_3, and Ta_2O_5 (which are typically pH-sensitive materials) are used to fabricate this layer. The various methods used for fabrication of these layers are plasma enhanced chemical vapor deposition (PCVD), plasma anodic oxidation, evaporation by electron beam, sputtering, etc. The gate of the ISFET is now exposed to the electrolyte solution. The pH variations in the electrolyte serve as an input for the device.

In the case of BioFET, a biological sensing layer is deposited on the top of the pH-sensitive gate oxide layer of the ISFET. This layer is formed in such a way that after deposition, it does not move about in the electrolyte and retains its activity for a longer duration. There are many different techniques of achieving this, such as physical adsorption [17, 18], entrapment [19, 20] covalent binding [21], and crosslinking [22].

14.5.2 ISFET and BioFET Fabrication Using a Chemical Solution Process

The conventional technique used for ISFET fabrication requires a huge set of laboratory equipment, making the process complex and expensive. To minimize the problems of traditional fabrication technology, the chemical solution process can be used [23–25]. This process uses the electrochemical deposition (ECD) technique for deposition of different layers of ISFET from the bottom to top, offering numerous advantages such as easy detection, low cost, low-power requirements, inherent miniaturization, portability, and high degree of compatibility with advanced micromachining technologies

The setup for deposition is a three-electrode system, as shown in Figure 14.9, which consists of a counter electrode (CE), reference electrode (RE), and working electrode (WE). A voltage is applied between the WE and CE. The potential difference between the RE and WE is fixed to a specified value and monitored. The role of the RE is to act as a reference in measuring and controlling the working electrode's potential and at no point does it pass any current. The CE passes all the current needed to balance the current observed at the WE. The deposition occurs on the WE.

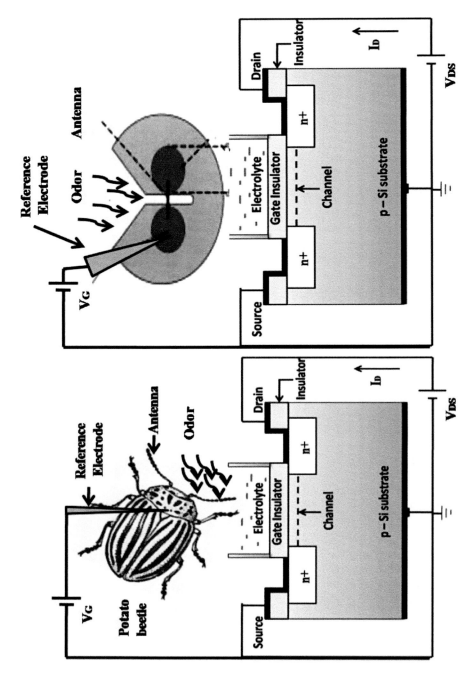

FIGURE 14.8 (a) BioFET consisting of the whole beetle and (b) BioFET formed from isolated antenna.

FET-Based Biosensors (BioFETs)

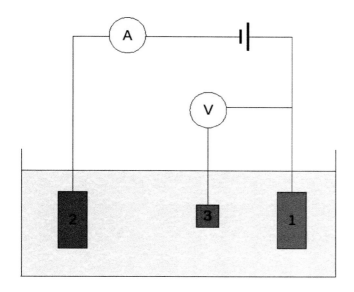

FIGURE 14.9 Basic setup for a three-electrode system (1) working electrode; (2) counter electrode; (3) reference electrode.

14.6 DETERMINATION OF THE CHARACTERISTICS OF BIOFETS

14.6.1 SENSITIVITY

Sensitivity is the measure of the smallest detectable change in the input. As discussed earlier, in the case of BioFETs the pH variations in the electrolyte serve as the input parameter. Therefore, theoretically, the sensitivity of a BioFET can be determined using the R.E.G. van Hal model [7] in terms of potential change due to pH change in the bulk of the electrolyte. The general expression of sensitivity obtained from this model is given by

$$\frac{\Delta \psi_0}{\Delta pH_{bulk}} = -2.3 \, \alpha \, \frac{kT}{q} \quad (14.13)$$

The Nernstian sensitivity of ~ 59.2 mV/pH is achieved at 298 K by the ISFET when $\alpha = 1$. The higher the sensitivity, the better the device performance.

Looking from the experimental point of view, the sensitivity can be determined from surface potential (ψ_0) versus pH curves as $\psi_o = f(pH)$. However, ψ_o cannot be directly experimentally determined because it is an interface potential. So, its experimental values can be determined as follows:

The threshold voltage of MOSFET is given by

$$V_{TH(MOSFET)} = \frac{\Phi_M}{q} - \frac{\Phi_{Si}}{q} - (Q_i + Q_{SS} + Q_B)/C_i + 2\Phi_f \quad (14.14)$$

where Φ_M is the work function of the gate metal and other terms are described in Equation (14.9). Now substituting the parameter $V_{TH(ISFET)}$ of Equation (14.9) with $V_{TH(BioFET)}$ and then comparing it with Equation (14.14) we get the following equation:

$$V_{TH(BioFET)} = V_{TH(MOSFET)} - \psi_0 + \text{constant} \quad (14.15)$$

It is important to note that $V_{TH(MOSFET)}$ is identical to the threshold voltage of BioFET fabricated by replacing the metal gate of MOSFET with a biological layer. Hence, during BioFET fabrication, an additional step, that is replacement of BioFET into a MOSFET, is required to analyze the behavior of the BioFET device following the gate dependency test of the MOSFET device. The value of the constant in Equation (14.15) is given by:

$$\text{Constant} = E_{ref} + \chi_{sol} - \frac{\Phi_M}{q} \quad (14.16)$$

Now from the transfer characteristic curve (V_{GS} vs. I_{DS} curve) of MOSFET, $V_{TH(MOSFET)}$ can be determined using the extrapolation in linear region (ELR) method. The ELR method is mostly used for threshold voltage extraction. It consists of finding the gate-voltage axis intercept (i.e., $I_{DS} = 0$) of the linear extrapolation of the $I_{DS} - V_{GS}$ curve at its maximum first derivative (slope) point (i.e., the point of maximum transconductance, g_m).

In a MOSFET device, the threshold voltage is constant (it is kept constant through a well-controlled fabrication process). Its transfer characteristics (V_{GS} vs. I_{DS}) are drawn, keeping V_{DS} at a constant value. Unlike MOSFET, BioFET is a pH- or concentration-dependent threshold voltage ($V_{TH(BioFET)}$) controlled device. It means that at different pH values, we get different characteristic curves. Its transfer characteristics can therefore be experimentally determined using the following method:

Keeping pH (and hence, concentration of analyte being measured) and V_{DS} constant, I_{DS} can be recorded by varying V_{GS} (i.e., V_{GS} vs. I_{DS} curve).

Let us now consider a BioFET for which the transfer characteristic curves have been plotted experimentally for pH 7, 6.5, and 5.75, keeping $V_{DS} = 0.2$ V (fixed at a value up to which linearity is observed in the output characteristic) (Figure 14.10). The variation of the BioFET's threshold voltage with a change in pH can be determined from these V_{GS} vs. I_{DS} curves using the ELR method.

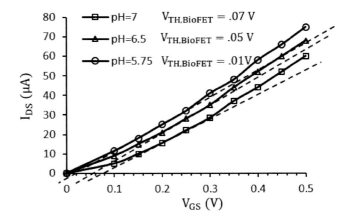

FIGURE 14.10 Transfer characteristics of the BioFET for different levels of pH showing different threshold voltages using ELR technique.

TABLE 14.2
Effect of pH on the Surface Potential of BioFET

pH	Experimental $V_{TH, BioFET}$ (in V)	Experimental ψ_0 (in V)
5.75	0.016	0.01
6	0.025	0.001
6.25	0.04	−0.014
6.5	0.055	−0.029
6.75	0.065	−0.039
7	0.077	−0.051

These experimentally obtained threshold voltages ($V_{TH, BioFET}$) at different values of pH are used in Equation (14.15) to determine the values of ψ_0, as shown in Table 14.2.

These experimentally obtained values can now be compared with the theoretical values, as shown in Figure 14.11.

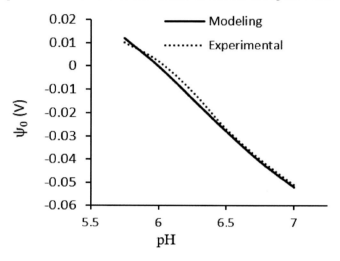

FIGURE 14.11 Surface potential vs. pH plot for determining sensitivity.

From the slope of these curves, the pH sensitivity can be determined theoretically, using either Equation (14.10) or (14.11), as well as experimentally. In this example, it is about 58 mV/pH from the theoretical model and 57.5 mV/pH from the experimental results.

14.6.2 DRIFT AND HYSTERESIS

Drift and hysteresis are important factors that influence the output accuracy of the pH-sensitive BioFETs. These factors are known as memory effects that limit the accuracy of pH response measurement of the device. BioFET is a pH-dependent, threshold voltage-controlled device; therefore, drift can be defined as the threshold voltage variation with time at constant pH. It is caused by a change in gate capacitance due to the penetration of ions contained in the electrolyte into the sensing membrane (insulating layer). It can therefore be measured in terms of threshold voltage shift given by:

$$\Delta V_{TH} = V_{TH}(t) - V_{TH}(0) \qquad (14.17)$$

For instance, Figure 14.12 shows the threshold voltage shift in pH 5.0, 7.0, and 9.0 buffer solutions after immersion of the device in buffer solution for 5 hours. It showed drift rates of 0.576 mV/h, 0.52 mV/h, and 0.317 mV/h in pH 5.0, 7.0, and 9.0 solutions and 0.734 mV/h, 0.683 mV/h, and 0.547 mV/h in pH 5.0, 7.0, and 9.0 solutions for HfO_2 and ZrO_2 gate-based BioFETs, respectively. The higher drift rate in the acidic solution could be due to easy transport of hydrogen ions into the buried surface sites through the sensing film.

Hysteresis is affected by the slow response of the pH-sensitive devices that can be interpreted through a surface site model. The hysteresis width is the voltage deviation when the device is measured many times at the same pH value. It can be measured in terms of the initial and final

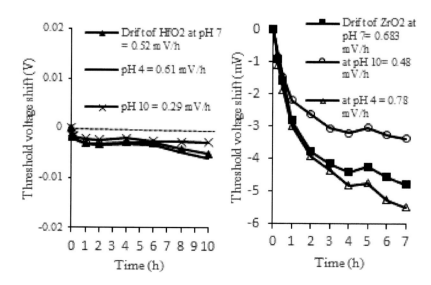

FIGURE 14.12 Drift characteristics of HfO_2 and ZrO_2 gate-based BioFETs.

FET-Based Biosensors (BioFETs)

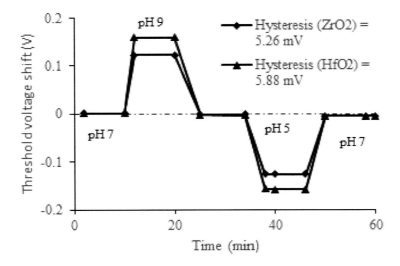

FIGURE 14.13 Hysteresis characteristics of the BioFETs measured in the loop of pH 7 → 9 → 7 → 5 → 7.

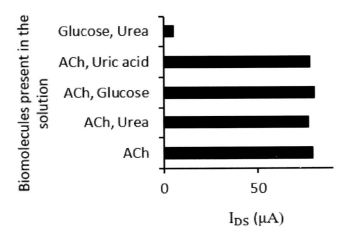

FIGURE 14.14 Interference of acetylcholine (ACh) with other biomolecules.

threshold voltage difference by subjecting the device to a pH cycle over a period. For instance, let us consider a pH cycle of pH 7 → 9 → 7 → 5 → 7 over a period of 60 minutes, as shown in Figure 14.13. The device showed hysteresis of 5.88 mV and 5.26 mV for HfO_2 and ZrO_2 gate-based BioFETs.

14.6.3 Selectivity

As discussed earlier, one of the most important features of the bioreceptors used in BioFETs is their selective nature. The device selectivity for the target molecule is evaluated experimentally under similar conditions by analyzing the interference signal caused by the other molecules present in the sample. For instance, the plot in Figure 14.14 shows the level of interference, which is ~ 0.4%, caused by the compounds such as uric acid, glucose, and urea present in the sample during detection of the target acetylcholine.

14.6.4 Linearity

The linearity test is carried out to find out the output range of the BioFET device. This test helps to fix the input voltage or any other input parameter to a particular value for performing other tests. For example, Figure 14.15 shows output current versus reference voltage plot for BioFET. This plot shows that the device behaves linearly up to 0.6 V and then saturates. So, the reference voltage can be fixed at 0.6 V for other measurements. Similarly, the linearity test can be performed for concentrations of analyte or any other biological substance under examination.

14.6.5 Limit of Detection

Limit of detection (LoD) is the lowest quantity of a substance that can be detected and it is an important parameter for all biosensors as discussed in Chapter 1. Mathematically, the LoD can be defined as:

$$\text{LoD} = \frac{3 \times \sigma}{S} \quad (14.18)$$

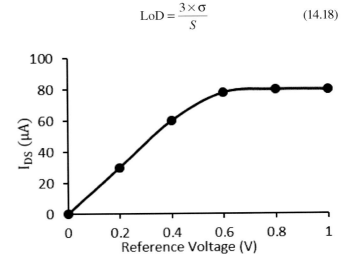

FIGURE 14.15 Linearity test.

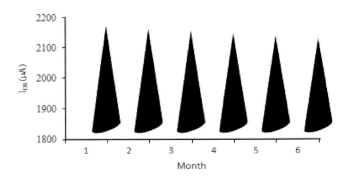

FIGURE 14.16 Stability test.

where σ is the standard deviation and S is the slope of the curve drawn between the substrate concentration and their responses. σ can be calculated using the following equation:

$$\delta = \sqrt{\frac{\sum (x-\bar{x})^2}{n-1}} \qquad (14.19)$$

where x and $\bar{x}$ are the sample value and average value of the samples, respectively, and n is the number of samples to be detected.

14.6.6 Stability

Stability deals with the degree to which sensor characteristics remain constant over time. Changes in stability occur on account of components aging, leading to the decrease in sensitivity. For instance, Figure 14.16 shows the pH stability test of a BioFET device. Here the degradation of the device has been tested for the same device after every 1 month up to 6 months, taking a solution of pH 7.0. The plot shows an average of 98% stability after 6 months storage. The plot has been drawn for V_{GS} V_s I_{DS} curves at pH 7 for every month and then taking I_{DS} as a parameter with respect to month. This can also be determined for other parameters to be measured.

14.6.7 Repeatability

Repeatability deals with how consistent a particular sensor is against itself. It can be used to describe the ability of a sensor to provide the same result, under the same environments, over and over again. For instance, Figure 14.17 shows the comparison of five tests performed on the same BioFET under similar conditions.

14.6.8 Reproducibility

Reproducibility is a characteristic of a device which tests whether a fabricated device can be mass-produced or not. In this test, at least five devices are developed using the same procedure under similar conditions, and their performance in terms of their output characteristics is measured. For instance, Figure 14.18 shows the comparison of five similar

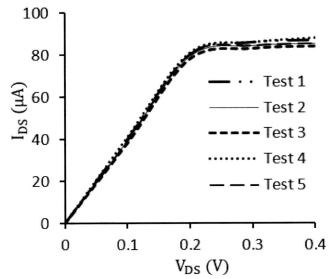

FIGURE 14.17 Repeatability test.

devices fabricated under similar conditions. The results show the reproducibility of the device.

14.7 INSTRUMENTATION

The major advantage of the use of FETs in the development of biosensors was miniaturization in order to integrate biosensors on a chip along with the signal processors. A measuring circuit is required in order to get the desired result of the BioFET sensor. For measuring the result, the basic circuit of a source and drain follower is used, as shown in Figure 14.19. The operational amplifier in the circuit helps in maintaining the zero-input difference voltage. All other voltages in this circuit are kept constant along with the input reference voltage of the BioFET. As a result of this, the changes in the oxide

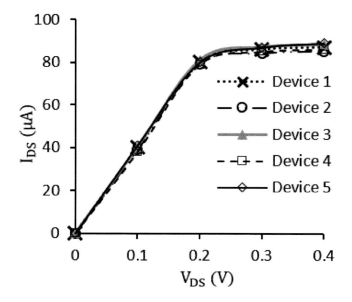

FIGURE 14.18 Reproducibility test.

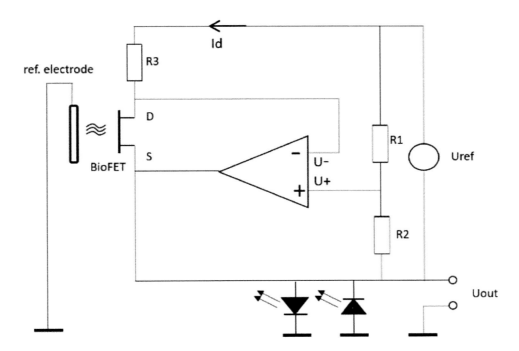

FIGURE 14.19 BioFET measurement circuit [26].

potential of the BioFET are very much accurately reflected in the output.

In practice, for exact measurement of the sample, a pH ISFET/BioFET differential arrangement can be employed [8].

14.8 APPLICATIONS

BioFETs have wide variety of applications in different fields such as medical, biotechnology, environmental, food quality, agriculture, industry, and defense. Different BioFETs have different applications depending on their ability to transduce the biochemical signals as discussed elsewhere

- *Medical:* BioFETs have played a dominant role in the medical field. For example, ENFETs are mostly used in the medical field for the detection of various biomolecules such as glucose, cholesterol, urea, creatinine, acetylcholine, etc. ImmunoFETs are excellent clinical diagnostic tools for quantifying how well the human immune system is functioning.
- *Biotechnology:* In this field, GenFET can be used for the sequence-selective detection of polymerase chain reaction (PCR) products, for diagnosis of genetic diseases, the detection of infectious agents, drug screening, etc. Currently about 400 diseases are diagnosable by molecular analysis of nucleic acids, and this number is increasing daily. Cell-based BioFETs can be used in cell biology and toxicology.
- *Environmental:* In this field, BioFETs can be used to monitor the presence of pesticides, herbicides, chemical fertilizers, and several water pollutants. They can serve as an early warning system to protect the environment.
- *Food quality:* In this field, BioFETs can be used to measure the protein contents of food, bacteria contaminating the foods, freshness of fish, etc.
- *Agriculture:* In this field, BioFETs can be used for measurement of the degree of ripeness of several products, odors, etc.
- *Industry:* BioFETs can be used in canning and fermentation industries.
- *Defense:* In this field, BioFETs can be used to monitor the presence of nerve gases in the environment.

14.9 CONCLUSIONS

Since the first invention of ISFET by Bergveld, various types of BioFETs have been developed using different biological recognition elements. The first BioFET that appeared was an ENFET for the measurement of penicillin. After this development, many efforts were made to create other BioFETs using bioreceptors such as immunological species through DNA molecules to single cells and even living organisms. While ENFETs are at a well-developed stage, most of the other BioFETs, especially GenFET and "beetle chip" FETs, are at their initial stages. Nevertheless, an effort has been made to give a fundamental concept of BioFETs through mathematical expressions and analysis. Also some important experiments that need to be performed during BioFET development have been discussed. Details of methodology used to deduce the mathematical expressions governing the characteristics of BioFETs such as Equations (14.8) to (14.15) have been discussed in many reference papers, books, and PhD theses. Readers can study these [27–29] to learn more.

REFERENCES

1. L. C. Clark, and C. Lyons, Electrode systems for continuous monitoring in cardiovascular surgery, Ann N Y Acad Sci., vol. 102, pp. 29–45, Oct 1962.
2. P. Bergveld, Development of an ion-sensitive solid-state device for neurophysiological measurements, IEEE Trans. Biomed. Eng., vol. BME-17, no. 1, pp. 70–71, January 1970.
3. J. Janata and S. Moss, Chemically sensitive field effect transistors, Biomed. Eng., vol. 6, pp 241–245, 1976.
4. S. Caras and J. Janata, Field effect transistor sensitive to penicillin, Anal. Chem., vol. 52, pp. 1935–1937, 1980.
5. D. R. Thevenot, K. Toth, R. A. Durst and G. S. Wilson, Electrochemical biosensors: recommended definitions and classification, Biosens. Bioelectron., vol. 16, pp. 121–131, 2001
6. D.E.Yates, S.Levine and T.W.Healy, Site binding model of the electrical double layer at the oxide/water interface J. Chem. Soc. Faraday Trans., vol. 70, pp. 1807–1819, 1973.
7. R. E. G. Van Hal, J. C. T. Eijkel and P. Bergveld. A novel description of ISFET sensitivity with the buffer capacity and double-layer capacitance as key parameters. Sens. Actuators B: Chem. vol. 24, nos. 1–3, pp. 201–205, 1995.
8. Michael J. Schoning and Arshak Poghossian, Recent advances in biologically sensitive field-effect transistors (BioFETs), Analyst, vol. 127, pp. 1137–1151, 2002.
9. J. F. Schenck, Technical difficulties remaining to the application of ISFET devices, In Theory, Design and Biomedical Applications of Solid State Chemical Sensors, ed. P. W. Cheung, CRC Press, Boca Raton, FL, pp. 165–173, 1978.
10. S. Schütz, B. Weissbecker, H. E. Hummel, M. J. Schöning, A. Riemer, P. Kordos and H. Lüth, Microscaled living bioelectronic systems – coupling beetles to silicon transducers, Naturwissenschaften, vol. 84, pp. 86–88, 1997.
11. P. Fromherz, A. Offenhäusser, T. Vetter and J. Weis, A neuronsilicon junction: A retzius cell of the leech on an insulated-gate fieldeffect transistor, Science, vol. 252, pp. 1290–1293, 1991.
12. M. Bove, S. Martinoia, M. Grattarola and D. Ricci, The neuron-microtransducer junction: linking equivalent circuit models to microscopic description Thin Solid Films, vol. 285, pp. 772–775, 1996.
13. S. Belli and G. Rechnitz, Prototype potentiometric biosensor using intact chemoreceptor structures Anal. Lett., vol. 19, pp. 403–405, 1986.
14. E. Souteyrand, J.P. Cloarec, J. R. Martin, C. wilson, I. Lawrence, S. Mikkelsen and M.F. lawrence, Direct detection of the hybridization of synthetic homo-oligomer DNA Sequences by field effect, J. Phys. Chem. B, vol. 101, pp. 2980–2985, 1997.
15. M. J. Schöning, S. Schütz, P. Schroth, B. Weissbecker, A. Steffen, P. Kordos, H. E. Hummel and H. Lüth, A BioFET on the basis of intact insect antennae, Sens. Actuators, B, vol. 47, pp. 235–238, 1998.
16. V. K. Khanna, Fabrication of ISFET microsensor by diffusion-based Al gate NMOS process and determination of its pH sensitivity from transfer characteristics. Indian J. Pure App. Phy., vol. 50, pp. 199–207, 2012.
17. M. A. Ali, P. R. Solanki, M. K. Patel, H. Dhayani, V. V. Agrawal, R. John, and B. D. Malhotra, A highly efficient microfluidic nano biochip based on nanostructured nickel oxide, Nanoscale, vol. 5, no. 7, pp. 2883–2891, 2013.
18. A. L. Sharma, R. Singhal, A. Kumar, Rajrsh, K. K. Pande, and B. D. Malhotra, Immobilization of glucose oxidase onto electrochemically prepared poly (aniline–co–fluoroaniline) films, J. Appl. Polym Sci., vol. 91, no. 6, pp. 3999–4006, 2004.
19. M. A. Gerard, A. Chaubey and B. D. Malhotra, Application of conducting polymers to biosensors, Biosens. Bioelectron., vol. 17, no. 5, pp. 345–359, 2002.
20. S. K. Sharma, Suman, C. S. Pundir, N. Sehgal, and A. Kumar, Galactose sensor based on galactose oxidase immobilized in polyvinyl formal, Sens. Actuators B: Chem., vol. 119, no. 1, no. 15–19, 2006.
21. M. D. Trevan, Enzyme immobilization by covalent bonding. Methods Mol. Biol., vol. 3, pp. 495–510, 1988.
22. S. Singh, P. R. Solanki, M. K. Pandey and B. D. Malhotra, Cholesterol biosensor based on cholesterol esterase, cholesterol oxidase and peroxidase immobilized onto conducting polyaniline films, Sens. Actuators B: Chem., vol. 115, no. 1, pp. 534–541, 2006.
23. M. A. Barik and J. C. Dutta, Fabrication and characterization of junctionless carbon nanotube field effect transistor for cholesterol detection, Appl. Phys. Lett., vol. 105, 053509 2014.
24. M. A. Barik, R. Deka and J. C. Dutta, Carbon nanotube-based dual-gated junctionless field-effect transistor for acetylcholine detection, IEEE Sens. J., vol. 16, no. 2, pp. 280–286, 2016
25. M. A. Barik, M. K. Sarma, C. R. Sarkar, and J. C. Dutta, Highly sensitive potassium-doped polypyrrole/carbon nanotube-based enzyme field effect transistor (ENFET) for cholesterol detection, Appl. Biochem. Biotechnol., vol. 174, no. 3, pp. 1104–1114, 2014.
26. P. Bergveld, Thirty years of ISFETOLOGY what happened in the past thirty years and what may happen in the next thirty years, Sens. Actuators B, vol. 88, pp. 1–20, 2003.
27. Massimo Grattarola and Giuseppe Massobrio, Bioelectronics Handbook: MOSFETs, Biosensors, and Neurons, McGraw Hill, New York, NY, 1998.
28. J. C. Dutta, Ion Sensitive Field Effect Transistor for Applications in Bioelectronic Sensors: A Research Review, IEEE National Conference on Computational Intelligence and Signal Processing (CISP), 185–191, 2012. DOI: 10.1109/NCCISP.2012.6189704
29. Purnima Kumari Sharma, Electrochemical modeling and validation of high-κ dielectric and nanomaterial based enzyme field effect transistors (ENFETs) for biomolecule detection, A thesis submitted in partial fulfillment of the requirements for award of the degree of Doctor of philosophy, www.tezu.ernet.in

Index

1-Acetylnapthalene 241
(3-Aminopropil) triethoxysilane 37
4-Cyano bi phenyl 241
2-Cyanofluorene 241
2,6-Dichloro-1,4-benzoquinone (DCBQ) 199
1,4-Dicyanobenzene 241
^{1}H NMR 115
4-Methoxy-(2-ethylhexoxy)-2,5-polyphenyl-enevinylene (MEH-PPV) 242
1,4-Naphthoquinone-2-sulfonate (NQS) 199
5-, 10-, 15-, 20-Tetra (4 pyridyl) porphyrin (TPyP) molecules 192
1-Thianthrenecarboxylic acid (1-THCOOH) 241
2-Thianthrenecarboxylic acid (2-THCOOH) 241
2,4,6-Trinitrotoluene (TNT) 132

A

Absorbance 12, 13, 14, 33, 40, 122, 149, 256, 258, 272
ABTS (2,2′-azino-bis-(3-ethyl-benzothiazoline-6-sulfonic acid)) 153, 155, 156, 158
Acceptor 15, 81, 167, 176, 188, 193, 203–205, 216–217, 235, 244–246,
Acetic acid 266
Acrylonitrile 269
Adenine 6, 9, 113, 221, 246, 288
Adenosine 15, 81, 128, 169, 178, 193, 255
Adsorption 18–19, 36, 178, 255, 267
Affinity Sensors 188
Aflatoxin 195
Agarose 19, 20, 30, 77
Aggregation 81, 98, 100, 130, 164, 169
Aggregation-induced emission (AIE) 128
Alcohol oxidase 33, 223, 266
Alcohol sensor 223
Algae 131, 196, 215, 258
Alginate 31, 34–35, 266, 269
Algorithms 22, 275–279
Alzheimer 126–127
Amino group 20, 153–154, 200
Aminonhexanethiol 203
Amperometric 2, 6, 9–10, 23, 186–190, 192–194, 221, 223, 258–259, 268–269
Amplex red 145, 150
Analyte 2–6, 21–22, 31, 130
Analytical tool 2, 220
Aniline 236
Annihilation 124, 167, 234–236, 247
Anodization 35, 37, 187
Anthracene 178, 235, 237, 241
Anthraquinone 193–194, 258
Antibacterial agents 133
Antibody 3, 5–7, 6, 31, 38, 57, 70, 100, 189, 247, 257, 288
Antifouling properties 30
Apoferritin 145, 159
Aptamer 38, 56, 70–85
Aptasensor 7, 11, 70, 80–82, 84–85, 245–248
ARAS BOD sensor 279
Arc discharge method 171
Arginine 266
Arsenic ions 131
Arsenite 224, 274
Artificial enzymes 133, 144
Artificial neural networks 22, 276–278, see ANN 276–277
Ascorbic Acid (AA) 95, 112, 133, 202, 287
Asparagine 117, 266
ATP 15, 169, 178
ATP synthase 197, 209
Autofluorescence 167, 170, 234
Avidin 20, 37
A-wire 199

B

Bacillus stearothermophilus 270
Bacteria 3, 6, 22, 38, 98, 101, 131, 168
Bacteriochlorophyll 197–198
Bacteriophages 255, 258
Bacteriopheophytin 197
BamH1 248
Band gap 109, 122
Bandgap energy 178, 184, 186, 189
Bayesian networks 276
Beers 257
Beetle FET 289, 295
Benzene 172, 268, 273
Benzoic anhydride (BA) 155
Benzophenone 241
Beverage 259, 261, 269
Bilirubin 9, 101, 129, 133, 147, 158
Bimetallic 48, 119,130, 156
Bimetallic nanoclusters 119
Bimolecular interaction 76, 101
Bio imaging 111
Bio labeling 111
Bioassays 220, 225, 234, 243, 245, 253
Biocatalyst 3, 181, 212–215, 224
Biochemical oxygen demand 12, 225, 267
Biocompatibility 22, 30, 43, 56, 64, 121, 133, 164, 253
Bioconjugate 243, 246–247
Bioelectrochemical reactions 223
Bioelectrochemiluminescence 233, 234
Bioelectrochemistry 211, 216
Bioelectrode 10, 198–200, 215, 218
BioFET 283–295
Biofilm 213–214, 217–219, 227
Biofuel cells 211–212
Bioinspired 31, 38, 268
Bioinspired coatings 268
Biointerface 48
Biolayer Interferometry 69, 80
Biological oxygen demand 265
Bioluminescence 15, 266, 271
Bioluminescent 266, 271–272, 279
Biomarkers 21–22, 85, 100–101, 125, 157, 176
Biomaterials 23–24, 56
Bio-MEMS 48
Biomolecules 61, 111–112, 125, 127, 218
Bionanomaterials 57
Bio-NEMS 48
Biophotovoltaics 196, 213
Biopolymers 6, 215, 218
Bioreceptor 3, 19–20, 31, 287, 293
Biorecognition 3, 5, 37–38
Bio-reduction 107, 114
Biosensing 4, 22, 33, 36, 80, 211

Bio-sensitive 53
Biosensor 2, 3, 21, 29, 36,
Biosynthesis 114
Bis-aniline 202
Bleaching 234, 255
Bleeding 31
BMPO(5-tertbutoxycarbonyl-5-methyl-1-pyrroline-N-oxide) 150
BOD 225, 270, 272, 277–279
Bohr radius 243
Bonding 60–62
Bottom-up 58, 111, 134, 165
Bovine serum albumin (BSA) 112, 125, 131
Brevibacterium ammoniagenes 269
BrPE (2-bromo-1-phenylethanone) 155
Brust-Schiffrin method 118
BSA 112, 125, 131
BSA stabilized gold nanoclusters 124, 130, 158
BTE 268
Buckminsterfullerene (C_{60}) 144–145, 158
Bulk metal 95, 109–110
Butler-Volmer equation 217
B-Wire 199

C

C. vulgaris 269
Cadmium 239, 242, 273–274, 170
Cadmium telluride (CdTe) 242
Calorimetry 17, 79, 234, 272
Cancer 38, 43, 73, 98, 111, 127, 138, 157, 170
Cancer cell detection 258
Cancer cells 43, 73, 114, 133, 170
Cancer therapy 111
Cantilever 34, 84
Capillary 56, 62, 71, 78, 245, 253
Capillary action 56, 253
Capillary electrophoresis 71, 78, 245
Capping agent 95, 98, 112, 114
Carbodiimide 20, 178, 200, 256
Carbohydrates 12, 36, 58, 98, 125, 165, 211
Carbon 11, 52, 155, 164, 172, 244
Carbon black 165, 257
Carbon dots 155, 165, 192
Carbon nanodots (CNDs) 164, 223
Carbon nanofibers (CNF) 52
Carbon nanomaterials 148, 164, 257
Carbon nanotube 52, 155, 171, 176, 244
Carbonaceous biochemical demand 270
Carbon-based 52, 155, 163, 238
Carbonization 165, 167
Carboxyl 19, 155, 200, 218, 255
Carboxylic groups 112, 116, 175, 190
Carcinoembryonic antigen (CEA) 100, 258, 189, 192
Catalase-like 144, 146, 148, 154–155
Catalyst 143, 148, 171, 211, 221, 247
Cation 168, 212, 237, 240
CCD cameras 271
Cell phone cameras 259
Cell SELEX 69, 73
Cell voltage 216
Cell-based FET 289
Cells 3, 7, 31, 34, 73, 78, 98, 101, 133, 211
Cellulose 12, 56, 62, 221, 252, 255
Cellulosic materials 30

297

Cephalosporin 270
CE-SELEX 71–72
Chalcogen 55
Charge recombination 189, 200
Charge transfer coefficient (α) 10
Chemical nose/ tongue 127
Chemical reduction 95, 98, 112, 114
Chemical vapour deposition 52, 57–59, 172
Chemiluminescence 167, 233
Chemisorption 58, 60
Chemometric 266
Chitin 30
Chitosan 30, 102, 219, 221
Chlorella vulgaris 269
Chloroauric acid 58, 112, 152
Chlorophyll A_0 201
Cholesterol 21, 34, 133, 157, 160, 223
Chromatography 34, 72, 78, 123, 251
Chromatography paper 253, 255
Chromium ions 131
Chromogenic materials 49
Chronoamperometry 11, 200, 203, 206, 258
Circular dichroism (CD) 78, 115, 123
Circulating tumor cells (CTC) 21, 38, 101
Citrate 98, 101, 113, 122, 152
Clark electrode 268
Clinical 2, 34, 43, 75, 170, 259
Clostridium butyricum 225, 266
CMOS 259, 272
CNT 40, 52, 171, 175–177, 218
CO_2 electrode 270
Cobalt 148, 154, 168, 187, 274
Cofactors 4, 129, 196
Collagen 19, 30, 34, 178, 265
Colloids 63, 97, 111, 123
Colorimetry 80, 149
Commercial 2, 25, 218, 234, 245, 254, 259
Competitive assay 100
Concanavalin A 101, 260
Conducting hydrogels 31
Conducting inks 252
Conduction band 109, 166, 179, 184
Conductive hydrogels 30, 54
Conductive polymers 40, 53, 207, 223
Conductivity 8, 30, 54, 95, 109, 174, 215, 238
Conductometric 7, 265, 269
Continuous-band 124
Copolymer 199
Copolymerization 219
Copolymerized 33
Copper ions 130
Copper nanoclusters 117
Co-reactant 124, 233, 236, 247
Core LH1 197
Core reaction center (RC) 196
Cortical neuron culture 144, 158
Cortisol 257
Co-sensitization 207
Cottrell equation 11
Covalent binding 19, 31, 289
Creatinine 21, 257, 287, 295
Crosslinking 20, 31, 192, 202
Crowding 248
Cryogelation 31
CTPO (3-carbamoyl-2, 2, 5, 5-tetramethy-3-pyrroline-1-yloxyl) 150, 152
Current 5, 8, 187, 216, 269, 289
Cyanide 131, 224
Cyanobacteria 196, 227, 258
Cyclic voltammetry (CV) 10, 177
Cyclodextrins 144

Cysteine 112, 125, 190
Cytochrome b_6f 199
Cytochrome bc1 Complex 197
Cytochrome c2 197
Cytochrome c6/plastocyanine 201
Cytochrome P450 34, 101
Cytokine 38
Cytometry 78, 85
Cytosine 113, 121, 126
Cytotoxicity 76, 95, 145–146, 170

D

DAB (3,3'-diaminobenzidine) 144, 150, 153
Data mining 274
de Feijter formula 13
Defense and security 283
Dendrimers 95, 112, 116, 120
Dengue virus 40
Dental plaque 266
Depletion 286, 235, 259
DEPMPO (5-diethoxyphosphoryl-5-methyl-1-pyrroline-N-oxide) 150
Deposition 38, 58, 172, 204
Deprotonated 116, 284
Detection 80, 82, 84, 100–101, 125, 129, 132, 157, 158, 176–177, 225, 258, 293
DHR (Dihydrorhodamine 123) 150
Di imides 242
Diagnosis 3, 38, 100, 129, 157, 168, 259
Diagnostic 6, 66, 85
Dialysis 76, 266
Dielectric function 92, 108
Differential pulse voltammetry (DPV) 11, 40, 123
Diketopyrrolopyrrole (dpp) 206–207
Dimethyl silicone sponge 269
Dimethylformamide (DMF) 119, 237
Dimyristoylphosphatidylcholine 199
Diphenyl anthracene (DPA) 235, 237
Diphenylanthracene-2-sulfonate (DPAS) 241
Dipole 17, 40, 92, 245, 286
Direct plasmon enhanced electrochemistry (DPEE) 101
Discrete dipole approximation (DDA) 92–95
Disposable 214, 228, 239, 253, 272
Dissolved oxygen 279
Dithiocarbamate 168, 248
DMPO(5,5,-dimethypyrroline N-oxide) 150–151
DNA biosensors 6, 31
DNA damage 272, 131
DNAzymes 56, 144, 246
Donor 9, 81, 167, 176, 188, 203, 216, 246
Dopamine (DA) 129
Dopant 30, 185, 244
Doping 30, 53–54, 175, 219, 244
Drift and Hysteresis 292
Drugs 34, 56, 100, 133, 152, 213, 258, 288

E

E. coli 24, 38, 101, 227, 266, 273
Early warning systems 34, 267, 280
E-bandages 56
EcoRI 248
EDC [1-ethyl-3-[3-dimethylaminopropyl] carbodiimide] – NHS [N-hydroxysuccinimide] 176–177, 200, 238
Elastomers 50
Electrical 3, 7, 9, 40, 52, 153, 184, 211, 218, 225
Electrical double layer model 180, 284–285
Electrocatalyst 257

Electrochemical 7, 32, 38, 52, 82, 100, 118, 179, 183, 215, 258
Electrochemical impedance spectroscopy, 11, 270
Electrochemically active 9, 40, 192, 214, 225
Electrochemiluminescence (ECL) 236–237, 239, 241–242, 245–248
Electrochromic 49
Electrochromism 49
Electro-decomposition 241
Electrodes 7, 9, 65, 183, 213, 218, 237–239
Electrogenic microorganisms 270
Electromagnetic 13, 78, 91
Electron beam (e-beam) 63
Electron flux 183, 196, 204
Electron mean free path 108–109
Electron spin resonance (ESR) 149–152, 154
Electron transfer kinetics 197–198, 203, 215, 225
Electron transfer 9, 54, 82, 145, 152, 197, 211
Electron transfer rate (k_{ET}) 203, 215–216
Electron-hole pair 175, 186, 194, 199
Electronic transitions 109, 123, 242
Electronics 24, 48, 65, 251
Electrophoresis 63, 71, 85
Electrophoretic mobility 71, 72, 77
Electropolymerization 35
Electro-reduction 114
Electrospinning 33, 132
Electrospray ionization mass spectroscopy (ESI MS) 112, 114–115, 123
Electrostatic 40, 58, 60, 97, 286
Electrosynthesis 202, 204, 215
Eley-Rideal mechanism 148–149, 152
ELISA 38, 100, 157, 247
Emission 15, 110, 234, 242
Emission energy 110, 166
Emission peak 14, 110, 115
Emit 14, 109, 122, 165, 234
Encapsulated 31, 51, 126, 145, 159
Endonucleases 128, 248
Energy levels 51, 63, 109, 122, 184, 235
Energy transfer 15, 81, 110, 167, 196, 245
Energy-deficient 235, 237
Energy-sufficient 235, 237
ENFET 287, 295
Entrapment 20, 256
Environmental 6, 34, 53, 130, 215, 260
Environmental monitoring 34, 261, 274, 283
Environmental risks 267
Enzymatic 11, 32, 117, 144, 186, 273, 252
Enzymatic biofuel cells 212
Enzymatic photoelectrochemical sensor 186
Enzyme 5, 18, 24, 133, 144, 148, 196, 212, 221, 247, 283, 287
Enzyme electrode 9, 196, 199, 221, 268
Enzyme mimics 133, 144, 152
Enzyme-like activity 144, 145, 152–155
Enzymes 5, 24, 31, 37, 128, 144, 212
E-Route 235
Escherichia coli 21, 31, 73, 157, 168, 213
E-Skins 56
ESR oximetry 150, 152
Etching 60–61, 114, 184, 220, 225, 243
Ethanol 65, 111, 172, 190, 228, 257, 287
Ethanol detection 227, 257
Ethylbenzene 268
Excimer 235, 248
Exciplex 235
Excitation 13, 101, 124, 126, 150, 170, 245
Excitation peak 110
Exciton 175, 186, 196, 243 Exonucleases 248
External quantum efficiency 197

Index

F

Fabrication 35, 60, 91, 189, 218, 254, 283, 289
Fabrication technology 289
FADH 9, 187, 202
Faraday method 111
Faradic 270
Femtomolar 11, 177, 240
Fenton reaction mechanism 148
Fermentation 21
Fermi energy level 165, 185
Fermi wavelength 108–109
Ferredoxin 201, 204, 215
Ferricyanide 9, 41, 205, 214, 228, 255, 266, 269, 279
Ferrocene 2, 9, 83, 101, 172, 215, 246
FeS_X, FeS_A/FeS_B 201
FET 283, 288–289
Fibrinogen 34
Field effect transistor 8, 64, 82, 180, 284
Filter paper 56, 221, 252, 254, 257
Fingerprinting 279
Flexographic printing 255
Flexography 62–63, 252
Flow cytometry 78, 85
Flow injection analysis (FIA) 245
FluMag-SELEX 72
Fluorene 241–242
Fluorescein 132, 150, 192
Fluorescein isothiocyanate 132
Fluorescence 12, 15, 78, 80, 85, 122, 124, 133, 167, 196, 248
Fluorescence correlation spectroscopy (FCS) 15
Fluorescence emission 110, 120, 132, 165, 175, 180, 196, 272
Fluorescence lifetime 15, 124, 168–169
Fluorescence lifetime imaging (FLIM) 15
Fluorescence quenching 125
Fluorescence spectroscopy 121–122, 168, 272
Fluorine-doped tin oxides (FTO) 187, 195–197, 238–239
Fluorophore 14, 80, 167, 176, 194, 272
Food chemistry 279
Food security 261
Foodborne pathogens 21, 85, 261
Formaldehyde detection 225, 227
Forster distance 237
Förster resonance energy transfer (FRET) 15, 81, 167, 176, 180, 196, 229, 245
Fourier-transform infrared spectroscopy 123
Free energy 186, 216, 224, 235
Fresnel diffraction 64
FRET 15, 81, 167, 176, 180, 196, 229, 245
Fullerene 144, 155, 191, 201
Functional materials 48, 57, 268
Functionalization 36, 54, 85, 102, 111, 125, 166, 174, 192, 205, 218, 243

G

Gas phase synthesis 111
Gas sensitive 53
Gasochromic 49
Gelatin 19, 270
GEMs 279
Genetically engineered 7, 24, 198, 272
Genetically engineered microorganisms 267
GENFET 288, 295
Genosensors 288, 258
Genotoxicity 95, 279
Geometry 36, 93, 102, 221

Germanium crystals 243
GFP 272–274
Gibbs free energy 10, 186, 216, 225
Glassy carbon electrodes (GCE) 11, 196, 238
Gluconobacter suboxydans 266
Glucose 2, 21, 34, 129, 144, 152, 221, 259, 287, 295
Glucose oxidase 2, 6, 132, 144, 170, 224, 257, 287
Glutamine 266
Glutathione 72, 112, 125, 145, 190
Glycoproteins 3, 6, 36, 101, 128
Gold colloids 111
Gold nanoclusters 111
GOx 2, 6, 132, 144, 170, 224, 257, 287, *see* Glucose oxidase
Gram-positive and gram-negative bacteria 103, 133
Grana 196
Graphene 11, 55, 101, 155, 179–180, 215, 257
Graphene oxide 11, 32, 101, 134, 179, 244, 257
Graphene quantum dots (GQDs) 164, 192
Graphite 35, 50, 165, 171, 218
Graphite paper 252
Graphite pens 258
Green energy technology 12
Green fluorescent protein 272–274, *see* GFP
Green synthesis 98, 112, 121
G-rich DNA 125, 127
Growth of clusters 112
Guanine 6, 113, 176, 288

H

Haber-Weiss reaction mechanism 147
Hairpin DNAs 113
Haloaromatic compounds 268
Healthcare 27, 67, 95, 104, 156, 282
Heating-assisted 113
Heavy metal 30, 102, 129, 136, 164, 170, 260, 274
Heavy-metal ions 27, 102, 129, 260, 269
HeLa cells 114, 133, 136
Heme 25, 101, 145, 146, 187, 215
Hemin 126, 144, 157, 161, 196, 236, 246, 250
Hemoglobin 120, 169, 181, 258
Hepatitis B virus (HBV) 18, 126
Hepatitis C virus (HCV) 18, 84, 258
Herbicides 208, 213, 266, 295
Hexagonal full-shell 110
High resolution TEM 123
High-performance liquid chromatography (HPLC) 69, 78, 87, 123, 240, 241, 245
Histidine 19, 21, 112, 117, 130, 142
Holes 50, 122, 166, 167, 176, 180, 184–187, 191, 193, 203
Hollow nanosphere 96
HOMO 122, 123, 124, 184, 241, 242
Homogenous 105, 168, 236
Horseradish peroxidase 6, 15, 113, 146, 187
HRP, *see* Horseradish peroxidase
HSA stabilized gold nanoclusters 158
Human cells 34, 159
Human chorionic gonadotropin 242, 259, 263
Human IgG 43
Human immunodeficiency virus 126
Human papillomavirus 258, 263
Human saliva 257, 262
Human serum 11, 134, 169, 259
Human serum albumin 113, 135, 158, 160, 161
Humidity sensitive 53
Hyaluronic acid 30, 43
Hybrid 15, 24, 38, 47–49, 51, 56–58, 64, 66, 101, 136, 157, 194, 205, 284, 289
Hybrid orbitals 244

Hybridization 6, 31, 38, 52, 56, 101, 127, 169, 179, 188, 288
Hydrazine (N_2H_4) 95, 236, 240
Hydrogels 29–35, 43, 44, 47, 53–56, 67
Hydrogen bonding 48, 60, 131, 155, 255
Hydrogen peroxide (H_2O_2) 17, 108, 132, 145, 146, 178, 187, 236, 242, 259
Hydrogen sulfide 270, 281
Hydrogenase 204, 206, 210, 215
Hydrophilic 20, 29–31, 53, 56, 62, 76, 133, 170, 179, 202, 221, 255
Hydrophilicity 36, 62, 197, 253
Hydrophobic 18, 53, 56, 62, 76, 144, 169, 198, 215, 218, 221, 255, 258
Hydrophobicity 31, 36
Hydroxysuccinimide 37, 200, 256

I

IDE 269
iGEM 274
Immobilization 1–3, 7, 9, 18–20, 22, 31, 36–40, 48, 56, 58, 72, 85, 203, 216, 219–221, 229, 245, 251, 255–260, 265, 267, 268, 272, 296
Immobilized bioreceptor 18, 20, 30, 214
Immunoassay 22, 100, 101, 153, 157, 192
Immunochromatographic test strip 101
ImmunoFET 283, 288
Immunosensors 1, 6, 21, 44, 257, 259
Impedance 11, 36, 38, 40, 82, 83, 258, 270
Impedance analysis, 268
Impedimetric detection 34, 270
Implantable 1, 22–24, 30, 31, 43, 56, 60, 64
Implantable biosensors 1, 23, 31, 43
In vivo SELEX 69, 73, 74, 86
Incubation 35, 42, 71, 72, 112, 113, 225, 268
Indium-doped tin oxides (ITO) 238
Industrial 49, 54, 132, 168, 225
Influenza A virus 86, 125
Inkjet printing 221, 255
Instrumentation 2, 252, 270, 279, 283, 294
Insulates 237
Insulin 113, 121, 177, 182, 259
Intelligent biosensors 275, 276
Intelligent devices 274
Inter-band transitions 110, 123
Interdigitated electrodes 269
Interdisciplinary 2, 48, 66, 164
Interferometric reflectance spectroscopy 38
Internal quantum efficiency 198, 209
Intraband (sp←sp) 108, 123, 124, 134
Invertase 2, 41, 287
Ion sensitive 2, 53
Ionic liquids (IL) 47, 55, 67
Iridium 233, 239, 241, 242, 249
Iron-oxide magnetic nanoparticles 148
Irradiation 92, 111, 116, 117, 165, 195, 199
ISFET 2, 8, 283–289, 291, 295, 296
Isothermal titration calorimetry 69, 88
I-V curves 198

J

Jellium model 107, 109, 110, 134
Joule effect 50

K

Kalman-Bucy filter 279
kidney cells 272

Kinetically stable 111
Kinetics 15, 49, 76–78, 197, 198, 215, 217, 222, 225, 238
Kits 234, 259, 279

L

Lab-on-a-chip 13, 43, 44, 231
β-Lactam 270
Lactate oxidase 32, 223
Langmuir-Blodgett film 198
Laser ablation method 163, 171
Lateral flow 157, 253, 254, 255, 262, 263
Lattice structure 154, 184, 185
Layer-by-layer assembly 60
Leaching 9, 18, 20, 203, 215, 234, 256
Lead ions 130
Leukemia cell 258, 260
Library 70–75
Ligand 7, 15, 31, 37, 38, 41, 70, 72, 78–81, 107, 108, 111–125, 127, 134
Ligands to the metal cluster core charge transfer (LMCCCT) 124
Light harvesting complex 198
Light matter interaction 91, 92
Light scattering 31, 33
Limit of detection (LOD) 3, 7, 38, 99, 100, 127, 222, 269, 283, 293
Limit of quantification (LOQ) 3
Linearity 1, 4, 222, 269, 283, 286, 291, 293
Lipopolysaccharide (LPS) 178
Liquid electrolysis method 163, 172
Liquid metal 47, 55, 63
Liquid phase synthesis 107, 111
Listeria monocytogenes 21, 43
Lithography 33, 60, 61, 63, 220
Living biosensors 31, 34, 35
Localized surface plasmon resonance (LSPR) 13, 51, 91
Locked nucleic acid 7, 76
Logarithmic response 270
Loop-mediated isothermal amplification 258
Low-cost 8, 32, 33, 43, 52, 60, 62, 119, 177, 248, 256
LSPR based sensors 91, 98
Lucigenin 233, 241, 242
Luminescence 12–16, 109, 110, 119–124, 167, 234, 237, 244, 271
Luminescent Materials 47, 51
Luminol 100, 233, 242, 244–246
Luminometer 279
Luminophore 233, 235–239, 241–245, 247
LUMO 122–124, 184, 241, 242, 245
Lux 258, 271, 274, 279
Lux genes 271, 279
Lyophilized 254, 261, 269
Lysogenic 266

M

Machine learning 22, 265, 274–277, 279, 282
Magic number 107, 109, 110
Magnetic field 50, 74, 219, 233, 237
Magnetic gels 50
Magnetic materials 47, 50
Magnetic nanoparticles 144, 148, 153, 219, 259
Magnetoresistors 50
Magnetorheological 50
Magnetostriction 50
Magnetostrictive materials 50
Marcus theory 216
Marcus-Hush-Chidsey equations 216
Mass spectroscopy (MS) 108, 119, 121, 122
Material science 22, 48, 221, 228
Matrix-assisted laser desorption ionization mass spectrometry (MALDI-MS) 112, 113, 119, 123
Mechanism 6, 51, 53, 112–115, 121–134, 143, 144, 146–149, 152–155, 176, 211, 213–215, 227, 233–237, 244–246
Mediators 57, 180, 197, 199–205, 214–217, 227, 269
MEMS 22, 48, 51, 61
Mercaptoacetic acid (MAA) 119
Mercaptohexanoic acid 203
Mercaptohexanol 203
Mercury ions 129
Mesoporous/microporous electrode 199, 207
Metabolic loss 217, 225
Metal aggregates 111
Metal center 111
Metal colloids 111
Metal complex 167, 194
Metal nanoclusters 107–109, 111, 119, 120, 123, 124, 132
Metal nanoparticles 51, 91, 95, 99, 108–111, 134, 148, 150, 152
Metal oxides 47, 54, 60, 184, 189
Metal precursor 98, 107, 111, 114, 117, 118
Metal-ligand complex 111
Metallophilic 129, 158
Metal-metal bonds 110
Methacrylamide 33
Methacrylate 30, 53, 98
Methyl parathion 272
Methyl viologen 203–205, 215
MFC, *see* microbial fuel cells
Microdox 279
Microarray 34
Microbial biosensors 34, 265–279
Microbial consortium 225, 266, 269
Microbial fuel cells 213, 226, 255, 265, 270
Microchannel cantilever 84
Microcystin 192
Microelectrodes 30, 270
Microfabrication 30, 47, 60–63, 220, 225
Microfluidic 33, 60–62, 72, 220, 221, 251–254, 258
Microgel 116
Microgravimetric transducers 34
Micromachining 60, 61, 220, 289
Microorganism detection 163, 177
Microplate reader 272
MicroRNA (miRNA) 126, 127, 248
Microscale thermophoresis 69, 79
Microwave-assisted 113, 118
Miniature 48
Miniaturized 22, 32, 48, 211, 212, 219, 223, 283
Mismatched DNA 126
Modification 30, 36, 47, 57–60, 69, 75, 174–178, 233, 242, 248, 255–258
Molecular orbital theory 184
Molecular self-assembly 47, 60
Monolayer 20, 37, 57, 63, 176, 177–179, 198, 202, 203, 245
Monolayer protected clusters (MPCs) 122
Moraxella sp. 269
Morpholino-DNA complex 38
MOSFET 284, 286, 289, 291, 296
Motif 74, 75, 196
Multiple signal sensor (MSS) 246
Multiplex detection 99, 133
Multiplexed 33, 43, 99, 126, 258, 259, 271
Multiwall carbon nanotube (MWCNT) 52, 259
Myocardial infarction 234

N

N,N,N′,N′-tetramethyl-p-phenylenediamine (TMPD) 235, 237
NADPH 168, 199, 201, 204
Nafion 53, 203, 224, 225, 244
Nanocage 97
Nanoceria 145, 154, 155, 157
Nanochannels 38
Nanocluster beacon (NCB) 125
Nanoclusters 97, 107, 111, 119, 121–124, 133
Nanocomposite 131, 133, 194, 257
Nanofabrication 9, 48, 60, 62
Nanoflower 132, 159
Nanomaterials 22, 47, 52, 54, 58, 118, 133, 145–147, 218, 233, 242, 251, 256
Nanoparticle 22, 58, 91, 92, 93, 95, 96, 108, 110, 122, 143, 152, 153–155
Nanoparticle geometry 91
Nanoplate 94, 95
Nanorod 95, 96
Nanosphere 96
Nanostar 95, 97, 101
Nanostructured materials 35
Nanotechnology 22, 265
Nanotubes 52, 155, 163, 172, 174, 218, 233, 238, 244
Nanowires 65, 66, 187, 192, 214, 219
Nanozyme 133, 143, 144, 147, 148, 153, 155, 156
Naphthalene 172, 266, 273
Natural hydrogels 30
Near infrared (NIR) 111
NEMS 48
Nernst equation 8
Neurotoxic organophosphate 269
Next generation sequencing (NGS) 74
Nitroaromatic explosives 108, 132
Nitrocellulose 56, 58, 69, 71, 252
Nonfaradic 270
Nonradiative 97, 124
N-type and p-type 54, 185
Nucleation 35, 95, 112
Nucleic acid 1, 5, 6, 7, 31, 38, 49, 78, 91, 98, 108, 125, 143, 157, 213, 233, 248, 251, 258
Nucleic acid sensing 143, 157
Nucleobases 113

O

ODA (o-dianisidine) 150, 156
OH · (hydroxyl radical) 146, 147, 150, 154
Oligonucleotide 6, 7, 47, 48, 56, 70, 108, 113, 114, 121
Onion bulb scales 272
OPD (o-phenylenediamine) 144, 145, 150, 153
Open circuit potential 205, 216
Optoelectronics 55
Organic molecules 43, 196, 242
Organic polymers 50, 233, 242
Organic semiconductors 47, 52
Organic thin film transistors (OTFT) 47, 52
Organophosphorus hydrolase 269
Organosilanes 37
Origami paper 256
Osmium 200, 201, 203, 205, 239
Osmium bipyridine complex 203
Overpotentials 217
Oxidant 167, 240
Oxidase-like 134, 144, 145, 148, 149, 155, 157, 158
Oxidation 145, 146, 150, 233, 236
Oxidative stress 145, 146

Index

Oxidoreductase 215, 144, 156
Oxygen evolution complex 196

P

P. putida 269, 273
μPAD 169, 221–222, 253–255, 259–260
PANI 20, 31, 66, 219, *see* Polyaniline
Paper-based analytical devices 251, 254
Pathogen detection 34, 101
Pathogens 18, 21, 38, 49, 100, 177, 261
Pattern 60, 61, 63, 265, 277
Pattern recognition 265, 275, 277
PCR 70, 85, 295
PDMS 63, 65, 220
PEDOT 31
PEG 30, 31, 33, 34, 165
Pentachlorophenol 189
Pentafluorophenyldimethylchlorosilane 37
Pepsin 113, 130
Peptides 47, 56, 107, 112, 117
Peripheral LH2 197
Peroxidase mimic 143, 152–155
Peroxidase-like 144, 147, 152–155, 157
Peroxidase-like activity 129, 134, 144, 147, 153, 155, 157
Persulfate ($S_2O_8^{2-}$) 236
Perylene 233, 236, 241
Perylene dicarboxylic imide (PI) 242
Perylene tetracarboxylic di imide (PDI) 242
Pesticides 8, 16, 18, 24, 261, 268, 269
pH electrode 7, 8, 266, 270
pH meter 132
pH sensitive 53, 287, 289, 292
Phages 255
Phenol 269, 270, 273
Phenyl hydrazine (PH) 155
Phenylalanine 57, 257
Pheophytin 199
Photocatalysis 165, 189
Photocalorimetry 272
Photochromic 49
Photochronoamperometry 187, 198, 200
Photocorrosion 188, 192
Photo-crosslinking 30
Photoelectrochemical biosensing 183, 185
Photoelectrochemistry 183, 184
Photoelectrode 185, 186, 188, 191, 195
Photoinduced electron transfer (PET) 167
Photolithography 60–63
Photoluminescence 15, 43, 119, 122, 124
Photopolymerization 20, 32
Photo-reduction 114
Photostability 81, 108, 111, 116, 125, 165, 170, 178
Photosynthesis 183, 196
Photosynthetic reaction centers 183
Photosystem 680 196
Photosystem 720 196
Photosystem 870 196
Phylloquinone A_1 201
Physicochemical 3, 49, 65, 167, 284,
Physisorption 58, 60
Piezoelectric materials 47, 49, 50
Ping-pong mechanism 143, 148, 155
Plasmon absorption 108, 115
Plasmon resonance 2, 13, 40, 51, 69, 78, 91, 92, 109, 194
Plasmonic materials 47, 51
Plasmonics 92
Plastoquinone 199
Platform 29, 47, 48, 51, 57, 69, 80, 211, 218, 265, 274
p-Nitrophenol 228, 272, 196

POBN(α-(4-pyridyl-1-oxide)-N-tert-butylnitrone) 150
Point-of-care (PoC) 176, 251
Point-of-need (PoN) 12, 176, 251, 279
Polarization curve 216, 222
Poly adenine 113
Poly(amidoamine) (PAMAM) 112, 116
Poly [dimethyldiallylammonium] chloride/poly[4-styrenesulfonate] 199
Poly(9,9-dioctylfluorene) 242
Poly(3-hexyl-thiophene) 242
Poly L-arginine 256
Poly-L-lysine 37
Poly lysine benzoquinone (PBQ) 204
Poly(methacrylic acid) (PMAA) 116, 121
Poly(N-isopropylacrylamide) 53, 54
Poly[triethyleneoxy-p-(diethoxy) phenylenevinylene] 242
Poly-(vinyl-9,10-diphenylanthracene) 242
Poly[vinyl]imidazole Os-[bipy]$_2$Cl 203
Polyacrylamide 30, 53
Polyacrylamide gel 19
Polyacrylamide gel electrophoresis (PAGE) 85, 123
Polyacrylate-based compounds 30
Polyacrylic acid 30
Polyamide-epichlorohydrin 255
Polyaniline 20, 30, 31, 66, 219,
Polyaromatic compounds 30
Polybenzyl viologen [PBV^{2+}] 204
Polycytosine 113, 128
Polyelectrolyte 53, 60, 268
Polyenes 30
Polyethylene glycol 30
Polyhydroxyethyl 30
Polymeric dots (PDs) 164
Polymerization 20, 32, 57
Polymers 47, 51, 53, 57, 107, 116, 121, 233, 242
Polymorphism 126, 157, 176
Polyphenylene vinylene (PPV) 242
Polypyrrole 20, 31, 66, 219
Polysaccharides 30, 53, 56
Polythiophene 30, 31
Polyvinyl alcohol 30, 58
Porous carbon 52
Portable 4, 12, 56, 66, 223, 254
Post-translational modifications (PTMs) 128
Potential 1, 24, 143, 292
Potentiometric 7, 265, 270
Potentiometric biosensor 7, 8
Potentiostatic 82, 258
Pregnancy test 259
Promoter gene 258, 272
Prostate 73, 257
Protein film voltammetry (PFV) 10
Protein stabilized 113
Proteins 53, 60, 107, 112, 117, 121, 127
Proton-exchange membrane 212, 270
Prussian blue 9, 40, 223, 257
PSA 257, *see prostate specific antigen*
Pseudomonas 213, 227, 268, 269
Pseudomonas aeruginosa 270
Pseudomonas fluorescens 266
Pseudomonas strains 268
Pyrroloquinoline quinone 202
Pyruvate 223, 236

Q

Quantum dots (QDs) 58, 164, 190, 243
Quantum efficiency 234, 236, 241
Quantum yield (QY) 111

Quartz crystal microbalance 18, 84
Quaterrylenecarboxylic di imide (QDI) 242
Quencher 16, 180, 237, 246
Quinone A 197, 199
Quinone B 197, 200

R

Radiation 15, 63, 166
Radioactive 234
Raman spectroscopy 41, 95
Randles−Sevcik equation 11
Reaction temperature 113
Reactive oxygen species (ROS) 129, 144, 146, 147, 178
Recombinant organisms 273
Redox cofactors 196, 199
Redox mediator 187, 203, 215
Redox polymer 199, 200, 203, 214
Redox proteins 9, 214, 215
Redox reaction 11, 49, 113, 131, 176, 184, 242
Reducing agent 54, 58, 95, 98, 111, 112, 114, 118, 121, 240
Reducing strength 112, 113
Reduction 58, 96, 107, 112, 114, 145, 146, 179, 233, 236
Reflectivity 36
Reflectometric interference spectroscopy 41
Refractive indices 108
Renewable energy 211, 221
Reorganization energy 10, 216
Repeatability 283, 294
Reporter 80, 81, 101, 158, 169, 258, 271, 272, 273
Reporter protein 258, 273
Reporter systems 80, 271
Reproducibility 1, 4, 216, 238, 261, 283, 294
Resonance 2, 13, 15, 33, 40, 51, 69, 78, 91, 143, 149, 194
Resonance energy transfer (RET) 237,
Resorufin 145, 150, 157
Respirometric 268
Responsive 21, 48, 49, 53, 185
Rhodococcus ruber 269
Ribosomal-binding sites 274
RNA 6, 38, 53, 70, 75, 99, 126, 258, 288
Rolling circle replication 127
Ronald Breslow 144
Rotaprint 252
Rotating ring disk electrode (RRDE) 201
RT-PCR 70
Rubrene 235, 236, 237
Ruthenium 233, 234, 240
Ruthenium metal complex 194, 241

S

S. cerevisiae 101
Salicylate 266
Saliva 22
Salt mediated aggregation 98
Sample matrix 4, 7, 267
Sandwich PEC immunosensor 189
Sauerbrey equation 18
Scaffolds 107, 117
Screen-printing 169, 221, 224
Selectivity 1, 3, 283, 293
Sensitivity 1, 3, 4, 233, 273, 283, 291, 292
SELEX 56, 69, 70–75
Self-assembled monolayer (SAM) 245
Self-assembly 47, 60, 128
Self-healing Materials 47, 51
Self-powered biosensors 221

Self-powered sensing 183, 194
Semiconductor 47, 52, 54, 185, 233, 242, 243
Sensitive 2, 53, 69, 84, 287
Sensor 6, 47, 48, 51, 57, 64, 82, 83, 91, 98, 146, 163, 167, 176, 179, 180, 183, 190, 193, 194
Sensor arrays 266
Sensor characteristics 294
Sensor platform 47, 48, 51, 57
Sequence 69, 74, 75
Shape memory alloys 49, 50
Shape memory materials 47, 50
Shape memory polymer 51
Sickle cell anemia 126
Silanization 37, 255
Silicon 55, 58, 60, 61, 65, 220, 239, 243, 285, 289
Silk fibroin (SF) 47, 56
Silole 233, 244
Silole-containing polymer (SCP) 245
Silver ions 114, 117, 131
Silver nanoclusters 114
Silylene 245
Single nucleotide polymorphism 126
Single signal sensors (SSS) 246
Single wall carbon nanotubes (SWCNT) 163, 172, 173, 174, 177
Single-stranded DNAs 113
Singlet 146, 147, 166, 191, 235, 242
Singlet oxygen ($^1O_2$) 146, 147
Site binding theory 287
Sludge 225, 279
Smart biosensors 22, 265, 277
Smart materials 47, 48, 49, 51, 54, 55, 56, 57, 64
Smartphone 48, 66
Sodium borohydride 98, 112, 114
Soft lithography 61, 220
Solar conversion efficiency 199
Sol-gel materials 47, 57
Sol-gel procedures 256
Sol-gels 245
Solid phase synthesis 107, 111
Solid shell 96
Sols 111
Soluble mediators 269
Solvatochromic effect 124
Sonication 111, 179
Sono-chemical 116, 130
SOS response 272
Special chlorophyll dimer 196
Special chlorophyll pair 196, 199, 201
Species 11, 13, 85, 129, 143, 146, 147, 148, 178, 265, 272
Specificity 3, 5, 6, 7, 38, 49, 56, 84, 188, 223, 247, 258
Sphingomonas sp. 272
Spin labelling 149
Spin trapping 149
Square wave voltammetry (SWV) 11
S-route 235, 237
Stability 1, 4, 65, 145, 147, 174, 283, 294
Stabilizer 58, 107, 117
Standard bioparts 274
Staphylococcus aureus 21, 258
Steric stabilization 111
Stigmatellin 197
Stimuli 48, 49, 52, 53, 57
Stokes Shift 14, 120, 125
Streptavidin 37
Streptococcus faecium 266
Strokes shift 108, 117
Stroma Lamellae 196
Structure 29, 35, 61, 111, 123, 285, 287, 288
Sugars 15, 78, 213, 221, 224, 266

Sulfide 131
Sulfonated poly(p-phenylene) 242
Superoxide anion ($O_2^{·-}$) 145, 146, 147, 178
Superoxide dismutase-like 145
Support vector machines 276
Surface acoustic wave 18, 84
Surface coverage area (Γ) 10
Surface plasmon resonance (SPR) 2, 13, 40, 51, 69, 81, 91, 92, 109, 194
Swelling properties 30
Synthetic biology 265, 274
Synthetic hydrogels 30

T

T7 exonuclease 248
Tafel plot 217
Tannic acid 195
Target 3–6, 49, 56–58, 65, 69, 95, 147, 213, 224
Techniques 2, 10, 18, 30, 40, 57–58, 60–62, 76, 121–123, 165, 270
TEMP (2,2,6,6-tetramethylpiperidine) 150
Terrylenetetracarboxylic di imide (TDI) 242
Test strips 66, 101, 253–254, 259, 261
Tetracyclines 273
Tetrasphaera duodecadis 269
Therapeutics 97, 144, 152, 158–170
Thermistor 16, 266, 269
Thermochromic 49, 54–55
Thermochromic hydrogels 54–55
Thermometer 16, 132
Thioaniline 202
Thiobacillus thioparus 270
Thioflavin T 117, 133
Thiol groups 108, 112–113, 116
Thiolate-gold ions 112
Thiols 111–112, 116–117, 119–121, 129, 134
Thrombin 80, 82–84, 125–127, 224, 245–247
Thylakoid membrane 196, 215
Thymine 6, 113, 116, 118, 158, 288
Titanium dioxide 184, 188, 194
TMB (3,3′,5,5′-tetramethylbenzidine) 131, 144–145, 150, 152–158
TMB (tri-methyl benzoate) 129
TMPD (N,N,N′,N′-tetramethyl p-phenylenediamine) 9, 197, 235, 237
Top-down 58, 111, 114, 128, 134, 165, 244
Toxic 4, 12, 22, 34, 49, 70, 129–131, 159, 168, 177, 224, 244, 258, 267–269, 279–280
Toxicity 34, 97, 108, 114, 125, 131, 152, 155, 164, 168, 214, 219, 225, 266–267, 269, 273, 279
Toxicity tests 279
Toxins 18, 22, 34, 85, 213, 271, 274
Training data 276
Transducer 2, 3, 5–13, 16, 19, 23, 30, 32, 34, 50, 64, 80, 84,183, 186, 189, 192, 213, 221, 259, 266–277, 284
Transducing elements 109–141
Transduction 2, 7–9, 22, 34, 36, 57, 80, 85, 129, 181, 256, 258, 266, 267, 268–273, 273, 287
Transition metal dichalcogenides (TMDC) 55
Transmission electron microscopy (TEM) 123, 175
Transmittance 36, 40, 55
Tributyltin 279
Trichlorethylene 270
Triethanolamine (TEOA) 190
Tri-n-propylamine (TPrA) 236, 240
Triplet 146, 147, 161, 191, 201, 235, 237, 240
Tris(2,2′-bipyridine) ruthenium (II)(Ru(by)$^{32+}$) 234
tris-Malonic acid (C_3) 144

Troponin I 234
Troponin T 234
T-route 235
Tryptophan 113
Tumors cells 114
Tungsten 188, 239
Turnover frequency 200
Two-photon absorption 124
Tyrosine 112, 113
Tyrosine Z 199

U

Ubiquinone 197–198, 203–204
Ultrafiltration 76
Ultraviolet-visible 259
Unsupervised learning 276
Urea 2, 12, 128, 129, 259, 269, 287, 293, 295,
Urease 129, 259, 287
Uric acid 152
UV-vis spectroscopy 114, 121, 122

V

Valence band 54,109, 155, 166, 179, 180, 184–187, 205
Villari effect 50
Vinylimidazole-allylamine backbone 199
Viruses 22, 43, 131
Vitamin B_{12} 269
Voltage 17, 35, 49, 269, 286, 289, 291–294
Voltammetry 10, 11, 38, 40, 82, 89, 123, 177,194, 199, 258

W

Wastewater 131, 211, 219, 225, 269, 272, 276, 279
Water 18, 20, 29, 31,34, 53, 65, 95, 98, 101, 116, 117, 120, 125, 130–134, 146, 155, 172, 177, 180, 184, 186, 187, 189, 193, 196, 198–200, 204–207, 212, 225, 238, 242, 256, 258, 260, 270, 279
Water splitting reaction 186, 201, 206
Wavefunction 243
Waveguides 40
Wavelength
Wax 63, 258
Wax dipping 255
Wax printing 62, 63, 221, 224
Wearable sensors 12, 64, 66
Whatman filter 56, 257
Whole-cell biosensors 279
Wireless 22–23, 64, 66
Wurster's blue cation 237

X

X-ray 60, 63,
X-ray crystallography 121
X-ray lithography 63
X-ray photoelectron spectroscopy (XPS) 112, 122

Y

Yeast 21, 101, 266, 269, 279

Z

Z scheme pathway 196, 199, 204
Zero-valent 110
Zinc 50, 228, 257, 273–274